D1287329

ANIMAL PHYSIOLOGY:

Adaptation and environment

SECOND EDITION

ANIMAL PHYSIOLOGY:
Adaptation and environment

KNUT   SCHMIDT-NIELSEN

James B. Duke Professor of Physiology, Department of Zoology, Duke University

CAMBRIDGE UNIVERSITY PRESS
CAMBRIDGE   LONDON   NEW YORK   MELBOURNE

Published by the Syndics of the Cambridge University Press
The Pitt Building, Trumpington Street, Cambridge CB2 1RP
Bentley House, 200 Euston Road, London NW1 2DB
32 East 57th Street, New York, NY 10022, USA
296 Beaconsfield Parade, Middle Park, Melbourne 3206, Australia

© Cambridge University Press 1975, 1979

First published 1975
Reprinted with corrections 1975
Reprinted 1976, 1977, 1978
Second edition 1979

Printed in the United States of America
Typeset by Vail-Ballou Press, Inc., Binghamton, N.Y.
Printed and bound by The Maple Press Co., York, Pa.

Library of Congress Cataloging in Publication Data
Schmidt-Nielsen, Knut, 1915–
    Animal physiology.
    Includes bibliographies and index.
    1. Physiology.   I. Title.
QP31.2.S363   1979   591.1   78-56822
ISBN 0 521 22178 1 hard covers
ISBN 0 521 29381 2 paperback
(First edition: ISBN 0 521 20551 4 hard covers
              ISBN 0 521 29075 9 paperback)

Contents

Preface to the second edition

The reason for a second edition of a book should be to make it better. This may be done by improving existing material and by updating the information it contains, and I have tried to do both. Every chapter is revised, important new information is added, and some chapters cover their subjects in greater depth than in the first edition.

As before, I consider that the understanding of principles is more important than the mere accumulation of facts that can smother in boredom the curiosity of interested young people. I have selected new material with this in mind.

Those readers who teach physiology may wish to see a listing of changes and improvements. The respiration chapters contain new material on fish respiration, on lungless salamanders, and on the oxygen supply to birds' eggs. The chapters on blood and circulation are expanded and clarified. The chapter on food and feeding contains new material on marine waxes and their importance in the food chain, on nitrogen fixation in termites and in corals, and a substantial expansion of the discussion of noxious plant compounds and chemical defenses.

In the chapter on metabolism, new material includes discussions of the oxygen minimum layer in the ocean and of the effects of high pressure and high altitude. The treatment of problems of scaling and body size is completely revised.

The chapters on temperature and heat problems contain new material on thermal tolerance and heat death. New concepts on the biological importance of fever are discussed, and recent research on the temperature regulation of birds and bees is included. The chapters on osmoregulation and excretion have been clarified, and recent changes in our concepts of how urine is concentrated in the mammalian kidney are clearly explained.

The three chapters toward the end of the book, those on muscle, nerve, and hormones, are completely revised and much new material is added. The expanded treatment ranges from new concepts of ameboid movement and the function of flagella to the molecular events in muscle contraction. Animal locomotion and biomechanics are given a prominent place, for moving about is an important characteristic of living animals.

The material on sensory perception is updated, especially in regard to electric and magnetic stimuli and the function of the lateral line of fishes. The explanation of the nature of nerve impulses and action potentials is completely rewritten and expanded to meet current requirements.

The last chapter, on physiological integration, has also been thoroughly revised. The close connection between the central nervous system and endocrine function, which was stressed in the first edition, is clarified through discussions of exciting new developments in neuroendocrine function. Instead of an enumeration of nearly endless numbers of hormones – to be memorized by students with retentive brains – there are clear tables that outline the important principles of modern endocrinology.

As in the first edition, some essential background material that already should be known to the students is placed in appendixes, not because it is considered peripheral, but because it is so important that it must be available to those who have forgotten and need a concise restatement of basic facts.

The International System of Units (the SI system) is clearly and accurately presented in Appendix A (as it was in the first edition). In addition, the inevitable transition to the common use of SI units is helped by the side-by-side use in the text of traditional units and the corresponding SI units.

Two important fields, vitamins and reproduction, are treated very lightly or not at all. The simple reason is that much of this material is already familiar. Lists of vitamins and their deficiency symptoms are of little use; they are found in books the student has met in courses on health, home economics, and introductory biology. A deeper understanding of the metabolic roles of vitamins requires a background in biochemistry that is beyond the scope of this book. The basics of human reproduction should already be familiar, and animal reproduction is a vast field that includes so much morphology and developmental biology that it is best treated as a separate subject.

I hope that the changes increase the usefulness of the book, and that both students and colleagues will let me know what I should have done better.

KNUT SCHMIDT-NIELSEN

To my students

This book was written in anger and frustration – frustration because I was unable to give my students a book that in simple words says what I find exciting and important in animal physiology, that deals with problems and their solutions, that tells how things work.

For some 20 years I have thought about writing such a book, well knowing that physiology is a field too vast for one person to know, understand, and handle adequately. I looked for and failed to find a coauthor on whom I could unload some responsibility. One day this excuse made me very angry at myself, and I decided to do the job alone. Perhaps the disadvantages of a task too big for the author might be offset by some coherence in style and viewpoint. A sabbatical leave gave me the needed time. Returning to my students, I found myself knowing more physiology than I did when I set out on the agonizing and rewarding task of writing a textbook out of my own heart.

About this book

This book is about animals and their problems. It is not only about how things are; it is about the problems and their solutions. It is also about aspects of physiology I happen to find particularly fascinating or interesting. It is written for the student who wants to know how things work, who wants to know what animals do and how they do it.

The book deals with the familiar subjects of physiology: respiration, circulation, digestion, and so on. These subjects are treated in 13 chapters, arranged according to major environmental features: oxygen, food and energy, temperature, and water. I consider this arrangement important, for there is no way to be a good physiologist, or a good biologist for that matter, without understanding how living organisms function in their environment.

The book is elementary and the needed background is minimal. I have assumed that the student is familiar with a few simple concepts, such as are obtained in a good high school course or in introductory biology at the college level. Otherwise, there will be few demands on prerequisite knowledge. I have included in the text sufficient background information to make physiological principles understandable in terms of simple physics and chemistry. In some cases, a more rigorous treatment has been placed in an appendix (e.g., concerning solutions and osmosis). This makes it available to the student who wants to acquire a better understanding and to the teacher who wants to make such information required knowledge.

The quantity and complexity of scientific information are steadily increasing, and students are already overburdened with material to remember. Furthermore, the mere recital of facts does not increase one's understanding of general principles. I have therefore tried to present information that can provide a reasoned background for my statements or conclusions. The student will find that many problems can be understood once a few fundamental principles are familiar. I also feel that clear concepts are more important than the learning of terms, but because concepts cannot be conveyed without words, terms are necessary. However, terms should clarify and help, and must be clearly and consistently defined.

To avoid overburdening the student with information, a textbook must necessarily be selective, and many of the omissions are intentional. For example, most students will be familiar with vitamins and with the physiology of reproduction, described in terms that have become household words, and there is no need to repeat these endlessly. But mere familiarity with common household words does not automatically confer an understanding of how living organisms work. It is more important to acquire coherent concepts, consistent with available information, consistent with the rules of chemistry and physics, and consistent with what the organism needs in order to live and function in its environment.

Much of this book explores how animals can live where the environment seems to place insurmountable obstacles in their way. The book tries to compare the possible approaches and the solutions found by different animals. The study of animals with anatomical or physiological specializations can contribute much to our understanding of general principles. However, unless we look for these general principles, comparative physiology is apt to become a description of functions peculiar to un-

common animals – uncommon not because they are rare, but because they are outside our daily experience with ourselves and with well-known pets and laboratory animals such as dogs, cats, rats, and frogs. Instead, we want to put information together into general concepts that help us understand how all animals function.

The text contains literature references. These are arranged at the end of each chapter, not only to tell where I obtained some of the facts, but also to help the student satisfy his curiosity without having to search for information that often is hard to come by. The vast quantity of scientific information made it necessary to be highly selective, and opinions about the proper selection will differ.

To bring more specialized and advanced information within the reach of the reader, I have arranged a short list entitled Additional Reading at the end of each chapter, following the main list of references. These can serve as a key to further study. To spare the student from a feeling of helplessness, I have made these lists short. They include titles that vary from brief and simple essays to large, comprehensive treatises. Except for a few older works, I have restricted these lists to reasonably recent and up-to-date material.

Like most authors, I hope that friendly, and perhaps not so friendly, readers will let me know about errors I have made and what they think I should have done better.

What is physiology?

Physiology is about the functions of living organisms – how they eat, breathe, and move about, and what they do just to keep alive. To use more technical words, physiology is about food and feeding, digestion, respiration, transport of gases in the blood, circulation and function of the heart, excretion and kidney function, muscle and movements, and so on. The dead animal has the structures that carry out these functions; in the living animal the structures work. Physiology is also about how the living organism adjusts to the adversities of the environment – obtains enough water to live or avoids too much water, escapes freezing to death or dying from excessive heat, moves about to find suitable surroundings, food, and mates – and how it obtains information about the environment through its senses. Finally, physiology is about the regulation of all these functions – how they are correlated and integrated into a smooth-functioning organism.

Physiology is not only a description of function; it also asks "why?" and "how?" To understand how an animal functions, it is necessary to be familiar both with its structure and with some elementary physics and chemistry. For example, we cannot understand respiration unless we know about oxygen. Since ancient times breathing movements have been known as a sign of life or death, but the true meaning of respiration could not be understood until chemists had discovered oxygen.

The understanding of how living organisms function is helped enormously by using a comparative approach. By comparing different animals and examining how each has solved its problems of living within the constraints of the available environment, we gain insight into general principles that otherwise might remain obscure. No animal exists, or can exist, independently of an environment, and the animal that utilizes the resources of the environment must also be able to cope with the difficulties it presents. Thus, a comparative and environmental approach provides insight into physiology.

Examining how an animal copes with its environment often tends to show what is good for the animal. This may bring us uncomfortably close to explanations that suggest evidence of purpose, or teleology, and many biologists consider this scientifically improper. However, we all do tend to ask "why?" or "what good is it for the animal?" Anyway, the animal has to survive, and there is nothing improper or unscientific in finding out how and why it succeeds. If it did not arrive at solutions to the problem of survival, it would no longer be around to be studied.

This book follows an environmental approach to comparative physiology. It begins with a description of how animals obtain oxygen from the environment, whether from water or air. Next it describes the role of blood in the transport of oxygen to the tissues and how the blood is pumped around in the organism. The energy supply (food) is dealt with in a chapter on feeding and digestion, followed by a discussion of energy metabolism in general. An important environmental factor, temperature and its effects, is discussed in two chapters. Then the equally important role of water for the organism is described. One chapter deals with movements and locomotion, another with the ways an animal obtains information about its environment (senses). The last chapter of the book discusses how all these functions with the aid of the hormonal and nervous systems are controlled, correlated, and integrated into a smoothly functioning whole organism.

PART ONE

OXYGEN

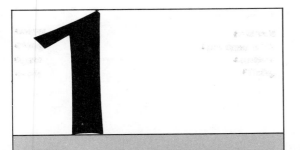

CHAPTER ONE

Respiration in water

All living organisms use energy, which they must obtain from outside sources. Most plants capture the energy of sunlight and use carbon dioxide from the atmosphere to synthesize sugars and eventually all the other complex compounds that make up a plant. Animals, on the other hand, use energy from chemical compounds they obtain from plants, either directly by eating the plants or indirectly by eating other animals that in turn depend on plants. The chemical energy animals use is therefore in the end derived from the energy of solar radiation.

Most animals satisfy their energy requirements by oxidation of food materials. A small number of animals can, in the absence of oxygen, utilize chemical energy from organic compounds, but the complete oxidation of these compounds makes available roughly 10 or 20 times as much energy. Most animal food consists of three major groups of compounds: carbohydrates, fats, and proteins. The oxidation of carbohydrates and fats yields carbon dioxide and water as the only end products; the oxidation of protein yields small amounts of other end products in addition to carbon dioxide and water.

The uptake of oxygen and release of carbon dioxide constitute *respiration*, a word that applies both to the whole organism and to the processes in the cells. Animals take up oxygen from the medium they live in and give off carbon dioxide to it. Aquatic animals take up oxygen from the small amounts of this gas dissolved in water; terrestrial animals from the abundant oxygen in atmospheric air. Many small animals can take up sufficient oxygen through the general body surface, but most animals have special respiratory organs for oxygen uptake. As the cells utilize oxygen for oxidation of foodstuffs, carbon dioxide is formed and follows the opposite path, being released through the general body surface or the respiratory organs. The water formed in the oxidation processes merely enters the general pool of water in the body and presents no special problems.

The most important and sometimes the only physical process in the movement of oxygen from the external medium to the cell is *diffusion* (i.e., movement of the gas, as a dissolved substance, from a higher to a lower concentration). The movement of carbon dioxide in the opposite direction also follows concentration gradients. The diffusion is often aided by a bulk movement (such as the circulation of blood), but this does not change the basic fact that concentration gradients provide the fundamental driving force in the movement of the respiratory gases. To understand respiration, it is therefore necessary to have a basic knowledge of the respiratory gases, their solubility, and the physics of diffusion processes.

Life presumably originated in the sea, and it is convenient to discuss aquatic respiration first and afterward deal with respiration in the air. The air-breathing animals (primarily vertebrates and insects) are among the most complex organisms; however, the largest number of animals, especially many of the less highly organized invertebrates, are aquatic. After a brief review of the respiratory gases and a bit of basic physics, this chapter deals with the problems of aquatic respiration.

GASES IN AIR AND WATER

Composition of dry atmospheric air

The physiologically most important gases are oxygen, carbon dioxide, and nitrogen. They are present in atmospheric air in the proportions shown in Table 1.1. In addition, the atmosphere contains water vapor in highly variable amounts.

What physiologists usually call nitrogen is actu-

TABLE 1.1 Composition of dry atmospheric air. All atmospheric air contains water vapor in highly variable amounts. The less common noble gases (helium, neon, krypton, and xenon) together make up only 0.002% of the total. [Otis 1964]

Component	%
Oxygen	20.95
Carbon dioxide	0.03
Nitrogen	78.09
Argon	0.93
Total	100.00

ally a mixture of nitrogen with about 1% of the noble gases, and for accuracy these should be listed as well. However, in physiology it is customary to lump these gases with nitrogen, the main reason being that in most physiological processes, nitrogen and the noble gases are equally inert to the organism. Another reason is that the analysis of respiratory gases is usually carried out by determining oxygen and carbon dioxide values and calling the remainder "nitrogen." To the physiologist the amount of "nitrogen" in air is, therefore, 78.09 + 0.93%, or 79.02%. The nearly 1% argon is of physiological interest only in some quite special circumstances: for example, in connection with the secretion of gases into the swimbladder of a fish. The complete analysis of all the gases in an air sample can be carried out with the aid of a mass spectrometer, an expensive and rather elaborate instrument that is unavailable to most physiologists.

The composition of the atmosphere remains extremely constant. Convection currents cause extensive mixing to a height of at least 100 km, and no discernible changes in the percentage composition have been demonstrated, although the pressure of the air is greatly reduced at high altitudes. The statement that the lighter gases, notably hydrogen and helium, are enriched in the outer reaches of the atmosphere applies to the very outermost layers, which are of no physiological interest whatsoever. For our purposes, the open atmosphere has a constant gas composition, except for its water vapor (Spitzer 1949).

The composition of the air is maintained as a balance between the use of oxygen in oxidation processes (primarily oxidation of organic compounds to carbon dioxide) and the assimilation of carbon dioxide by plants, which in the process release oxygen.

The fear that our use of fossil fuels – oil, coal, and natural gas – may deplete the atmosphere of oxygen and add large amounts of carbon dioxide is probably unfounded. In 1910 an extremely accurate oxygen analysis showed the value of 20.948%, and during 1967 to 1970 repeated measurements gave a value of 20.946% ± 0.006. The investigators who made these very accurate analyses then calculated that if all known recoverable fossil fuel reserves were depleted, there would still be 20.8% oxygen left in the atmosphere (Machia and Hughes 1970). Physiologically this change would be of no consequence.

The slight increase in carbon dioxide caused by the combustion of all the fuel would likewise have negligible physiological effects, but this is not to say it would be harmless. Even a slight change in carbon dioxide alters the absorption of solar radiation in the atmosphere and may have an unpredictable *greenhouse effect* that over the years may drastically change climatic conditions on the earth's surface. The atmosphere is more transparent to incoming short-wave radiation than to the long-wave radiation emitted by the earth. The outgoing long-wave radiation is absorbed in the atmosphere mainly by carbon dioxide and water vapor. It is estimated that a doubling of the atmospheric carbon dioxide content would increase world temperature by 1.3 °C if atmospheric water remained constant. However, at higher temperature the atmosphere can hold more water vapor, which enhances the blanketing effect and causes further temperature rise. On the other hand, increased water vapor in the atmosphere may augment formation of clouds, which in turn reflect more of the incoming solar radiation, thus having the opposite effect. The complexity of these relationships makes predictions about the greenhouse effect of increased carbon dioxide highly uncertain (Sawyer 1972; Baes et al. 1977).

Having stressed the constancy of the atmospheric composition, we must add a few words about special cases. For example, microenvironments, such as burrows occupied by animals, have more variable air composition, with the oxygen as low as 15% or even less (Darden 1972). The carbon dioxide content is increased, but not necessarily to the same extent. However, carbon dioxide may rise to above 5%, an amount that has considerable physiological effects.

The air contained in soil – in open spaces between the soil particles – is often low in oxygen. The reason is that the soil may contain oxidizable material that can deplete the oxygen severely. Not only organic matter, but also substances such as iron sulfide, can consume oxygen until practically all free oxygen has been removed. These oxidation processes depend on temperature, humidity, and other factors, as well as on the amount of exchange with the atmosphere. Rain, for example, may block the surface porosity of the soil and at the same time provide humidity for increased oxidation, and the microatmosphere may then change drastically.

Water vapor in air

The preceding information about the percentage composition of the atmosphere referred to dry air, and we must now turn to air's water content. The pressure of water vapor over a free water surface changes with temperature (Table 1.2). At the freezing point the vapor pressure is 4.6 mm Hg

TABLE 1.2 Water vapor over a free water surface at various temperatures.

Temperature (°C)	Water vapor			
	mm Hg	kPa	% of 1 atm	mg H$_2$O per liter air
0	4.6	0.61	0.6	4.8
10	9.2	1.23	1.2	9.4
20	17.5	2.34	2.3	17.3
30	31.7	4.24	4.2	30.3
40	55.1	7.38	7.3	51.1
50	92.3	12.33	12.2	83.2
100	760.0	101.33	100.0	598.0
37	46.9	6.28	6.2	43.9

(0.61 kPa).* It increases with increasing temperature, and reaches 760 mm Hg (101.3 kPa) at 100 °C. For this reason water boils at 100 °C if the atmospheric pressure is 760 mm Hg. If the atmospheric pressure is lower, water boils at a lower temperature – for example, at 20 °C if the pressure is reduced to 17.5 mm Hg (2.34 kPa).

Any mixture of gases, such as atmospheric air, that is in equilibrium with free water contains water vapor at a pressure corresponding to the temperature, and the fraction of the air sample that is made up of water vapor therefore increases with the temperature (column 4 of Table 1.2). At 37 °C, the usual body temperature of mammals, the water vapor pressure is about 47 mm Hg (6.28 kPa), and water vapor then makes up 6.2% of the air volume (see also Figure 9.9, page 308).

The lung air of man and of other air-breathing vertebrates is always saturated with water vapor at body temperature, but the outside atmospheric air usually is not. When air is saturated with water vapor, we say that the *relative humidity* (r.h.) is 100%. If the air contains less water vapor, the humidity can be expressed as a percent of the amount required for saturation at that temperature; for example, 50% relative humidity means that the air contains half the water it would contain if saturated with water vapor at that temperature.

For some purposes relative humidity is a convenient expression, but when we want to know the total *amount* of water vapor in the air, this can be expressed as milligrams of water per liter of air. Because cold air has a very low water vapor content, even at 100% r.h., the absolute amount of water in cold air is small. Therefore, if saturated outside air in winter enters our houses and is heated, the indoor relative humidity will be extremely low (although the absolute humidity of the air is unchanged), and we say that "the air is very dry." This dryness causes moist surfaces and mucous membranes to dry out, often to the great discomfort of sensitive persons.

Altitude and atmospheric pressure

Climbing in high mountains or ascent to high altitude in nonpressurized aircraft has serious

* The unit millimeters of mercury (mm Hg) is traditionally used in physiology. It is derived from the use of mercury manometers, and 1 mm Hg at 0 °C is also known as 1 torr. In the International System of Units (the SI System) the pressure unit is the pascal (Pa), defined as 1 newton per square meter (N m^{-2}). Thus, 1 mm Hg = 133.3 Pa or 0.133 kPa; 1 atm or 760 mm Hg = 101.3 kPa. Further details on the use of the SI System are given in Appendix A.

physiological effects. At an altitude of 3000 m humans begin to feel the effects of altitude as a reduction in physical performance, and at 6000 m (about 20 000 ft) most humans can just barely survive. This is because of lack of oxygen, although the air still contains the usual 20.95% oxygen.

At sea level, where atmospheric pressure is 760 mm Hg and 20.95% of this is oxygen, the *partial pressure* of oxygen in dry air is 159 mm Hg (21.2 kPa).* At 6000 m the atmospheric pressure is half that at sea level, or about 380 mm Hg (50.7 kPa). The partial pressure of oxygen is also half that at sea level, or about 80 mm Hg or 10.6 kPa (20.95% of 380 mm Hg or 50.7 kPa). It is this decrease in partial pressure of oxygen that produces such severe effects.

The relation between altitude and atmospheric pressure is shown in Figure 1.1. The lower scale of the abscissa shows the partial pressure of oxygen of inhaled air. The zero of this scale does not coincide with zero atmospheric pressure, but instead falls where atmospheric pressure is slightly less than 50 mm Hg (6.67 kPa). The reason is simple. At the body temperature of man (37 °C) the water vapor pressure is 47 mm Hg (6.28 kPa). Therefore, if a man were placed at an atmospheric pressure of

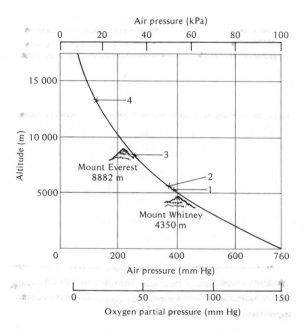

FIGURE 1.1 The relationship between altitude and atmospheric pressure. 1. Altitude where most unacclimatized persons will lose consciousness because of oxygen lack. 2. Highest permanent human habitation 3. Highest altitude where acclimatized humans can survive for a few hours when breathing air. 4. Highest altitude possible for humans breathing pure oxygen. [Dejours 1966]

47 mm Hg (19 000 m or 63 000 ft altitude), his lungs would be filled with water vapor and no air or oxygen could enter his lungs.

We shall later return to some of the effects the low oxygen partial pressure has on animals at high altitude.

Solubility of gases in water

Gases are soluble in water. If a sample of pure water is brought into contact with a gas, some gas molecules enter the water and go into solution. This continues until an equilibrium has been established and an equal number of gas molecules enters and escapes from the water per unit time. The amount of gas that is then dissolved in the water depends on (1) the nature of the gas, for the solubility is not the same for all gases; (2) the pressure of the gas in the gas phase; (3) the tempera-

---

* In a mixture of gases, the total pressure is the sum of the pressure each gas would exert if it were present alone. In dry atmospheric air at standard barometric pressure (760 mm Hg), the partial pressure of oxygen (P_{O_2}) is 159.2 mm Hg (20.95% of 760 mm Hg), of nitrogen (P_{N_2}) 600.6 mm Hg (79.02% of 760 mm Hg), and of carbon dioxide (P_{CO_2}) 0.2 mm Hg (0.03% of 760 mm Hg).

Atmospheric air is never completely dry, and its water vapor exerts a partial pressure (P_{H_2O}) corresponding to the water vapor content in the air. The partial pressure of the other gases is then reduced in exact proportion. If the air at 760 mm Hg contains 5% water vapor (P_{H_2O} = 38 mm Hg), the total pressure of the remaining gases is 722 mm Hg, and their individual partial pressures are in the proportion of their relative concentrations to make up the total of 722 mm Hg.

TABLE 1.3 Solubilities of gases in water at 15 °C when the gas is at 1 atm pressure. The solubility coefficient α is defined as the volume of gas (in milliliters STPD) dissolved in 1 liter of water, when the pressure of the gas itself, without the water vapor, is 1 atm. The amount of gas is expressed as the volume this gas would occupy if the dry gas were at 0 °C and 1 atm pressure; this is designated *standard temperature and pressure dry* (STPD).

Oxygen	34.1 ml O_2 per liter water
Nitrogen	16.9 ml N_2 per liter water
Carbon dioxide	1019.0 ml CO_2 per liter water

TABLE 1.4 Amount of oxygen dissolved in fresh water and in sea water in equilibrium with atmospheric air. [Krogh 1941]

Temperature (°C)	Fresh water (ml O_2 liter^{-1} water)	Sea water (ml O_2 liter^{-1} water)
0	10.29	7.97
10	8.02	6.35
15	7.22	5.79
20	6.57	5.31
30	5.57	4.46

ture; and (4) the presence of other solutes. For an understanding of physiology it is necessary to be familiar with the fundamentals of these relationships.

The solubilities of different gases in water are very different; those of the physiologically most important gases are listed in Table 1.3. Because solubility depends on both temperature and gas pressure, the conditions of these two variables must be specified.

We immediately see that nitrogen is only about half as soluble as oxygen, but carbon dioxide is roughly 30 times as soluble as oxygen or 60 times as soluble as nitrogen. This high solubility of carbon dioxide in water makes possible soda water or champagne. In an unopened bottle of soda water, which usually is at between 2 and 3 atm pressure, the small amount of gas at the top of the neck is under the pressure, and the remainder of the carbon dioxide is dissolved in the liquid. As we know, this must be a substantial amount, for large numbers of bubbles continue to rise to the surface for a long time. If nitrogen were dissolved at 3 atm pressure, only a few bubbles would be formed, for the amount of dissolved nitrogen would be only about 1/60 of the carbon dioxide.

Effect of pressure and temperature

The amount of gas dissolved in a given volume of water depends on the pressure of the gas. If the gas pressure is doubled, twice as much gas will be dissolved. The proportionality between gas pressure and the amount dissolved is known as *Henry's law* and can be expressed as follows:

$$V_g = \alpha \frac{P_g}{760} \cdot V_{H_2O}$$

α is the solubility coefficient (see Table 1.3), and the equation tells us the number of milliliters of the gas (V_g, at STPD) which is dissolved in the water at the pressure (P_g) given in millimeters of mercury.

The solubility of a mixture of gases depends on the partial pressure of each gas present in the gas phase. Any one of the gases is dissolved according to its own partial pressure in the gas phase, independently of the presence of other gases.

The solubility of gases decreases with increasing temperature. Most of us know this from our own experience. When we open a bottle of warm soda water or beer (God forbid!), the liquid has a much greater tendency to foam and overflow than does cold liquid. Also, if we watch a pot of water being heated on a stove, small bubbles begin to form on its walls long before boiling begins; these bubbles consist of gas driven out of solution as temperature rises. The solubility of gases in water is thus exactly the reverse of the solubility of solids, which for the most part are more soluble in hot water than in cold (sugar, for example).

We can now examine the solubility of a gas in greater detail. As an example, the solubility of oxygen is given in Table 1.4. First, we must note that this table refers to the amount of oxygen dissolved in water in equilibrium with atmospheric air (not with 1 atm pure oxygen as in Table 1.3). We can see that the solubility of oxygen decreases to about half as the temperature is raised from the freezing point to 30 °C. This decreased solubility is quite important for many aquatic animals.

Table 1.4 also gives the solubility of atmospheric oxygen in sea water, which on the whole is some 20% lower than in fresh water. This is because salts reduce the solubility of gases. This effect is characteristic for dissolved solids, but does not apply to dissolved gases, which do not affect the solubility of other gases (under conditions with which we deal in physiology).

Partial pressure and tension

We have discussed the solubility of gases in terms of the amount of a gas that enters into solution when the gas has a certain pressure. Let us look at it the other way. Take a sample of water that has a certain gas dissolved in it; the amount of gas in the water sample must correspond to one specific gas pressure in the gas phase. This pressure is called the *tension* of this gas in the water sample.* If a water sample has several gases dissolved in it, the tension of each gas corresponds to the partial pressure of that particular gas in the atmosphere with which the water is equilibrated. The tension of a gas in solution is thus defined as the partial pressure of that gas in an atmosphere in equilibrium with the solution.

When the gas pressure over a water sample is

---

* Physiologists often refer to the *tension* of a gas in a liquid, rather than its partial pressure, and define tension of a gas in solution as the pressure of this gas in an atmosphere with which that particular liquid sample is in equilibrium. The major reason for using the word *tension* is that a dissolved gas as such exerts no measurable pressure, and the term *partial pressure* is therefore conceptually somewhat misleading. However, we frequently find the two terms used interchangeably.

reduced, gases tend to leave the solution. If we reduce the gas pressure to about half the original value, gas leaves the solution until equilibrium is reached when the amount of dissolved gas has reached half its original value. If the gas pressure is reduced to zero, which is the same as exposing the water sample to a vacuum, all the gas leaves or is extracted from the water. Such vacuum extraction is one way of removing all dissolved gas from a liquid; in fact, it is a commonly used method in the analysis of the gas content of blood samples.

Because the gas dissolved in a liquid is in equilibrium with a given partial pressure in the gas phase, we can say that the gas in the liquid is under that particular partial pressure. If we have a container of water that has been equilibrated with atmospheric air, and we introduce a tiny bubble of, for example, pure nitrogen into the water, oxygen (as well as carbon dioxide) will diffuse from the water into the bubble, equilibrium being reached when the bubble contains 20.95% oxygen. Some nitrogen will initially dissolve into the water, for the initial nitrogen pressure in the bubble is 1 atm and the tension in the water only 0.79 atm. However, as oxygen enters, the loss of nitrogen subsides, and the final concentrations within the bubble are those of the initial equilibration atmosphere. (This argument depends on the bubble's volume being so small, relative to the volume of water, that it does not materially influence the gas concentrations in the water.)

Solubility of carbon dioxide and its rate of diffusion

The solubility of carbon dioxide in water is some 30 times as high as that of oxygen (see Table 1.3). However, because the amount of carbon dioxide in the atmosphere is very small (0.03%), the total quantity dissolved in water is very small. The amount can be calculated from the solubility coefficient and the fractional concentration of carbon dioxide in the atmosphere as follows:

$$\text{volume dissolved } CO_2 = \frac{1019 \times 0.03}{100}$$
$$= 0.3 \text{ ml } CO_2 \text{ per liter water}$$

In addition to the dissolved carbon dioxide, natural water contains a variable amount of carbon dioxide found in bicarbonates and carbonates. The total amount of carbon dioxide present in water in nature therefore can be quite high and varies with the cations present. Hard water, for example, contains large amounts of dissolved calcium bicarbonate, $Ca(HCO_3)_2$, which adds to the total amount of carbon dioxide. In seawater, which is slightly alkaline (pH = ca. 8.2), the total amount of carbon dioxide may range between 34 and 56 ml CO_2 per liter sea water (Nicol 1960), yet the amount of carbon dioxide present as dissolved gas is still that which is in equilibrium with the atmosphere, about 0.3 ml CO_2 per liter; the remainder is primarily in the form of bicarbonate ion. Thus, in spite of the high carbon dioxide *content* of sea water (and many other natural waters), sea water still has a carbon dioxide tension close to that in the atmosphere, 0.23 mm Hg or 30.7 Pa (photosynthesis and respiration may cause minor changes).

In the respiratory organs of animals (e.g., gills, lungs) gases diffuse between the environment and the organism, oxygen enters and carbon dioxide leaves the animal. It is therefore of interest to know how fast the gases diffuse, the *rate of diffusion*. This subject needs special attention, for many biologists have been led to believe that carbon dioxide diffuses much faster than oxygen. This is not so.

The rate of diffusion of a gas is inversely proportional to the square root of its molecular weight; carbon dioxide is heavier than oxygen and therefore diffuses more slowly. The molecular weight of carbon dioxide is 44 and that of oxygen 32; the square roots of these numbers are 6.6 and 5.7. Because the diffusion rates are inversely proportional to the square roots, the diffusion rates of carbon dioxide and oxygen are in the proportion of 5.7/6.6, or 0.86. In other words, carbon dioxide diffuses at a rate which is 0.86 that of oxygen.

When carbon dioxide diffuses between air and water, its high solubility in water makes carbon dioxide appear to diffuse faster. The situation can be explained by reference to Figure 1.2. Let us assume we have an atmosphere that contains oxygen at 100 mm Hg and carbon dioxide at 100 mm Hg pressure (i.e., the partial pressures, or concentrations, of the two gases are equal). Each gas dissolves in the surface water independently of the other, in proportion to its concentration in the gas phase and the solubility coefficient. The amount of the two gases dissolved at the surface is therefore 4.5 ml O_2 per liter and 134 ml CO_2 per liter (at 15 °C). In other words, at the surface the concentration of carbon dioxide in solution is 30 times as high as that of oxygen (29.8 times, to be exact), although in the gas phase their concentrations are equal. Carbon dioxide and oxygen now diffuse from the surface into the body of water. The amount of carbon dioxide dissolved at the surface is much higher, and more carbon dioxide therefore diffuses into the water. The carbon dioxide molecules as such diffuse at a rate 0.86 times that of the oxygen molecules, and the amount of carbon dioxide diffusing, relative to oxygen, is $29.8 \times 0.86 = 25.6$.

We now understand that, when carbon dioxide and oxygen diffuse from equal concentrations in air into water, the total amount of carbon dioxide diffusing is about 25 times higher. Likewise, if the carbon dioxide and oxygen tensions in the water are equal (in our case, 100 mm Hg at the surface of

FIGURE 1.2 Diffusion between air and water of oxygen and carbon dioxide. The amount of carbon dioxide dissolved at the surface of water (at 15 °C) is 29.8 times as high as the amount of oxygen when the two gases are at the same partial pressure in the gas phase. Carbon dioxide molecules diffuse more slowly (at a rate 0.86 times the rate for oxygen), but because of their higher concentration in the surface layer, the amount of carbon dioxide diffusing from the same pressure in the gas phase into the water is 25.6 times the amount of oxygen ($29.8 \times 0.86 = 25.6$).

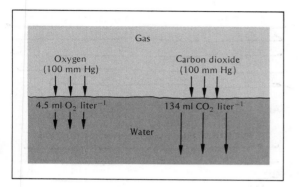

the water), diffusion into the atmosphere is also higher for carbon dioxide. It is important to note that this apparently faster diffusion of carbon dioxide is applicable only when diffusion takes place in water or between gas and water, and we refer to *the tension or partial pressure of the gases rather than to their molar concentrations*. This fast diffusion is attributable to the higher solubility of carbon dioxide; as a molecular species carbon dioxide still diffuses according to the laws of physics (i.e., somewhat slower than oxygen).*

With a knowledge of the physical basis of gas diffusion, we can now move on to the question of how animals live and function within these bounds.

AQUATIC RESPIRATION

Some of the simplest arrangements for respiratory gas exchange are found in aquatic animals. Many small organisms obtain oxygen by diffusion through their body surfaces, without having any special respiratory organs and without circulating

---

* In a gas the molar concentrations are directly proportional to the pressures and no confusion can arise as to the rate of diffusion.

blood. Larger and more complex animals often have specialized surfaces for gas exchange and a blood system that transports oxygen more rapidly than diffusion alone can provide.

Animals without specialized respiratory organs

The simplest geometrical shape of an organism is a sphere. For our considerations it is important that a sphere has the smallest possible surface corresponding to a given volume; any deviation from the spherical shape gives a relative enlargement of the surface area. If we assume that a spherical organism is to be supplied with oxygen by diffusion through the surface and into every part of the body, the longest diffusion distance is from that surface to the center.

For oxygen to reach the center, the oxygen concentration at the surface must be of a certain magnitude, for as oxygen diffuses inward it is consumed by the metabolism of the organism. The necessary oxygen tension at the surface, sufficient to supply the entire organism with oxygen by diffusion, can be calculated from an equation developed by E. Newton Harvey (1928), who was well known for his studies of luminescent organisms in the sea:

$$F_{O_2} = \frac{\dot{V}_{O_2} r^2}{6 K}$$

In this equation F_{O_2} is the concentration of oxygen at the surface expressed in fractions of an atmosphere, $\dot{V}_{O_2}$ is the rate of oxygen consumption as milliliters of oxygen per gram per minute, r is the radius of the sphere in centimeters, and K is the diffusion constant in square centimeters per atmosphere per minute. (K signifies the milliliters of oxygen which will diffuse per minute through an area of 1 cm^2 when the gradient is 1 atm cm^{-1}.

For a further discussion of diffusion, see Appendix B.)

Taking as a hypothetical example a spherical organism with a radius of 1 cm, an oxygen consumption of 0.001 ml O_2 g^{-1} min^{-1}, and a diffusion constant of 11×10^{-6} cm^2 atm^{-1} min^{-1} (the same as for connective tissue and many other animal tissues), we find that the required oxygen concentration at the surface, necessary to supply the entire organism to the center by diffusion, is 15 atm. This shows clearly that the organism cannot be supplied by diffusion alone if it has the postulated metabolic rate, which, incidentally, is rather low even for an invertebrate. The conclusion is that the organism, to be supplied with oxygen, must either be much smaller or have a much lower metabolic rate. Let us consider a smaller organism, choosing a radius of 1 mm. We then find that the required oxygen concentration at the surface is 0.15 atm. Because well-aerated water is in equilibrium with the atmosphere, which contains 0.21 atm oxygen, such an organism could obtain enough oxygen by diffusion only, and would be quite feasible.

If we consider real animals we find that these calculations have given quite reasonable orders of magnitude. Organisms that are supplied with oxygen by diffusion only (e.g., protozoans, flatworms) are mostly quite small, less than 1 mm or so, or have very low metabolic rates, as jellyfish do. Although a jellyfish can be very large, it may contain less than 1% organic matter; the rest is water and salts. It has a very low average rate of oxygen consumption, and the actively metabolizing cells are located along the surfaces where the diffusion distances are relatively short.

An organism that deviates from the spherical shape has a larger relative surface and shorter diffusion distance than a sphere. This holds for a variety of relatively simple organisms. They are flattened or threadlike, have pseudopodia, or have very large and complex surfaces (such as corals and sponges) and can thus obtain enough oxygen by diffusion, although some may be much larger than the sizes used in the preceding calculations.

Animals with respiratory organs

Although a small organism can get enough oxygen by diffusion through the surface, this is usually not true for larger organisms. Of course, any shape deviating from the sphere has a larger surface, and the diffusion distances are also reduced. However, in most cases this does not suffice, and we find specialized respiratory organs with greatly enlarged surfaces. Often these organs also have a thinner cuticle than other parts of the body, thus facilitating gas exchange.

If the respiratory surface is turned out, forming an evagination, the resulting organ is usually called a *gill*. Secondarily, the gill may be enclosed in a cavity, such as in a fish, but this does not change the fact that gills fundamentally are evaginations.

If the general body surface is turned in, or invaginated, the resulting hollow is called a *lung*. Our own lungs are a good example, although secondarily they are finely subdivided and have a quite complex structure. Simpler lungs exist; a pulmonate land snail, for example, has a lung that is little more than a simple saclike invagination in which gas exchange takes place. The term lung is used whether the respiratory medium is water or air.

Insects have a special form of respiratory system. Small openings on an insect's body surface connect to a system of tubes (*tracheae*) that branch and lead to all parts of the body. In this case the respiratory organ combines a distribution system (the tubes) with the gas-exchange system, for most of the gas passes through the walls of the finest branches of this system and diffuses directly to the cells.

In general, gills mostly serve for aquatic breathing and lungs for breathing in air. There are exceptions: Sea cucumbers have waterlungs in which most of the gas exchange seems to take place. Gills may also be modified for use in air, but on the whole they are rather unsuited for atmospheric respiration. For example, most fish when taken out of water rapidly become asphyxiated, although there is far more oxygen in air than in water. The reason is that in water the weight of the gills is well supported, but the gills lack the mechanical strength and rigidity to support their own weight in air.

Requirements for an effective respiratory organ are (1) a large surface and (2) a thin cuticle. Both these demands make it difficult to provide the mechanical rigidity for support of a gill in air. Furthermore, in air the surfaces of the fish gill tend to stick together because of surface adhesion. Therefore, the surface area exposed to air is reduced to a minute fraction of what is exposed to water, severely impeding oxygen uptake.

Ventilation of gills

If a gill removes oxygen from completely still water, the immediately adjacent boundary layer of water will soon be depleted of oxygen. Renewal of this water is therefore important in supplying oxygen, and various mechanical devices serve to increase the flow of water over the gill surface. Increased flow can be achieved in two ways: by moving the gill through the water or by moving water over the gill.

Moving the gill through the water is practical only for small organisms. Some aquatic insect larvae – mayfly larvae (Ephemeridae), for example – ventilate their gills in this way. The difficulty is that, if a gill is moved through the water with its base serving as a pivot point, the force needed to overcome the resistance to the movement is too great. This resistance increases with the square of the linear velocity of the organ, and therefore the energy needed to move the gill increases in the same proportion. The mechanical strength of the gill would also need to be increased, again with the square of the linear velocity, as would the force applied to the base of the gill to make it move through the water. The large aquatic salamander known as the mudpuppy (*Necturus*) does move its gills, but the movements are very slow.

Moving water over the respiratory surface is a much more feasible solution. The movement may be achieved by ciliary action, as in protozoans and in the gills of mussels and clams. Sponges move water through their ostia by the action of flagella.

Moving the water with a mechanical pumplike device is more common. Fish and crabs, for example, move water over their gills in this way. As a matter of principle, it is less expensive to move water slowly over a large surface than to move water fast over a smaller surface.

For some animals their own locomotion contributes to the movement of water. This is true of many pelagic fish; the large, fast-swimming tunas have practically immobile gill covers and obtain the required high water flow over the gills by swimming rapidly through the water. They probably cannot survive if kept from swimming forward, and when these fish are maintained in aquaria it is common to keep them in large circular tanks so that they can keep moving without meeting obstacles.

Also in squid and octopus there is a close correlation between locomotion and water flow over the gills, but in these animals the relationship is in a way reversed. A squid or octopus ventilates its gills by taking water into the mantle cavity, and by ejecting the water through the siphon, it propels itself through the water by jet propulsion. In this

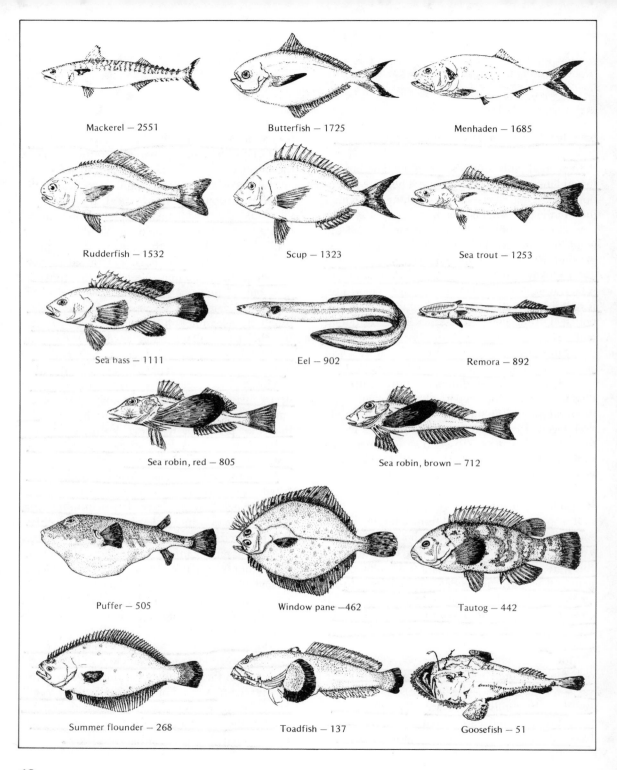

Mackerel — 2551

Butterfish — 1725

Menhaden — 1685

Rudderfish — 1532

Scup — 1323

Sea trout — 1253

Sea bass — 1111

Eel — 902

Remora — 892

Sea robin, red — 805

Sea robin, brown — 712

Puffer — 505

Window pane —462

Tautog — 442

Summer flounder — 268

Toadfish — 137

Goosefish — 51

case the ventilatory system has been modified for locomotion, but as is true of fish, as increased swimming increases the call for oxygen, oxygen is automatically provided in greater amounts.

Other functions of gills

Gills may have functions other than the respiratory gas exchange, and sometimes it is difficult to decide what is the primary or only function of a gill. Some so-called anal gills of mosquito larvae function in osmotic regulation: They absorb ions from the surrounding water, and it is doubtful that they play any major role in respiration. The gills of both fish and crabs serve in osmotic regulation, but in these cases it is quite clear that the gills also serve a primary function in respiration.

To evaluate the role in gas exchange of a gill or other suspected respiratory organ, we must have information about the amount of oxygen taken up, and preferably also about the carbon dioxide given off, through this organ. By comparing this with the oxygen consumption of the entire organism, we can see whether the organ is responsible for virtually all oxygen uptake, a large part, a small part, or just a trivial amount.

If an organism has definite and well-developed respiratory organs, there is usually a need for a circulatory system as well to carry oxygen to the various parts of the body. (This does not hold for the tracheal system of insects, which in principle is independent of a circulatory system.)

The gills of mussels and clams were mentioned above as establishing water currents by ciliary action. In the filter-feeding bivalves the gills are arranged so that they act as a sieve, retaining particles suspended in the water, which afterward are carried to the mouth and ingested. The gills therefore have a primary function in food uptake; whether they also are of importance in gas exchange is less certain. Bivalves on the whole have quite low metabolic rates, and it is possible that the surface of the mantle is sufficient to provide the required gas exchange.

Gas exchange and water flow

The fact that gill surface area must be large enough to provide adequate gas exchange is well expressed in fish. Highly active fish have the largest relative gill areas (Figure 1.3). The fast-swimming mackerel's gill surface area, expressed per unit body weight, is some 50 times as high as the sluggish, bottom-living goosefish's.

For the gas exchange to be adequate, a high rate of water flow and a close contact between the water and the gill are necessary. This is achieved by the anatomical structure of the gill apparatus. The gills are enclosed in a gill cavity, which provides protection for these rather fragile organs and, equally important, permits water to be perfused over the gills in a most effective way. A special advantage is gained by an arrangement that makes the stream of water flowing over the gill and the stream of blood running within the gill flow in opposite directions to each other. This we call a *countercurrent flow*. To understand the importance of this arrangement, it is necessary to know the structure of the gill (Figure 1.4).

Fish gills consist of several major *gill arches* on each side. From each gill arch extend two rows of *gill filaments*. The tips of these filaments from adjoining arches meet, forcing water to flow between the filaments. Each filament carries densely packed, flat *lamellae* in rows. Gas exchange takes place in these lamellae as water flows between them in one direction and blood within them in the opposite direction.

This countercurrent type of flow has an important consequence. Just as the blood is about to leave the gill lamella, it encounters water whose oxygen has not yet been removed. Thus, this blood

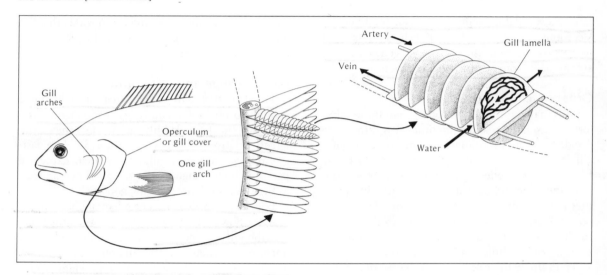

takes up oxygen from water which still has the full oxygen content of inhaled water, and this permits the oxygen content of the blood to reach the highest possible level. As the water runs further between the lamellae, it meets blood with a lower and lower oxygen content, and it therefore continues to give up more oxygen. Thus, the lamella, along its entire length, serves in taking up oxygen from the water, and the water may leave the gill having lost as much as 80 or 90% of its initial oxygen content (Hazelhoff and Evenhuis 1952). This is considered a very high oxygen extraction; for example, mammals remove only about one-quarter of the oxygen present in the lung air before it is exhaled.

We can express the effect of the countercurrent type of gas exchange in a diagram (Figure 1.5). This figure shows how blood, as it flows through the gill lamellae, takes up more and more oxygen and approaches the oxygen tension of the incoming water. The outflowing water has had most of its oxygen removed and has a tension far lower than

that of the blood leaving the gill. If the flows of water and blood were in the same direction, this would be impossible, for the blood could at best reach the oxygen tension of the outflowing water. Because the pumping of water over the gills requires energy, the countercurrent flow through the increased oxygen extraction also reduces the energetic cost of pumping.

It was previously believed that sharks have no countercurrent flow in their gills, although they are excellent swimmers and might be expected to have superbly functioning respiratory organs. However, sharks do indeed have countercurrent flow in their gills (Grigg 1970). Countercurrent flow is found also in the gills of some crabs, but in these the efficiency of oxygen removal from the water is far less than in fish. In part this may be because the water flow is less effective, but probably it is mainly because the gill–blood diffusion barrier is greater. The latter explanation seems to be the case in the European shore crab (*Carcinus*), in which the oxygen extraction, in spite of coun-

FIGURE 1.5 The countercurrent flow in the fish gill permits the blood to leave the gill with an oxygen tension almost as high as that in the incoming water. The water gives up oxygen all along the lamella, and when it leaves the gill may have lost nearly all its oxygen.

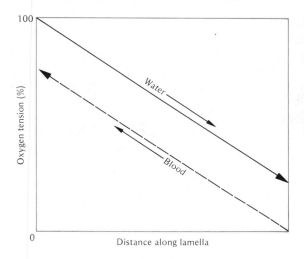

FISH GILL. The tip of a single gill filament from a trout. Water flows between the regularly arranged lamellae in a direction perpendicular to the plane of the paper. [Courtesy of G. M. Hughes, University of Bristol]

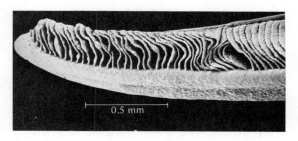

0.5 mm

that in all cases the calculated ventilation volumes could easily be greater than those measured in the living fish.

It is interesting that the highest water flows are found in some Antarctic fish, the icefish, which are very unusual in the sense that there is no hemoglobin in their blood. We shall return later to these interesting fish and how they can live without the normal oxygen-carrying mechanism of the blood.

To move water over the gills, teleost fish use a combined pumping action of the mouth and the opercular covers, aided by suitable valves to control the flow. A model of this pump is shown in Figure 1.6. The system actually consists of a double set of pumps. The volume of the first, the oral cavity, can be enlarged by lowering the jaw and especially the floor of the mouth. The volume of the second, the opercular cavity, can be increased by movements of the opercular covers while backflow of water around the edges is prevented by a skin flap acting as a passive valve. The diagram shows only one opercular pump; in reality there are two opercular chambers, one on each side.

The action of the two pumps is such that a flow of water through the gills is maintained throughout nearly the entire respiratory cycle. This flow continues although the pressure in the mouth during part of the cycle may be less than that in the surrounding water; the reason is simply that the pres-

tercurrent flow, ranges between no more than 7 and 23% (Hughes et al. 1969).

Because the lamellae in the fish gill are very close, it has been suggested that there is a high resistance and virtually no flow of water between them, for the space may be no more than 0.02 mm. The pressure that drives the water through the gills is often less than 10 mm H_2O, and such a low pressure appears insufficient to drive water through the narrow space. A careful analysis of this problem has been carried out by Hughes (1966), based on the width of the passageway between the lamellae, its height, and its length. Using a modified form of Poiseuille's equation (see chapter 4), Hughes calculated that the flow through the gills of a 150-g tench (*Tinca*), for a pressure of 5 mm H_2O, would be 10.1 ml s^{-1}. The normal volume of water pumped through the gills of this fish is about 1 to 2 ml s^{-1}; the obvious conclusion is that the gill lamellae do not offer much resistance to flow, as they would allow a greater flow than actually occurs in the living fish. Similar calculations for more than a dozen other species of fish showed

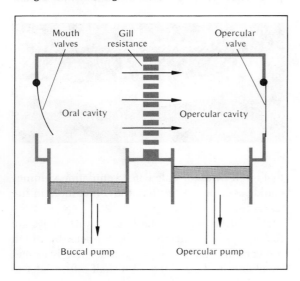

FIGURE 1.6 Water is pumped over the gills of a fish by a dual pumping system. With the aid of suitable valves, the pumps provide a unidirectional flow of water over the gill surface. [Hughes 1960]

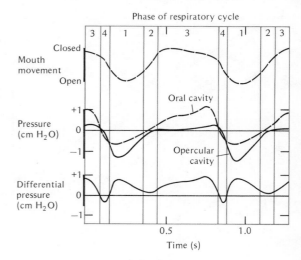

FIGURE 1.7 Record of pressure changes in the respiratory pump of a carplike fish, the roach (*Rutilus*). The lower curve shows the difference between the pressures in the oral and the opercular cavities. A pressure of 1 cm $H_2O \approx 0.1$ kPa. [Hughes and Shelton 1958]

sure in the opercular cavities is maintained even lower than in the mouth.

This is evident from pressure recordings made in the mouth and the opercular cavities during the respiratory cycle (Figure 1.7). The graph shows that the pressure changes are synchronized with the movements of the mouth and the opercula. Furthermore, the difference between the pressures in the mouth and in the opercular cavities (as shown by the curve at the bottom of the graph) remains positive almost throughout the cycle and provides the pressure that drives water through the gills. Only during a brief moment is there a slight pressure reversal. The details of the pressure curves differ from fish to fish, but all those studied so far are, in principle, the same: During almost the entire respiratory cycle the pressure in the oral cavity remains higher than in the operacular cavities, providing for a virtually continuous flow of water over the gills.

Some fish are unable to breathe this way. Fish biologists have long known that the large tunas cannot be kept alive in captivity unless they can swim continuously; this can be arranged by keeping them in large ring-shaped tanks where they can cruise without stopping. The fish swims with its mouth partly open, there are no visible breathing movements, and water flows continuously over the gills; this is called *ram ventilation*.

Ram ventilation is not restricted to large, fast-swimming pelagic fish. Many fish breathe by pumping at low speed and change to ram ventilation at higher speeds. The transition takes place around speeds of 0.5 to 1 m s^{-1}, and above this speed active breathing movements cease (Figure 1.8).

The change to ram ventilation does not mean that the gills are ventilated for free; it means only that the work of breathing is transferred from the muscles of the opercular pumps to the swimming muscles of the body and tail. The open mouth causes increased drag, and this has to be paid for in

FIGURE 1.8 When the swimming speed of a mackerel increases to between 0.5 and 0.8 m s⁻¹, opercular pumping ceases and the fish breathes entirely by ram ventilation of the gills. The records in this graph were obtained from five individuals weighing about 70 g each. [Roberts 1975]

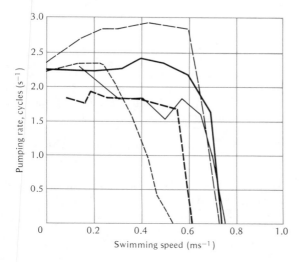

increased muscular work. However, the continuous flow during ram ventilation is probably more economical in energy than opercular pumping at the high rates required for fast swimming.

The degree of opening of the mouth seems to be adjusted to give just the right flow of water over the gills, keeping the drag to the lowest possible level compatible with the ventilation requirement. This conclusion is based on experiments in which mackerel were kept swimming at constant speeds, but subjected to a progressive lowering of the oxygen content in the water. This resulted in a graded increase in mouth gape so that the lowered oxygen supply was compensated for by increased water flow (Brown and Muir 1970).

As water flows over the gills, solid particles suspended in the water tend to get caught in the gills. Such material can be dislodged by a sudden reversal of the flow, which is brought about by enlarging the oral cavity with closed lips, causing a sudden lowering of the pressure in the mouth. This maneuver is analogous to the use of coughing to remove material from the respiratory passageways of mammals.

Crabs keep their gills clean in a somewhat similar way. The flow of water over the gills is unidirectional practically all the time, but at intervals the pumping is stopped and there is a sudden reversal of the flow for a few seconds. The frequency of the abrupt reversal varies greatly. It may happen once a minute or once in ten minutes, or even less frequently (Hughes et al. 1969). We assume that such abrupt reversals of water flow serve to remove particles that have become lodged in the gills.

The physiological basis for respiratory gas exchange as outlined in this chapter depends on some very simple physical principles. We have seen how these principles are used, and how animal function is adapted to the requirements of the physical environment. We shall now move on to the examination of how equally simple physical principles apply to the physiology of respiratory gas exchange in animals that breathe air.

REFERENCES

Baes, C. F., Jr., Goeller, H. E., Olson, J. S., and Rotty, R. M. (1977) Carbon dioxide and climate: The uncontrolled experiment. *Am. Sci.* 65:310–320.

Brown, C. E., and Muir, B. S. (1970) Analysis of ram ventilation of fish gills with application to skipjack tuna (*Katsuwonus pelamis*). *J. Fish. Res. Bd. Can.* 27:1637–1652.

Darden, T. R. (1972) Respiratory adaptations of a fossorial mammal, the pocket gopher (*Thomomys bottae*). *J. Comp. Physiol.* 78:121–137.

Dejours, P. (1966) *Respiration*. New York: Oxford University Press. 244 pp.

Gray, I. E. (1954) Comparative study of the gill area of marine fishes. *Biol. Bull.* 107:219–225.

Grigg, G. C. (1970) Use of the first gill slits for water intake in a shark. *J. Exp. Biol.* 52:569–574.

Harvey, E. N. (1928) The oxygen consumption of luminous bacteria. *J. Gen. Physiol.* 11:469–475.

Hazelhoff, E. H., and Evenhuis, H. H. (1952) Importance of the "counter-current principle" for the oxygen uptake in fishes. *Nature, Lond.* 169:77.

Hughes, G. M. (1960) A comparative study of gill ventilation in marine teleosts. *J. Exp. Biol.* 37:28–45.

Hughes, G. M. (1966) The dimensions of fish gills in relation to their function. *J. Exp. Biol.* 45:177–195.

Hughes, G. M., Knights, B., and Scammell, C. A. (1969) The distribution of P_{O_2} and hydrostatic pressure changes within the branchial chambers in relation to gill ventilation of the shore crab *Carcinus maenas* L. *J. Exp. Biol.* 51:203–220.

Hughes, G. M., and Shelton, G. (1958) The mechanism of gill ventilation in three freshwater teleosts. *J. Exp. Biol.* 35:807–823.

Krogh, A. (1941) *The Comparative Physiology of Respiratory Mechanisms.* Philadelphia: University of Pennsylvania Press. 172 pp.

Machia, L., and Hughes, E. (1970) Atmospheric oxygen in 1967 to 1970. *Science* 168:1582–1584.

Nicol, J. A. C. (1960) *The Biology of Marine Animals.* New York: Interscience. 707 pp.

Otis, A. B. (1964) Quantitative relationships in steady-state gas exchange. In *Handbook of Physiology*, sect. 3, *Respiration*, vol. I (W. O. Fenn and H. Rahn, eds.), pp. 681–698. Washington, D.C.: American Physiological Society.

Randall, D. J. (1968) Fish physiology. *Am. Zool.* 8:179–189.

Roberts, J. L. (1975) Active branchial and ram gill ventilation in fishes. *Biol. Bull.* 148:85–105.

Sawyer, J. S. (1972) Man-made carbon dioxide and the "greenhouse" effect. *Nature, Lond.* 239:23–26.

Spitzer, L. (1949) The terrestrial atmosphere above 300 km. In *The Atmospheres of the Earth and Planets* (G. P. Kuiper, ed.), pp. 211–247. Chicago: University of Chicago Press.

ADDITIONAL READING*

Altman, P. L., and Dittmer, D. S. (eds.) (1971) *Biological Handbooks: Respiration and Circulation.* Bethesda: Federation of American Societies for Experimental Biology. 930 pp.

Bartels, H., Dejours, P., Kellogg, R. H., and Mead, J. (1973) Glossary on respiration and gas exchange. *J. Appl. Physiol.* 34:549–558.

Dejours, P. (1975) *Principles of Comparative Respiratory Physiology.* Amsterdam: North-Holland. 253 pp.

Hoar, W. S., and Randall, D. J. (eds.) (1969–1971) *Fish Physiology*, 6 vols. New York: Academic Press.

Hughes, G. M. (1963) *Comparative Physiology of Vertebrate Respiration.* Cambridge, Mass.: Harvard University Press. 145 pp.

Hughes, G. M., and Morgan, M. (1973) The structure of fish gills in relation to their respiratory function. *Biol. Rev.* 48:419–475.

Johansen, K. (1971). Comparative physiology: Gas exchange and circulation in fishes. *Annu. Rev. Physiol.* 33:569–612.

Jones, J. D. (1972). *Comparative Physiology of Respiration.* London: Edward Arnold, 202 pp.

Krogh, A. (1939). *The Comparative Physiology of Respiratory Mechanisms.* Philadelphia: University of Pennsylvania Press, 172 pp. Reprinted by Dover Publications, New York, 1968.

The following book is a useful source of references in all fields of an animal physiology:

Prosser, C. L. (ed.) (1973) *Comparative Animal Physiology.* Philadelphia: Saunders. 1011 pp.

* See also Additional Reading list at end of Chapter 2.

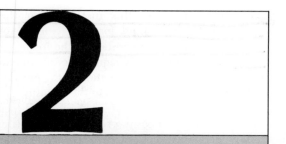

CHAPTER TWO

Respiration in air

Successful, large-scale evolutionary adaptation to air breathing and terrestrial life has occurred only in arthropods and vertebrates. In addition, some snails are well adapted to terrestrial life (some even live in deserts), and a small number of other invertebrates lives in various terrestrial microhabitats.

The atmosphere provides a high and constant oxygen concentration, available practically everywhere. The greatest drawback to breathing in air is the evaporation of water.

The easy and trouble-free access to oxygen permits a high rate of metabolism and a high degree of organizational development, both structurally and physiologically. Another consequence of life in air is that the low heat conductivity and heat capacity of air, compared with water, permits an animal to maintain a considerable difference between its own body temperature and that of the surroundings. This is nearly impossible for aquatic animals. Life in air permits the evolution of warm-blooded animals, such as birds and mammals. However, there are schemes through which some aquatic animals, especially large and highly active fish, maintain their bodies at temperatures substantially above that of the water (see chapter 8).

Let us compare how much oxygen is available in water and in air. Water in equilibrium with atmospheric air at 15 °C contains 7 ml O_2 per liter (1000 ml). This 7 ml O_2 weighs 0.01 g, and this amount is found in a weight of water 100 000 times as great. To obtain a given amount of oxygen, we must therefore move 100 000 times its weight* of water over the respiratory organs.

In contrast, 1 liter of air contains 210 ml O_2,

---

*More correctly, mass. Because in daily speech we say that we weigh an object when we determine its mass, it is very common to say weight when we should say mass. See footnote on page 32.

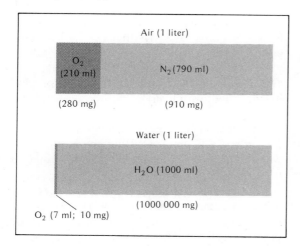

which weighs 280 mg. The remainder of the air, 790 ml N_2, weighs 910 mg. To obtain the oxygen, we must move only 3.5 times its mass of inert gas. This difference is illustrated in Figure 2.1.

This tremendous difference in the mass of the inert medium has one important consequence: In aquatic respiratory systems the movement of water is almost always unidirectional. If the flow of water were in and out or back and forth, a large mass of water would have to be accelerated, then stopped, and again accelerated in the opposite direction. A high expenditure of energy would be required for the continuous changes in kinetic energy of the water. For respiratory organs that use air, such as the lung, an in-and-out flow is not very expensive, for the inert mass will be only a few times more than the mass of oxygen used. However, air-breathing animals may use a unidirectional flow of air, as we shall see later in this chapter when we discuss bird and insect respiration.

In addition to the mass of the respiratory medium, another factor contributes to the amount of work required to move it. Water has a higher vis-cosity than air, and this increases the work required to pump the fluid, for the driving pressure must be increased in proportion to the higher vis-cosity. Because the energy required for moving the fluid increases in direct proportion to the pressure, it also increases in proportion to the viscosity.

The viscosity of water at 20 °C is 1 cP (centipoise), and the viscosity of air is 0.02 cP (i.e., water is about 50 times as viscous as air). The work required for pumping increases accordingly.

Another advantage of air respiration is the high rate of diffusion of oxygen in air, which, at the same partial pressure or tension, is some 10 000 times as rapid as in water (see Appendix B). This fast diffusion in air permits very different dimensions in the respiratory organs. The distances over which a gas can diffuse, for example in a lung, may be several millimeters, but the diffusion distance in the fish gill is a minute fraction of a millimeter.

To prevent undue evaporation from the respiratory surfaces, these should not be freely accessible to the outside air. The gas-exchange surfaces are usually located in specialized respiratory cavities (lungs), and this greatly limits the access of air. The renewal of air in the cavity is often very carefully regulated, being no greater than dictated by the requirement for oxygen.

Gas exchange across the general body surface is usually possible only in a moist habitat. Earthworms, for example, in which the entire respiratory gas exchange takes place through the body surface, are very susceptible to water loss. They live in moist habitats, and if they remain exposed on the surface of the earth they rapidly dry out and die.

With regard to water and the need for gas exchange, plants are really in a much worse situation than animals. Plants need carbon dioxide for photosynthesis, and they must obtain this gas from the air, which, as we have seen, contains only

FIGURE 2.2 For plants it is difficult to obtain by diffusion the necessary carbon dioxide from the atmosphere without incurring a relatively tremendous water loss. For animals to obtain oxygen, the water loss presents less of a problem because of the high partial pressure of oxygen in air (ca. 700 times as high as the carbon dioxide pressure).

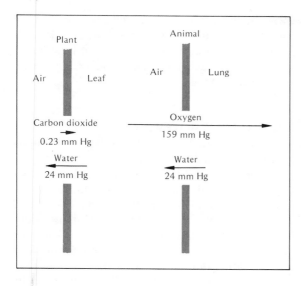

the daytime the stomata are kept closed, thus reducing water loss, and the now available sunlight is used for photosynthesis of the stored carbon dioxide (Joshi et al. 1965; Szarek et al. 1973).

The situation for plants is compared with that for animals in Figure 2.2. In animals the inward driving force for oxygen is the 21% O_2 in the air, or 159 mm Hg. At 25 °C the outward driving force for loss of water to dry air is 24 mm Hg, or only a small fraction of the inward driving pressure for oxygen. Even at high temperatures, say 37 °C, the water vapor pressure is small compared with the pressure for inward movement of oxygen. There is thus a great contrast between the unfavorable water balance plants must cope with and the favorably high oxygen concentration animals enjoy (at least, hopefully so).

0.03% CO_2 (0.23 mm Hg). The diffusion gradient for carbon dioxide into the plant is therefore extremely small. Even if the plant can maintain a zero concentration of carbon dioxide inside its tissues, the driving force for carbon dioxide diffusion is no more than 0.23 mm Hg. The plant tissues' water vapor pressure varies with temperature; at 25 °C it is 24 mm Hg. The driving force for diffusion of water to dry air is, therefore, 24 mm Hg. In moist air it is less; at 50% r.h., for example, it is half as much. In any event, the magnitude of the vapor pressure that drives the outward diffusion of water is some 100 times as high as the inward driving pressure for carbon dioxide diffusion. As a consequence of this unfavorable situation, plants have a very high requirement for water.

One scheme for alleviating this situation is used by the pineapple. This plant has open stomata during the night when the air humidity is high; carbon dioxide diffuses in and is stored in the plant (it cannot be photosynthesized in the dark, of course). In

RESPIRATORY ORGANS

Air-breathing animals have three major types of respiratory organs: gills, lungs, and tracheae.

Gills. On the whole, gills are rather poorly suited to respiration in air and are used only by few animals – mostly some that relatively recently have invaded the terrestrial habitat, taking with them the remnants of their previous mode of aquatic respiration. Land crabs are good examples. The coconut crab (*Birgus latro*), which has adopted an almost completely terrestrial existence (including climbing coconut palms), has gills that are sufficiently rigid to remain useful for respiration in air. Another land crab, *Cardiosoma*, is particularly interesting because it can survive indefinitely either in air *or* in water, whereas the coconut crab will drown if it is kept submerged (Cameron and Mecklenburg 1973). Another crustacean group with air-breathing gills is the terrestrial isopods (commonly known as sow bugs, pill bugs, or wood lice); these animals usually prefer to live in moist surroundings, and those that are most successful in the terrestrial habitat have their gills within cavities that can be regarded as functional lungs.

Among fish that can breathe air, functional gills have been maintained in some, but not in all. The common eel (*Anguilla vulgaris*) survives quite well in air if it is kept reasonably cool and moist. Much of the oxygen is then taken up through the skin and less through the gills, for the filaments tend to stick together and expose only a small surface to the air in the gill chamber. As a consequence, the eel does not obtain its normal oxygen requirement, and the oxygen uptake in air is reduced to only about one-half of that in water (Berg and Steen 1965).

Lungs. We can distinguish two types of lungs: *diffusion lungs* and *ventilation lungs*. Diffusion lungs are characterized by the fact that air exchange with the surrounding atmosphere takes place by diffusion only. Such lungs are found in relatively small animals such as pulmonate snails, scorpions, and some isopods. Ventilation lungs are found only among vertebrates. Substantial and regular renewal of the air in the lung is necessary for a large body size combined with a high metabolic rate. Vertebrate respiratory systems are ventilated by an in-and-out, or tidal, flow of air. The respiratory system of birds, however, which is far more complex than that of mammals, is arranged so that the air can flow unidirectionally through the lung during both inspiration and expiration (see under Bird Respiration, later in this chapter).

Tracheae. This type of respiratory organ is characteristic of insects. It consists of a system of tubes that supply oxygen directly to the tissues, thus obviating the need for circulation of blood for the purpose of gas transport. Exchange of gas in the tracheal system may take place by diffusion only, but in many large, and especially in highly active

insects, there is active unidirectional pumping of air through parts of the tracheal system. The advantage of unidirectional flow is that it permits a far better gas exchange than is obtained by pumping air in and out.

RESPIRATORY MOVEMENTS

Vertebrate lungs are ventilated by active pumping of air. (Tracheal ventilation will be discussed later, for insects differ from other animals in so many respects that it is convenient to treat them separately.) Ventilation of the vertebrae lung can be achieved in two different ways. Filling of the lung can take place with the use of a pressure pump, as in amphibians, or a suction pump, as in most reptiles, birds, and mammals.

A frog fills its lungs by taking air into its mouth cavity, closing its mouth and nostrils, and pressing air into its lungs by elevating the floor of its mouth. The detailed sequence is somewhat more complex, as explained in Figure 2.3. As a result of this filling mechanism, a frog can continue to take in repeated volumes of air several times in sequence without letting air out, and thus can blow itself up to a considerable size.

Nevertheless, Bentley and Shield (1973) have shown that, contrary to commonly accepted opinion, at least some amphibians can also breathe with the aid of suction-type pumping. Whether this is universal for amphibians is not known.

Positive pumping, similar to the amphibian mechanism, is found in some reptiles. The chuckawalla (*Sauromalus*), a desert lizard from southwestern North America, often hides in a rock crevice where, by inflating its lungs, it lodges itself so firmly that it cannot be pulled out. One way to dislodge the animal, used by Indians who want it for food, is to puncture it with a pointed stick.

The normal mechanism for filling the lungs in

FIGURE 2.3 Breathing cycle in the frog (*Rana*): *a*, air is taken into the buccal cavity by lowering the floor of the mouth; *b*, air is permitted to escape from the lungs, passing over the buccal cavity; *c*, the external nares are closed and air forced into the lungs; *d*, while air is retained in the lungs by the closed glottis, the cycle can be repeated by again taking air into the mouth. [Gans 1969].

FIGURE 2.4 Oxygen uptake through the skin of frogs is nearly constant throughout the year. The increased oxygen consumption during summer is covered by a greatly increased oxygen uptake through the lung. [Dolk and Postma 1927]

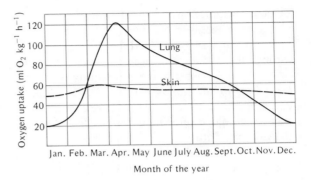

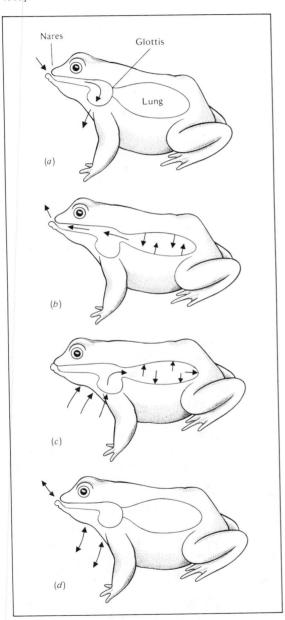

reptiles is the same as in birds and mammals: The lungs are filled by suction. Exhalation can be passive, following inhalation by elastic recoil, or it can be actively aided by muscular contraction. A suction-type pump requires a closed thoracic cavity where the pressure during inhalation is less than the surrounding atmosphere. In mammals inhalation is aided by contraction of the muscular diaphragm. Birds have a membranous diaphragm attached to the body wall by muscles, but its function differs from the mammalian diaphragm's. The common statement that birds have no diaphragm is incorrect.

ROLE OF THE SKIN IN RESPIRATION

Gas exchange through the skin is normal and important for amphibians, which have moist and well-vascularized skin. In fact, some small salamanders (plethodont salamanders) have no lungs, and all gas exchange takes place through the skin surface, except for a small contribution by the oral mucosa.

In frogs the relative roles of skin and lungs change through the year (Fig. 2.4). In winter, when the oxygen uptake is quite low, the skin takes up more oxygen than the lungs. In summer, when oxygen consumption is high, the uptake through the lungs increases several-fold and far exceeds the cutaneous uptake. The fact that oxygen uptake through the skin remains nearly constant throughout the year is related to the constant oxygen concentration in the atmosphere, which provides a constant diffusion head. If the oxygen concentration in the blood remains uniformly low through the year, the diffusion through the skin should not change much, for diffusion rates change very little with temperature. Because the need for oxygen increases greatly in summer, the increase could not be handled by the skin and must be covered by additional uptake in the lung, as is indeed the case.

Is the change with the seasons a result of the changing temperature? This question has been studied with toads (*Bufo americanus*), which were kept at three different temperatures: 5, 15, and 25 °C (Fig. 2.5). At the two higher temperatures, the pulmonary oxygen uptake exceeds that through the skin, but at the lowest temperature the cutaneous oxygen uptake is greater. This is similar to the situation in the frog. For carbon dioxide exchange the skin is more important at all temperatures. At the lowest temperature the skin is therefore more important than the lungs, both for carbon dioxide and for oxygen.

Salamanders of the family Plethodontidae are unusual because they have neither lungs nor gills. These salamanders are by no means uncommon – they constitute about 70% of existing salamander species. They are quite small and occur both in terrestrial and aquatic habitats. Gas exchange and the role of the blood in gas transport have been studied in *Desmognathus fuscus*, which as adult weighs about 5 to 7 g.

This animal is essentially a skin breather, although about 15% of the total gas exchange takes place across the mucosa of the mouth and pharynx. The blood has no exceptional character-

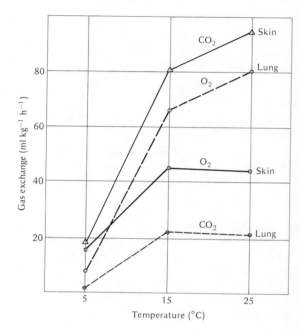

FIGURE 2.5 Pulmonary and cutaneous gas exchange in the toad *Bufo americanus* at different temperatures. [Hutchison et al. 1968]

istics: Both its hemoglobin content and its oxygen affinity are in the same range as in other salamanders, both aquatic and terrestrial. There is no special blood vessel carrying the oxygenated blood from the skin to the heart, and oxygen-rich blood leaving the skin is mixed with the general venous blood. In other amphibians there is a partial separation of oxygenated and venous blood in the heart, but such separation appears to be absent in the plethodont salamanders, and as a result their arterial blood is never fully saturated with oxygen. Nevertheless, their common occurrence indicates that the plethodont salamanders are quite successful with what from our viewpoint appears a rather inadequate respiratory apparatus (Gatz et al. 1974).

Reptiles, in contrast to amphibians, are often believed to have nearly impermeable skin. They breathe with lungs, and most of them are terrestrial. However, the true sea snakes are marine and some even bear living young at sea. They are excellent swimmers and are able to dive to a depth of at least 20 m. When the sea snake *Pelamis platurus* is submerged, it can take up oxygen through the skin at rates up to 33% of its total standard oxygen uptake and excrete carbon dioxide at rates up to 9% of the total. Although the lung is the primary organ of gas exchange, the skin obviously is of considerable aid in gas exchange as this snake pursues juvenile fish on which it preys (Graham 1974).

In mammals gas exchange through the skin is trivial. There is an oft-repeated legend about some children who for a religious procession in Italy were painted with gold paint; the story goes that they all died of asphyxiation because the skin could not "breathe." Death from asphyxiation is out of the question, for oxygen uptake through the skin is barely measurable and carbon dioxide loss from the skin is less than 1% of that from the lung (Alkalay et al. 1971). The gold-painted children must have died from other causes. A plausible explanation is that the gold paint was made by amalgamating gold and mercury and suspending the amalgam in oil, a common paint base. Mercury emulsifies readily in oil and is then rapidly absorbed through the skin, and the children may well have died from acute mercury poisoning.

Bats have a relatively much larger skin surface than other mammals; the large, thin, hairless wing membranes are highly vascularized and may contribute to gas exchange. There is, in fact, some carbon dioxide loss from the wing membrane. In the bat *Eptesicus fuscus* 0.4% of the total carbon dioxide production is lost from the wing skin at 18 °C. The amount increases with temperature, and at an air temperature of 27.5 °C as much as 11.5% of the total carbon dioxide is lost this way

(Herreid et al. 1968). The uptake of oxygen through the wing membranes, however, is not sufficiently great to be of any significance; as discussed earlier, the diffusion between water and air is some 25-fold slower for oxygen than for carbon dioxide.

MAMMALIAN LUNGS

As we move up through the vertebrate classes, the lungs become increasingly complex. In amphibians the lung is a single sac, subdivided by a few ridges that give an increased surface. The mammalian lung is much more finely divided into small sacs, the *alveoli*, which vastly increase the surface area available for gas exchange. Measurements of the surface of the frog lung indicate that 1 cm^3 lung tissue has a total gas-exchange surface of 20 cm^2; the corresponding figure for the human lung is 300 cm^2 surface per 1 cm^3 lung tissue. The large surface area is essential for the high rate of oxygen uptake required for the high metabolic rate of warm-blooded animals.

Lung volume

The lung volume of a mammal constitutes about 6% of the body volume, irrespective of the body weight (Fig. 2.6). If the lung volume were to remain the exact same proportion of body size, the slope of the regression line in Figure 2.6 would be exactly 1.0. The best fitting regression line has a slope of 1.02 (i.e., the deviation from strict proportionality is insignificant). As expected, the individual points do not fall exactly on the line, but there are no large characteristic deviations. This means that small mammals, which have high specific metabolic rates, obtain sufficient oxygen with lungs of the same relative size as in large animals.

It is worth noting that diving animals, such as porpoise, manatee, and whale, follow the com-

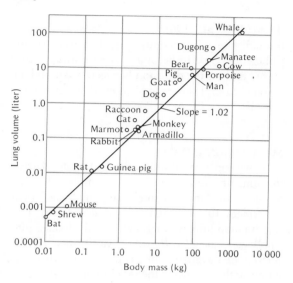

mon mammalian pattern in regard to lung size. One might expect that these animals, in order to stay under water longer, would have very large lungs that they could fill with air before a dive. This is not the case; as we shall see later, diving animals do not depend for diving on the oxygen reserves in their lungs.

The regression line in Figure 2.6 can be described by the equation:

$$V_1 = 0.0567 \cdot M_b^{1.02}$$

where V_1 is the lung volume in liters and M_b the body mass in kilograms.* For a mammal of 1 kg

* Mass is a fundamental property of matter, and weight refers to the force exerted on a given mass by a specified gravitational field. In the gravitational field of the earth mass and weight, if both are expressed in kilograms, numerically have the same value. In a different gravitational field the mass is unchanged but the weight is different. For example, on the moon a 70-kg man still has a mass of 70 kg, but his weight is only one-sixth of that on the earth, or about 12 kg (more correctly, kilogram force

body mass the expected lung volume would be 0.0567 liter, or 56.7 ml. Assuming that the body volume in liters equals the body mass in kilograms (this is for practical purposes correct, for we know that man is close to neutral buoyancy in water and therefore has a density near 1.0), the lung volume is 5.67% of the body volume. (For those unfamiliar with the arithmetic manipulation of exponential equations, a brief discussion is included in Appendix C.)

Inhalation and exhalation

Gas exchange in the lung takes place in the alveoli; the trachea, the bronchi, and their branches are only connecting tubes. At the end of an exhalation these tubes are filled with "used" air from the lung, and when inhalation follows, this air is pulled back into the lung before fresh outside air enters. The volume of air in the passageways reduces the amount of fresh air that enters the lung and is called the *dead space*. The volume of air inhaled in a single breath is the *tidal volume*. A normal man at rest has a tidal volume of about 500 cm³. Because the dead space is about 150 cm³, only 350 cm³ of fresh air will reach the lungs. The dead space thus constitutes about one-third of the tidal volume at rest. In exercise, the relative role of the dead space is less. For example, if a man who breathes heavily inhales 3000 cm³ air in a single breath, a dead space of 150 cm³ is now only about one-twentieth of the tidal volume. Therefore, the dead space is a substantial fraction of the tidal volume at rest, but in exercise it is relatively insignificant.

or kgf). Because of the similarity between mass and weight on earth, it is very common not to distinguish between them. However, the correct statement is that the mass of our man is 70 kg, whether he is on the earth, on the moon, or in space; but his weight will be 686 N on earth, about 110 N on the moon, and zero in space.

An important aspect of respiration is that the lungs are never completely emptied of air. Even if a man exhales as much as possible, there is still about 1000 cm³ of air left in his lungs. It is therefore impossible for a man to fill his lungs completely with "fresh" air, for the inhaled air is always mixed with air that remained in the lungs and the dead space.

In respiration at rest a man may have about 1650 cm³ of air in the lungs when inhalation begins. During inhalation, 350 cm³ fresh air reach the lungs and are mixed with the 1650 cm³ already there. The renewal of air is therefore only about one part in five. The result is that the composition of the alveolar gas remains quite constant at about 15% oxygen and 5% carbon dioxide. This composition of alveolar air remains the same during exercise; in other words, the increased ventilation during exercise is adjusted to match accurately the increased use of oxygen.

Surface tension

Every person who has blown soap bubbles knows that, when the connection to the atmosphere is open, the bubble tends to contract and expel air until it collapses. The vertebrate lung is somewhat similar; the bubble-like shape and high curvature of the alveoli mean that the surface tension of the moist inner surface tends to make the "bubbles" contract and disappear. The surface tension should make the lung collapse, but this tendency is minimized by the presence on the inner surface of the alveoli of substances that greatly reduce the surface tension.

These substances are phospholipids, and their effect on surface tension has given them the name *surfactants*. Surfactants are found in the lungs of all vertebrates – mammals, birds, reptiles, and amphibians. The amount of surfactant present in the vertebrate lung seems always to be above the mini-

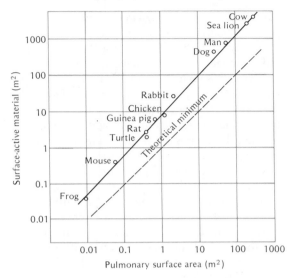

FIGURE 2.7 The amount of surfactant that can be extracted from the lungs of various vertebrates plotted against the surface area of the lung. The extracted amount is uniformly greater than the theoretical minimum amount needed to cover the lung surface with a monomolecular layer. [Clements et al. 1970]

mum required to cover the pulmonary surface with a monomolecular layer (Figure 2.7).

Mechanical work of breathing

The movement of air in and out of the lung requires work, and it is of interest to compare the cost of pumping with the amount of oxygen provided. The simplest way is to compare the amount of oxygen it takes to run the pump with the amount of oxygen the organism obtains in the same period of time.

Determinations of the work of breathing are rather difficult and have a substantial margin of error. Most determinations indicate that the cost of breathing in a man at rest is 1.2% of the total resting oxygen consumption. At rest the lung ventilation is about 5 liters per minute and the cost of respiration about 0.5 cm³ O_2 per liter ventilated air.

With increasing ventilation, however, the cost of breathing increases. If the ventilation increases to 10 liters per minute, the cost increases to 1 cm³ O_2

per liter air, and in very heavy respiration (50 liters per minute) the cost increases to 2 cm³ O_2 per liter ventilated air (Otis 1954). Other determinations indicate that the maximum work of breathing during exercise is no more than 3% of the total oxygen consumed, or slightly less than the figure just mentioned (Margaria et al. 1960). Cost of respiration has also been determined with some success in dogs; the figures are of the same magnitude as those for man.

What is the cost of pumping for fish, which must move a much heavier and more viscous medium? Accurate determinations are difficult to make, and the question remains somewhat controversial. It has been suggested that the cost of breathing in fish may be as high as 30% or even 50% of the oxygen obtained (Schumann and Piiper 1966). These figures seem unrealistic, and recent determinations indicate that the cost of breathing is much lower – a few percent of the total oxygen uptake. If the mechanical work of breathing is calculated from the pressure drop across the gill and the volume of water flow, the calculated work comes out to less than 1%, a much more reasonable figure. Without a unidirectional flow of water, such a low cost of ventilation would probably be impossible.

REGULATION OF RESPIRATION

If the need for oxygen increases, the ventilation of the respiratory organs must be increased accordingly. Likewise, if the oxygen concentration in the medium falls, there must be compensation by increasing the ventilation, by increasing the amount of oxygen removed from the respired air, or by both methods.

In warm-blooded vertebrates – mammals and birds – the ventilation of the lungs is very precisely adjusted to the need for oxygen, but interestingly, the primary agent responsible for the regulation is

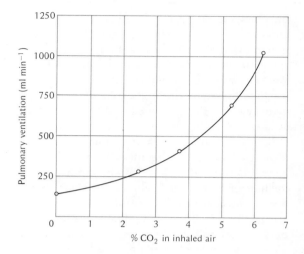

FIGURE 2.8 The effect on pulmonary ventilation volume of increased carbon dioxide content of the inhaled air in chickens. [Johnston and Jukes 1966]

the carbon dioxide concentration in the lung air. This is readily demonstrated by adding carbon dioxide to the inhaled air. This causes a rapid increase in pulmonary ventilation (Figure 2.8). Normal atmospheric air contains virtually no carbon dioxide (0.03%), and if 2.5% CO_2 is added to the inhaled air, the ventilation volume is approximately doubled. The effect is similar in mammals and birds.

This 2.5% CO_2 is really not much, for the lung air of mammals already contains about 5% CO_2. If the carbon dioxide concentration in inhaled air is increased to what is normally found in the lung, the increase in respiratory ventilation volume is sevenfold. In much higher concentrations carbon dioxide becomes narcotic and therefore induces abnormal responses.

Oxygen has a much smaller effect on ventilation. If we reduce the oxygen concentration in inhaled air by 2.5%, from 21% to 18.5% O_2, there is virtually no change in respiration.

The sensitivity of respiration to the carbon dioxide concentration is often used by swimmers who wish to stay under water for long periods. By breathing deeply for some time, a person can increase the loss of carbon dioxide from the lungs and from the blood. This removes the usual stimulus to respiration, and as a consequence, a person who has hyperventilated can remain under water longer before he is forced to the surface by the urge to breathe. This practice is extremely dangerous. As a person swims under water, the oxygen in his blood is gradually depleted, but in the absence of the usual concentration of carbon dioxide the urge to breathe is not very strong. He therefore remains submerged, and as the blood oxygen falls, he may lose consciousness without even being aware of the danger. If in this state he is not immediately discovered and rescued, he will drown. This sequence of events has in fact been the cause of many drowning accidents, particularly in swimming pools, when good swimmers competitively attempt long underwater swimming feats (Craig 1961a, b).

It has been stated that seals and whales are less sensitive to carbon dioxide than other animals and therefore are able to stay under water longer. However, the duration of a dive is probably limited by the amount of oxygen available, and it is unlikely that a dive could be prolonged merely by a decreased sensitivity to carbon dioxide.

If we want to examine the response to inhaled carbon dioxide in various animals, it is not enough to measure the respiration frequency. In some animals the respiration frequency increases considerably in response to carbon dioxide, but in others there may be little or no change in frequency and yet a considerable increase in tidal volume. This has, for example, been found in the spiny anteater (the echidna, *Tachyglossus*) (Bentley et al. 1967). The information we need is the *ventilation volume* (i.e., the product of respiration frequency and tidal volume).

If we determine the increase in ventilation volume in response to carbon dioxide, it turns out

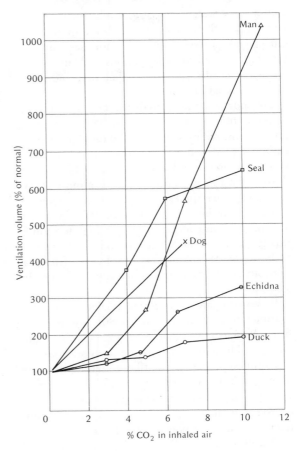

FIGURE 2.9 An increase in the carbon dioxide content of inhaled air causes a several-fold increase in respiration volume. Note that the seal is more sensitive to carbon dioxide than the dog. [Bentley et al. 1970]

that the seal, at carbon dioxide concentrations up to 6%, is even more sensitive to carbon dioxide than nondiving animals such as man and dog (Figure 2.9). This contradicts the statements about diving seals being insensitive to carbon dioxide.

Comparison of air and aquatic respiration

In many aquatic invertebrates the regulation of respiration is poor or even absent. This is true especially in marine species that normally live in well-aerated water with a relatively constant oxygen supply. Some lower animals are quite tolerant of lack of oxygen; bivalve molluscs can keep the shells closed for long periods and, in the absence of ventilation, utilize anaerobic metabolic processes (see Chapter 6).

For most aquatic animals the primary stimulus to respiration is lack of oxygen. This is usual in crustaceans, octopus, fish, and so on. The effect of carbon dioxide on aquatic invertebrates is never very pronounced and may be absent. The carbon dioxide tension in natural water is almost always low, and, as we shall see in Chapter 3, because of the high solubility of carbon dioxide in water, aquatic animals cannot build up a high tension of this gas. If they were to depend on an increase in carbon dioxide tension for stimulation of respiration, an adequate supply of oxygen could not be assured.

If we compare two closely related forms, the marine lobster (*Homarus*) and the fresh-water crayfish (*Astacus*), we notice a characteristic difference. The lobster, a marine animal, does not show much change in ventilation with a decrease of oxygen in the water (Thomas 1954). The fresh-water crayfish, in contrast, responds to a decrease in oxygen with an increase in ventilation. The difference is readily understood. The lobster lives in cold waters where the oxygen is always high, and an elaborate mechanism for regulation of ventilation would be superfluous. The crayfish, on the other hand, may readily encounter fresh-water environments where oxygen is depleted, and a ventilatory response is needed.

Fish in general respond to a decrease in oxygen, and their response to changes in carbon dioxide is minimal. In this regard they resemble other aquatic animals rather than the air-breathing vertebrates. Insects are for the most part highly sensitive to carbon dioxide.

Can we generalize these differences? We have already mentioned that aquatic animals cannot readily base their respiratory regulation on carbon dioxide, not only because the carbon dioxide tension in natural waters usually is low, but also because it is an unreliable measure of the oxygen content. Sea water is highly buffered, so that the carbon dioxide tension never builds up to any appreciable extent. Stagnant fresh water, on the other hand, may have high carbon dioxide concentrations, usually associated with low oxygen. Aquatic animals could not possibly depend on something as unreliable as the carbon dioxide concentration; when their respiration is regulated, it is in response to oxygen concentration.

The question now becomes: Why have terrestrial air-breathing animals abandoned regulation on oxygen and gone over to carbon dioxide as the primary stimulus? The answer is probably that, as oxygen becomes easily available in the atmosphere, the carbon dioxide concentration tends to build up in the respiratory organs. For example, if a mammal reduces the oxygen in the inhaled air from 21% to 16%, this entails a simultaneous buildup of carbon dioxide to about 5%, an amount that greatly affects the acid–base balance of the organism. A decrease in oxygen to 16% does not have any profound physiological effect, but a change in 1% of the carbon dioxide concentration, say from 4% to 5%, constitutes a 25% increase in carbonic acid. Also, it may be easier, physiologically speaking, to design a precise control system based on carbon dioxide and the detection of small changes in hydrogen ion concentration than to devise a system sensitive to small changes in oxygen concentration. Be this as it may, we universally find that air-breathing animals are far more sensitive to changes in carbon dioxide than to changes in oxygen.

AIR-BREATHING FISH

Everybody has heard about lungfish, but many other fish also can breathe air. Many of these resort to air only when the oxygen content in the water is low; relatively few depend on air to such an extent that they drown if kept submerged. There are two main ecological reasons for using accessory or exclusive air breathing: (1) depletion of oxygen in the water and (2) the occurrence of periodic droughts. A lungfish, for example, during dry periods burrows deep into the mud, encases itself in a cocoon, and remains inactive until the next flood.

Most of the air-breathing fish are tropical freshwater or estuarine species; few if any are truly marine. Oxygen-deficient fresh water is much more common in the tropics than in temperate climates. This is because there is much decaying organic matter in the water, the temperature is high and speeds up bacterial action, small bodies of water are often heavily shaded by overhanging jungle (which reduces photosynthesis and oxygen production in the water), and there is little temperature change between day and night and therefore minimal thermal convection to bring oxygen-rich surface water to deeper layers. Not all air-breathing fish belong in the tropics, however. The well-known bowfin (*Amia calva*) is found in the northern United States even where the lakes are frozen over in winter. During such periods they manage well without air breathing because the low temperature reduces their oxygen consumption.

It was explained in Chapter 1 that fish gills are not well suited for respiration in air. They lack the necessary rigidity and tend to stick together, but even so, some oxygen can be taken up through gills in air. Any other moist surface will supplement the gas exchange, provided it is a surface that has access to air and is supplied with blood. Some gas exchange can always take place through the skin and the surface of the mouth cavity, but in addition there may be other anatomically more specialized organs that aid in gas exchange. Organs commonly utilized in air breathing are gills, skin, mouth, opercular cavities, stomach, intestine, swimbladder, and lung.

Some air-breathing fish are listed in Table 2.1, with a few comments to place them in a familiar frame of reference. It is worth noting that all except the last five are ordinary higher bony fishes (Actinopterygii). The bowfin (*Amia*) is a representative of the primitive group Holostei, and the bichir (*Polypterus*) is peculiar because it has a lung, but both are considered primitive representatives of the Actinopterygii. Only the last three fish listed are the true lungfish (Dipnoi), which presumably are closely related to the Crossopterygii (the subclass that contains the famous coelacanth, *Latimeria*). There is one genus of true lungfish in each of three continents: Australia, Africa, and South America.

The need for air breathing depends on the amount of oxygen in the water and the temperature, for the rate of oxygen consumption increases with temperature. For this reason it is not always possible to say whether a given fish does or does not depend on air breathing. However, some fish are so dependent on air breathing that they cannot survive even in well-aerated water. They are obligatory air breathers and will "drown" if deprived of access to air. The fish listed in Table 2.2 are such obligatory air breathers; interestingly, the Australian lungfish, *Neoceratodus*, is not included.

We shall now discuss how some of the various mechanisms function and how effective they are by describing some fish that have been studied by physiologists.

Common eel

The common European and North American eel (*Anguilla vulgaris*) supposedly is able to crawl

TABLE 2.1 Fish that can use accessory air breathing or depend completely on air. Most are modern teleosts; only the last three are true lungfish.

Organ used for respiration in air	Fish	Habitat	Comment
Gill	*Symbranchus*	South America, fresh water	An eel-shaped fish without any common English name
Skin	*Anguilla*	North America, Europe	The common eel; breeds in the sea; larva migrates to fresh water
Skin	*Periophthalmus*	Tropical estuarine beaches	A common fish, often called mud skipper
Mouth and opercular cavities	*Electrophorus*	South America, fresh water	The electric eel
Mouth and opercular cavities	*Anabas*	Southeast Asia, fresh water	Called climbing perch, but not really a perch; related to betta, the Siamese fighting fish
Mouth and opercular cavities	*Clarias*	Southeast Asia, (Florida, introduced) fresh water	A catfish, known also as the walking catfish
Mouth and opercular cavities	*Gillichthys*	Pacific Coast of North America	Also called the mudsucker
Stomach	*Plecostomus*	South America fresh water	A small catfish common in home aquaria
Stomach	*Anicistrus*	South America fresh water	An armored catfish, protected by heavy spines and bony plates
Intestine	*Hoplosternum*	South America, fresh water	An armored catfish
Swimbladder	*Arapaima*	South America, rivers	The world's largest freshwater fish
Swimbladder	*Amia*	North America, fresh water	The bowfin; range extends north to areas where lakes remain ice-covered through winter; belongs to primitive group Holostei
Swimbladder	*Lepisosteus*	North America, fresh water	The garpike; belongs to the primitive group Holostei
Lung	*Polypterus*	Africa, fresh water	The bichir; has a lung, but is not a true lungfish (see text)
Lung	*Lepidosiren*	South America, fresh water	A true lungfish
Lung	*Protopterus*	Africa, fresh water	A true lungfish
Long	*Neoceratodus*	Australia, fresh water, rivers	A true lungfish

TABLE 2.2 Fish that are obligatory air breathers.

Fish	Respiration organ	Habitat
Protopterus	Lung	Africa
Lepidosiren	Lung	South America
Arapaima	Swimbladder	South America
Hoplosternum	Intestine	South America
Ophiocephalus	Pharyngeal cavities	South Asia and Africa
Electrophorus	Mouth	South America

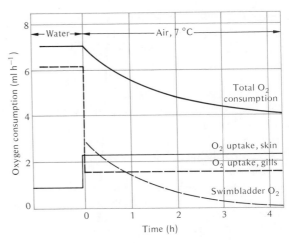

FIGURE 2.10 When an eel is transferred from water to air (at 7 °C), its rate of oxygen consumption gradually decreases and stabilizes at a lower level (top curve). Initially, oxygen is taken from the swimbladder, but when this is used up, the combined oxygen uptake from skin and gills suffices to sustain the lowered rate of oxygen consumption. [Berg and Steen 1965]

considerable distances over land and cross from one watercourse to another, especially at night and through moist grass. Fishermen often keep live eels in a box for days, merely covered by a wet sack to keep them from drying out.

When an eel is out of water, it keeps its gill cavities inflated with air. About once a minute it renews the air in the gill cavity; in contrast, in water at 20 °C an eel respires at a frequency of about 20 per minute.

The oxygen consumption of an eel in air is about half what it is in water at the same temperature. When an eel is moved to air, oxygen taken from the air is initially supplemented with oxygen removed from the swimbladder; this amount may, during the first hour, nearly equal that taken from the air. However, when an eel is kept in air longer, it gradually accumulates a substantial oxygen debt, and the lactic acid concentration in the blood increases. When returned to water, recovery is rapid, and in 2 hours the blood lactic acid is normal again.

At a lower temperature, 7 °C, the rate of oxygen consumption is much lower and there is no rise in lactic acid in the blood and apparently no oxygen debt. The permeability of the skin to oxygen should be nearly independent of temperature, and at the low temperature the combined oxygen up-take through skin and gills seems sufficient to cover the metabolic rate (Figure 2.10).

The relative importance to the eel of gills and skin differs in water and in air. In water (at 20 °C) about 90% of the total oxygen uptake is via the gills; the remainder through the skin. In air, only about one-third of the total oxygen uptake is via the gills; the remaining two-thirds via the skin. Although the gills are still of some use in air, they are far from adequate, and even with the aid of the skin the total available oxygen is insufficient to avoid an oxygen debt, except at the lowest temperatures.

Mud skipper

Another fish that, like the eel, depends primarily on cutaneous respiration is the well-known mud skipper (*Periophthalmus*), which is common on tropical beaches, especially in muddy estuarine areas. These fish are up to about 20 cm long; they prefer to remain in air for extended periods and move about supported on their pectoral fins. When disturbed, a mud skipper rapidly disappears into a mud hole that usually has some water at the bottom. When kept submerged in water, however, the fish displays circulatory reactions that indicate it suffers from partial asphyxia.

Symbranchus

This South American swamp fish has no common English name. It is notable because it seems able to breathe equally well in water and air. It uses primarily the gills for respiration, but the walls of the gill chamber are highly vascularized and contribute to gas exchange. When the fish is in air, the gill chamber is periodically inflated, and the air is retained for some 10 to 15 minutes while about half its oxygen is extracted.

When the fish is in water, the arterial oxygen saturation is always rather low, never more than 50 or 60%. Inflation of the gill chamber with air is invariably followed by a substantial increase in the arterial oxygen, which then approaches saturation (Johansen 1966).

When Symbranchus breathes air, the arterial carbon dioxide concentration increases, a change characteristic of all air breathers, and when the fish is returned to water, there is a prompt fall in the carbon dioxide again. If the carbon dioxide concentration in the water is increased, the respiration rate decreases; thus carbon dioxide inhibits rather than stimulates respiration, in contrast to the case in other air breathers. The decreased respiration caused by a high carbon dioxide level in the water leads to hypoxia, and this stimulates air breathing. The primary stimulus in the respiratory regulation of Symbranchus is therefore a decrease in oxygen, which makes the fish turn to air respiration.

Electric eel

The South American electric eel (*Electrophorus electricus*) is known for its powerful electric dis-

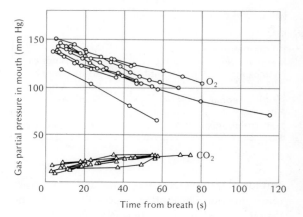

FIGURE 2.11 Air kept in the oral cavity of the electric eel gradually loses oxygen as it is used by the fish. The carbon dioxide in the mouth air gradually increases, but not to the same degree. When about one-third of the oxygen has been removed, the fish seeks to surface for renewal of the air. [Johansen et al., 1968]

charges, which have been measured at 550 V, enough to stun or kill other fish and perhaps even a man. The electric eel is a mouth breather and drowns if denied access to air. The oral cavity, which is immensely vascularized, has multiple foldings and papillations that greatly enlarge its surface area.

The electric eel takes air into the mouth at intervals from a few seconds to several minutes. The oxygen in this air gradually decreases while the carbon dioxide increases, although not in the same proportion (Figure 2.11). When the oxygen has fallen by about one-third, the fish slowly moves to the surface for renewal of the air.

In an air breather in which the entire gas exchange takes place between the blood and a body of air, the buildup of carbon dioxide should correspond roughly to the oxygen depletion. As this is not the case for the electric eel, carbon dioxide must be lost elsewhere, perhaps through the rather rudimentary gills or, more likely, through the skin.

The regulation of respiration in the electric eel can be examined by changing the carbon dioxide and oxygen concentrations in the atmosphere it is allowed to breathe. The result is that a substantial increase in the carbon dixoide content in the air gives only a moderate increase in the rate of breathing, but a drop in the oxygen content causes a several-fold increase. Likewise, increasing the oxygen pressure above that in normal atmospheric air has a somewhat depressing action on the respiration (Johansen 1968). It is interesting that this fish is an obligatory air breather, but has retained the aquatic mode of regulating respiration on oxygen. Evidently, this is associated with the fact that carbon dioxide escapes through the relatively permeable skin, thus not building up to sufficient concentrations to be physiologically important.

Garpike

The garpike (*Lepisosteus osseus*) is a relic of the ancient group Holostei, which presently is confined to North America. It is a long-nosed fish with hard scales and not popular as a game fish. Often it can be seen to break through the water surface as it comes up to breathe air. It is a large fish that may exceed 10 kg in weight, and is widely distributed and occurs in the northern United States even where the lakes are frozen over for several months in winter.

The garpike has gills and also uses its modified swimbladder for respiration. The frequency with which it comes to the surface to take fresh air into the swimbladder is a function of water temperature. At 22 °C the interval between breaths averages 8 minutes in an undisturbed fish, but in winter, when the water is covered by ice, the fish obviously cannot breathe air and depends entirely on the gills. At 20 to 25 °C, air breathing meets about 70 to 80% of the total oxygen requirement; the remainder is met by the gills. Carbon dioxide elimination from the swimbladder is, of course, zero in winter when the fish depends entirely on the gills.

AIR-BREATHING FISH. A garpike (*Lepisosteus osseus*) breathing air at the water surface. A bubble of air is visible at the posterior end of the head; it has just been expelled from the gill slit, and the fish is ready to take in air by lowering the floor of the mouth. The whole sequence of breathing requires only slightly more than 0.5 second before the fish returns below the surface. [Courtesy of Katharine Rahn, Buffalo, New York]

In summer, carbon dioxide elimination from the swimbladder approaches 10%, and the bulk of the carbon dioxide is eliminated via the gills and skin (Rahn et al. 1971). This is characteristic of many air-breathing fish: Although air breathing is important for the uptake of oxygen, carbon dioxide, because of its high solubility, is eliminated mostly to the water.

Lungfish

The African and South American lungfish (*Protopterus* and *Lepidosiren*) were listed in Table 2.2 as obligatory air breathers. They live in stagnant bodies of water and in lakes where long droughts may cause complete drying out of their habitat. They estivate until the next wet period, when they come out from their cocoons in the mud and resume normal life. The Australian lungfish (*Neoceratodus*) lives in rivers and slow streams. It also estivates in dry periods, but it depends much less on the lung and is primarily a gill breather. If it is kept in a tank from which the water is drained slowly, it becomes frantic and seeks to keep itself submerged in what little water remains. It does fill its lung with air, however, which contributes substantially to the gas exchange.

FIGURE 2.12 The relative roles of gills and lungs in respiratory gas exchange in three kinds of lungfish when kept in water and with access to air. [Lenfant et al. 1970]

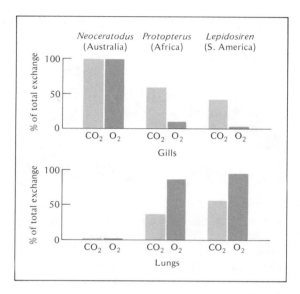

FIGURE 2.13 When African and South American lungfish are kept out of water, their rates of oxygen consumption remain nearly unchanged. The Australian lungfish shows a precipitous decrease in oxygen consumption when kept out of water. In this species the oxygen saturation in the arterial blood shows a corresponding drop. [Lenfant et al. 1970]

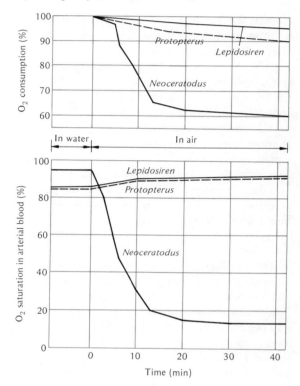

When the three kinds of lungfish are kept in water with access to air, the relative roles of gills and lungs are as shown in Figure 2.12. In the Australian lungfish the gills are the primary organs of gas exchange. In the African lungfish most of the oxygen is taken up by the lung and only a small fraction through the gills. In the South American lungfish virtually all the oxygen uptake is via the lung and only a minute amount via the gills. With carbon dioxide the situation is different; both the African and the South American lungfish exchange about half through the lung, the remainder through the gills. Hence, gills are important in carbon dioxide exchange, and the lung plays the dominant role in oxygen uptake.

The changes that take place as the lungfish are removed from water to air show a similar relationship among the three species (Figure 2.13). The oxygen consumption of the African and South American lungfish remains nearly un-

changed in air. In contrast, the Australian lungfish shows a precipitous drop in oxygen consumption, because normally it is an aquatic breather and in air it cannot obtain sufficient oxygen.

The oxygen saturation of the arterial blood (lower part of Figure 2.13) shows that the two obligatory air breathers maintain or slightly increase the oxygen saturation of the blood when they are removed to air; the Australian lungfish's arterial oxygen content drops to nearly zero, a clear indication that the fish is becoming asphyxiated.

The lungfish are among the best studied of the air-breathing fish. The information on many others is inadequate or virtually nil. The study of these could be very rewarding, although often technically difficult, for many are quite small.

BIRD RESPIRATION

Structure of the respiratory system

The respiratory organs of birds are very different from their counterparts in mammals. The small, compact bird lungs communicate with voluminous, thin-walled *air sacs* and air spaces that extend between the internal organs and even ramify into the bones of the skull and the extremities. This extensive and intricate respiratory system has been considered as an adaptation to flight. We can immediately say, however, that it is not *necessary* for flight, because bats (which have typical mammalian lungs) are good fliers and at times even migrate over long distances.

We may imagine that flying birds should have a very high oxygen consumption and that the avian respiratory system should be seen in this light. However, the rates of oxygen consumption of resting birds and mammals of equal body sizes are very similar, and although normal flight requires an 8- or 10-fold increase in oxygen consumption, many mammals are capable of similar increases. Finally, during flight bats have an oxygen consumption very similar to that of birds of the same body mass. On the other hand, although the avian-type respiratory system is not a prerequisite for flight, it may still have considerable advantages.

The presence of air spaces in the body of a bird can be said to make the bird lighter, but only in a very limited sense. The bird needs a digestive system, a liver, kidneys, and so on, and merely adding large sacs of air to the abdominal cavity does not make the bird any lighter. If we removed this air, or doubled its volume, the bird would still have exactly the same weight to carry during flight.

TABLE 2.3 Volumes of the respiratory systems of typical birds and mammals of 1 kg body size. [Estimated from data collected by Lasiewski and Calder 1971].

	Bird	Mammal
Lung volume (ml)	29.6[a]	53.5
Tracheal volume (ml)	3.7	0.9
Air sac volume (ml)	127.5	—
Total respiratory system volume (ml)	160.8	54.4
Tidal volume (ml)	13.2	7.7
Respiratory frequency (min^{-1})	17.2	53.5

[a] This is the total volume of the bird lung, which contains only 9.9 ml air.

On the other hand, if the marrow of a bone is replaced by an equal volume of air, the bone weighs less. The air-filled bones, therefore, do contribute to making a bird lighter, but the other large air spaces do not.

If we compare the volumes of the respiratory systems of birds and mammals, we find some conspicuous differences (Table 2.3). The lung volume of a typical bird is only a little more than half that of a mammal of the same body size. In contrast, the tracheal volume of a bird is much larger than that of a mammal. This can easily be understood in view of the bird's long neck, but we shall later see that it has other implications as well. The air sacs of a bird are large, several times as large as the lung, and mammals have no air sacs at all. Therefore, the total volume of the respiratory system of a bird is some three times as large as that of a mammal.

The difference between birds and mammals is not restricted to the air sacs; in structure the avian lungs differ radically from those of mammals. In mammals the finest branches of the bronchi terminate in saclike alveoli (Figure 2.14). In birds the finest branches of the bronchial system (known as *parabronchi*) permit through passage of air. Air can thus flow *through* the bird lung and continuously

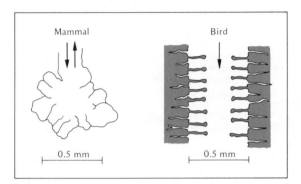

FIGURE 2.14 The smallest units of the mammalian lung are the saclike alveoli. In the bird lung the finest branches are tubes that are open at both ends and permit through flow of air. [Schmidt-Nielsen 1972]

FIGURE 2.15 The body of a bird contains several large, thin-walled air sacs. The paired lungs are small and located along the vertebral column. The main bronchus, which runs through the lung, has connections to the air sacs as well as to the lung. Below the bird is a diagram of the system, simplified by combining all anterior sacs into one single space and all posterior sacs into another. [Schmidt-Nielsen 1972]

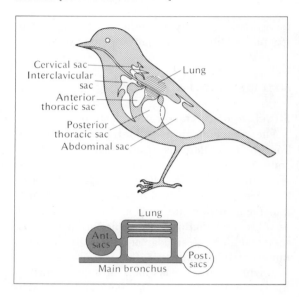

past the exchange surface; in mammals air must flow in and out. This is the most important difference between the respiratory systems of birds and mammals, and it has profound physiological consequences.

To understand how the avian respiratory system works, we must know a few additional facts about its complex anatomy; a diagram of the lungs and major air sacs will help (Figure 2.15). It is not necessary to learn the names of all the sacs, particularly as different investigators often use different names that merely confuse the nonspecialist. It is important, however, to know that the air sacs anatomically and functionally form two groups: a *posterior group* that includes the large abdominal sacs, and an *anterior group* that consists of several somewhat smaller sacs.

The trachea divides into two bronchi; each bronchus runs to and then actually through one of the lungs, and terminates in the abdominal sac. The anterior sacs connect to this main bronchus in the anterior part of the lung; the posterior sacs connect to the posterior part of the main bronchus. The main bronchus also connects to the lung, and furthermore, some of the air sacs connect directly to the lung tissue.

Function of the respiratory system

What is the function of the air sacs? Do they function in gas exchange, or do they serve as bellows to move air in and out?

The morphology of the air sacs does not indicate that they have any major role in gas exchange between air and blood. Their walls are thin, flimsy, poorly vascularized, and there are no foldings or ridges to increase the surface area. A simple experiment that excluded a direct role in gas exchange was made in the last century by a French investigator. He plugged the openings from the large abdominal air sacs to the rest of the respiratory system and then introduced carbon monoxide into the sacs (Soum 1896). Birds are quite sensitive to carbon monoxide, but these birds showed no signs of carbon monoxide poisoning. Therefore, carbon monoxide had not been taken up by the blood from the air sacs. This conclusion can be general-

BIRD LUNG. In cross section a bird lung shows cylindrical tubes (parabronchi) that allow the unidirectional through flow of air characteristic of bird lungs. The diameter of each tube in this picture, which is from a chick of domestic fowl, is slightly less than 0.5 mm. [Courtesy of Professor H.-R. Duncker, Giessen, Germany]

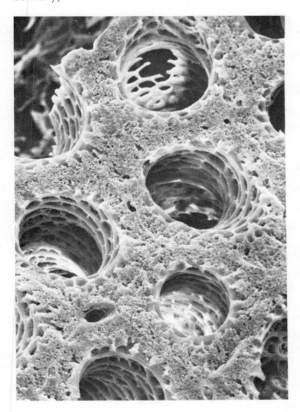

FIGURE 2.16 When an ostrich inhales a single breath of pure oxygen and then breathes ordinary air, the oxygen acts as a tracer gas. It is first found in the posterior air sacs. Only during the second or third respiratory cycle (indicated at top), does the gas show up in the anterior sacs. Respiratory rate about 6 cycles per minute. The usual gas composition in the air sacs is given at right. [Schmidt-Nielsen et al. 1969]

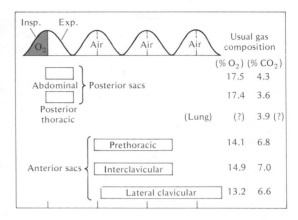

	(% O$_2$)	(% CO$_2$)
Abdominal } Posterior sacs	17.5	4.3
Posterior thoracic	17.4	3.6
(Lung)	(?)	3.9 (?)
Prethoracic	14.1	6.8
Interclavicular	14.9	7.0
Lateral clavicular	13.2	6.6

ized; it is valid for other gases that have similar diffusion properties – oxygen, for example.

A more plausible hypothesis for the function of the air sacs is that they serve as bellows to move air in and out. As an inspiration begins, there is a simultaneous pressure fall in both the anterior and posterior sacs as they expand (see diagram in Figure 2.15). This means that during inspiration air flows into all sacs (but, as we shall see later, not all sacs fill with outside air). During exhalation the pressure in the sacs increases, and air flows out again.

To follow the flow of gas in greater detail, we could use a gas mixture of a composition different from the usual air and follow the movements of this gas through the respiratory system. A convenient gas is pure oxygen, for it is harmless and can easily be measured with an oxygen electrode. For such experiments the ostrich has the advantage that it breathes quite slowly (a single respiration lasts about 10 seconds), and changes in gas composition can therefore be followed readily.

When an ostrich is permitted to inhale a single breath of pure oxygen instead of air, the oxygen shows up in the posterior air sacs toward the end of this breath (Figure 2.16). This must mean that the inhaled oxygen reaches these sacs directly through the main bronchus. In the anterior sacs, on the other hand, the oxygen concentration never increases during the inhalation of oxygen; however, these sacs do expand during inhalation, and this means that they receive air from somewhere else. Toward the end of the second inspiration, when the bird again breathes ordinary air, the oxygen in the anterior sacs begins to increase. This can only mean that the oxygen which now appears has, in the intervening time, been located elsewhere in the respiratory system, presumably in the posterior sacs, and has passed through the lung to reach the anterior sacs.

The gas concentrations found in the air sacs are interesting. The posterior sacs contain some 4% carbon dioxide (see Figure 2.16), and relative to atmospheric air the oxygen is depleted by a similar amount, from 21% to about 17%. In the anterior sacs, however, the carbon dioxide concentration is higher, between 6 and 7%, and the oxygen is correspondingly reduced to some 13 or 14%. These differences in gas composition might suggest that the posterior sacs are better ventilated and the anterior sacs contain more stagnant air, which therefore reaches a higher carbon dioxide concentration. This conclusion is incorrect.

A marker gas can also serve to determine how rapidly air is renewed in a sac. After introduction directly into a sac, the concentration of the marker decreases stepwise in synchrony with each respiratory cycle, and the rate of washout indicates the extent of air renewal. The time required for the marker gas to be reduced to one-half the initial level can be designated as the *half time*. In the ostrich this half time was between two and five respiratory cycles for both anterior and posterior air sacs; this means that both sets of sacs are about equally well ventilated. Similar determinations on ducks also gave about equal half times for anterior and posterior sacs (Bretz and Schmidt-Nielsen 1972). Because tracer gas experiments indicate that air flows into the anterior sacs from the lung (their carbon dioxide concentration is consistent with this conclusion), and because their air renewal is high, the evidence indicates that the air sacs serve as a holding chamber for air from the lung, to be exhaled on the next exhalation.

How air flows can also be determined by placing small probes, sensitive to air flow, in the various

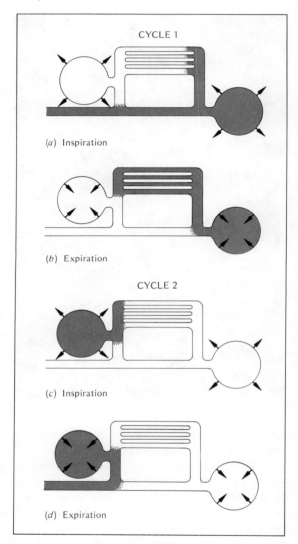

FIGURE 2.17 The movement of a single inhaled volume of gas through the avian respiratory system. It takes two full respiratory cycles to move the gas through its complete path. [Bretz and Schmidt-Nielsen, 1972]

CYCLE 1

(a) Inspiration

(b) Expiration

CYCLE 2

(c) Inspiration

(d) Expiration

passageways. Studies on ducks indicate the flow patterns illustrated in Figure 2.17, which shows how a single bolus of air would flow. During inhalation (a) most of the air flows directly to the poste-

FIGURE 2.18 Gas exchange in the bird lung at high altitude. This highly simplified diagram shows the flow of blood and air through the lung, each represented by a single stream with opposite flow directions. This flow pattern permits the oxygenated blood to leave the lung with the highest possible oxygen tension. [Schmidt-Nielsen 1972]

FIGURE 2.19 The blood in the avian lung does not flow in parallel capillaries, but rather in an irregular, complex network. This simplified diagram shows that blood leaving the lung is a mixture of blood flowing through different parts of the lung and having different degrees of oxygenation. This pattern can be described as crosscurrent flow. [Scheid and Piiper 1972]

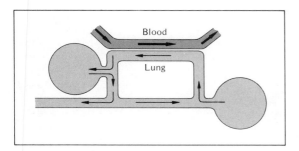

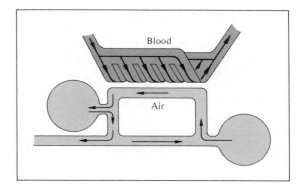

rior sacs. Although the anterior sacs expand on inhalation, they do not receive any of the inhaled outside air; instead, they receive air from the lung. On exhalation (*b*), air from the posterior sacs flows into the lung instead of out through the main bronchus. On the following inhalation (*c*), air from the lung flows to the anterior sacs. Finally, on the second exhalation (*d*), air from the anterior sacs flows directly to the outside. Two full respiratory cycles are required to move a single bolus of gas through the respiratory system. This does not mean that the two cycles differ in any way: They are completely alike, each bolus of gas being followed by another similar bolus, in tandem, on the next cycle.

The most notable characteristics of this pattern are that air always flows through the lung from the posterior to the anterior and that air moves through the lung during both inhalation and exhalation.

This flow pattern has an important consequence for gas exchange between air and blood that in principle is similar to the countercurrent flow in the fish gill (see Figure 1.4). It allows the oxygenated blood that leaves the lung to have a higher oxygen tension than the oxygen partial pressure in exhaled air. A highly simplified diagram explains this (Figure 2.18). Blood that is just about to leave the lung (right side of the diagram) is in exchange with air that has just entered the lung as

it comes directly from the posterior sacs with a high oxygen partial pressure. As the air flows through the lung (toward the left in the diagram), it loses oxygen and takes up carbon dioxide. All along, this air encounters blood with a low oxygen tension and therefore gives up more and more oxygen to the blood. Due to this type of flow, the blood can become well saturated with oxygen, and yet be able to extract more oxygen from the pulmonary air and deliver more carbon dioxide to it, than is the case in mammals.

The flow of air in the bird lung is in fact not an ideal countercurrent exchange system, but rather a *cross-current* type of flow (Figure 2.19). The result, with regard to arterial gas tensions, is similar to that described above, although a cross-current system is not as effective as a true countercurrent system in achieving maximum advantage in gas exchange.

The effectiveness of the unidirectional air flow in the avian lung is particularly important at high altitude. In experiments in which mice and sparrows were exposed to an atmospheric pressure of 350 mm Hg, corresponding to 6100 m or 20 000 ft altitude (slightly less than 0.5 atm), the mice were lying on their bellies and barely able to crawl, while the sparrows were still able to fly (Tucker

1968). Mice and sparrows have the same body weight, their blood has the same affinity for oxygen, and their metabolic rates are similar, so the difference cannot be explained in terms of their rates of oxygen consumption or the chemistry of their blood. The flow pattern in the bird lung is the most plausible explanation, for it allows blood to take up oxygen from air that has a higher oxygen concentration than would be found in the mammalian system, and in addition, because of the unidirectional and continuous flow through the lung, to extract more oxygen from the air. It is consistent with these experiments that birds in nature have been seen in the high Himalayas, flying overhead at altitudes where mountain climbers can barely walk without breathing oxygen.

Respiration in bird eggs

The problems of supplying oxygen to the growing embryo and chick within the hard shell of a bird's egg form an interesting chapter in respiration physiology. Most of the work in this field has been concerned with hens' eggs, but presumably the general principles apply to all birds' eggs, except for differences that result from the enormous differences in size. The smallest eggs, those of hummingbirds, may weigh less than 0.3 g; those of the ostrich weigh over 1 kg. The largest bird eggs known are those of the extinct *Aepyornis*, which average about 10 kg. Thus, there is more than a 40 000-fold difference in weight between the smallest and largest bird egg.

The shell of an egg consists of a hard outer layer of calcium carbonate, which on the inside has two soft membranes, called the outer and the inner shell membrane. The hard shell is much less permeable to gases than the shell membranes, which are thinner and much more permeable to gases. At the blunt end of the egg is an *air cell* that lies between the inner and outer shell membranes. This

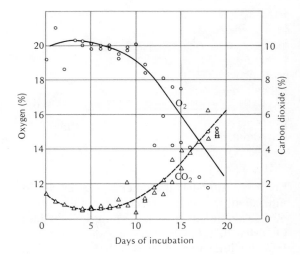

FIGURE 2.20 As the incubation of a hen's egg progresses, the oxygen concentration in the air cell gradually decreases as the carbon dioxide shows a corresponding increase. [Romijn and Roos 1938]

air space increases in size during incubation because water is lost from the egg by evaporation.

The hard shell of a typical hen's egg is perforated by about 10 000 pores; because the surface area of the egg is about 70 cm², there is an average of 1.5 pores per square millimeter of shell. The pore diameter is about 0.017 mm, so that the total pore area is 2.3 mm² (Wangensteen et al. 1971). All gas exchange between the embryo and the surrounding atmosphere must take place through these pores.

The oxygen consumption and the carbon dioxide production increase throughout the incubation period, which lasts about 21 days. Because the diffusion characteristics of the eggshell remain constant, or very nearly so, there must be a progressive increase in the gradients of gases between the ambient air and the embryo; that is, there must be a progressive fall in the oxygen partial pressure and a concomitant rise in carbon dioxide pressure within the egg during incubation (Figure 2.20).

During the last day of incubation, things change rapidly. About 28 hours before hatching the chick perforates the air cell membrane, lung respiration begins, and the chick rebreathes the gas of the air space (which because of evaporation has increased to more than 10 cm³ in volume). About 12 hours later the chick begins to break through the eggshell, the stage known as *pipping,* and the main barrier to gas exchange is now broken. During the following hours, while the chick struggles to break out completely, pulmonary respiration has taken over the whole task of gas exchange.

At high altitude, where the oxygen pressure is reduced, the oxygen supply to the embryo is severely challenged. This problem was studied in a colony of white leghorns that had been maintained for many years at the White Mountain Research Station in California at 3800 m altitude (barometric pressure = 480 mm Hg, or 64 kPa). When the colony was started, only 16% of the fertile eggs hatched, compared with the usual 90% at sea level. As it became established the hatching rate gradually increased and after eight generations reached a maximum of 60%.

One consequence of a reduced air pressure is an increase in the diffusion coefficient for gases. At 3800 m the increase is slightly more than 1.5-fold, and this increase is beneficial to the delivery of oxygen to the embryo. However, the higher diffusion coefficient at low atmospheric pressure applies also to water vapor, and this heightens the danger of desiccation.

Eggs laid at high altitude are small, and the incubation period is longer than at sea level. The pore area per unit shell surface is decreased, though the thickness of the shell is unchanged. The reduction in pore area in combination with the increased gas diffusion coefficient results in an overall gas conductance that is similar at sea level and at high altitude. We can therefore see that the reduction in total pore area of an egg laid at high altitude is a necessary adaptation because the water

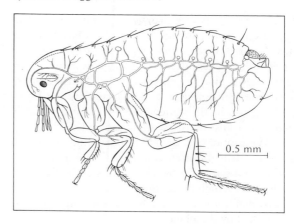

FIGURE 2.21 The tracheal system of an insect (e.g., the flea) consists of fine air-filled tubes (tracheae) that convey respiratory gases to and from all parts of the body. The tracheae connect to the outside air through the spiracles. [Wigglesworth 1972]

0.5 mm

loss otherwise would be excessive (Wangensteen et al. 1974).

INSECT RESPIRATION

Terrestrial life poses a continual conflict between the need for oxygen and the need for water. Conditions that favor the entry of oxygen also favor the loss of water. Insects, the most successful terrestrial animals, have a hard cuticle, which is highly impermeable to gases and whose covering wax layer makes it virtually impermeable to water as well. Gas exchange takes place through a system of internal air-filled tubes, the *tracheae,* which connect to the outside by openings called *spiracles* (Figure 2.21). The spiracles usually have a closing mechanism that permits accurate control of the exchange between the air in the tracheal system and the outside atmosphere. The tubes branch and ramify and extend to all parts of the body. The finest branches, the *tracheoles,* are about $1\,\mu m$ or less in diameter and can even extend into individual cells, such as muscle fibers.

The tracheal system conveys oxygen directly to the tissues, and carbon dioxide in the opposite direction. This makes insect respiration independent of a circulatory system. Insect blood has no direct role in oxygen transport, in contrast to the role of blood in vertebrates. The comparison with vertebrates is by no means irrelevant, for the rates of oxygen consumption of a large moth and a humming bird in flight are similar; the oxygen supply mechanisms, however, differ radically.

We cannot conclude that the circulatory system in insects is unimportant; it has many other functions. One role is obvious: During flight the muscles have a high power output, and while oxygen is provided through the tracheal system, fuel must be supplied by the blood.

The tracheal system

Tracheal systems are suitable primarily for respiration in air and must have evolved in air. Insects that secondarily have become aquatic have in principle maintained air respiration, but their tracheal systems have many interesting modifications that make them suitable for gas exchange in water.

The tracheal system may show various modifications from a basic pattern, in which there characteristically are 12 pairs of spiracles, 3 pairs on the thorax and 9 on the abdomen. Often there are fewer spiracles, and there may be none. Some of the possible variations are shown in Figure 2.22.

In a typical pattern the larger tracheae connect so that the system consists of tubes that all interconnect (Figure 2.22a). This pattern is commonly modified by the addition of enlarged parts of the tracheae, or air sacs, which are compressible (b). The volume of these sacs can be altered by movements of the body, thus pumping air in and out of the tracheal system. This is important, for in large and highly active insects diffusion alone does not provide sufficient gas exchange. The tracheae themselves have internal spiral ribs and are rather incompressible; the large air sacs are therefore necessary for ventilation. It is possible to achieve a unidirectional stream of air through the larger tracheal stems by having the spiracles open and close in synchrony with the respiratory movements, but out of phase with each other.

A modification characteristic of many aquatic insects is shown in Figure 2.22c. Most of the spiracles are nonfunctional, and only the two hindmost open to the outside. They are located so that the insect, by penetrating the water surface with the tip of the abdomen, can make contact with the atmosphere and obtain gas exchange, either by diffusion, or more effectively, by respiratory movements.

The tracheal system may be completely closed, without any opening to the surface, although it is filled with air (Figure 2.22d). In this event gas exchange must be entirely by diffusion through the cuticle. Many small aquatic insects can obtain sufficient gas exchange in this way; their cuticle is relatively thin, and they can remain submerged without the need to make contact with the atmosphere. Yet, the tracheal system is necessary for gas transport inside the animal; otherwise diffusion through the tissues would be too slow. As we saw earlier, diffusion of oxygen in air is some 300 000 times faster than in water; this clearly points out the advantage of a tracheal system in conveying gases when the circulation of blood is unsuited for this purpose.

A further development on the same theme again represents an aquatic insect (Figure 2.22e). The closed tracheal system extends into abdominal appendages or "gills," which, with their relatively large surface and thin cuticle, permit effective gas exchange between the water and air within the tracheal system. Such tracheal gills are found, for example, in mayfly larvae (Ephemeroptera).

Another variation is shown in Figure 2.22f. The tracheal gills are located within the lumen of the

FIGURE 2.22 (Top) The basic pattern of the insect tracheal system (a) may be modified into a variety of other patterns (b-f). For details, see text. [Wigglesworth 1972]

INSECT TRACHEOLES. Tracheoles in the tymbal muscle of a cicada (*Tibicen* sp.). The tracheoles (Tr) are cut in cross section, except for one which is cut tangentially and shows the spirally ribbed wall. The diameter of these tracheoles is about 0.5 µm. The distance between the two labeled tracheoles is about 6.5 µm (i.e., about the diameter of a human red blood cell). The light areas (M) are muscle fibers in cross section; the dark areas are mitochondria. [Courtesy of David S. Smith, University of Miami]

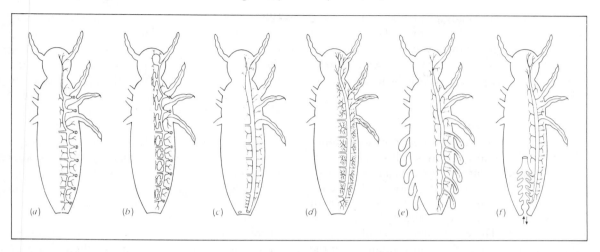

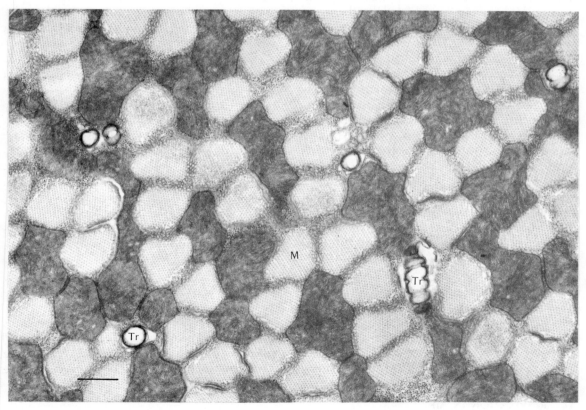

rectum, and ventilation takes place by moving water in and out of the rectum. This system is found in some dragonfly larvae. These larvae may also use the water in the rectum for locomotion by expelling it rapidly, thus moving by jet propulsion like squid and octopus.

Diffusion and ventilation

Diffusion alone suffices for gas exchange within the tracheal system of many small insects and in relatively inactive large insects. As an example we will use a large wood-boring larva of the genus *Cossus* (goat moth), which was studied by Krogh (1920). This larva weighs about 3.4 g and is 60 mm long. There are nine pairs of spiracles along the body, and careful measurements of the tracheae leading from these showed that the aggregate cross-sectional area of all the tracheae combined with 6.7 mm². The average length of the tracheal system was 6 mm. Curiously, as the tracheae branched and subdivided, the aggregate cross-sectional area of the system remained constant, not changing much with the distance from the spiracles. Therefore, diffusion through the entire tracheal system can be represented by a single cylindrical tube of 6.7 mm² cross-sectional area and a length of 6 mm. The oxygen consumed by the animal, 0.3 μl O_2 s^{-1}, will diffuse through a tube of these dimensions if the partial pressure difference between the two ends is 11 mm Hg (1.5 kPa). This means that, with an oxygen pressure in the atmosphere of 155 mm Hg (20.7 kPa), the tissues could still have an oxygen tension of 144 mm Hg (19.2 kPa). Obviously, for the *Cossus* larva an adequate oxygen supply is secured by diffusion alone, even if the metabolic rate during activity should be increased several-fold.

Intuitively it may seem that the diffusion through 6 mm of air would be slow, but because the diffusion of oxygen in air is 300 000 times faster than in water, the diffusion through 6 mm of air is as fast as the diffusion through a water layer of 0.02 μm. The greatest barrier to oxygen reaching the tissues is therefore likely to be between the finest branches of the tracheoles and the cells. In very active tissues, such as insect flight muscles, electron micrographs reveal that the tracheoles are as close as 0.07 μm to a mitochondrion.

A quantitative consideration of the branched tracheal system gives information about another aspect of gas exchange. As the tracheae divide and subdivide, the aggregate wall area of the system increases. In the *Cossus* larva the largest tracheae are about 0.6 mm in diameter and the finest tracheoles 0.001 mm. The total cross-sectional area of the tubes remains constant, and the wall area of the tracheoles therefore increases to 600 times that of an equal length of the largest tracheae. Thus, the reduction in diameter alone increases the wall surface area 600-fold. This immediately tells us that virtually the entire area available for gas exchange is in the finest branches. Consider now that the wall thickness in the tracheoles is less than one-tenth of the wall thickness of the large tracheae, and it becomes evident that diffusion through the walls of the larger branches must be insignificant.

The spiracles

The openings of the tracheal system to the outside, the spiracles, are highly complex structures that can be opened or closed to allow a variable amount of gas exchange. Their accurate control helps impede the loss of water.

The spiracles open more frequently and more widely at high temperature and when there is increased activity, in accord with the increased need for oxygen. The spiracles do not necessarily all open simultaneously; they are under control of the central nervous system, and out-of-phase opening

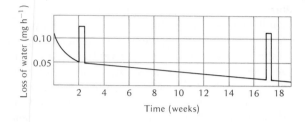

FIGURE 2.23 The water loss from a mealworm kept in dry air. During periods when the spiracles were made to remain open by adding carbon dioxide to the air, the water loss increased several-fold. [Mellanby 1934]

and closing permit the control of air flow through the tracheal system.

The ventilation of the tracheal system, and especially the function of the spiracles, is influenced by both carbon dioxide and lack of oxygen. Carbon dioxide seems to be a primary stimulus for opening of the spiracles. If a tiny stream of carbon dioxide is directed toward one spiracle, only this spiracle will open; this shows that the single spiracle may respond independently. The necessary concentration of carbon dioxide is fairly small; in the cockroach, for example, 1% carbon dioxide in the air has a perceptible effect; 2% keeps the spiracles open; and 3% makes them remain widely open.

The effect of open spiracles on water loss is considerable. In the mealworm, which spends its entire life in the dry environment of flour, the control of the spiracles is very important. If the spiracles are made to remain open by adding carbon dioxide to the air, the water loss immediately increases several-fold (Figure 2.23).

Carbon dioxide is not the only agent that controls the spiracles; they can be made to open also by pure nitrogen (i.e., oxygen deficiency is a stimulus). This could be interpreted as an effect of carbon dioxide, for when there is a lack of oxygen there is often an accumulation of acid metabolites (e.g., lactic acid), and acids increase the carbon dioxide tension. The anoxic effect could therefore, be an indirect carbon dioxide effect. However, even a moderate oxygen deficiency stimulates respiration, and it is most likely that oxygen has a direct effect.

Ventilation

The tracheae are relatively rigid and do not easily collapse, but the spiral folding of the wall permits some accordion-like shortening. This can serve to reduce the volume sufficiently to contribute to the active ventilation of the system. If the tracheae are oval or flat in cross section, they are more compressible. However, effective ventilation depends on thin-walled dilations or air sacs that are connected with the larger tracheae, for when these are compressed, a large volume of air can be expelled. Expiration is usually the active phase, inspiration being passive.

In many insects the major respiratory movements are made by the abdomen. During flight, synchronous pressure changes and movements may increase the ventilation, and even the head may have some active ventilation owing to transmission of blood pressure. The ventilation volume may be quite large, as much as one-third or one-half of the total capacity of the air system being emptied in one expiration, giving a renewal of about half the volume of the respiratory system. This is much greater than in a mammal at rest, where the renewal for one breath is about one-tenth of the air contained in the system, although the maximum possible renewal in a mammal is closer to two-thirds of the total volume of the respiratory system.

The breathing movements are synchronized with the opening and closing of spiracles and are similarly controlled by lack of oxygen or excess carbon dioxide. Excess oxygen may cause complete arrest of respiratory movements; as high oxygen should have no influence on the carbon dioxide production, it probably affects the nervous system directly.

The breathing movements are directed by centers in the segmented ventral nerve chain, and isolated segments may perform respiratory movements. Coordination between the segments is brought about by higher centers located in the prothorax. The head, apparently, is not involved. Insects in general have fewer higher centers in the head, and decapitation has little effect on respiration and many other functions.

Respiration in aquatic insects

Insects that live in water have evolved from terrestrial forms and have retained many characteristics of their terrestrial ancestors. Almost all are fresh-water forms; very few live in brackish water, and virtually no insect is truly marine.

Some aquatic insects have retained air breathing through spiracles. The spiracles are then usually reduced in number and located at the hind end of the body, which is brought into contact with the atmosphere. The spiracular openings are hydrophobic and often surrounded by hydrophobic hairs. The water surface is therefore easily broken and contact established with the air.

Small insects (e.g., mosquito larvae) depend on diffusion exchange between the atmosphere and the tracheal system, but larger larvae use respiratory movements to aid in the replacement of the tracheal air. The tracheae are often large and voluminous so they can contain enough air to serve as an oxygen store, permitting submersion for longer periods. There are strict limits to this solution, however, because an unlimited increase of air stores gives too much buoyancy and the insect will be unable to submerge.

External oxygen stores

Many insects carry air attached to the outside of the body, and the spiracles then open into this air mass. The air is held in place by nonwettable sur-faces, often helped by hydrophobic hairs. The adult *Dytiscus* beetle has a large air space under the wings; when the beetle makes contact with the water surface, this air space is ventilated, and on submersion it serves as an air store into which the spiracles open. A similar means is used by the back swimmer, *Notonecta*, which carries air attached to the ventral surface of the abdomen. The rather large air bubble gives it positive buoyancy, so that it must swim vigorously to descend from the surface and must attach itself to vegetation or other solid objects in order to remain submerged.

A simple experiment can demonstrate that the air carried by *Notonecta* has a respiratory role and is not related to buoyancy problems. If *Notonecta* is kept submerged in water that has been equilibrated with pure nitrogen instead of pure air, it will live for only 5 minutes. If it is kept submerged in water saturated with air, it will live for 6 hours. If it is submerged in water equilibrated with 100% O_2 instead of air, and is allowed to fill its air space with oxygen before submerging, it will live for only 35 minutes (Ege 1915). This seemingly paradoxical situation comes about because the bubble serves two purposes: It contains a store of oxygen that is gradually used up, but in addition it serves as a *diffusion gill* into which oxygen can diffuse from the surrounding water.

Let us examine what happens in each case. In the first case, the insect is allowed to carry with it a bubble of atmospheric air into oxygen-free water. The oxygen in the bubble disappears because the animal uses some; in addition, more oxygen diffuses rapidly into the surrounding oxygen-free water, and after 5 minutes all the oxygen is gone.

In the second case, the insect carries with it a bubble of air. After a while the animal's oxygen consumption has reduced the oxygen in the bubble to, say 5%, while the carbon dioxide has increased to 1%. The remainder of the gas in the

AIR-BREATHING AQUATIC INSECTS. Many aquatic insects breathe air. The mosquito larva (left) penetrates the surface film with a spiracular tube that opens at the tip of the abdomen. The leaflike structures to the right are anal papillae, whose primary function is in osmoregulation. The larva remains hanging still at the surface, while gas exchange takes place by diffusion.

The water beetle, *Dytiscus* (right), makes contact briefly with the surface, only long enough to pull air into the space beneath its wings. The spiracles open into this air space, and gas exchange in the tracheae continues after the beetle has submerged again. [Courtesy of Thomas Eisner, Cornell University]

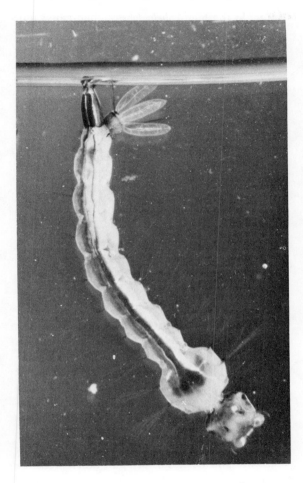

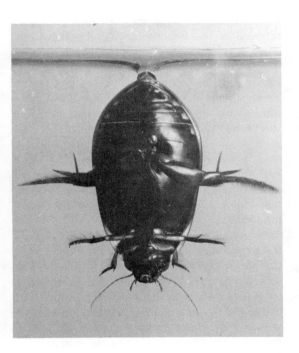

bubble (94%) is nitrogen. Because the oxygen partial pressure in the bubble is lower than in the water, oxygen diffuses from the water into the bubble. The nitrogen in the bubble, on the other hand, is at 94%, and the water is in equilibrium with the normal 79% nitrogen in the atmosphere. The partial pressure of nitrogen in the bubble is, therefore, higher than in the water; nitrogen diffuses out; and the volume of the bubble gradually diminishes. In the end the bubble is lost, but in the case described it lasted for 6 hours.

In the third case, both bubble and water contained only oxygen. The bubble disappears at the rate the animal uses oxygen. Although the water is saturated with oxygen, this is of no help, for the partial pressures of oxygen in bubble and water are the same and no additional oxygen diffuses in from the water. Consequently, the bubble lasts a much shorter time than if nitrogen is also present, as in the second case.

An air bubble has one liability: How long it lasts depends on the depth to which the insect dives. With increasing depth in the water, the total pressure within the bubble increases, while the partial pressures of dissolved gases in the water remain unchanged. The bubble, therefore, is lost faster.

For a 1 m increase in depth, the partial pressure of nitrogen in the bubble increases by about one-tenth, accelerating the rate of loss.

At a depth of 10 m, the total pressure within a submerged air mass is 2 atm. This high pressure causes a rapid loss of gas to the water, and an air mass at this depth could function only for a very short time. Therefore, there are limitations on the depth to which an insect with an air mass can dive. It has to return to the surface very often for renewal of the air; the distance to be traversed requires additional work and use of oxygen; exposure to predators is increased; and the total time spent at the bottom would be short.

For an insect that carries an air mass as a diffusion gill, the total obtainable oxygen far exceeds the initial supply carried in the mass when the insect submerges. Nitrogen is lost more slowly than oxygen diffuses in, primarily because nitrogen, because of its lower solubility, diffuses only half as fast as oxygen between air and water. Because of the initial nitrogen supply, an air mass will last long enough to obtain eight times as much oxygen by diffusion from the surrounding water as was originally in the bubble. This 8-fold ratio remains constant and independent of the metabolic rate of the insect, of the exposed surface area, of the initial volume of the air mass, and of the thickness of the boundary layer of the water (meaning that turbulence in the water in the form of active ventilation will nòt influence the amount of obtainable oxygen) (Rahn and Paganelli 1968).

Plastron

Another type of respiratory device is the *plastron*. In a plastron-breathing insect part of the body surface is densely covered with hydrophobic hairs that provide a nonwettable surface, the plastron, where air remains permanently. This device permits the insect to remain submerged without time limits. In the bug *Aphelocheirus*, there are 2 million single hairs per square millimeter of plastron surface, each single hair being 5 μm long (Thorpe and Crisp 1947). The plastron breathers have a tremendous advantage above those that carry a single bubble as a diffusion gill, for water will not penetrate between the hydrophobic hairs. As a result the plastron serves as a noncompressible gill into which oxygen diffuses from the water. Because the volume remains constant, the net diffusion of nitrogen in this gill must be zero, and the total gas pressure within the air mass is negative relative to the gas tensions in the surrounding water. This is possible because the surface film of the water is supported by the hydrophobic hairs, and to force water into the air spaces between the hairs takes a pressure of 3.5 to 5 atm.

Internal oxygen store

The hemipteran insects, *Buenoa* and *Anisops*, which are relatives of the back swimmer *Notonecta*, are remarkable for their ability to remain poised in midwater for several minutes, almost effortlessly. This is most unusual among insects, but it permits these predaceous animals to exploit the midwater zone in oxygen-poor waters. In small temporary bodies of water where fish are absent, the midwater zone is relatively free from predators. The bottom is dangerous because of voracious dragonfly and beetle larvae, and other predators hunt at the surface, but in midwater there is safety as well as little competition for food.

Both *Buenoa* and *Anisops* have large hemoglobin-filled cells in the abdomen. These serve to store oxygen, which is consumed during the dive. The insects carry with them only a small air mass, but this external supply is depleted only slowly while oxygen from the hemoglobin is being used. When the oxygen of the hemoglobin has been exhausted, the external oxygen is rapidly consumed, and the insects lose buoyancy and become heavier than water.

The long-lasting internal oxygen store confers an advantage on these animals. After they have replenished their oxygen store at the surface, they swim actively to descend and then remain poised in midwater where they are in near-neutral buoyancy for an extended period while the oxygen stores are being used. Not until they begin to use the external oxygen store and begin to sink do they return to the surface to replenish the oxygen supply (Miller 1964).

If *Anisops* is placed in air-free water, the duration of the dive is not much shorter. This shows that the relatively small air mass does not act as a gill, as it does in *Notonecta*. However, if the hemoglobin is poisoned with carbon monoxide, the dive is much shortened, for the hemoglobin now carries no oxygen. Even if the insect is permitted to take normal air with it, the situation is unchanged; the oxygen store is absent, and the external air mass does not help appreciably.

Discontinuous or cyclic respiration

Many insects show a peculiar phenomenon characterized by a periodic or cyclic release of large amounts of carbon dioxide. Oxygen is taken up at a more or less constant rate, but carbon dioxide is released during brief periods or bursts. The carbon dioxide bursts may last for a few minutes and alternate regularly with long periods (several hours or even days). A record of a single carbon dioxide burst from the pupa of a silkworm is shown in Figure 2.24. Such bursts may occur from once every week to many times per hour, depending on temperature and metabolic rate.

Insect groups in which cyclic respiration has been observed include roaches, grasshoppers, beetles, larvae and pupae of butterflies and moths (Lepidoptera), and diapausing adults of Lepidop-

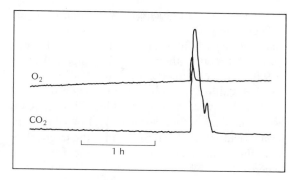

FIGURE 2.24 A record of the gas exchange of the pupa of a *Cecropia* silkworm (at 20 °C) shows steady oxygen uptake. The concurrent release of carbon dioxide is minimal until a major burst occurs after several hours. [Punt et al. 1957]

tera. The phenomenon is rather common, but seems to be characteristic of insects with low metabolic rates. In some it can be produced by lowering the temperature, thus lowering the metabolic rate (Schneiderman 1960).

Between the bursts, during the interburst period, a minute amount of carbon dioxide is being released, but the concurrent oxygen uptake is many times greater, in some cases 20 or even 100 times as high. It is clear that gas exchange takes place through the spiracles and not through the cuticle, for if all the spiracles are sealed with wax, there is no detectable oxygen uptake or carbon dioxide release (i.e., the cuticle is highly impermeable to gases). This means that during the interburst period the spiracles permit inward passage of oxygen while virtually no carbon dioxide escapes. How is this possible? This and many other questions need answers.

Amount of carbon dioxide released

In a *Cecropia* pupa that weighs 6 g, more than 0.5 cm^3 CO_2 is released during a single burst. This volume exceeds the total volume of the tracheal system, so that even if the tracheal system were filled with carbon dioxide only, the carbon dioxide would still have to come from the blood or tissues during the short duration of the burst.

Accurate measurements of the carbon dioxide released from the pupa of another moth, *Agapema*, showed the tracheal volume to be 60 mm³. Just before a burst, the tracheal air contained about 5.9% CO_2. This means that 3.5 mm³ CO_2 was present within the tracheal system when the burst started. However, during the burst 30 mm³ CO_2 was released; in other words, only 10% of the carbon dioxide could have come from the tracheal air and the remainder must have come from elsewhere.

Is metabolism a cyclic event?

The record in Figure 2.24 shows that the oxygen uptake, although continuing between the bursts, was intensified during the burst. This periodic nature of both carbon dioxide release and oxygen uptake raises the question of whether the metabolism of the pupa is cyclic, with short periods of sudden or cataclysmic carbon dioxide production interspersed with long quiet periods of virtual metabolic standstill.

This question can be answered by a simple experiment. If a single spiracle is kept permanently open by inserting a fine glass tube through the opening, the cyclic phenomenon disappears. We can, therefore, discount the hypothesis of cyclic biochemical events in the metabolism. It then becomes necessary to examine the function of the spiracles in relation to the cycles.

Movements of the spiracles

Direct observation of the spiracles with a microscope shows that they are widely open during the carbon dioxide burst. Shortly after the burst, the valve of each spiracle flutters for a short while, and then closes completely and remains closed for as long as an hour. Following this period the valve begins to pulse or flutter faintly, slightly enlarging the spiracular opening. The flutter continues,

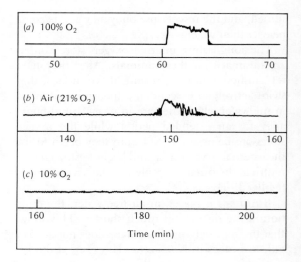

FIGURE 2.25 Record of the movements of the spiracular opening of a *Cecropia* pupa. With the pupa in pure oxygen (a), there is no flutter before the burst. With the pupa in atmospheric air (b), the burst is preceded by flutter. With the pupa in 10% oxygen (c), the spiracle flutters continuously. [Schneiderman 1960]

often for several hours, until the next period of wide opening and burst of carbon dioxide.

The movements of the spiracles of a *Cecropia* silkworm pupa are shown in Figure 2.25. The top record (a) was obtained with the insect in pure oxygen. After a long period closed, the spiracles opened, remained open for about 4 minutes, and then closed again. In atmospheric air (b), after a long closed period, the spiracles fluttered for about 10 minutes before they opened completely; then they fluttered for some time and gradually closed and remained closed until the next fluttering period started. The third graph (c) shows that 10% oxygen caused the spiracles to flutter continuously.

During the long period of flutter, when the spiracles are slightly open, the release of carbon dioxide is minimal, but oxygen uptake is almost continuous. We therefore are faced with the perplexing fact that, when the spiracle flutters, oxygen enters at a rate up to 100 times as high as that with which carbon dioxide leaves.

How can the spiracles, which are the only site of

gas exchange, allow oxygen to enter but effectively prevent carbon dioxide from leaving? To answer this question we need information about the gas concentrations inside the tracheal system.

The insects we have discussed usually have no visible respiratory movements, and we might therefore assume that gas exchange between the tracheal system and outside air takes place by diffusion. However, the gas concentrations actually found in the tracheal system cannot be explained by diffusion processes only.

Schneiderman and his collaborators have succeeded in removing minute gas samples for analysis (Figure 2.26). As expected, they found the lowest carbon dioxide concentration, about 3%, immediately after a burst, and during the next 6 hours the carbon dioxide concentration rose very slowly to 6.5%. During the burst the oxygen concentration increased to nearly atmospheric concentrations, 18 to 20%, but after the burst it declined rapidly again and remained constant at 3.5% throughout the interburst period. These amazing figures cannot be explained if diffusion is the only driving force for the movement of gases through the spiracles. However, gases can also be moved by pressure, and it was suggested by Buck (1958) that a slightly negative pressure within the tracheal system would cause a mass flow of air in through the spiracle.

It is possible to measure the pressure within the tracheal system by sealing a small glass capillary into one spiracle while leaving the others intact, and then connecting the tube to a manometer. During periods of flutter the pressure within the tracheal system is very close to atmospheric, although slightly negative. This slightly negative pressure causes an inward mass flow of air. When the spiracles close completely, the pressure in the tracheal system drops precipitously. As the spiracles remain closed, a considerable negative pres-

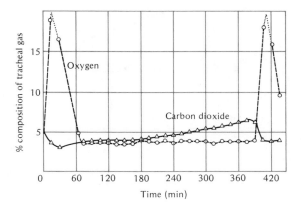

FIGURE 2.26 Composition of air obtained from the tracheal system at various stages of cyclic respiration. A sudden rise in oxygen concentration occurs when the spiracles are fully open during the burst. Pupal weight, 8.4 g; temperature, 25 °C. [Levy and Schneiderman 1958]

sure develops, corresponding to the rapid decrease in oxygen concentration within the tracheal system immediately after the closing of the spiracles (Figure 2.26). When flutter begins again, the pressure rises stepwise, and when flutter continues regularly, the pressure has again returned to near-atmospheric (Levy and Schneiderman 1966b).

The rapid drop in pressure when the spiracles close is caused by the rapid consumption of oxygen from the closed tracheal system. Most of the carbon dioxide produced remains dissolved or buffered in blood and tissues, and therefore a negative pressure must develop. If the pupa is placed in pure oxygen before the spiracles close, so that the tracheal system is filled with oxygen only, the negative pressure becomes even greater. This is precisely what would be expected when the oxygen is consumed and no nitrogen is present.

If we combine all the information at hand, the facts are consistent with the following interpretation (Figure 2.27). The top graph (Figure 2.27a) shows flutter, followed by a period of open spiracles, and then a period of completely closed spiracles, after which the flutter is resumed. The carbon dioxide output (b) is low during the closed and flutter periods; most of the carbon dioxide output

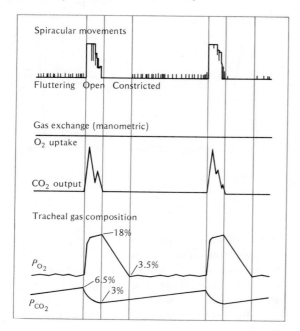

FIGURE 2.27 Summary graph, showing the synchronous events in spiracular movements during cyclic respiration. [Levy and Schneiderman 1966a]

Spiracular movements

Fluttering Open Constricted

Gas exchange (manometric)

O_2 uptake

CO_2 output

Tracheal gas composition

—18%

3.5%

P_{O_2}

6.5%

3%

P_{CO_2}

takes place in a short burst during the open period. The oxygen uptake is recorded as being continuous; this corresponds to the actual consumption of oxygen by the tissues and can be measured by suitable manometric methods (Schneiderman and Williams 1955).

The composition of the tracheal air (Figure 2.27c) shows that the carbon dioxide concentration falls during the burst from 6.5% to 3%, and again rises slowly during the interburst period. The changes in the oxygen concentration are more spectacular. When the spiracles open the oxygen rises rapidly to 18%, not far from atmospheric concentration. During the ensuing constricted period, the tracheal oxygen concentration falls steadily as the tissues use oxygen at a constant rate. When flutter begins, a negative pressure has already developed within the tracheal system, and inward

bulk flow of air begins. This partly replenishes the oxygen, but nitrogen also enters and gradually increases. If there is, say, 4% carbon dioxide and 4% oxygen in the tracheal system, the remainder must be nitrogen in a concentration of 92%.

Significance of cyclic respiration

Terrestrial animals are vulnerable to desiccation and insects often live in very dry environments. The diapausing pupa, which may remain encased for nearly a year from one summer to the next, has no possibility for replenishing lost water. It must be extremely economical in the use of water while permitting sufficient gas exchange for its metabolic processes. The spiracles must therefore be governed by a compromise between permitting oxygen uptake and preventing water loss.

We can now see a rationale for the periodic opening of the spiracles. The water loss depends on the vapor pressure within the tracheae (a function of temperature), the humidity in the outside atmosphere, and the spiracular opening and closing. If the spiracles are closed, water loss from the tracheal system is nil; when they are fully open, water loss is maximal. As we have seen, partial closing (flutter) permits a sufficient oxygen uptake, for in addition to the inward mass flow of air, the high oxygen concentration in the atmosphere gives a steep inward diffusion gradient. However, as the carbon dioxide concentration in the tracheal system builds up and reaches a limit of 6% or so, the spiracles open widely and permit this gas to escape (the carbon dioxide burst). If, instead, the carbon dioxide were permitted to leave continuously through open spiracles, as fast as it is formed, water would also be lost continuously. The cyclic nature of the gas exchange, which at first glance appears as a puzzling deviation, can be considered a clever device for water conservation.

We have now seen how gas exchange takes place in the respiratory organs and how special adaptations serve to provide for an adequate uptake of oxygen and the concomitant discharge of carbon dioxide. In the next chapter we shall discuss how the blood serves in the transport of these two important gases between the respiratory organs and the metabolizing tissues.

REFERENCES

Alkalay, I., Suetsugu, S., Constantine, H., and Stein, M. (1971) Carbon dioxide elimination across human skin. *Am. J. Physiol.* 220:1434–1436.

Bentley, P. J., Herreid, C. F., and Schmidt-Nielsen, K. (1967) Respiration of a monotreme, the echidna, *Tachyglossus aculeatus. Am. J. Physiol.* 212:957–961.

Bentley, P. J., and Shield, J. W. (1973) Ventilation of toad lungs in the absence of the buccopharyngeal pump. *Nature, Lond.* 243:538–539.

Berg, T., and Steen, J. B. (1965) Physiological mechanisms for aerial respiration in the eel. *Comp. Biochem. Physiol.* 15:469–484.

Bretz, W. L., and Schmidt-Nielsen, K. (1972) Movement of gas in the respiratory system of the duck. *J. Exp. Biol.* 56:57–65.

Buck, J. (1958) Cyclic CO_2 release in insects. 4. A theory of mechanism. *Biol. Bull.* 114:118–140.

Cameron, J. N., and Mecklenburg, T. A. (1973) Aerial gas exchange in the coconut crab, *Birgus latro,* with some notes on *Gecarcoidea lalandii. Respir. Physiol.* 19:245–261.

Clements, J. A., Nellenbogen, J., and Trahan, H. J. (1970) Pulmonary surfactant and evolution of the lungs. *Science* 169:603–604.

Craig, A. B., Jr. (1961a) Causes of loss of consciousness during underwater swimming. *J. Appl. Physiol.* 16:583–586.

Craig, A. B., Jr. (1961b). Underwater swimming and loss of consciousness. *J.A.M.A.* 176:255–258.

Dolk, H. E., and Postma, N. (1927) Ueber die Haut- und die Lungenatmung von *Rana temporaria. Z. Vergl. Physiol.* 5:417–444.

Ege, R. (1915) On the respiratory function of the air stores carried by some aquatic insects (Corixidae, Dytiscidae, and *Notonecta*). *Z. Allg. Physiol.* 17:81–124.

Gans, C., De Jongh, H. J., and Farber J. (1969) Bullfrog (*Rana catesbeiana*) ventilation: How does the frog breathe? *Science* 163:1223–1225.

Gatz, R. N., Crawford, E. C., Jr., and Piiper, J. (1974) Respiratory properties of the blood of a lungless and gillless salamander, *Desmognathus fuscus. Respir. Physiol.* 20:33–41.

Graham, J. B. (1974) Aquatic respiration in the sea snake *Pelamis platurus. Respir. Physiol.* 21:1–7.

Herreid, C. F., Bretz, W. L., and Schmidt-Nielsen, K. (1968) Cutaneous gas exchange in bats. *Am. J. Physiol.* 215:506–508.

Hutchison, V. H., Whitford, W. G., and Kohl, M. (1968) Relation of body size and surface area to gas exchange in anurans. *Physiol. Zool.* 41:65–85.

Johansen, K. (1968) Air breathing in the teleost *Symbranchus marmoratus. Comp. Biochem. Physiol.* 18:383–395.

Johansen, K. (1968) Air-breathing fishes. *Sci. Am.* 219:102–111.

Johansen, K., Lenfant, C., Schmidt-Nielsen, K., and Petersen, J. A. (1968) Gas exchange and control of breathing in the electric eel, *Electrophorus electricus. Z. Vergl. Physiol.* 61:137–163.

Johnston, A. M., and Jukes, M. G. M. (1966) The respiratory response of the decerebrate domestic hen to inhaled carbon dixoide–air mixture. *J. Physiol* 184:38–39P.

Joshi, M. C., Boyer, J. S., and Kramer, P. J. (1965) Growth, carbon dioxide exchange, transpiration, and transpiration ratio of pineapple. *Botan. Gaz.* 126:174–179.

Krogh, A. (1920) Studien über Tracheenrespiration. 2 Ueber Gasdiffusion in den Tracheen. *Pflügers Arch.* 179:95–112.

Lasiewski, R. C., and Calder, W. A., Jr. (1971) A preliminary allometric analysis of respiratory variables in resting birds. *Respir. Physiol.* 11:152–166.

Lenfant, C., Johansen, K., and Hanson, D. (1970) Bimodal gas exchange and ventilation–perfusion relationship in lower vertebrates. *Fed. Proc.* 29:1124–1129.

Levy, R. I., and Schneiderman, H. A. (1958) An experimental solution to the paradox of discontinuous respiration in insects. *Nature, Lond.* 182:491–493.

Levy, R. I., and Schneiderman, H. A. (1966a) Discontinuous respiration in insects. 2. The direct measurement and significance of changes in tracheal gas composition during the respiratory cycle of silkworm pupae. *J. Insect Physiol.* 12:83–104.

Levy, R. I., and Schneiderman, H. A. (1966b) Discontinuous respiration in insects. 4. Changes in intra-tracheal pressure during the respiratory cycle of silkworm pupae. *J. Insect. Physiol* 12:465–492.

Margaria, R., Milic-Emili, G., Petit, J. M., and Cavagna, G. (1960) Mechanical work of breathing during muscular exercise. *J. Appl. Physiol.* 15:354–358.

Mellanby, K. (1934). The site of loss of water from insects. *Proc. R. Soc. Lond. B,* 116:139–149.

Miller, P. L. (1964) The possible role of haemoglobin in *Anisops* and *Buenoa* (Hemiptera:Notonectidae). *Proc. R. Entomol. Soc. Lond.* A39:166–175.

Otis, A. B. (1954) The work of breathing. *Physiol. Rev.* 34:449–458.

Punt, A., Parser, W. J., and Kuchlein, J. (1957) Oxygen uptake in insects with cyclic CO_2 release. *Biol. Bull.* 112:108–119.

Rahn, H., and Paganelli, C. V. (1968) Gas exchange in gas gills of diving insects. *Respir. Physiol.* 5:145–164.

Rahn, H., Rahn, K. B., Howell, B. J., Gans, C., and Tenney, S. M. (1971) Air breathing of the garfish (*Lepisosteus osseus*). *Respir. Physiol.* 11:285–307.

Romijn, C., and Roos, J. (1938) The air space of the hen's egg and its changes during the period of incubation. *J. Physiol.* 94:365–379.

Scheid, P., and Piiper, J. (1972) Cross-current gas exchange in avian lungs: Effects of reversed parabronchial air flow in ducks. *Respir. Physiol.* 16:304–312.

Schmidt-Nielsen, K. (1972) *How Animals Work.* London: Cambridge University Press. 114 pp.

Schmidt-Nielsen, K., Kanwisher, J., Lasiewski, R. C., Cohn, J. E., and Bretz, W. L. (1969) Temperature regulation and respiration in the ostrich. *Condor* 71:341–352.

Schneiderman, H. A. (1960) Discontinuous respiration in insects: Role of the spiracles. *Biol. Bull.* 119:494–528.

Schneiderman, H. A., and Williams, C. M. (1955) An experimental analysis of the discontinuous respiration of the cecropia silkworm. *Biol. Bull* 109:123–143.

Schumann, D., and Piiper, J. (1966) Der Sauerstoffbedarf der Atmung bei Fischen nach Messungen an der narkotisierten Schleie (*Tinca tinca*). *Pflügers Arch.* 288:15–26.

Soum, J. M. (1896) Recherches physiologiques sur l'appareil respiratoire des oiseaux. *Ann. Univ. Lyon* 28:1–126.

Szarek, S. R., Johnson, H. B., and Ting, I. P. (1973) Drought adaptation in *Opuntia basilaris*: Significance of recycling carbon through crassulacean acid metabolism. *Plant Physiol.* 52:539–541.

Tenney, S. M., and Remmers, J. E. (1963) Comparative quantitative morphology of the mammalian lung: Diffusing area. *Nature, Lond.* 197:54–56.

Thomas, H. J. (1954) The oxygen uptake of the lobster (*Homarus vulgaris* Edw.). *J. Exp. Biol.* 31:228–251.

Thorpe, W. H., and Crisp, D. J. (1947) Studies on plastron respiration. 1. The biology of *Aphelocheirus* [Hemiptera, Aphelocheiridae (Naucoridae)] and the mechanism of plastron retention. *J. Exp. Biol.* 24:227–269.

Tucker, V. A. (1968) Respiratory physiology of house sparrows in relation to high-altitude flight. *J. Exp. Biol.* 48:55–66.

Wangensteen, O. D., Wilson, D., and Rahn, H. (1971) Diffusion of gases across the shell of the hen's egg. *Respir. Physiol.* 11:16–30.

Wangensteen, O. D., Wilson, D., and Rahn, H. (1974) Respiratory gas exchange of high altitude adapted chick embryos. *Respir. Physiol.* 21:61–70.

Wigglesworth, V. B. (1972) *The Principles of Insect Physiology,* 7th ed. London: Chapman & Hall. 827 pp.

ADDITIONAL READING

Altman, P. L., and Dittmer, D. S. (eds.) (1971) *Biological Handbooks: Respiration and Circulation.* Bethesda:

Federation of American Societies for Experimental Biology. 930 pp.

Buck, J. (1962) Some physical aspects of insect respiration. *Annu. Rev. Entomol.* 7:27–56.

Comroe, J. H., Jr. (1965) *Physiology of Respiration: An Introductory Text.* Chicago: Year Book. 245 pp.

Duncker, H.-R. (1972) Structure of avian lungs. *Respir. Physiol.* 14:44–63.

Fenn, W. O., and Rahn, H. (eds.) (1964, 1965) *Handbook of Physiology*, sect. 3, *Respiration*, vol. 1, pp. 1–926 (1964); vol. 2, pp. 927–1696 (1965).

Heath, A. G., and Mangum, C. (eds.) (1973) Influence of the environment on respiratory function. *Am. Zool.* 13:446–563.

Hughes, G. A. (1963) *Comparative Physiology of Vertebrate Respiration.* Cambridge, Mass.: Harvard University Press. 145 pp.

Johansen, K., et al. (1970) Symposium on cardiorespiratory adaptations in the transition from water breathing to air breathing. *Fed. Proc.* 29(3):1118–1153.

Jones, J. D. (1972) *Comparative Physiology of Respiration.* London: Edward Arnold. 202 pp.

Miller, P. L. (1964) Respiration: Aerial gas transport. In *The Physiology of Insecta*, vol. III (M. Rockstein, ed.), pp. 557–615. New York: Academic Press.

Piiper, J., and Scheid, P. (1977) Comparative physiology of respiration: Functional analysis of gas exchange organs in vertebrates. *Int. Rev. Physiol.* 14:219–253.

Weibel, E. R. (1973) Morphological basis of alveolar-capillary gas exchange. *Physiol. Rev.* 53:419–495.

3

CHAPTER THREE

Blood

The preceding two chapters dealt with respiratory gases and the gas exchange between the organism and the environment. They focused on the two most important gases, oxygen and carbon dioxide.

We learned that in all but very small organisms diffusion alone does not suffice to distribute the gases within the organism and that a mechanical transport system is necessary. Nearly every large animal, except some that have very low oxygen demands, has a distribution system designed around the mechanical movement in the organism of a fluid: blood. Insects do not use blood for gas transport; they use air-filled tubes. Yet insects do have blood that is pumped around in the body, for many other substances need to be transported at rates faster than diffusion alone can provide.

Blood, in fact, has many functions that may not be immediately apparent. This becomes clear when we list the major functions served by this important fluid (Table 3.1).

We tend to think of gas transport as the primary function of blood, but the list notes six major categories of substances (in addition to heat) that are transported by blood. A very important although often overlooked function is the transmission of force. Of the two last items, one is concerned with a highly specific inherent characteristic of blood, the ability to coagulate; the other is concerned with the general chemical composition of the blood fluid.

Most of these functions can be carried out by nearly any aqueous medium. For example, no special characteristics of the blood are necessary to carry metabolites or excretory products. Exceptions are gas transport and coagulation, both of which are associated with highly complex biochemical properties of the blood.

In this chapter we shall be concerned primarily with the role of blood in the transport of gases and those properties of blood that serve this purpose.

TABLE 3.1 The most important functions of blood.

1. *Transport of nutrients* from digestive tract to tissues; to and from storage organs (e.g., adipose tissue, liver)
2. *Transport of metabolites* (e.g., lactic acid from muscle to liver), enabling metabolic specialization
3. *Transport of excretory products* from tissues to excretory organs; from organ of synthesis (e.g., urea in liver) to kidney
4. *Transport of gases* (oxygen and carbon dioxide) between respiratory organs and tissues; storage of oxygen
5. *Transport of hormones* (e.g., adrenaline [fast response]; growth hormone [slow response])
6. *Transport of cells* of nonrespiratory function (e.g., vertebrate leukocytes); insect blood lacks respiratory function, yet carries numerous types of blood cells
7. *Transport of heat* from deeper organs to surface for dissipation, (essential for large animals with high metabolic rates)
8. *Transmission of force* (e.g., for locomotion in earthworms; for breaking shell during molting in crustaceans; for movement of organs such as penis, siphon of bivalve, extension of legs in spider; for ultrafiltration in capillaries of the kidneys)
9. *Coagulation,* inherent characteristic of many blood and hemolymph fluids; serves to protect against blood loss
10. *Maintenance of "milieu interieur"* suitable for cells in regard to pH, ions, nutrients, etc.

OXYGEN TRANSPORT IN BLOOD

Respiratory pigments

In many invertebrates oxygen is carried in the blood or hemolymph in simple physical solution. This aids in bringing oxygen from the surface to the various parts of the organism, for diffusion alone is too slow for any but the smallest organisms.

The amount of oxygen that can be carried in simple solution is small, however, and many highly organized animals (vertebrates almost without exception) have blood that can bind larger quantities of oxygen reversibly, thus greatly increasing the amount of oxygen carried. In mammalian blood, the amount of physically dissolved oxygen is about 0.2 ml O_2 per 100 ml blood, and the amount bound reversibly to hemoglobin is up to some 100 times as great, about 20 ml O_2 per 100 ml blood. Dissolved oxygen is therefore of minuscule importance compared with the oxygen in hemoglobin.

The substances we know as oxygen carriers in blood are proteins that contain a metal (commonly iron or copper). Usually they are colored, and therefore they are often called *respiratory pigments*. The commonest ones are listed in Table 3.2.

Blood corpuscles

In some animals the respiratory pigment occurs dissolved in the blood fluid; in others (such as vertebrates) it is enclosed in cells and the blood fluid contains no dissolved respiratory pigment. The distribution between cells and blood fluid for the most common respiratory pigments is listed in Table 3.3, which in addition gives the molecular weights of the pigments in representative animals.

This table immediately reveals that when the pigments occur enclosed in cells, their molecular weights are relatively low, ranging from 20 000 to 120 000; if the pigments occur dissolved in plasma, their molecular weights are much higher, from 400 000 to several million. The sole exception is the hemoglobin in the insect *Chironomus*.

The large molecules of respiratory pigments dissolved in the plasma are, in fact, aggregates of smaller molecules. They increase the total amount of the pigment without increasing the number of

TABLE 3.2 Common respiratory pigments and examples of their occurrence in animals.

Pigment	Description	Molecular weight	Occurrence in animals
Hemocyanin	Copper-containing protein, carried in solution *blue when oxygenated*	300 000– 9 000 000	**Molluscs**: chitons, cephalopods, prosobranch and pulmonate gastropods; not lamellibranchs **Arthropods**: crabs, lobsters **Arachnomorphs**: *Limulus, Euscorpius*
Hemerythrin	Iron-containing protein, always in cells, nonporphyrin structure *purple when oxygenated*	108 000	**Sipunculids**: all species examined **Polychaetes**: *Magelona* **Priapulids**: *Halicryptus, Priapulus* **Brachiopods**: *Lingula*
Chlorocruorin	Iron-porphyrin protein, carried in solution *green*	2 750 000	Restricted to four families of Polychaetes: Sabellidae, Serpulidae, Chlorhaemidae, Ampharetidae Prosthetic group alone found in starfish (*Luidia, Astropecten*)
Hemoglobin	Iron-porphyrin protein, carried in solution or in cells; most extensively distributed pigment	17 000– 3 000 000	**Vertebrates**: almost all, except leptocephalus larvae and some Antarctic fish (*Chaenichthys*) **Echinoderms**: sea cucumbers **Molluscs**: *Planorbis*, Pismo clam (*Tivela*) **Arthropods**: insects (*Chironomus, Gastrophilus*); crustacea (*Daphnia, Artemia*) **Annelids**: *Lumbricus, Tubifex, Arenicola, Spirorbis* (some species have hemoglobin, some chlorocruorin, others no blood pigment), *Serpula* (both hemoglobin and chlorocruorin) **Nematodes**: *Ascaris* **Flatworms**: parasitic trematodes **Protozoa**: *Paramecium, Tetrahymena* **Plants**: yeasts, *Neurospora*, root nodules of leguminous plants (clover, alfalfa)

dissolved protein molecules in the blood plasma. A large increase in the number of dissolved particles would raise the colloidal osmotic pressure of the plasma, which in turn would influence many other physiological processes such as the passage of fluid through capillary walls and ultrafiltration (the initial process in the formation of urine in the kidney).

It also has been said that enclosing a high concentration of hemoglobin within cells has the advantage of decreasing the viscosity of the blood as compared with a hemoglobin solution with the same oxygen-carrying capacity. Mammalian plasma contains about 7% protein, and the red blood cells contain about 35% hemoglobin. If all the hemoglobin from the red cells were carried dissolved in the blood instead of enclosed in the red blood cells, the total protein concentration would be about 20%, or three times as high as the typical mammalian plasma protein concentration.

The argument goes that a 20% protein solution is highly viscous or syrupy – far more viscous than blood. However, if we first measure the viscosity of a blood sample, and then disrupt the red cells with ultrasound so that the hemoglobin is set free, the resulting hemoglobin solution is less than half as

TABLE 3.3 Molecular weights and location (in cells or in plasma) of respiratory pigments in various animals.

	In cells		In plasma	
Pigment	Animal	Molecular weight	Animal	Molecular weight
Hemoglobin	Mammals	ca. 68 000[a]	Oligochaetes	
	Birds	ca. 68 000[a]	*Lumbricus*	2 946 000
	Fish	ca. 68 000[a]	Polychaetes	
	Cyclostomes		*Arenicola*	3 000 000
	Lampetra	19 100	*Serpula*	3 000 000
	Myxine	23 100	Molluscs	
	Polychaetes		*Planorbis*	1 539 000
	Notomastus	36 000	Insects	
	Echinoderms		*Chironomus*	31 400
	Thyone	23 600		
	Molluscs			
	Arca	33 600		
	Insects			
	Gastrophilus	34 000		
Chlorocruorin			Polychaetes	
			Spirographis	3 400 000
Hemerythrin	*Sipunculus*	66 000		
	Phascolosoma	120 000		
Hemocyanin			Molluscs	
			Helix	6 680 000
			Cephalopods	
			Rossia (squid)	3 316 000
			Octopus	2 785 000
			Eledone	2 791 000
			Arthropods	
			Limulus	1 300 000
			Crustacea	
			Pandalus	397 000
			Palinurus	447 000
			Nephrops	812 000
			Homarus	803 000

[a]The molecular weight of muscle hemoglobin (myoglobin) is 17 000.

viscous as blood with the hemoglobin carried in the cells (Figure 3.1). Nevertheless, this is not the full answer, for the viscous properties of blood are very complex and will be discussed further in Chapter 4.

Aside from viscosity, other advantages are gained by enclosing hemoglobin in cells. The most important is probably that the chemical environment within the red cell can differ from that of the blood plasma. The reaction between oxygen and hemoglobin is greatly influenced by inorganic ions as well as by certain organic compounds, no-

FIGURE 3.1 Blood viscosity increases with increasing hemoglobin content. Blood viscosity is here expressed relative to the viscosity of pure water (water = 1.00). The viscosity of goat blood containing intact red cells is shown in the upper curve. The lower curve shows the relative viscosity of hemolyzed blood. The observed plasma viscosity of the same blood samples is shown by the dashed line. A comparison of points A, B, and C shows that a hemoglobin solution has a relative viscosity about half that of intact blood with the same total hemoglobin content. [Schmidt-Nielsen and Taylor 1968]

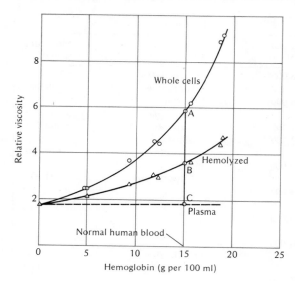

tably organic phosphates, and hemoglobin located within cells can be provided with a separate well-adjusted environment different from the plasma.

Size and shape of the red cell

The red blood cells of mammals are round, slightly biconcave discs. There is one exception: All members of the camel family have oval red cells. The diameter of the red cell is characteristic for each species, the range for mammals being from about 5 to 10 μm. There is no particular relationship between the size of the red cell and the size of the animal: The smallest mammalian red cell is found in a small Asiatic deer, but a mouse and an elephant have red cells of approximately the same size.

Mammalian red cells lack nuclei, although they have a life span of about 100 days and carry out complex metabolic functions. The red cells of all other vertebrates – birds, reptiles, amphibians, and fish – have nuclei and are almost universally oval. Frequently they are much larger than mammalian cells; the red cells of salamanders, for example, have 100 or 200 times as great a volume as a mammalian red cell. The importance, if any, of a nucleated versus a non-nucleated red cell is not understood. Neither is it clear whether the size of the red cell is of functional importance.

The only vertebrates that lack hemoglobin and red cells are the larvae of eels (leptocephalus larvae) and a few Antarctic fish of the family Chaenichthyidae.

Oxygen dissociation curves

The property of blood that is of greatest importance to oxygen transport is the reversible binding of oxygen to the hemoglobin molecule. This binding can be written as an ordinary reversible chemical reaction:

$$Hb + O_2 \rightleftarrows HbO_2$$

At high oxygen concentration the hemoglobin (Hb) combines with oxygen to form oxyhemoglobin (HbO_2) and the reaction goes to the right. At low concentration, oxygen is given up again and the reaction proceeds to the left. If the oxygen concentration is reduced to zero, the hemoglobin gives up all the oxygen it carries.

Each iron atom in the hemoglobin molecule binds one oxygen molecule, and when at high oxygen concentration (or pressure) all available binding sites are occupied, the hemoglobin is fully saturated and cannot take up any more oxygen. At any given oxygen concentration there is a definite proportion between the amounts of hemoglobin and oxyhemoglobin. If we plot the amount of oxyhemoglobin present at each oxygen concentration, we obtain an oxygen–hemoglobin dissociation curve, which describes how the reaction depends on the oxygen concentration or partial pressure. Such a dissociation curve is depicted in Figure

FIGURE 3.2 Oxygen dissociation curve for pigeon blood: *a*, curve determined for normal temperature (41 °C), blood P_{CO_2} (35 mm Hg), and pH (7.5) for this bird; *b*, curve determined after shifting the pH from 7.5 to 7.2 with unchanged P_{CO_2}. [Lutz et al. 1973]

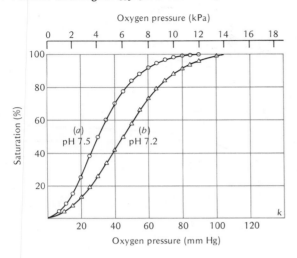

3.2. This curve shows that pigeon blood is fully saturated with oxygen (100% HbO_2) at or above about 80 to 100 mm Hg (10 to 13 kPa). At still higher oxygen pressure, the hemoglobin takes on no more oxygen, and we say it is fully saturated. At lower oxygen pressure, however, it gives off oxygen, and at $Po_2 = 30$ mm Hg (4.0 kPa), half the hemoglobin is present as oxyhemoglobin, the other half as hemoglobin. This particular oxygen pressure is referred to as the *half-saturation pressure* (P_{50}). As the oxygen pressure is decreased further, more oxygen is given off, until the hemoglobin is fully deoxygenated at zero oxygen pressure.

The oxygen dissociation curves of other animals have a similar shape, but are not identical to the one shown in Figure 3.2. The blood of some animals has a higher affinity for oxygen (it gives up oxygen less readily), and the dissociation curve is located further to the left. The blood of others gives up oxygen more readily (the affinity for oxygen is lower), and the dissociation curve is located further to the right. In addition, other parameters (pH, ions, organic phosphates, temperature) influence the binding of oxygen and thus the dissociation curve. These variations are of great functional significance.

Effect of temperature

Increased temperature weakens the bond between hemoglobin and oxygen and causes an increased dissociation of the bond. As a consequence, at higher temperature the hemoglobin gives up oxygen more readily and the dissociation curve is shifted to the right. This is of physiological importance because increased temperature is usually accompanied by an increased metabolic rate (or need for oxygen); therefore, it is an advantage that hemoglobin delivers oxygen more readily at higher temperature. Even in warm-blooded animals this may be of some importance — in fever or in exercise, for example, when there is a rise in the rate of oxygen consumption.

Effect of carbon dioxide and pH

Another important influence on the dissociation curve is the pH of the blood plasma. Increase in carbon dioxide or other acids lowers the pH of the plasma, and this shifts the dissociation curve to the right (Figure 3.2). The significance of this shift is that a high carbon dioxide concentration causes more oxygen to be given up at any given oxygen pressure. This effect is known as the *Bohr effect*. In the capillaries of the tissues, as carbon dioxide enters the blood, the hemoglobin gives up a larger amount of oxygen than would be the case if there were no effect of carbon dioxide on the binding. The Bohr effect can therefore be said to facilitate the delivery of additional oxygen to the tissues.

The Bohr effect is not of equal magnitude in all mammals. We can express the Bohr effect as the shift in the P_{50} that occurs for a unit change in pH. We then find that the Bohr effect is dependent on body size and that the shift in P_{50} is greater in a

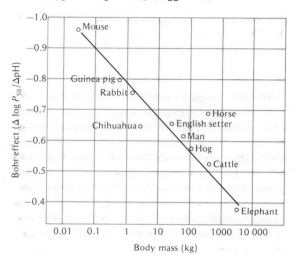

FIGURE 3.3 The Bohr shift of hemoglobin in relation to body size. The hemoglobin of small mammals has a greater Bohr shift (i.e., is more acid-sensitive) than the hemoglobin of large mammals and, therefore, releases more oxygen at a given P_{O_2}. [Riggs 1960]

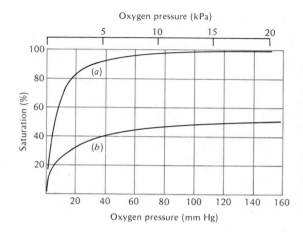

FIGURE 3.4 Oxygen dissociation curves of toadfish blood at 15 °C: a, curve for whole blood in the absence of CO_2; b, curve for whole blood in the presence of 25 mm Hg (3.3 kPa) CO_2. [Root et al. 1939]

mouse than in an elephant (Figure 3.3). In other words, mouse hemoglobin is more acid-sensitive than elephant hemoglobin. We shall later see that this is of importance to the delivery of oxygen to the tissues of small animals, which have a higher metabolic rate per gram than larger animals (Schmidt-Nielsen 1972).

The effect of acid on the dissociation curve is much more pronounced in some fish than in mammals; this stronger effect is known as the *Root effect* (Figure 3.4). The most significant consequence of the Root effect has to do with the function of the blood in the secretion of oxygen into the swimbladder (Chapter 11).

The peculiarity of the Root effect is that the blood is not fully saturated with oxygen at low pH, however high the oxygen pressure. This is explained by the fact that fish blood contains two different hemoglobin species, one highly sensitive to acid and the other acid-insensitive. In the presence of high carbon dioxide the former carried no oxygen, while the latter presents the usual dissocia-

tion curve for this hemoglobin species (Gillen and Riggs 1973).

Why do fish have two different hemoglobins? It is easy to see the value of the acid-sensitive species, but why have another kind that is acid-insensitive? Isn't this unnecessary?

The explanation is that when a fish in an emergency is making a maximal swimming effort, it produces a great deal of lactic acid. If the presence of this acid in the blood made all the hemoglobin incapable of binding oxygen, the fish would simply die of asphyxia. However, the presence of an acid-insensitive hemoglobin avoids this calamity.

Effect of organic phosphates

The function of the red cell in gas transport is intimately related to the metabolic activities of the cell itself. Due to the discovery of the importance of organic phosphates, we can now understand many peculiarities of the oxygen dissociation curve that formerly seemed extremely confusing.

The mammalian red cell has no nucleus, and for many years this cell was considered little more than an inert sac packed full of hemoglobin. The red cell, however, has an active carbohydrate metabolism that not only is essential to the viability of the cell, but also has profound effects on its function in oxygen transport. The red cell also has a high content of adenosine triphosphate (ATP) and an even higher level of 2,3-diphosphoglycerate (DPG). The presence of these organic phosphates helps explain why a purified solution of hemoglobin has a much higher oxygen affinity than whole blood (a hemoglobin solution has a dissociation curve far to the left of the curve for whole blood). If organic phosphates, notably DPG, are added to a hemoglobin solution, its oxygen affinity is greatly decreased and approaches that of intact cells. The decrease is attributable to a combination of the hemoglobin with the DPG, which alters the oxygen affinity (Chanutin and Curnish 1967; Benesch et al. 1968a,b).

The discovery of the DPG effect explains many observations that previously were quite puzzling. It also explains why, if we examine whole blood, we can observe many characteristics of the dissociation curve that remain unnoted if hemoglobin solutions are studied. For example, the DPG level of the red cell is higher in people living at high altitude than in people living at sea level. As a consequence, there is a shift to the right of the oxygen dissociation curve at high altitude. This shift to the right has been well known, but the mechanism remained unclear until it was discovered that the level of DPG increases in response to hypoxia (Aste-Salazar and Hurtado 1944; Lenfant et al. 1968).

The role of DPG and other phosphates explains the long-known fact that the oxygen affinity of blood stored in a blood bank increases with time. It also helps explain why purified fetal and adult hemoglobins, when stripped of DPG, have similar oxygen affinities, although fetal blood has a higher

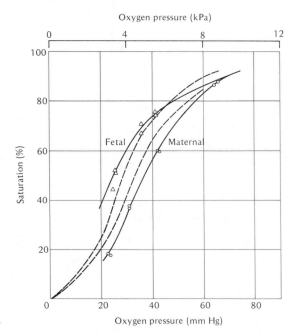

FIGURE 3.5 Oxygen dissociation curves of fetal and maternal blood of goat. The higher oxygen affinity of the fetal blood helps in the transfer of oxygen to the fetal blood in the placenta. Broken lines indicate limits of nonpregnant adult goats. [Barcroft 1935]

oxygen affinity. Furthermore, a shift to the right of the maternal dissociation curve relative to the curve for a nonpregnant adult can now be understood, although the pregnant animal (as opposed to the fetus) has the same type of hemoglobin throughout.

Dissociation curve of fetal hemoglobin

For many mammals, including man, the dissociation curve of fetal blood is located to the left of that of maternal blood (Figure 3.5). This is related to how the fetus obtains its oxygen by diffusion from the maternal blood. Because fetal blood has a higher affinity for oxygen than maternal blood, it can take up oxygen more readily. At a given oxygen pressure fetal blood contains more oxygen (has a higher percent saturation) than maternal

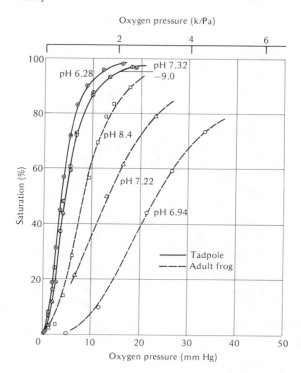

FIGURE 3.6 The oxygen dissociation curve for tadpole blood is located to the left of that for adult frogs, (i.e., tadpole blood has a higher oxygen affinity). Furthermore, tadpole blood is nearly acid-insensitive, whereas adult frog blood has a pronounced Bohr shift. [Riggs 1951]

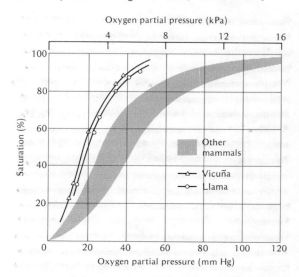

FIGURE 3.7 Oxygen dissociation curves of llama and vicuña blood are located to the left of curves for other mammals (shaded area). The higher oxygen affinity of the blood of these animals aids the oxygen uptake at the low pressure of high altitude. [Hall et al. 1936]

blood at the same oxygen pressure; this facilitates the uptake of oxygen by the fetal blood in the placenta. The difference in dissociation curve between fetal and maternal blood is in part attributable to the fact that fetal hemoglobin is slightly different from maternal hemoglobin, and in part to a difference in the organic phosphate within the red cell. After birth fetal hemoglobin gradually disappears and is replaced by adult-type hemoglobin.

Displacement of the dissociation curve to the left, so that the blood more readily takes up oxygen, is found also in the chick embryo and frog tadpole. Tadpole hemoglobin has a higher oxygen affinity than adult-type frog hemoglobin and, particularly interesting, it is insensitive to acid,

whereas adult-type hemoglobin has a pronounced Bohr effect and is highly acid-sensitive (Figure 3.6). If we consider the normal habitat of tadpoles, we can readily see the advantage; tadpoles often live in stagnant pools with low oxygen content and, at times, also high carbon dioxide content in the water. A high oxygen affinity therefore is an advantage. Because low oxygen and high carbon dioxide are likely to occur at the same time, a pronounced Bohr effect would be undesirable, for it would decrease the oxygen affinity of the blood when a high affinity is most needed.

Effect of altitude

The low atmospheric pressure at high altitude means that the partial pressure of oxygen is lower than at sea level. Animals that normally live in this environment of relative oxygen deficiency have corresponding adaptations in the oxygen dissociation curves of their blood. For example, the dissociation curve of the llama is located to the left of the usual range for mammals (Figure 3.7). The llama lives in the high Andes of South America, often above 5000 m, and the high oxygen affinity of its hemoglobin enables its blood to take up oxygen at the low pressures. This adaptation to altitude is an inborn characteristic, for the oxygen dissociation curve of a llama brought up in a zoo at sea level shows the same high oxygen affinity.

If a man moves to high altitude, he gradually becomes more adapted to the low oxygen pressure; he becomes acclimated to the altitude and performs better, both physically and mentally. However, the adaptation does not include an increased oxygen affinity of the blood. Paradoxically, the dissociation curve moves slightly to the right as altitude tolerance is improved. This is a result of other, more complex physiological adaptations (Lenfant et al. 1969).

The shift to the right of the dissociation curve of humans at high altitude (owing to increased DPG levels) is exactly the opposite of the change in the blood of the llama and other animals native to high altitude. This may be difficult to reconcile, but we should realize that the affinity of hemoglobin for oxygen is important at two different locations in the body: (1) when the hemoglobin takes up oxygen in the lung and (2) when the hemoglobin gives up oxygen to the tissues. For uptake in the lung, a high oxygen affinity (shift to the left) increases the effectiveness of the uptake; in the tissues, on the other hand, a low oxygen affinity (shift to the right) facilitates the delivery of oxygen. It appears that the left-shift of the llama's dissociation curve is primarily an adaptation to improved oxygen uptake at the low pressures of high altitude, and that the right-shift in humans under similar conditions serves to augment the oxygen delivery to the tissues. Thus, a given dissociation curve can be regarded as a trade-off between a desirable high oxygen affinity in the lung and an equally desirable low affinity in the tissues.

The red cells of various mammals do not have equally high DPG levels, nor are their hemoglobins equally sensitive to the DPG level. The red cells of man, horse, dog, rabbit, guinea pig, and rat all have high levels of DPG, and their hemoglobins, when stripped of DPG, have high oxygen affinities. In contrast, sheep, goat, cow, and cat have low concentrations of DPG, and their hemoglobins interact only weakly with DPG (Bunn 1971). At present, these differences are difficult to understand. They seem to follow neither phylogenetic relationships nor any characteristic differences in the biology of the animals.

Dissociation curve in diving animals

It might be expected that diving mammals, such as seals, which are normally exposed to a relative oxygen deficiency when they dive, could gain an advantage by having blood with a high oxygen affinity. This is not so, however, and a little thought will clarify why they would gain no special advantage in this way. Because the seal breathes ordinary atmospheric air, the uptake of oxygen into the blood in the lung takes place under circumstances similar to those in other mammals, and the loading of the blood in the lung does not require any particularly high oxygen affinity. During diving, the unloading of oxygen in the tissues (primarily the central nervous system and the heart muscle) would not be aided by a high oxygen affinity of the blood; on the contrary, perhaps, a high oxygen affinity of the hemoglobin would tend to impede delivery of oxygen to the most vital organs. Accordingly, oxygen dissociation curves of diving animals generally differ little from those of similar nondiving animals.

Dissociation curve and body size

If the dissociation curves for various mammals are plotted together, we see that mammals of large

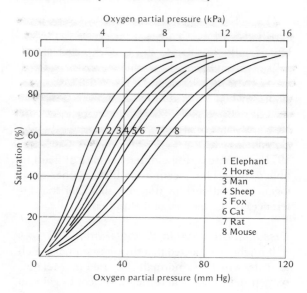

FIGURE 3.8 Oxygen dissociation curves of blood of mammals of various sizes. Small mammals have have a lower oxygen affinity. This helps in the delivery of oxygen in the tissues to sustain the high metabolic rate of a small animal. [Schmidt-Nielsen 1972]

1 Elephant
2 Horse
3 Man
4 Sheep
5 Fox
6 Cat
7 Rat
8 Mouse

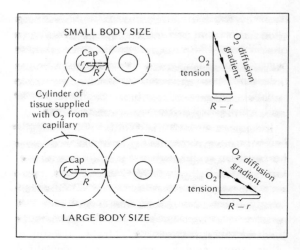

FIGURE 3.9 Each capillary is surrounded by an approximately cylindrical mass of tissue, which is supplied with oxygen by diffusion from the capillary. The steepness of the diffusion gradient depends on the radius of the tissue cylinder (R), the radius of the capillary (r), and the unloading pressure (tension) for oxygen.

body size have curves located to the left, and small animals have curves to the right (Figure 3.8).

As stated before, the loading–unloading reactions of hemoglobin are of importance at two locations: in the lung when oxygen is taken on and then in the tissues when oxygen is unloaded. Figure 3.8 shows that in all mammals the blood is practically 100% saturated at the normal oxygen pressure in the mammalian lung, which is about 100 mm Hg (13 kPa). The blood becomes fully loaded with oxygen in the lung, and there is no great difference among the species in this regard. Let us therefore examine the unloading of oxygen in the tissues.

The blood of a small mammal releases oxygen more readily because of its lower affinity for oxygen. Usually blood does not give up all its oxygen in the tissues; as an approximation we can say that on the average, about half the oxygen is given up. This corresponds to half-saturation of the hemoglobin (50% oxyhemoglobin and 50% hemoglobin), which is a convenient way of describing the location of the dissociation curve. The corresponding oxygen pressure, which we can consider an average unloading pressure, is about 22.5 mm Hg (3.0 kPa) for elephant blood, 28 mm Hg (3.7 kPa) for human blood, and about 45 mm Hg (6.0 kPa) for mouse blood.

These different unloading pressures of the blood can be related to the metabolic need for oxygen. Small animals have a higher rate of oxygen consumption per gram body weight than large animals. For example, the oxygen consumption of a horse weighing 700 kg is 1.7 μl O_2 g^{-1} min^{-1}, and that of a 20-g mouse is 28 μl O_2 g^{-1} min^{-1}, i.e. the mean oxygen consumption of 1 g mouse tissue is about 16 times as high as that of 1 g horse tissue.

It follows that mouse tissue must be supplied with oxygen at a rate 16 times higher than horse tissue, and this can only be achieved by a diffusion gradient from capillary to cells 16 times as steep. This is shown diagrammatically in Figure 3.9. Each capillary supplies the surrounding tissue with oxygen, and the distance over which oxygen must diffuse is determined by the distance to the next capillary, the halfway point being the most distant point to which oxygen must diffuse. Horses and mice have approximately the same size red cells; therefore, their capillaries are also of approximately the same size, and this simplifies the analysis. To achieve a diffusion gradient in the mouse that is 16 times as steep as that in the horse, two variables can be adjusted. First, the diffusion distance can be reduced by reducing the intercapillary distance (i.e., an increase in capillary density in the tissue), and second, the unloading pressure for oxygen can be increased (i.e., a dissociation curve shifted to the right).

The capillary density is indeed larger in the small animal. In a mouse, for example, a 1 mm² cross section of a muscle may contain about 2000 capillaries; 1 mm² cross section of a horse muscle contains less than 1000 capillaries (Schmidt-Nielsen and Pennycuik 1961). Figure 3.8 shows that the unloading pressure (P_{50}) for mouse blood is about 2.5 times higher than for horse blood. We can therefore conclude that the higher metabolic rate in the mouse, and the necessary steep diffusion gradient for oxygen from capillary to tissue cells, is sustained by a combination of a reduced diffusion path and a higher unloading pressure for oxygen.

Respiratory function of fish blood

When we compare the oxygen affinities of the blood of various fishes, we immediately meet some problems that are difficult to resolve. For example, at what temperature should the comparison be made? If we compare blood from two different fish, should they be examined at the same temperature, or should they be examined at the temperature at which each fish normally lives? The question is not easy to answer if we compare a tropical and an Arctic fish, but with fish from similar habitats we can use the same temperature.

The dissociation curves for fish hemoglobin show a wide range in oxygen affinity with characteristic differences related to the activity of the fish (Hall and McCutcheon 1938). The mackerel, an active, fast-swimming fish that normally moves in well-oxygenated water, has a dissociation curve to the right of the other fish. This is in accord with what we have just learned: A low oxygen affinity facilitates the delivery of oxygen to the tissues, and in well-aerated water the arterial blood will still be 100% saturated. The toadfish, whose curve lies to the extreme left, is a slow-moving and relatively sluggish bottom-living fish (Randall 1970). It is often found in less well oxygenated water, and it is highly tolerant of oxygen deprivation. The hemoglobin of the toadfish has a particularly high oxygen affinity, and this is in accord both with the environment in which the fish normally lives and with its relatively low metabolic rate, the demands on the loading mechanism apparently being more important than the unloading in the tissues.

It might seem disadvantageous to a fish that the oxygen affinity of its blood changes with temperature, for it is exposed to considerable temperature changes from season to season, especially if it lives in fresh water. The change in dissociation curve as a fish becomes acclimated to different temperatures is therefore interesting. The dissociation curve for blood of the brown bullhead, *Ictalurus nebulosus* (a catfish), depends on the acclimation temperature (Figure 3.10). The blood of the warm-acclimated fish has a higher oxygen affinity than the blood of a cold-acclimated fish, when measured at the same temperature. Instead, compare the curves measured at the temperature to which each fish has been acclimated (the two curves in the middle of the group of four). The fish acclimated at 9 °C has a curve (when measured at

FIGURE 3.10 Oxygen dissociation curves of warm-acclimated and cold-acclimated bullheads measured at the temperatures marked on each curve; pH 7.8. [Grigg 1969]

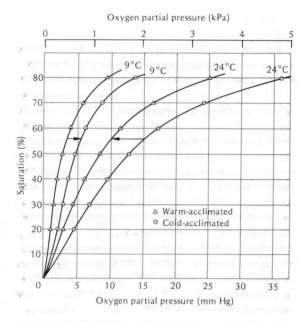

coordinate system. The hemoglobin from a bivalve mollusc, *Phacoides*, for example, has a P_{50} for oxygen of about 0.2 mm Hg (27 Pa) (Read 1962).

The hemoglobin of the *Chironomus* larva also has an extremely high oxygen affinity. The dissociation curve is influenced by both temperature and pH, and within a temperature range of 5 to 24 °C, the P_{50} varies from 0.1 to 0.6 mm Hg (13 to 80 Pa). By mammalian standards this is an extremely high oxygen affinity. Characteristically, these animals live in environments that may periodically have extremely low oxygen concentrations, and it seems that their hemoglobins are of importance primarily when there is a lack of oxygen.

This suggestion is further supported by the fact that those invertebrates that do have hemoglobin often develop much higher hemoglobin concentrations when they are kept in water low in oxygen. This holds for *Daphnia*, the brine shrimp (*Artemia*), *Chironomus* larvae, and many others.

Hemerythrin differs chemically from hemoglobin in that it contains no porphyrin group. It is found only in a few marine forms (see Table 3.2), which show an interesting relationship between oxygen affinity and the ecological conditions under which the animals normally live. The two sipunculid worms, *Dendrostomum* and *Siphonosoma*, have hemerythrin both in their blood and in their coelomic fluid. In *Dendrostomum* the tentacles around the mouth are richly supplied with blood; they are kept above the surface of the sand in which the animal lives and serve respiratory purposes. Characteristically, in *Dendrostomum* the oxygen affinity of the coelomic fluid is higher than that of the blood (Figure 3.11). This means that oxygen taken up by the blood in the respiratory organ is readily transferred to the pigment of the coelomic fluid. In *Siphonosoma*, the blood and coelomic fluid do not differ much; in fact, the blood has a slightly higher oxygen affinity. This is

9 °C) which is located to the left of the curve of the blood from the fish acclimated at 24 °C (and measured at 24 °C). Acclimation brings the two sets of curves closer together, but the warmer fish, which has a higher metabolic rate, is still shifted to the right relative to the cold-acclimated fish.

The differences in the blood and the seasonal changes depend on the characteristics of the red cell, not the hemoglobin. If solutions of hemoglobin from warm- and cold-acclimated fish are examined, there is no measurable difference in the oxygen affinities of the two hemoglobins.

Dissociation curves of invertebrate respiratory pigments

We saw earlier (Table 3.2) that invertebrates have a variety of respiratory pigments. In many, the oxygen affinity is very high, and the dissociation curve is located far to the left in the customary

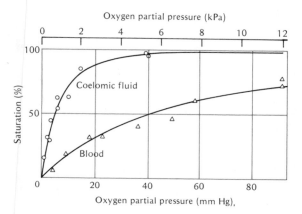

FIGURE 3.11 Oxygen dissociation curves for hemerythrin from blood and coelomic fluid of the sipunculid worm *Dendrostomum zostericolum*. [Manwell 1960]

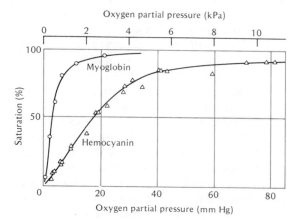

FIGURE 3.12 Oxygen dissociation curves of hemocyanin (from blood) and myoglobin (from radula muscle) of the amphineuran mollusc *Cryptochiton*. [Manwell 1958]

in accord with the habits of the worm: The oral tentacles are used for feeding but not for respiration; the worm remains in its burrow deep below the surface of the mud, and oxygen is taken up through the general body surface. Hence, both blood and coelomic fluids have a similar high oxygen affinity.

A similar relationship in the oxygen affinities of two different pigments is found in the amphineuran mollusc, *Cryptochiton*. This animal's blood contains hemocyanin, and the muscle of its tongue contains myoglobin. The blood has a relatively low oxygen affinity and the myoglobin a high oxygen affinity (Figure 3.12). This situation is completely analogous to what we find in vertebrates. In these, oxygen is transferred from the hemocyanin of the blood to the muscle hemoglobin (myoglobin), which has a far greater oxygen affinity and a dissociation curve displaced to the far left, with a P_{50} of about 0.5 mm Hg (0.07 kPa).

Facilitated diffusion in hemoglobin solutions

It is easy to see that hemoglobin carried in the blood vastly improves the transport of oxygen to

the tissues. But hemoglobin occurs not only in blood but also in various tissues, in higher animals especially in muscle. Can this hemoglobin, lodged inside the cells and removed from circulation, serve any function?

An increased oxygen capacity of tissues can help smooth fluctuations when there are rapid increases in the demand for oxygen. For example, during muscle contraction the blood flow to a muscle may be reduced or momentarily stopped due to mechanical compression of the blood vessels by the contracting muscle. In this situation the myoglobin gives up its oxygen and thus acts as a short-time oxygen store (Millikan 1937). It is unlikely, however, that muscle contractions regularly cause complete cessation of blood flow, on the contrary, the circulation in heavily working muscles is usually greatly increased.

The discovery that hemoglobin greatly accelerates the diffusion rate of oxygen throws a different light on the presence of hemoglobin in muscle, as well as its wide occurrence in animals that live in environments that at least periodically subject them to low oxygen tensions. Some such animals are mentioned in Table 3.2, and it is probable that

FIGURE 3.13 The diffusion of oxygen and nitrogen from air to a vacuum through a layer of blood plasma takes place at the constant ratio of 0.5. Through a hemoglobin solution the relative rate of oxygen diffusion is higher and increases with decreasing pressure. Crosshatched area at right indicates the water vapor pressure, which was identical on the two sides of the membrane. For further explanation, see text. [Scholander 1960]

these also enjoy the advantages of accelerated or *facilitated diffusion* of oxygen.

Scholander (1960) discovered that although nitrogen diffuses through a hemoglobin solution a little more slowly than through water, the rate of oxygen transfer through the hemoglobin solution is much faster than through water.

Scholander arranged the following experiment. The solution to be tested (hemoglobin solution, blood plasma, or water) was absorbed in a highly porous membrane. One side of this membrane was then exposed to ordinary air and the other to vacuum. Both oxygen and nitrogen would now diffuse from the air through the membrane to the vacuum. If the membrane contained water or blood plasma, the ratio between the amount of oxygen and nitrogen diffusing through the liquid layer would be about 0.5. (This particular ratio could be predicted and comes about because the air contains about four times as much nitrogen as oxygen, but nitrogen is only one-half as soluble as oxygen.)

If the air pressure above the membrane was lowered to 0.5 atm, only half as much air would diffuse across to the vacuum per unit time, but the ratio between the two gases remained unchanged at 0.5 volume oxygen per volume of nitrogen (Figure 3.13). If the plasma was replaced by a hemoglobin solution containing as much hemoglobin as normal blood (obtained by hemolyzing blood), the ratio of oxygen to nitrogen diffusion changed to nearly 1. This means that relative to the nitrogen the diffusion of oxygen was speeded by the presence of hemoglobin. With further lowering of the pressure in the air chamber, the ratio increasingly favored the oxygen above the nitrogen, until the ratio indicated that the diffusion of oxygen was speeded up about eightfold relative to its diffusion in the absence of hemoglobin.

Facilitated diffusion is associated with the ability of hemoglobin to bind oxygen reversibly. Hemo-

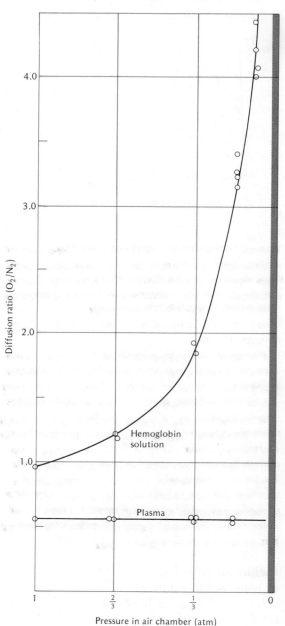

globin that has been changed to methemoglobin (achieved by oxidizing the divalent iron to trivalent iron) can no longer bind oxygen, and a solution of methemoglobin causes no facilitation of oxygen diffusion. However, myoglobin and various hemoglobins from vertebrates as well as invertebrates all show facilitated diffusion.

One important aspect of facilitated diffusion is that the enhanced diffusion occurs only when the oxygen pressure at the downhill end of the diffusion path is close to zero. In fact, maximal facilitation is found when the degree of saturation on the high and low sides of the hemoglobin layer are on the steepest part of the dissociation curve.

Do other respiratory pigments, such as hemocyanin, also facilitate diffusion? They do not. This seems related to the large molecular size of hemocyanin (several millions), which greatly reduces the motility of the hemocyanin molecule in solution. Although the exact mechanism of oxygen transfer is inadequately understood, it seems that the normal thermal motility of the hemoglobin molecules in solution plays a role. If this motility is slowed (e.g., by adding gelatin to the hemoglobin solution), the facilitation is reduced. Myoglobin, which has a lower molecular weight (17 000) than hemoglobin (68 000) shows greater facilitation, roughly in inverse proportion to the square root of the molecular weights of the two pigments. This indicates that the difference results from the thermal motion of the pigment in solution (Hemmingsen 1965).

The main points of facilitated diffusion of oxygen in hemoglobin solutions can be summarized as follows: (1) the net diffusion of oxygen is accelerated when the exit pressure is low, (2) the transfer equals the sum of simple diffusion and the enhancement attributable to the pigment, (3) the facilitation increases with increasing hemoglobin concentration, and (4) the transfer rate is inversely related to the square root of the molecular weight of the pigment, so that a small hemoglobin molecule gives greater facilitation.

Physiological implications

In vertebrates hemoglobin is present in high concentration primarily in the red blood cells and secondarily in red muscles. It is doubtful that facilitated diffusion has much importance in the erythrocytes, for facilitation requires a near-zero oxygen pressure at the receiving end of the diffusion path. The red blood cells, which already have a very short diffusion distance, are rarely if ever exposed to near-zero oxygen pressures, except in the capillaries of heavily working muscles.

For myoglobin the situation is different. At some distance from a capillary the oxygen tension in a working muscle may readily approach zero. This establishes an ideal situation for facilitated diffusion, the oxygen tension in the capillary being somewhere between arterial and venous tension, and in the tissue (the receiving end of the diffusion path) nearly zero. A several-fold diffusion facilitation, such as was observed in the laboratory experiments, would indeed be important in improving the oxygen supply to the muscle. Experiments on pigeon breast muscle confirm that myoglobin may transport a significant fraction of the oxygen consumed by the muscle (Wittenberg et al. 1975).

It now appears particularly noteworthy that those invertebrate organisms that have hemoglobin frequently live in environments that, at least at times, have low oxygen tensions. When the oxygen supply is poor, the oxygen tension in the tissues is particularly low, and facilitation is most important. Hemoglobin is especially important for these animals, and many of them adapt to prolonged exposure to oxygen deficiency by increasing the amount of hemoglobin in their bodies. It seems that the major role of hemoglobin in these animals is to facilitate oxygen diffusion.

Indeed, some parasitic nematodes that live in the intestines of sheep respire actively at considerably higher rates than the calculated maximum rate of inward diffusion of oxygen, in the absence of facilitation (Rogers 1949). However, these animals have highly concentrated intracellular hemoglobin that would serve to augment the rate of oxygen diffusion at the very low oxygen partial pressure in the intestine of their host.

The role of hemoglobin in plants has been the subject of much speculation, but is now reasonably clear. Nitrogen-fixing plants, such as soybeans and clover, have small nodules on the roots that contain bacteria that utilize atmospheric nitrogen to form nitrogen compounds (nitrogen fixation). These nodules contain a hemoglobin, called leghemoglobin, which is required for nitrogen fixation. It has a very high oxygen affinity ($P_{50} = 0.04$ mm Hg, 0.005 kPa), and the flux of leghemoglobin-bound oxygen exceeds the flux of dissolved oxygen by 10 000- to 100 000-fold. The oxygen delivered supports the formation of ATP, which is consumed in the complex reduction process involved in nitrogen fixation (Wittenberg 1977).

Evolutionary implications

It is curious that a molecule as complex as hemoblogin occurs so widely, yet in such a scattered or sporadic way. In addition to being nearly universally present in vertebrates, hemoglobin occurs here and there in a number of animal phyla, including protozoans, and in plants and bacteria. Often, it is found in only one or a few species within a large phylum, at times in a parasite and not in its free-living relative, and so on. Once the advantage of hemoglobin in facilitated diffusion has been established, however, the peculiar distribution of hemoglobin is easier to understand.

The basic structure of the heme part of the hemoglobin molecule is a porphyrin nucleus, which is universally present in plants and animals as part of cellular enzymes that belong to the cytochrome system. Relatively minor biochemical alterations can change this raw material into a hemoglobin-type pigment that binds oxygen reversibly; for such mutations to be permanently established requires some suitable evolutionary advantage that favors their maintenance.

The most obvious and immediate advantage of hemoglobin is the improvement in oxygen delivery due to facilitated diffusion. Once hemoglobin synthesis is established, further improvement in oxygen transport can be achieved by adding convective transport (i.e., blood and a circulatory system).

The establishment of a complete circulation system is a complex change that would require not only relatively minor biochemical changes in porphyrin, but also complex morphological and physiological alterations of the entire organism. Such drastic changes are, of course, possible, but can hardly be achieved as a single or few-step event. In contrast, an initial occurrence of hemoglobin in any tissue immediately endows the organism with the advantage of facilitated diffusion, giving a selective advantage to its maintenance. A later step would then be to permit this hemoglobin to enter an already existing blood system, thus further augmenting oxygen delivery by convection.

It follows from this hypothetical sequence of events that the scattered occurrence of hemoglobin is attributable to the universal presence of iron-porphyrin compounds, that facilitated diffusion gives the primary impetus to maintaining the biochemical mechanism for hemoglobin synthesis, and that adding hemoglobin to a pre-existing convective system then becomes a logical later step in the evolutionary history.

CARBON DIOXIDE TRANSPORT IN BLOOD

As blood passes through the tissues and gives off oxygen, it concurrently takes up carbon dioxide; as it passes through the lungs, the reverse takes place. Let us examine the events in the lung as carbon dioxide is given off.

The venous blood of a mammal, as it reaches the lungs, contains about 550 cm^3 CO_2 per liter blood and has a carbon dioxide tension of about 46 mm Hg (6.1 kPa). As the blood leaves the lungs, it contains about 500 cm^3 CO_2 per liter blood at a tension of about 40 mm Hg. Thus, the blood has given up only a small amount of the total carbon dioxide it contains, the arterio-venous difference being about 50 ml CO_2 per liter blood. To understand carbon dioxide transport, we need to know more about carbon dioxide and its behavior in solution.

The total amount of carbon dioxide carried in the blood far exceeds the amount that would be dissolved in water at the carbon dioxide tension of the blood. This is because the largest part of the carbon dioxide is carried in chemical combination, rather than as free dissolved carbon dioxide gas. The two most important such compounds are (1) bicarbonate ions and (2) a combination of carbon dioxide with hemoglobin. (Note that the carbon dioxide binds to terminal amino groups of the protein molecule, not to the binding site for oxygen.

When carbon dioxide is dissolved in water, it combines with water to form carbonic acid (H_2CO_3). This can be described as follows:

$$CO_2 + H_2O \rightleftarrows H_2CO_3 \rightleftarrows H^+ + HCO_3^-$$

$$\rightleftarrows (H^+ + CO_3^{2-})$$

The H_2CO_3 dissociates as an acid into hydrogen ion and bicarbonate ion (HCO_3^-). Carbonic acid (H_2CO_3) acts as a very weak acid with an apparent dissociation constant of 8.0×10^{-7} at 37 °C (pK = 6.1).* The further dissociation of bicarbonate ion into one additional hydrogen ion and the carbonate ion (CO_3^{2-}) is included for completeness, but at the pH of living organisms the amount of carbonate ion present is minuscule and can usually be disregarded.

At the normal pH of mammalian blood, most of the total carbon dioxide present is in the form of bicarbonate ion. This can readily be calculated from the Henderson-Hasselbalch equation in the following way:

$$\frac{[H^+][A^-]}{[HA]} = K$$

$$[H^+] = K \cdot \frac{[HA]}{[A^-]}$$

$$pH = pK - \log \frac{[HA]}{[A^-]}$$

for pK = 6.1 and at pH = 7.4

$$-\log \frac{[HA]}{[A^-]} = pH - pK = 7.4 - 6.1 = 1.3$$

$$\log \frac{[HA]}{[A^-]} = -1.3 = 0.7 - 2$$

$$\frac{[HA]}{[A^-]} = \frac{[H_2CO_3]}{[HCO_3^-]} = 0.05 = \frac{1}{20}$$

* Carbonic acid is actually a stronger acid with a true pK value of 3.8. The reason for the discrepancy in pK values is that only about 0.5% of the dissolved carbon dioxide combines with water to form the acid. The bulk of the carbon dioxide remains as dissolved gas, and only the fraction 0.005 is hydrated to form H_2CO_3. In physiology it has been customary to consider that all carbon dioxide is in the form of carbonic acid; hence, the carbonic acid appears to be weaker than the molecular species, H_2CO_3, in fact is. If to the true pK for carbonic acid (3.8) we add the negative log of the hydration constant (log

This means that at pH = 7.4 the ratio of carbonic acid to bicarbonate ion is 1:20, i.e., for each one part of carbon dioxide present as acid, 20 times as much carbon dioxide is present as bicarbonate ion.

The hydrogen ion formed as a result of the ionic dissociation of the carbonic acid is buffered by various substances in the blood, and the effect of carbon dioxide on the pH of the blood is therefore only moderate. For example, arterial blood may have a normal pH of 7.45 and venous blood a pH of 7.42; the difference in pH caused by the uptake of carbon dioxide by the blood in the tissues thus is no more than 0.03.

The most important buffering substances in the blood are the carbonic acid–bicarbonate system, the phosphates, and the proteins in the blood. Proteins act as excellent buffers because they contain groups that can dissociate both as acids and as bases, the result being that proteins can either take up or give off hydrogen ions. Hemoglobin is the protein present in the largest amount in blood, and it has the greatest role in buffering; the plasma proteins are second in importance in this regard.

Carbon dioxide dissociation curve

The amount of carbon dioxide taken up by blood varies with its partial pressure. It is therefore possible to construct a carbon dioxide dissociation curve for blood, analogous to the oxygen dissociation curve. This is done by plotting on the ordinate the total amount of carbon dioxide in the blood and on the abscissa the various carbon dioxide partial pressures. The result is a curve as in Figure 3.14. This graph shows a slightly different dissociation curve for oxygenated and deoxygenated blood.

$0.005 = -2.3$), the sum is 6.1, which is the apparent pK for carbon dioxide as used in physiology. In this book we will follow physiological convention and assume that all dissolved carbon dioxide forms H_2CO_3.

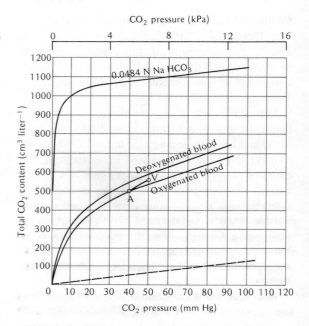

FIGURE 3.14 Carbon dioxide dissociation curves of mammalian blood. The curves for oxygenated and deoxygenated blood differ somewhat. The line A-V is the physiological dissociation curve in the body, describing the difference between arterial and venous blood. The dashed line indicates the amount of carbon dioxide held by physical solution in water. [Winton and Bayliss 1955]

This is because oxyhemoglobin is a slightly stronger acid than hemoglobin; hence, oxygenated blood binds slightly less carbon dioxide. This phenomenon is closely related to the Bohr effect, which merely shows the other side of the coin. When carbon dioxide (acid) is added to the blood, it pushes the equilibrium between oxyhemoglobin and hemoglobin in the direction of the weaker acid (hemoglobin). This tends to release more oxygen, which, in fact, is the Bohr effect.

In Figure 3.14 the point marked A stands for normal arterial blood, and the point marked V for mixed venous blood (venous blood is never completely deoxygenated). In the body the actual dissociation curve for carbon dioxide is the fully drawn line A–V, and this curve is said to be the *functional dissociation curve*.

Figure 3.14 also shows the dissociation curve for

a sodium bicarbonate solution with the same total available base as in blood. The slope of the upper part of this curve represents the solubility of carbon dioxide in an aqueous medium. As carbon dioxide pressure is reduced, this dissolved carbon dioxide is given up. When the carbon dioxide pressure falls to about 10 mm Hg, the bicarbonate solution begins to give up carbon dioxide from the bicarbonate ion, leaving carbonate behind.

By exposing a solution of sodium bicarbonate to a vacuum, half the total carbon dioxide of the bicarbonate can be removed, leaving a solution of sodium carbonate (Na_2CO_3) behind. Sodium carbonate, however, does not give off carbon dioxide to a vacuum, and the final solution retains half the total carbon dioxide that was originally present as bicarbonate. In contrast, whole blood exposed to a vacuum gives up all the carbon dioxide present. This is because other ions, primarily proteins, furnish the necessary anions. If separated plasma is subjected to a vacuum, it behaves more like a solution of sodium bicarbonate; the presence of red cells is necessary to obtain the normal dissociation curve shown in Figure 3.14.

Carbon dioxide in aquatic respiration

The amount of carbon dioxide produced in metabolism is closely related to the amount of oxygen consumed. The ratio between the respiratory exchange of carbon dioxide and oxygen, the *respiratory exchange ratio* (also called the *respiratory quotient*), normally remains between 0.7 and 1.0 (see Chapter 6). For the following discussion we shall consider an animal in which the exchange ratio is 1.0 (i.e., for each molecule of oxygen taken up, exactly one molecule of carbon dioxide is given off).

An air-breathing animal – a mammal, for example – with a gas exchange ratio of 1.0 has an easily predicted carbon dioxide concentration in exhaled air. If the 21% O_2 in inhaled air is reduced to 16% in the exhaled air, the exhaled air of necessity contains 5% CO_2. If, instead, we consider the change in partial pressure of the two gases, we can say that if oxygen in the respired air is reduced by 50 mm Hg, the carbon dioxide is increased by the same amount, 50 mm Hg (6.7 kPa). Each millimeter change in the partial pressure of oxygen in the respiratory air results in an equal and opposite change in the carbon dioxide.

If the exchange ratio of unity is applied to gas exchange in water, the amounts of oxygen and carbon dioxide are, of course, equal, but the changes in the partial pressures of the two gases are very different because the solubilities of the two gases are so different. For example, if a fish removes two-thirds of the oxygen present in the water, thus reducing the P_{O_2} from 150 to 50 mm Hg (20 to 6.7 kPa), the increase in P_{CO_2} is about 3 mm Hg (provided the water is unbuffered; a buffered medium may give an even smaller increase). Depending on the temperature, the 100 mm Hg reduction in P_{O_2} may mean that, say, 5 cm^3 O_2 was removed from 1 liter of water. Returning 5 ml CO_2 increases the P_{CO_2} by only about 3 mm Hg because the solubility of carbon dioxide is very high, roughly 30 times as high as the solubility of oxygen.

This relationship in the carbon dioxide tensions resulting from respiration in air and in water is expressed in Figure 3.15. This graph shows the P_{CO_2} that results from any given decrease in oxygen in the respired medium. Even if a fish were to remove all oxygen in the water, the carbon dioxide in the water would increase to only 5 mm Hg (0.67 kPa) partial pressure. Because carbon dioxide diffuses across the gill epithelium nearly as readily as oxygen, the blood P_{CO_2} of the fish can never exceed 5 mm Hg (in air-breathing vertebrates it is about 10 times as high). As a consequence, in regard to

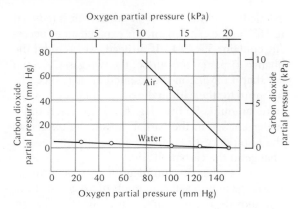

FIGURE 3.15 Relation between P_{O_2} and P_{CO_2} in air and in water, when the respiratory exchange ratio (*RQ*) between oxygen and carbon dioxide is unity (*RQ* = 1). For further explanation, see text. [Rahn 1966]

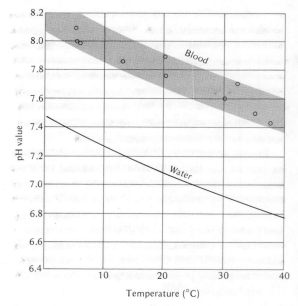

FIGURE 3.16 Blood pH values for vertebrates of all classes, determined at the normal body temperature for each animal. Shaded band indicates range of observed values; central fully drawn line is the pH of water at neutrality as a function of temperature. [Rahn 1966]

the effect on acid–base regulation, air respiration and aquatic respiration are very different. This is particularly important for animals that shift between air and water respiration, such as amphibians and air-breathing fish. For these animals a successful shift between the two media requires substantial physiological adjustments in their acid–base regulation.

Blood carbon dioxide and pH

We have seen that the tension or partial pressure of carbon dioxide in the arterial blood is very different in aquatic and terrestrial animals. The total amount of carbon dioxide present also varies greatly: It is usually low in fish (often less than 100 cm³ CO_2 per liter blood) and may be very high in turtles (in excess of 1000 cm³ CO_2 per liter blood). Clearly, there is uniformity neither in the carbon dioxide tension nor in the total carbon dioxide content of vertebrate blood. In contrast, we find a much greater uniformity in the pH of the blood of various vertebrates than would be expected from the large variation in carbon dioxide. This results from corresponding adjustments in the acid–base balance, achieved primarily through the amount

of sodium (often called *free bicarbonate*) present in the blood.

The pH of the blood of a variety of vertebrates, both warm- and cold-blooded, usually varies around values between about 7.4 and 8.2. If the data are plotted against the temperature of the animals, however, we obtain a more regular distribution than indicated by this rather wide pH range (Figure 3.16). At temperatures near 0 °C, the pH often reaches values well over 8.0, and at 40 °C it is closer to 7.5. At any given temperature, however, the blood pH of a variety of animals falls within a narrow range of about 0.2 pH units. It is important to remember that the dissociation of water, and therefore the pH of neutrality, changes with temperature. The neutrality point is at 7.0 only at room temperature (25 °C) and changes with temperature according to the line in the center of Figure 3.16, marked "water."

The blood pH of all vertebrates deviates from neutrality by about 0.6 pH units, always in the alkaline direction. Consequently, although vertebrates differ considerably in regard to blood pH value, they are amazingly uniform in the extent of deviation from neutrality when measured at the normal temperature for the animal. There is a similar uniformity in the maintenance of the intracellcular pH, which is of great importance in providing suitable conditions for the activity of metabolic enzymes, etc. (Malan et al. 1976).

The very different conditions for achieving an appropriate acid–base balance in water and in air cause great physiological difficulties for animals that live in a transition between the two media. A fish that normally lives in water has its acid–base balance adjusted to a carbon dioxide tension of 2 or 3 mm Hg. If the fish changes to air respiration, the blood carbon dioxide tension rises sharply; this requires an adjustment in the blood buffering system, primarily of the sodium ion concentration, or some auxiliary mechanism for the elimination of carbon dioxide. Amphibious animals lose substantial amounts of carbon dioxide through the skin, and this is of help to both lungfish and aquatic amphibians. If an animal emerges upon land and becomes completely terrestrial, it cannot retain a highly permeable skin because of the danger of desiccation. If it were to keep the P_{CO_2} low by lung ventilation, the ventilation would be vastly increased (relative to that needed for the oxygen supply), and evaporation would become a major problem. Thus, a radical readjustment in the acid–base regulatory system is a primary prerequisite for moving permanently onto land.

Carbonic anhydrase and velocity of carbon dioxide–water interaction

The events that take place when carbon dioxide is dissolved in water (see equations earlier in this chapter) are not instantaneous. When carbon dioxide enters water and becomes dissolved, the initial hydration of the carbon dioxide molecule to form H_2CO_3 is a relatively slow process. The reverse process – the release of carbon dioxide from H_2CO_3 – is also slow, requiring several seconds to a fraction of a minute. (The dissociation of H_2CO_3 into H^+ and HCO_3^- and the reverse process are, for physiological considerations, instantaneous.) If we consider that the time blood remains in a capillary is usually only a fraction of a second, how is it possible for the blood to take up carbon dioxide in the tissues as rapidly as it does, and again give off carbon dioxide during the short time the blood is in the lung capillary?

The problem appeared to be solved with the discovery of an enzyme that accelerates the formation of carbon dioxide from H_2CO_3 so that the process becomes very fast. The enzyme was given the name *carbonic anhydrase,* which is something of a misnomer because both the combination of carbon dioxide with water and the release of carbon dioxide from carbonic acid are speeded up (i.e., the process is catalyzed in both directions). The enzyme is not present in blood plasma, but is found in high concentration inside the red blood cells. It also occurs in other organs, notably in glands such as kidneys, the secretory epithelium of the stomach, pancreas, and salivary glands, but this is of no concern to us here. It should be noted, however, that carbonic anhydrase is highly specific for the hydration reaction of carbon dioxide and has no other known function in the organism.

Some years after the discovery of carbonic anhydrase, several highly potent carbonic anhydrase inhibitors were synthesized and became important as drugs, especially in the treatment of certain types of kidney malfunction. Much to the surprise of physiologists, it turned out that these enzyme inhibitors have a very low toxicity and that the

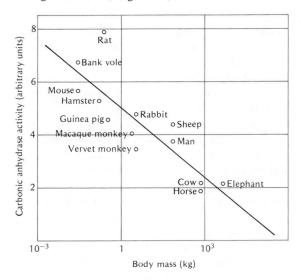

FIGURE 3.17 Carbonic anhydrase activity in mammalian blood is related to body size and is higher in small than in large mammals. [Magid 1967]

complete inhibition of blood carbonic anhydrase has only minor effects on carbon dioxide transport. In other words, the enzyme is not a critical factor in carbon dioxide transport as such, although its inhibition causes some changes in the acid–base balance.

An examination of the amount of carbonic anhydrase found in various animals might be helpful in examining its possible role. We know that small animals have higher metabolic rates, per unit body weight, than large animals. They use oxygen and produce carbon dioxide more rapidly, and carbon dioxide must therefore be released in a shorter time in the lungs of a small animal. An examination of the carbonic anhydrase of the blood in relation to body size shows that small animals indeed have significantly higher concentrations of the enzyme in their blood (Figure 3.17). This seems to be in accord with the need for carbon dioxide transport, but the enzyme is present in amounts far in excess

of what is needed. There is, however, another process where rapid hydration of carbon dioxide is important. As carbon dioxide in the tissue capillary enters the blood, the presence of carbonic anhydrase permits an immediate formation of carbonic acid, which in turn immediately affects the dissociation of the acid-sensitive oxyhemoglobin. This acid sensitivity, the Bohr effect, obviously would not have time to take effect while the blood still remains in the capillary unless the hydration of the carbon dioxide molecule were speeded up by the enzyme. In the absence of carbonic anhydrase, therefore, the Bohr effect could not be of much importance.

It seems, then, that carbonic anhydrase, an enzyme that specifically catalyzes the hydration of the carbon dioxide molecule, perhaps is of minor importance in carbon dioxide transport but contributes to the efficient delivery of oxygen to the tissues. As inhibition of the enzyme in the living animal does not cause great difficulties to the animal, the enzyme is probably important primarily under conditions of extreme demands on the gas transport mechanism.

In other organs where carbonic anhydrase occurs, it seems always to be related to functions where ion transport processes are important, such as in glands (kidneys, pancreas, salivary gland) and in the ciliary body of the eye. The effect of carbonic anhydrase inhibition on these processes is profound and is the basis for the widespread and important use of inhibitor drugs in clinical medicine as well as in physiological research.

After having discussed the physiological properties of blood and the way in which it carries gases, we can move on to the next chapter, in which we will be concerned with the movement of blood within the organism.

REFERENCES

Aste-Salazar, H., and Hurtado, A. (1944) The affinity of hemoglobin for oxygen at sea level and at high altitudes. *Am. J. Physiol.* 142:733–743.

Barcroft, J. (1935) Foetal respiration. *Proc. R. Soc. Lond. B.* 118:242–263.

Benesch, R., Benesch, R. E., and Enoki, Y. (1968a) The interaction of hemoglobin and its subunits with 2,3-diphosphoglycerate. *Proc. Natl. Acad. Sci. U.S.A.* 61:1102–1106.

Benesch, R., Benesch, R. E., and Yu, C. I. (1968b) Reciprocal binding of oxygen and diphosphoglycerate by human hemoglobin. *Proc. Natl. Acad Sci. U.S.A.* 59:526–532.

Bunn, H. F. (1971) Differences in the interaction of 2, 3-diphosphoglycerate with certain mammalian hemoglobins. *Science* 172: 1049–1050.

Chanutin, A., and Curnish, R. R. (1967) Effect of organic and inorganic phosphates on the oxygen equilibrium of human erythrocytes. *Arch. Biochem. Biophys.* 121:96–102.

Gillen, R. G., and Riggs, A. (1973) Structure and function of the isolated hemoglobins of American eel, *Anguilla rostrata. J. Biol. Chem.* 248:1961–1969.

Grigg, G. C. (1969) Temperature-induced changes in the oxygen equilibrium curve of the blood of the brown bullhead, *Ictalurus nebulosus. Comp. Biochem. Physiol.* 28:1203–1223.

Hall, F. G., Dill, D. B., and Barron, E. S. G. (1936) Comparative physiology in high altitudes. *J. Cell. Comp. Physiol.* 8:301–313.

Hall, F. G., and McCutcheon, F. H. (1938) The affinity of hemoglobin for oxygen in marine fishes. *J. Cell. Comp. Physiol.* 11:205–212.

Hemmingsen, E. A. (1965) Accelerated transfer of oxygen through solutions of heme pigments. *Acta Physiol. Scand. Suppl.* 246:1–53.

Lenfant, C., Torrance, J., English, E., Finch, C. A., Reynafarje, C., Ramos. J., and Faura, J. (1968) Effect of altitude on oxygen binding by hemoglobin and on organic phosphate levels. *J. Clin. Invest.* 47:2652–2656.

Lenfant, C., Ways, P., Aucutt, C., and Cruz, J. (1969) Effect of chronic hypoxic hypoxia on the O_2–Hb dissociation curve and respiratory gas transport in man. *Respir. Physiol.* 7:7–29.

Lutz, P. L., Longmuir, I. S., Tuttle, J. V., and Schmidt-Nielsen, K. (1973) Dissociation curve of bird blood and effect of red cell oxygen consumption. *Respir. Physiol.* 17:269–275.

Magid, E. (1967) Activity of carbonic anhydrase in mammalian blood in relation to body size. *Comp. Biochem. Physiol.* 21:357–360.

Malan, A., Wilson, T. L., and Reeves, R. B. (1976) Intracellular pH in cold-blooded vertebrates as a function of body temperature. *Respir. Physiol.* 28:29–47.

Manwell, C. (1958) The oxygen-respiratory pigment equilibrium of the hemocyanin and myoglobin of the amphineuran mollusc *Cryptochiton stelleri. J. Cell. Comp. Physiol.* 52:341–352.

Manwell, C. (1960) Histological specificity of respiratory pigments. 2. Oxygen transfer systems involving hemerythrins in sipunculid worms of different ecologies. *Comp. Biochem. Physiol.* 1:277–285.

Millikan, G. A. (1937) Experiments on muscle haemoglobin *in vivo*: The instantaneous measurement of muscle metabolism. *Proc. R. Soc. Lond. B.* 123:218–241.

Rahn, H. (1966) Gas transport from the external environment to the cell. In *Development of the Lung* (A. V. S. de Reuck and R. Porter, eds.), pp. 3–23. London: CIBA.

Randall, D. J. (1970) Gas exchange in fish. In *Fish Physiology,* vol. 4 (W. S. Hoar and D. J. Randall, eds.), pp. 253–292. New York: Academic Press.

Read, K. R. H. (1962) The hemoglobin of the bivalved mollusc, *Phacoides pectinatus* Gmelin. *Biol. Bull.* 123:605–617.

Riggs, A. (1951) The metamorphosis of hemoglobin in the bullfrog. *J. Gen. Physiol.* 35:23–44.

Riggs, A. (1960) The nature and significance of the Bohr effect in mammalian hemoglobins. *J. Gen. Physiol.* 43:737–752.

Rogers, W. P. (1949) On the relative importance of aerobic metabolism in small nematode parasites of the alimentary tract. 2. The utilization of oxygen at low partial pressures by small nematode parasites of the alimentary tract. *Aust. J. Sci. Res.* 2:166–174.

Root, R. W., Irving, L., and Black, E. C. (1939) The effect of hemolysis upon the combination of oxygen with the blood of some marine fishes. *J. Cell. Comp. Physiol.* 13:303–313.

Schmidt-Nielsen, K. (1972) *How Animals Work.* London: Cambridge University Press. 114 pp.

Schmidt-Nielsen, K., and Pennycuik, P. (1961) Capillary density in mammals in relation to body size and oxygen consumption. *Am. J. Physiol.* 200:746–750.

Schmidt-Nielsen, K., and Taylor, C. R. (1968) Red blood cells: Why or why not? *Science* 162:274–275.

Scholander, P. F. (1960) Oxygen transport through hemoglobin solutions. *Science* 131:585–590.

Winton, F. R., and Bayliss, L. E. (1955) Blood and the transport of oxygen and carbon dioxide. In *Human Physiology* (F. R. Winton and L. E. Bayliss, eds.), pp. 78–118. Boston: Little, Brown.

Wittenberg, B. A., Wittenberg, J. B., and Caldwell, P. R. B. (1975) Role of myoglobin in the oxygen supply to red skeletal muscle. *J. Biol. Chem.* 250:9038–9043.

Wittenberg, J. B. (1977) Facilitation of oxygen diffusion by intracellular leghemoglobin and myoglobin. In *Oxygen and Physiological Function* (F. F. Jöbsis, ed.), pp. 228–246. Dallas: Professional Information Library.

ADDITIONAL READING

Altman, P. L., and Dittmer, D. S. (eds.) (1971) *Biological Handbooks: Blood and Other Body Fluids* (3rd printing, with minor corrections). Bethesda: Federation of American Societies for Experimental Biology. 540 pp.

Bonaventura, J., Bonaventura, C., and Sullivan, B. (1977) Non-heme oxygen transport proteins. In *Oxygen and Physiological Function* (F. F. Jöbsis, ed.), pp. 177–220. Dallas: Professional Information Library.

Howell, B. J., Rahn, H., Goodfellow, D., and Herreid, C. (1973) Acid-base regulation and temperature in selected invertebrates as a function of temperature. *Am. Zool.* 13:557–563.

MacFarlane, R. G. (1970) *The Haemostatic Mechanism in Man and Other Animals.* Symposium of the Zoological Society of London, No. 27. London: Academic Press. 248 pp.

Reeves, R. B. (1977) The interaction of body temperature and acid–base balance in ectothermic vertebrates. *Annu. Rev. Physiol.* 39:559–586.

Riggs, A. (1965) Functional properties of hemoglobins. *Physiol. Rev.* 45:619–673.

Rørth, M., and Astrup, P. (eds.) (1972) *Oxygen Affinity of Hemoglobin and Red Cell Acid Base Status.* Proceedings, Alfred Benzon Symposium IV. Copenhagen: Munksgaard. 832 pp.

CHAPTER FOUR

Circulation

The main role of circulating fluids in the body is to provide rapid mass transport over distances where diffusion is inadequate or too slow. Circulation is therefore important in virtually all animals more than a few millimeters in size and is a necessity for all large animals with high metabolic activity. The transport of gases – oxygen and carbon dioxide – is an important, but far from the only function of circulation.

The main functions of blood (listed in Table 3.1) fall into three major categories: (1) mass transport of solutes and cells, (2) transport of heat, and (3) transmission of force. Functions that depend on the transmission of force are related mostly to the movement of organs, the locomotion of the entire animal, and providing pressure for ultrafiltration in the capillaries of the kidney; these functions will be discussed elsewhere: The transport of heat and solutes (including gases) is discussed in this chapter. A system that is adequate for the convective transport of oxygen invariably suffices for transport of other solutes as well. Therefore, our main emphasis will be on the role of circulation in gas, notably oxygen, transport.

Circulatory systems of vertebrates are well known and will be discussed first. The function of most invertebrate circulatory systems is less well known; in many cases the morphology has been adequately worked out, but information about function is often less than satisfactory. As a result, there is little possibility for generalizations, and the main emphasis in this chapter will be on the highly developed and better understood circulatory systems of vertebrates.

GENERAL PRINCIPLES

An adequate circulatory system depends on one or more pumps and on channels or conduits in which the blood can flow. The pump, or *heart*, is

FIGURE 4.1 The three types of pumps that move blood in a circulatory system. In the peristaltic pump (a) a constriction in a tube moves along the tube and pushes blood ahead of it. In the common chamber pump (b) rhythmical contractions of the walls force blood out. Valves prevent backflow, and the blood is therefore expelled in one direction only. In another form of chamber pump (c) blood is pushed out of a collapsible tube by pressure from surrounding tissues. In this diagram contracting muscles provide the pressure and valves prevent backflow. The venous pump in the legs of man is of this type.

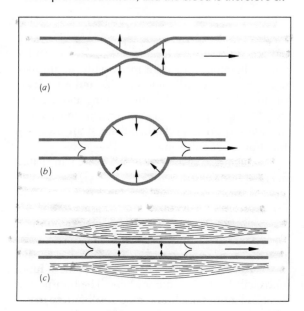

based on the ability of muscle to contract or shorten. By wrapping muscle around a tube or chamber, it is possible to achieve a reduction of volume. Two different types of pumps can be designed this way: *peristaltic pumps* or *chamber pumps* with valves (Figure 4.1). A chamber pump may have contractile walls (Figure 4.1b) or the reduction in volume may be achieved by external pressure from other body parts (Figure 4.1c).

Peristaltic hearts are found mostly in invertebrates; the vertebrate heart is almost without exception a chamber pump with contractile walls. A typical pump for which the force is provided by outside pressure is found in the larger veins in the human legs. The walls of these veins are relatively thin; the veins are provided with valves that prevent the backflow of blood, and when the leg muscles contract the veins are collapsed and the valves ensure that blood is forced in the direction of the heart. This pumping action aids greatly in moving the blood out of the leg veins against the force of gravity, which tends to make blood accumulate in the legs.

Because muscle can provide force only by shortening, the heart cannot pull blood unless special arrangements are made. The elasmobranch heart is an example. It is enclosed in a stiff capsule, and when one chamber contracts, the other chamber is filled with blood by "suction" while valves prevent backflow from the arterial side. This is described in further detail later in this chapter.

The circulating fluid or blood is carried from the heart in channels or conduits and eventually returns to the heart. Vertebrate blood is carried in a system of elastic tubes or pipes (arteries, capillaries, veins). Blood returns to the heart without leaving this system of tubes, and because the blood remains within this closed system, we refer to vertebrate circulation as a *closed circulation*. In many invertebrates (e.g., insects, most crustaceans, many molluscs) blood is pumped from the heart into the blood vessels, but these terminate and blood flows more or less freely between the tissues before it eventually returns to the heart. Such a system is called an *open circulation*. Some of the characteristics of open and closed circulatory systems are listed in Table 4.1.

Closed circulatory systems are found in vertebrates, cephalopods (octopus), and echinoderms; open systems are found in most arthropods, noncephalopod molluscs, and tunicates. The circulatory system of tunicates is interesting in several ways. It is an open system, and the tube-shaped heart pumps by means of a peristaltic wave passing along from one end to the other. The heart has no valves. After a series of contractions, say several dozen or a hundred, the heart gradually slows and stops. After a pause the beat reverses direction, and blood is now forced in the opposite direction

TABLE 4.1 Major differences between closed and open circulatory stems.

Closed systems	Open systems
Usually high pressure systems	Usually low pressure systems
Development of high pressure requires closed system and resistance.	High pressure not possible
Sustained high pressure between heart-beats requires elastic walls	Sustained pressure not possible
Blood conveyed directly to organs	Similar to closed systems
Distribution to different organs can be regulated	Distribution of blood not readily regulated
Blood return to heart rapid	Blood return to heart slow

(Krijgsman 1956). This is an unusual arrangement: In most circulatory systems the heart pumps, and the blood always flows, in one direction only.

VERTEBRATE CIRCULATION

Body water compartments and blood volume

Roughly two-thirds of the vertebrate body consists of water. The exact figure varies a great deal, particularly as a result of variations in fat content. Adipose tissue contains only some 10% water, and a very fat individual, who may have half or more of the body weight as fat, therefore has a low overall water content. If the water content is expressed as a percentage of the fat-free body weight, the figures shows much less variation. Most of the water is located inside the cells (*intracellular water*); a smaller fraction is outside the cell membranes (*extracellular water*). The extracellular water is located partly in tissue spaces (*interstitial fluid*) and partly in the blood (*plasma water*). We can therefore consider the body water as being distributed among three compartments: intracellular, interstitial, and plasma (Figure 4.2).

The volumes of the body fluid compartments can be measured in various ways, often without

harm to the organism. One of the most convenient ways is to introduce some suitable substance that is distributed uniformly within the compartment to be measured. The dilution of this substance can then be used for a calculation of the volume into which it is distributed. The quantities most readily measured in this way are plasma volume, extracellular volume, and total body water.

Total blood volume cannot be measured accurately by draining the blood from an animal, for a substantial amount of blood will remain in the vascular system. The dilution technique, however, gives reliable and reproducible results. One of the commonly used substances for this purpose is a blue dye, Evans blue (also known as T-1824). A measured small amount of this dye is injected as an aqueous solution into the bloodstream; the dye binds to the plasma proteins and remains within the vascular system. After complete mixing has taken place (after several minutes), a blood sample is withdrawn and the dye concentration in the plasma is measured. The *plasma volume* is then readily calculated from the dilution of the dye. The *total blood volume* can be calculated from a determination of the percent red cells in the blood (the *hematocrit*), which is obtained by centrifugation of a blood sample.

The extracellular fluid volume is determined by injecting a substance that penetrates the capillary

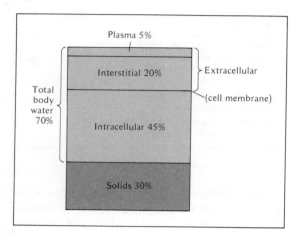

FIGURE 4.2 The approximate distribution of water and solids in the vertebrate body, expressed as percentage of the fat-free body weight. More detailed information is given in Table 4.2.

walls and becomes distributed in blood and interstitial spaces, but does not penetrate the cells. One such substance is the polysaccharide inulin (also used in renal function studies; see Chapter 10). Inulin, which has a molecular weight of about 5000, distributes itself uniformly in extracellular space, but is not metabolized and does not penetrate the cells. Its dilution, therefore, indicates the total extracellular volume. The volume of interstitial fluid is then calculated as the difference between total extracellular volume and plasma volume.

Total body water can, of course, be determined by killing the organism and determining the water content by drying. A dilution technique is, however, less dramatic and more convenient. Presently the commonest substance used is water labeled with the radioactive hydrogen isotope tritium. A small amount of tritiated water is injected; it diffuses through capillary walls and cell membranes without hindrance and rapidly becomes evenly distributed throughout all body water. A determination of the radioactivity of any sample of a body fluid can be used for a calculation of the volume

into which the injected tritiated water was diluted. In fact, it is not even necessary to take a blood sample, for if the bladder has been emptied before the experiment, a freshly voided urine sample will be as representative of the body water as a water sample obtained from any other source.

Table 4.2 gives the total body water and its distribution in intracellular and extracellular volumes for some selected vertebrates. The total body water in aquatic and terrestrial vertebrates is of the same magnitude, about 70%, but there are variations in both directions. The distribution between intracellular and extracellular volume again is rather uniform, but with some inconsistencies.

The blood volume of vertebrates is usually between 5 and 10% of the body weight, except that teleost fish in general have blood volumes only 2 to 3% of body weight, and elasmobranchs and cyclostomes have volumes 6 to 10% of body weight. Air-breathing vertebrates (amphibians, reptiles, birds, and mammals) have blood volumes between 5 and 10%. We can therefore say that vertebrates on the whole have blood volumes between 5 and 10%, with the exception of teleost fish, which have substantially smaller blood volumes. At the present time it is uncertain what functional significance this difference may have.

Circulation patterns

Each class of vertebrates has a quite uniform type of circulation, but differences among the classes are substantial. As vertebrate life changes from aquatic to terrestrial, the circulation becomes more complex.

There are two important consequences of the arrangement of mammalian circulation, both immediately apparent from Figure 4.3. First, the blood flow through the lungs must equal the blood flow through the entire remaining part of the body (except for minute transient variations that can be

TABLE 4.2 Body fluid compartments in representative vertebrates. All values in percent of body weight. The minor differences shown are of uncertain significance and may be due to different techniques, to differences in the fat content of the animal body, etc. The blood volume of teleost fish is, however, consistently much lower than in other vertebrates.

FIGURE 4.3 In the circulatory system of a fish, blood is pumped by the heart through the gills, from where it flows on to the tissues. In the mammalian system, blood is pumped through the lungs and then returns to the heart before being passed on to the tissues by a second pump. The mammalian arrangement allows a higher blood pressure in the tissues than in the lungs.

Animal	Total body water (%)	Intracellular water (%)	Extracellular water (%)	Blood volume (%)
Lamprey[a]	76	52	24	8.5
Dogfish[b]	71	58	13	6.8
Carp (fresh water)[c]	71	56	15	3.0
Red snapper (marine)[c]	71	57	14	2.2
Bullfrog[d]	79	57	22	5.3
Alligator[e]	73	58	15	5.1
Gopher snake[e]	70	52	17	6.0
Pigeon[f]	—	—	—	9.2
Great horned owl[f]	—	—	—	6.4
Dog[g]	63	37	19	9.1

[a] Thorson (1959).   [b] Thorson (1958).   [c] Thorson (1961).   [d] Thorson (1964).   [e] Thorson (1968).
[f] Bond and Gilbert (1958).   [g] Hopper et al. (1944).

caused by slight changes in heart volume during one or a few strokes). Second, because both halves of the heart contract simultaneously, as blood is ejected from the heart, the entire ejected volume must be taken up by changes in the volume of the elastic blood vessels.

Fish and mammals represent two extremes in vertebrate circulation. The gradual separation of the heart into two separate pumps as the vertebrates progress from aquatic life to fully terrestrial respiration is shown in Figure 4.4. Each vertebrate class has certain characteristics, and a few comments will help explain these differences.

Cyclostomes

The circulatory system of hagfish differs from that of all other vertebrates. It is partly an open system with large blood sinuses, rather than a closed system as in other vertebrates. Its notable characteristic is that in addition to the regular heart (the *branchial heart*), the hagfish has several accessory hearts, especially in the venous system (Figure 4.4a). There are three sets of such accessory hearts: the *portal heart*, which receives venous blood from

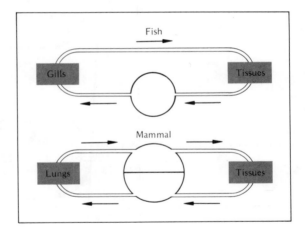

the cardinal vein and from the intestine and pumps this blood to the liver; the *cardinal hearts*, which are located in the cardinal veins and help to propel the blood; and the *caudal hearts*, which are paired expansions of the caudal veins. In addition to these accessory hearts, all located in the venous system, the gills take active part in the forward propulsion of the arterial blood. This is ac-

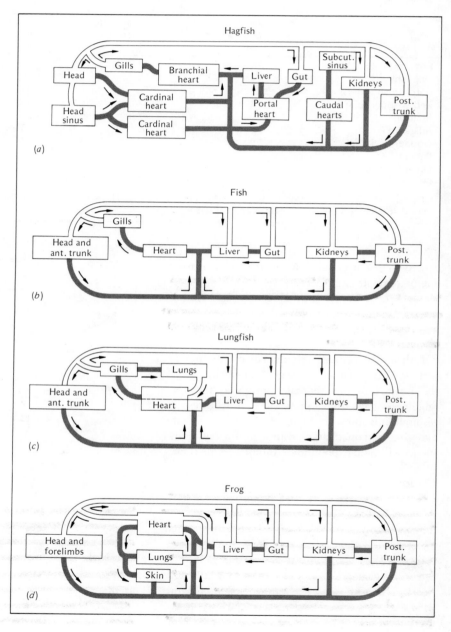

FIGURE 4.4 The main circulatory patterns of the vertebrate classes. The black portions represent venous blood; the white portions oxygen-rich arterialized blood.

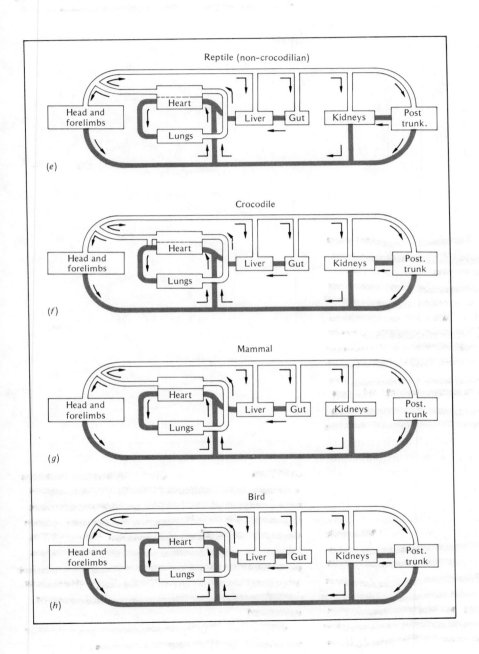

FIGURE 4.5 The caudal heart of hagfish differs from other vertebrate hearts. A central cartilage is flexed by alternate contractions of muscles on the two sides, which also serve as walls for the two chambers. As the muscles on one side contract, the chamber on this side is filled with blood, while blood is expelled from the other side. Valves help to ensure a unidirectional flow of blood [Gordon 1972]

FIGURE 4.6 The two-chambered heart of fish is preceded by an enlarged venous sinus that supplies blood for filling the atrium. The heart ejects blood into a thickened part of the artery, the bulbus arteriosus (in teleosts) or conus arteriosus (in elasmobranchs). [Randall 1970]

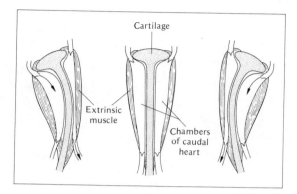

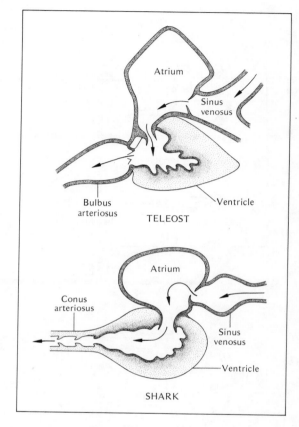

complished by contraction of striated muscular elements in the gills and gill ducts that helps propel the blood in the arterial system (Johansen 1960).

The caudal hearts of the hagfish are particularly interesting because they differ in design from all other hearts (Figure 4.5). A longitudinal rod of cartilage separates two chambers, and alternate contractions of muscles on the two sides cause the rod to be flexed. As the muscles on one side contract, those on the opposite side provide pressure for expulsion of the blood on that side. Simultaneously, the volume on the contracting side increases, so that this chamber becomes filled with blood. By alternate contractions, the two chambers fill and empty in opposing phase, while appropriate valving assures a unidirectional flow.

Fish

The circulation in fish, both teleosts and elasmobranchs, is shown in Figure 4.4*b*. The heart consists of two chambers in series, an atrium and a ventricle. On the venous side the heart is preceded by an enlarged chamber or sinus on the vein, the *sinus venosus*, which helps assure a continuous flow of blood to the heart. On the arterial side the teleost heart is immediately followed by a thickened muscular part of the ventral aorta, the *bulbus*

arteriosus (Figure 4.6). The elasmobranch heart has a similarly located thickened part, the *conus arteriosus*, developed from the heart muscle. It is fibrous and is equipped with valves that prevent backflow of blood into the ventricle. This is particularly important because the heart, located in a rigid chamber, can produce negative pressures. A negative pressure in the heart facilitates the filling by "suction" of the atrium from the large sinus venosus.

The sequence of events in a shark heart during one contraction cycle can be followed on the pres-

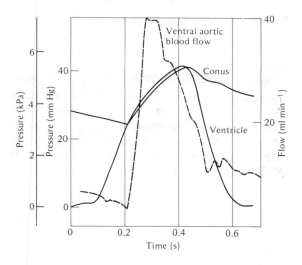

FIGURE 4.7 Pressure changes in the elasmobranch heart during a single contraction cycle in relation to ventral aortic blood flow. [Randall 1970]

sure tracing shown in Figure 4.7. During contraction of the ventricle, the blood pressure rises and this pressure is transmitted to the conus arteriosus. As the ventricle relaxes, backflow from the conus is prevented by valves, and the high pressure therefore persists in the conus after the ventricle begins to relax.

During contraction of the ventricle, the heart decreases in volume, and because the heart is located in a rigid chamber, the decrease in volume causes a negative pressure to develop in this chamber. In the Port Jackson shark the pressure in the ventricle during its contraction rises to above 30 mm Hg (4.0 kPa), and the concurrent negative pressure in the atrium and the sinus venosus may be as low as −4 mm Hg (−0.6 kPa) (Satchell 1970). Unless the large sinus and the adjoining veins could provide an equal volume of blood to flow into the atrium as the ventricle contracts, the negative pressure would only impede the contraction. With the inflow of blood, however, the negative pressure does not become excessive, but merely serves to fill the atrium. As the atrium af-

terward contracts, the now relaxed ventricle becomes filled with blood from the atrium, backflow into the sinus being prevented by valves. Thus, in this part of the cycle blood is merely shifted from atrium to ventricle, while the volume of the pericardial contents remains unchanged (Randall 1970).

In an air-breathing vertebrate, the sinus venosus and bulbus arteriosus decrease in importance, and in the mammalian heart they are not present as separate structures.

Lungfish

The major evolutionary change in lungfish is that, in addition to gills, they have lungs as respiratory organs (Figure 4.4c). The gills in part receive blood that has already passed through the lungs. If the gills were similar to the gills of ordinary fishes, this might be a disadvantage, for a lungfish swimming in oxygen-depleted water would then lose oxygen from the blood to the water that flows over the gills. The lungfish gills, however, have degenerated, and some of the gill arches permit a direct through-flow of blood (Johansen et al., 1970).

The atrium of the heart is divided into two chambers by a septum, and the ventricle is partially divided. In this way the lungfish heart somewhat resembles the completely divided heart of mammals, birds, and crocodiles. The lungfish heart, in fact, shows an amount of structural division greater than that of any amphibian (Foxon 1955). Blood from the lungs returns to the left atrium, and the right atrium receives blood from the general circulation. The partial division of the ventricle tends to keep the two bloodstreams separated, so that oxygenated blood tends to flow into the first two gill arches and supply the head with relatively oxygen-rich blood. The less well oxygenated blood from the right side of the heart flows through the posterior gill arches and passes on to the dorsal aorta and in part to the lungs. The lungfish represents the beginning of a complete separation between circulation to the lungs and to the remaining parts of the body (Johansen et al. 1968).

The anatomy of the lungfish does not explain how blood actually flows. Studies of live African lungfish (*Protopterus*) show that circulation to the lungs and circulation to the systemic vascular circuit have a high degree of functional separation, with preferential passage of oxygen-poor blood to the lungs and oxygen-rich blood to the systemic circulation. It is particularly significant that the functional separation is highest immediately after a breath of air when the oxygen in the lungs is highest, whereas later in the interval between breaths, the degree of separation diminishes. This is of obvious importance for the efficiency of gas exchange in a lung that is filled with fresh air only at intervals (Johansen et al. 1968).

Amphibians

The heart of the modern amphibians (frogs, toads, salamanders) has two completely separate atria, but only one undivided ventricle (Figure 4.4*d*). The left atrium receives oxygenated blood from the lungs, and the right atrium receives venous blood from the general systemic circulation. Although the ventricle is undivided, the two kinds of blood tend to remain unmixed, so that oxygenated blood enters the general circulation and oxygen-poor blood flows separately into the pulmonary circulation.

The pulmonary artery also sends branches to the skin; this is important because the moist amphibian skin is a major site of oxygen uptake. The anatomical arrangement of the heart includes a longitudinal ridgelike baffle in the bulbus arteriosus (known as the *spiral valve*), which seems important in keeping the bloodstreams separate (DeLong 1962).

Reptiles

In a noncrocodilian reptile the atria are completely separated, but the ventricle is only partially divided (Figure 4.4*e*). Even so, the streams of oxygenated and nonoxygenated blood are kept well separated so that there is very little mixing of the blood; there is, in effect, a well-developed double circulation. Thus, the incomplete division of the ventricle in amphibians and noncrocodilian reptiles cannot be interpreted simply on the basis of anatomical appearance, which by itself would suggest mixing of blood in the ventricle.

As a rule, a small animal has a higher rate of oxygen consumption per unit body weight than a remain much more separated than anatomical considerations would indicate, both in amphibians and in reptiles. The method used to evaluate the mixing is to determine the oxygen content in blood entering the heart and in blood in the vessels leaving the heart. Such studies have shown nearly complete separation of the bloodstreams (White 1959; Johansen 1962).

In crocodiles both chambers of the heart are completely divided (Figure 4.4*f*). This complete separation is confounded by the peculiar fact that the left aortic arch originates from the right ventricle and thus should receive venous blood. There is, however, a hole or foramen connecting the two aortic arches. On an anatomical basis this has been used to argue that crocodilian circulation allows mixing of oxygenated and nonoxygenated blood, but more careful study shows that both types of blood remain separate and that both aortic arches carry unmixed oxygenated blood (White 1956). During diving, however, the circulation changes: Blood flow to the lungs is decreased, and a major part of the output of the right ventricle is ejected into the left aortic arch. This shunt permits a rerouting of the blood during diving and a partial or complete bypass of the lungs.

Birds and mammals

The division of the heart and separation of pulmonary and systemic circulation are complete in birds and mammals (Figure 4.4g, h). This has one important consequence: The pressure can be different in the pulmonary and the systemic circulation. The resistance to flow in the pulmonary system is much lower than in the systemic circulation, and the blood pressure in the pulmonary circulation is only a small fraction of the pressure in the systemic part. Such a difference is, of course, not possible if separation of the heart is incomplete. Because incomplete separation has been retained in both amphibians and reptiles, this arrangement may have other advantages that are not well understood.

There are some differences between the circulation in birds and mammals that are of great significance in comparative anatomy. For example, mammals retain the left aortic arch, whereas birds retain the right. A difference of physiological importance is that the kidneys of all nonmammalian vertebrates receive venous blood from the posterior part of the body (the *renal portal circulation*). Birds have retained this renal portal circulation, but it is absent in mammals. This difference is important to the understanding of renal function (Chapter 10) (Johansen and Martin 1965).

Heart and cardiac output

As a rule, a small animal has a higher rate of oxygen consumption per unit body weight than a large animal, and thus the heart of the small animal must supply oxygen at a higher rate. We have already seen that the oxygen capacity of the blood of small and large mammals is similar; as a consequence the heart of the small mammals must pump blood at a higher rate. Is the necessary increase achieved by a larger pump, by a larger stroke volume, or by a higher frequency?

Size of vertebrate hearts

The heart sizes of a number of mammals are plotted in Figure 4.8. As expected, heart size increases with body size, but surprisingly, relative to body size, small and large mammals have about the same heart size. If the heart mass were exactly proportional to body mass, the slope of the regression line in Figure 4.8 would be 1.0; the actual slope is 0.98, which is statistically indistinguishable from exact proportionality. The equation for the line is $M_h = 0.0059 M_b^{0.98}$, which says that the average mass (M_h) of the mammalian

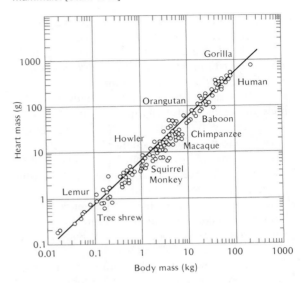

FIGURE 4.8 Heart size of mammals in relation to body size. The size of the heart is nearly proportional to body size and makes up approximately 0.6% of the body mass in small and large mammals alike. The heart size of a wide variety of primates falls within the range of other mammals. [Stahl 1965]

heart is 0.59% of the body mass (M_b in kilograms), irrespective of the size of the mammal.

The heart size of birds can be described by a similar equation: $M_h = 0.0082 \, M_b^{0.91}$ (Lasiewski and Calder 1971). This equation says that heart size is not strictly proportional to body size; the body mass exponent is significantly less than 1.0, which means that, relative to body size, larger birds tend to have slightly smaller hearts than birds of small body size. A 1-kg bird can be expected to have a heart of 8.2 g; for a mammal of the same size the expectation is a heart of 5.9 g.

The heart sizes of reptiles and amphibians are less well studied, but available data indicate that a reptile's heart size is about 0.51% of body weight; an amphibian's, 0.46%. These proportions are only slightly lower than in mammals, although the metabolic rates of reptiles and amphibians are about one-tenth the mammalian rate. Fish have smaller hearts again, about 0.2% of body weight.

To summarize, whether we compare the different classes, or animals of different body size within one class, the large differences in metabolic rate of vertebrates are not conspicuously reflected in the size of the heart. Differences in the need for oxygen must therefore be reflected primarily in pumping frequency, for the stroke volume depends on heart size, and the amount of oxygen contained in each volume of blood is not size-dependent.

Heartbeat frequency

The heartbeat frequency, or pulse rate, is usually given as the number of heartbeats per minute. The pulse rate for an adult human at rest is about 70 per minute; in exercise the rate increases several-fold.

The heart frequency is clearly inversely related to body size. An elephant that weighs 3000 kg has a resting pulse rate of 25 per minute, and a 3-g shrew, the smallest living mammal, has a resting

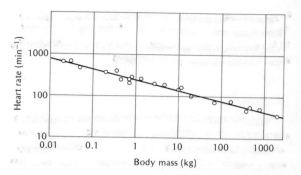

FIGURE 4.9 Relationship between heart rate and body size of mammals. The heart rate is higher in small than in large animals. If the observed data are plotted on logarithmic coordinates, they are grouped around a straight line, which can be represented by a logarithmic equation (see text). [Stahl 1967]

pulse rate of over 600 (Morrison et al. 1959). This means that in the shrew the heart goes through 10 complete contraction cycles per second, an almost unbelievable rate. During activity, the heart frequency increases further; as much as 1200 beats per minute have been measured in hummingbirds (Lasiewski et al. 1967) and in small bats in flight (Studier and Howell 1969).

If heart frequency is plotted relative to body mass on logarithmic coordinates, the points are located close to a straight line (Figure 4.9). The equation for the line is $f_h = 241 \, M_b^{-0.25}$. The slope of the regression line is negative; that is, the larger the body mass (M_b), the lower is the heart frequency (f_h).

In addition to the negative slope, the most significant information contained in this equation is the numerical value of the slope, which is 0.25. This is exactly the same as the slope of a regression line between body mass and resting oxygen consumption per unit body mass, the *specific oxygen consumption* (see Chapter 6).*

We now have two important pieces of information: (1) the size of the pump, the heart, remains a

---

* The word *specific*, as used here in front of a physical quantity, has the recommended meaning "divided by mass." For example, specific volume is the volume divided by the mass.

constant percentage of the body mass, and (2) the increase in the pumping rate (the heart rate) in the smaller mammal increases in exact proportion to the need for oxygen. Let us turn to the volume of blood pumped by the heart.

Cardiac output

The volume of blood pumped by the heart per unit time is usually called the *cardiac output* or the cardiac minute volume. (In hearts with a complete separation of the ventricles, the expression refers to the output of one side of the heart only.) The cardiac output can be determined in a number of different ways. One commonly used method is known as *Fick's principle*. It depends on the simple fact that all the oxygen consumed ($\dot{V}_{O_2}$)* by the animal is carried by the blood ejected from the heart. Therefore, if we know the difference in the oxygen content of arterial and venous blood, we can calculate cardiac output ($\dot{Q}_h$) from the following equation:

$$\dot{V}_{O_2} = \dot{Q}_h(C_{aO_2} - C_{vO_2})$$

To determine the cardiac output, we can determine (1) the oxygen consumption, (2) the oxygen content in arterial blood, and (3) oxygen content in venous blood. An arterial blood sample can be obtained from any artery, but the venous blood must be obtained as mixed venous blood (i.e., from the right side of the heart or from the pulmonary artery). Venous blood from some other place will not suffice, for different organs remove different

---

* In equations related to gas metabolism V usually stands for volume, and $\dot{V}$ (V-dot) for the time derivative of V (i.e., the volume per unit time). The subscript (in this case O_2) is chosen to aid easy recognition. Q stands for quantity of blood, $\dot{Q}$ for blood flow per unit time, and $\dot{Q}_h$ for blood flow from the heart (i.e., cardiac output). C stands for gas concentration in the blood, C_a for arterial blood and C_v for venous blood.

amounts of oxygen from the blood passing through them. A sample of mixed venous blood is most easily obtained from the right side of the heart by inserting a catheter in a vein and threading it in until it reaches the heart. This is a widely used and relatively harmless procedure.

During activity, when the oxygen consumption increases, the cardiac output also rises, but not in proportion to the increased oxygen consumption. Part of the increase is covered by an increase in the arteriovenous oxygen difference (i.e., during activity the amount of oxygen extracted from the blood increases).

Determinations of cardiac output according to Fick's principle can be carried out only in animals that have a complete separation between arterial and venous blood in the heart, as do mammals and birds. In amphibians and reptiles, where there is a possibility for mixing arterial and venous blood in the heart, this principle cannot be used. However, in fish the principle applies: The venous blood sample can be taken from or near the heart (but before the gills) and arterial blood from any artery after the gills.

Other methods for determining cardiac output are based on a variety of principles. One method consists of injecting an easily measured dye at a known moment in a vein leading to the heart. The dye will shortly afterward appear on the arterial side, and by integrating the curve representing the concentration of the dye as it rapidly increases and again disappears in the artery, it is possible to calculate the cardiac output.

In principle, the thermodilution method is similar. A known volume of saline is injected in the vein; if the saline is colder than the blood, there will be a transient decrease in arterial blood temperature. Again, the integrated curve gives information about the blood flow (i.e., the cardiac output).

TABLE 4.3 Blood flow to the major organs of a 70-kg man at rest. [Data from Folkow and Neil 1971]

Organ	Organ size (kg)		Blood flow (liter min⁻¹)		Blood flow (liter kg⁻¹ min⁻¹)
Kidneys	0.3		1.2		4.0
Liver	1.5	3.5	1.4	3.6	0.9
Heart	0.3		0.25		0.8
Brain	1.4		0.75		0.5
Skin	2.5		0.2		0.08
Muscle	29	66.5	0.9	2.0	0.03
Remainder	35		0.9		0.03
Total	70		5.6		

Distribution of blood flow

The blood is not evenly distributed to all parts of the body; some organs receive much more than others, both in relative and in absolute terms. The distribution of blood to the major organs of a man at rest is shown in Table 4.3.

The four most heavily perfused organs – kidneys, liver, heart, and brain – make up only 5% of the body mass, but these organs receive more than half the total cardiac output. The relative blood flow to the kidneys is higher than to any other major organ (although blood flow to some endocrine organs, especially the thyroid and adrenals, may be even higher). Per unit weight the kidneys receive more than 100 times as much blood as muscles at rest.

During exercise the cardiac output is increased and the distribution of blood flow is changed. This is primarily because of the vastly increased oxygen demands of the muscles and because the work load on the heart is augmented due to the increase in pumping.

The blood flow to the skin is also very variable, changing with the need for temperature regulation. In the cold there is vasoconstriction in the skin, and the blood flow to the skin is at a minimum; when there is a need to dissipate heat, as a result either of high ambient temperatures or of increased heat production during exercise, blood flow to the skin is vastly increased. This subject will be further discussed in Chapter 8.

Regulation of heartbeat

The heart has the inherent capacity to contract rhythmically without any external stimulus. Even if the heart is completely removed from the body, it may continue to beat for a considerable time: An isolated frog or turtle heart beats rhythmically for hours. Mammalian hearts do not continue to beat for as long, for they are more sensitive to temperature and an adequate oxygen supply. Since the isolated heart continues beating by itself, its ability to contract rhythmically must be inherent. Excellent proof of the independent contractility of the heart are the facts that the heart in a developing chick embryo begins to beat before any nerves have reached it and that heart muscle cells grown in tissue culture also contract rhythmically without any external stimulus.

The contraction in the mammalian heart begins in a small piece of embryonic-type muscle located where the vena cava enters the right atrium. It is the vestige of the sinus venosus and is known as the *sinus node*. The contraction spreads rapidly through the muscle of the two atria, and, after a slight delay, to the muscle of the ventricles. When the wave of contraction reaches the partition between the atria and the ventricles, a patch of tissue known as the *atrioventricular bundle* conducts the impulse to the ventricles, which then, after the minute delay resulting from conduction, contract simultaneously.

The sinus node, where the heart contraction originates, is also known as the *pacemaker* of the heart. It starts the beat and sets the rate by its inherent rhythmicity. As we know, the heart rate can vary by several-fold. Two different mechanisms are involved in the control of the rate: nerve impulses to the pacemaker and hormonal influences. A branch of the vagus nerve, when stimulated, slows the rate of heartbeat. The nerve is of parasympathetic origin and acts by release of *acetylcholine*. Another nerve, the accelerans nerve, is of sympathetic origin. When it is stimulated, the heart accelerates because *noradrenaline* is released from the nerve endings at the pacemaker.

Noradrenaline is also known as a hormone. When it is released from the adrenal medulla into the blood, it accelerates the heart. This effect on the heart is one of the major effects of adrenaline, well known as part of the fight-or-flight syndrome. There is no corresponding release of the decelerating substance, acetylcholine, except from the endings of the vagus nerve. If acetylcholine were released elsewhere in the body, it would not affect the heart for the simple reason that the blood contains an enzyme that rapidly splits acetylcholine by hydrolysis.

The amount of blood pumped by the heart can be changed, not only by changing the frequency, but also by changing the volume of blood ejected in a single beat, the *stroke volume*. When the heart contracts, it does not eject all the blood contained in the ventricles. A substantial but variable volume of blood remains in the ventricles at the end of contraction. Two major factors influence the stroke volume. One is the hormone adrenaline, which augments the force of the cardiac muscle contraction, thus forcing a larger amount of blood out of the heart in a single stroke. The other is the amount of blood in the ventricles when contraction begins. The latter results from an inherent characteristic of heart muscle, which will contract with greater force if it is more distended at the moment contraction begins.

If the return of venous blood to the heart is increased for some reason, the ventricles are filled with more blood, the muscle is stretched more than before, and the following contraction ejects a larger volume of blood. This relationship between increased venous return and increased cardiac output was discovered by the famous English physiologist Starling, and is known as *Starling's law of the heart*. In exercise the venous return to the heart increases, partly due to the increased pumping effect of the contracting muscles on the veins and partly to contraction in the venous system of the abdominal region, causing an increase in cardiac output.

In summary, the heartbeat originates in the sinus node or pacemaker. Cardiac output is under the influence of three major controlling systems: the nervous system, hormonal control, and the autoregulation attributable to the effect of the venous return on the heart muscle.

The conduits: blood vessels

The blood vessels are not merely tubes of different sizes. They have elastic walls, and a layer of smooth muscle within their walls enables them to change diameter.

There are characteristic differences among arteries, capillaries, and veins. Arteries have relatively thick walls that consist of heavy, strong layers of elastic fibers and smooth muscle. As the arteries branch and become smaller, the relative amount of muscle gradually increases in proportion to the elastic tissue in the wall. Capillaries, the smallest units of the system, consist of a single cell layer. Virtually all exchange of substances between blood and tissues takes place through the walls of the capillaries. Veins have thinner walls than arteries, but both elastic fibers and smooth muscle are found throughout the venous system.

Some measurements made on the vascular system of dogs are given in Table 4.4 (Burton 1972).

TABLE 4.4 Geometry of the blood vessels in the mesentery of the dog.

Kind of vessel	Diameter (mm)	Number	Total cross-sectional area (cm^2)	Length, approx. (cm)	Total volume (cm^3)[a]
Aorta	10	1	0.8	40	
Large arteries	3	40	3	20	
Arterial branches	1	2 400	5	5	190
Arterioles	0.02	40 000 000	125	0.2	
Capillaries	0.008	1 200 000 000	600	0.1	60
Venules	0.03	80 000 000	570	0.2	
Veins	2	2 400	30	5	
Large veins	6	40	11	20	680
Vena cava	12.5	1	1.2	40	

[a] The estimates of total volume are based on a more detailed analysis than indicated by the figures for length given in the preceding column.

Every time a larger vessel branches, the number of branches increases and the diameter decreases. As the number of branches increases, their total combined cross-sectional area increases, and at the level of the capillaries the aggregate cross section of all the capillaries in the mesentery alone is some 800 times that of the aorta. From that point on, the vessels converge into larger and larger veins, their number decreases, and so does the total cross-sectional area, until in the vena cava the cross section slightly exceeds that of the aorta. Another interesting fact is evident from the table: The amount of blood located in the venous system exceeds by several-fold the amount of blood in the arterial system.

In addition to the volume of the vascular system, we are interested in the pressure and the flow velocity in the various parts. Such information is given in Figure 4.10, which refers to measurements on man. The first column shows that most of the blood at any given moment is located on the venous side.

The last column shows that the velocity of the blood is highest in the aorta. As the total cross-sectional area increases, the velocity decreases drastically, until in the capillaries it is between one-hundredth and one-thousandth of the velocity in the aorta. On the venous side, the velocity increases again, but does not reach the high velocity of the arterial system.

The center column in Figure 4.10 shows the pressure throughout the system. The pressure must, of course, gradually decrease throughout, from the aorta to the larger veins. The greatest pressure drop takes place in the smallest arteries, called arterioles, which have a high proportion of smooth muscle in their walls and, by changing their diameter, are the most important factor in changing the resistance to flow and thus in regulating the distribution of blood flow to various organs.

The pulmonary circulation differs from the systemic in that the total resistance is less; the blood pressure can therefore be lower and the arteries have thinner walls. In cross section the pulmonary artery is not circular, but oval. During the contraction of the heart, as blood is injected into the pulmonary artery, the flattened vessel absorbs a considerable increase in volume as the cross section becomes circular; only then does the wall become stretched. In this way, much of the ejected blood is

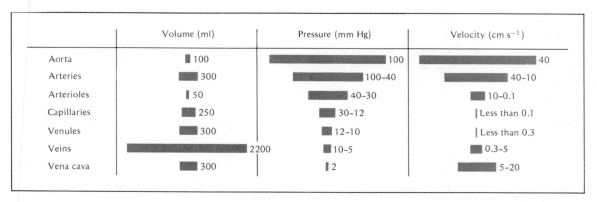

FIGURE 4.10 Distribution of blood volume, blood pressure, and velocity in the various parts of the vascular system of man (1 mm Hg = 0.13 kPa).

	Volume (ml)	Pressure (mm Hg)	Velocity (cm s⁻¹)
Aorta	100	100	40
Arteries	300	100–40	40–10
Arterioles	50	40–30	10–0.1
Capillaries	250	30–12	Less than 0.1
Venules	300	12–10	Less than 0.3
Veins	2200	10–5	0.3–5
Vena cava	300	2	5–20

taken up in the increased volume of the artery, which thus serves as a damping device so that the blood does not flow through the lung capillaries in spurts with each heartbeat (Melbin and Noordergraaf 1971).

The physics of flow in tubes

The flow of a fluid (whether gas or liquid) may be smooth and regular so that in a straight tube each particle of fluid moves in a straight line. This is called *laminar flow*. If in a curved tube the direction of flow at any given point remains constant, the flow is also said to be laminar. The path followed by a fluid particle in laminar flow is called a *streamline*. If the fluid particles flow irregularly and swirl about, however, the flow is *turbulent*. In a given tube the flow changes from laminar to turbulent if the velocity increases above a critical point. The following discussion will refer specifically to laminar flow; the fluid dynamics of turbulent flow is more complex and has little application to the flow in blood vessels.

Bernoulli's theorem

The steady flow in an ideal fluid along a streamline is described by *Bernoulli's theorem*, named after a Swiss physician (1700–1782) who at the age of 25 became professor of mathematics. First we must note that flow in a tube is called steady if the velocity at every point of the tube remains constant, even if the pipe gets narrower or wider and the fluid therefore moves faster or slower as it flows along the tube.

Bernoulli's theorem states that the total fluid energy (E) is the sum of (1) the potential energy attributable to internal pressure, (2) the potential energy attributable to gravity, and (3) the kinetic energy of the moving fluid.

$$E = (p \, v) + (m \, g \, h) + (\tfrac{1}{2} \, m \, u^2) \tag{1}$$

On the right side of the equation the first term is the pressure energy of the fluid or the product of pressure (p) and volume (v). The second term is the gravitational potential energy or the product of the mass (m), gravitational constant (g), and the height (h). The third term is the kinetic energy attributable to the velocity (u) of the fluid. The sum of these three is the total fluid energy. The energy content per unit volume (E') can be obtained by dividing both sides of the equation by the volume:

$$E' = p + \rho \, g \, h + \rho \frac{u^2}{2} \tag{2}$$

in which ρ (rho) is the density of the fluid.

FIGURE 4.11 (Top) When a fluid flows in a horizontal tube, the frictional resistance leads to a drop in pressure along the tube.

FIGURE 4.12 (Bottom) When a fluid flows through a narrow part of a tube, the velocity of the fluid is increased at this point. The increased velocity is accompanied by a decrease in pressure. For further explanation, see text.

Flow of a fluid in a horizontal tube permits a considerable simplification of the equation. Because there is no change in the gravitational potential energy from one end of the tube to the other, this term remains constant and can be dropped. Likewise, because the density of the fluid remains constant, this constant can be removed so that we obtain:

$$E' \propto p + \frac{u^2}{2} \tag{3}$$

If the flow is in a frictionless tube, the energy content of the fluid (E') remains constant, and if the velocity (u) changes due to a change in diameter, the pressure must change in the opposite direction, or:

$$\triangle \frac{u^2}{2} = -\triangle p \tag{4}$$

Let us next look at a tube of uniform diameter in which there is friction. With constant diameter, the velocity (u) remains constant. Because of resistance to flow, energy is used to drive the fluid through the tube, and the energy loss is therefore expressed as a decrease in pressure (p), as indicated in Figure 4.11. The energy loss from frictional forces is degraded as heat; this shows up as a temperature increase in the fluid, but in the bloodstream this is of no significance.

If the diameter of the tube changes, there must be a change in the velocity (u) as well. Again, in a horizontal tube the gravity term of equation 2 can be disregarded. Let us, for the moment, assume that no energy is dissipated by frictional forces and the energy content remains constant. We now see that velocity and pressure must change inversely according to expression 4, which states that if velocity increases, pressure in the fluid must decrease. This is illustrated in Figure 4.12 for a tube with frictional resistance.

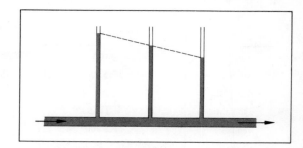

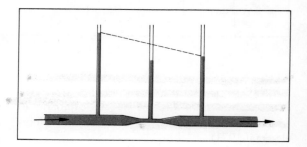

It is often stated that fluid always moves from higher pressure to lower pressure. Figure 4.12 shows that this is not necessarily so, for the pressure is lower in the narrow tube than further downstream. The correct statement is that fluid always moves from a point where the total fluid energy is higher to a point where it is lower.

In the preceding discussion the gravitational term in equation 2 was disregarded. If we have a U-tube, as in Figure 4.13, a liquid at rest has the same energy content per unit volume throughout the tube. Although the pressure at the bottom of the U is higher than in the arms, the fluid remains still. Reference to equation 2 shows that, as the height (h) decreases, the gravitational potential energy per unit volume ($\rho\,g\,h$) decreases, and the pressure energy (p) must increase accordingly. We refer to this increase in pressure energy as the *hydrostatic pressure* of a fluid, an expression used to

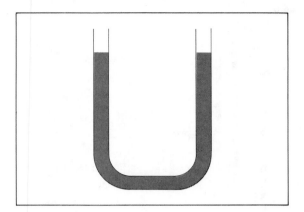

FIGURE 4.13 A fluid in a U-shaped tube remains at rest. At the bottom of the U the pressure is higher than in the arms, due to the effect of gravity on the fluid.

describe the increase in pressure with depth relative to the pressure at the surface.

Blood pressure

Let us see how these principles can be applied. In the foot of a man the venous pressure is increased due to the height of the column of blood (Figure 4.14). The arterial pressure at the foot actually exceeds the arterial pressure at the heart, again due to the weight of the column of blood. The high pressure in the venous system of the legs distends the veins, and contractions of the leg muscles and valves in the veins are main aids in returning the venous blood to the heart.

Another consequence of the gravitational term is that a certain arterial pressure is necessary to drive the blood to the head. Because of the gravity effect, the arterial blood pressure in the head of a man is reduced to about 50 mm Hg (7 kPa). For a giraffe, which carries its head about 2 m above the level of the heart, sufficient pressure to supply the brain with blood requires a much higher blood pressure than in man.

Measurements of blood pressures in the giraffe have shown a systolic blood pressure as high as 260 mm Hg (35 kPa) (Van Citters et al. 1968). (The normal systolic blood pressure in man is about 100 mm Hg or 13 kPa.) To withstand the high pressures, the giraffe's arterial system has exceptionally thick walls, and the venous system is equipped throughout with valves that facilitate the return of blood from the limbs. When a giraffe that stands 4.5 m tall lowers its head to drink, valves in the neck veins help impede backflow of blood to the head and prevent the increase in hydrostatic pressure in the brain that otherwise would be caused by swinging the head down.

Wall thickness and tube diameter

In a hollow cylinder the tension in the wall (T) equals the product of the pressure across the wall (p) and the radius (r):

$$T = p \times r$$

This relationship was derived by Bernoulli, but is usually known as *Laplace's law*. It tells us that, for a given pressure, the tension in the wall increases in direct proportion with increasing radius. For the wall to withstand the tension, the thickness must therefore be increased accordingly. This is the reason a large artery must have a thicker wall than a small artery.

A vein, which carries much lower blood pressure, has a thinner wall. Again, a smaller vein can withstand the venous pressure with a thinner wall than a larger vein. A capillary, of course, has a higher blood pressure than the veins, but nevertheless, because of its very small radius, a wall consisting of a single layer of cells has sufficient strength. Thus, the smallness of the capillary is a prerequisite for a capillary wall thin enough to permit rapid exchange of material between the blood and the tissues.

FIGURE 4.14 Arterial and venous pressures in a man as he assumes different postures. The figures indicate the pressures at various points in relation to the pressure in the right atrium of the heart. (1 mm Hg = 0.13 kPa). [Burton 1972]

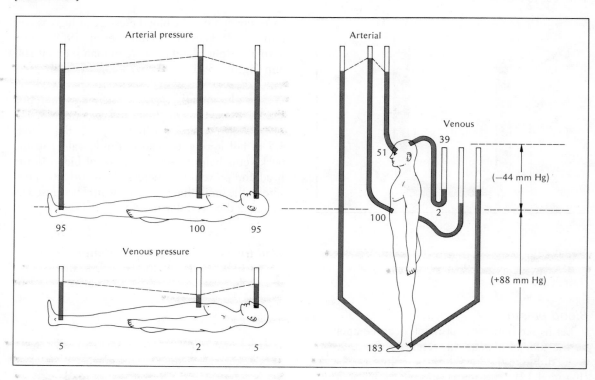

Viscosity

The resistance to flow in a tube results from inner friction in the fluid, the *viscosity* of the fluid. Everybody has some experience with viscosity – water and syrup do not flow equally fast out of a bottle. We say that water has a low viscosity and syrup a high viscosity.

The flow through a tube ($\dot{Q}$) is proportional to the pressure drop through the tube and inversely proportional to the resistance:

$$\dot{Q} = \frac{p}{R}$$

The resistance term (R) is a function of the dimensions of the tube and the nature of the fluid (viscosity). This relationship was clarified by the French physician Poiseuille (1799–1869), who became interested in the flow of blood in capillaries. Poiseuille worked with tubes of different diameters and found that the flow through a tube is proportional to the pressure and proportional to the fourth power of the radius of the tube, but inversely proportional to the length of the tube and to the viscosity of the fluid. This is formally expressed in *Poiseuille's equation*, which can be written as follows:

$$\dot{Q} = \triangle p \, \frac{r^4}{l\eta} \times \frac{\pi}{8}$$

in which $\dot{Q}$ = rate of blood flow, $\triangle p$ = pressure drop, r = radius, l = length of tube, and η (eta) = viscosity.*

The most striking aspect of this equation is the tremendous effect the radius of the tube has on the flow. For our purposes here, however, we shall be more concerned with the viscosity. The equation shows that as viscosity increases, the fluid flow decreases proportionately.

For many purposes it is convenient to express the viscosity of a fluid relative to that of water, assigning unity to the viscosity of water. Blood plasma is more viscous than water and has a *relative viscosity* of about 1.8, mostly as a result of the 7% dissolved protein. Assume that we measure the flow of water and of plasma through a given tube: If we assign the value of 1.0 to the flow of water, the flow of plasma will be 1/1.8. Let us now reduce the radius of the tube by one-half, keeping pressure and length of the tube constant. According to Poiseuille's equation the flow of water will be reduced to $(\frac{1}{2})^4$ or 1/16. So will the flow of plasma, and the ratio of the flows will still be 1/1.8; that is, the viscosity of plasma relative to water (the relative viscosity) is independent of the size of the tube.

If instead of plasma we use whole blood, we find that its viscosity is higher than the viscosity of plasma and increases with the concentration of red cells (Figure 3.1). Blood, however does not behave quite as expected. Its relative viscosity changes with the radius of the tube through which it flows. With decreasing radius, the relative viscosity of

* The viscosity of a fluid changes with temperature. A reduction in temperature from 37 to 0 °C causes the viscosity of water to increase about 2.6-fold.

blood decreases (i.e., it becomes more similar to water and flows more easily).

Such an anomalous fluid is called a *non-Newtonian fluid*; its behavior makes the flow of blood in the body difficult to describe, particularly in view of the fact that blood vessels are elastic and by no means maintain a constant radius. The end result, however, is important: Flow of blood through the capillary bed is easier, and the apparent viscosity of blood is lower than expected merely from the dimensions of the blood vessels.

We can now return to the question of blood viscosity that was raised in Chapter 3: What is the effect on blood viscosity of enclosing the hemoglobin in red cells? Because of its non-Newtonian behavior, whole blood flows more easily than corresponding hemoglobin solutions in the arterioles and capillaries (which are the major resistance elements of the circulatory system) (Snyder 1973).

Another peculiar aspect of blood flow through capillaries is that the capillary diameter is often considerably less than the size of a red blood cell. Unexpectedly, this seems to be no impediment to flow in the capillary: The red cell is easily deformed and passes readily through the capillary (Skalak and Branemark 1969). This gives rise to a very different type of flow (*bolus flow*) in which the red cell acts as a plug that causes a rapid renewal of liquid in the boundary layer along the capillary wall, thus facilitating the renewal of diffusible substances in this layer.

The capillary system

The total number of capillaries in the body is enormous. Merely in the mesentery of a dog the number of capillaries exceeds one thousand million (Table 4.4). Muscle is particularly well suited for an exact count of the number of capillaries, for here the capillaries run between and parallel to the muscle fibers. In a cross section of a muscle it is therefore relatively easy to count the number of capillaries per unit cross-sectional area. The number is very great, but the capillaries usually are not all open and filled with blood. In the resting muscle of a guinea pig, a cross section of 1 mm² contains about 100 open capillaries through which blood flows. If the muscle is performing work, however, the constricted arterioles open, more capillaries carry blood, and in maximal exercise a 1-mm² cross section may have more than 3000 open capillaries. In an ordinary pencil the cross section of the lead is about 3 mm². Imagine nearly 10 000 tiny tubes running in parallel down the lead of the pencil!

A mammal of small body size has a higher capillary density than a mammal of large body size. This is consistent with the higher metabolic rate and need for oxygen in the small animal, as discussed in Chapter 3.

Two major processes account for the exchange of material between the fluid within the capillary and the interstitial spaces of the tissues. The capillary wall is quite thin, about 1 μm thick, and consists of a single layer of cells. Water and dissolved substances of small molecular weight (gases, salts, sugars, amino acids, etc.) can diffuse relatively unhindered across the capillary wall, but in addition, pressure within the capillary forces fluid out through the wall by bulk filtration. Molecules as large as most proteins (molecular weight about 70 000 or larger), however, do not pass through the capillary wall. Proteins are withheld within the capillary, and a relatively protein-free filtrate is forced out through the wall. The concentrations of the small-size solutes in this filtrate are similar to those in the blood plasma, but not quite identical because the proteins have some influence on the distribution of ions (the Gibbs–Donnan effect).

If we regard the capillary wall as a semipermeable membrane, we can see that because the nonprotein solutes penetrate freely, they have no osmotic effect. The proteins, however, are retained inside the capillary and exert a certain osmotic effect, called the *colloidal osmotic pressure*. In mammalian plasma this pressure is about 25 mm Hg and tends to draw water back into the capillary from the surrounding tissue fluid. When the hydrostatic pressure (blood pressure) within the capillary exceeds the colloidal osmotic pressure, fluid is forced out through the capillary wall; when the hydrostatic pressure within falls below the colloidal osmotic pressure, fluid is drawn in. The blood pressure within the capillary is variable, but in the arterial end it is often higher and in the venous end lower than the colloidal osmotic pressure. As a result, fluid is filtered out at the arterial end and reenters at the venous end of the capillary (Figure 4.15). This theory for bulk filtration and reentry of fluid through the capillary wall was originally proposed by Starling (1896).

The amount of fluid filtered through the capillary wall, and the amount withdrawn again because of the colloidal osmotic pressure, fluctuate greatly. Usually outflow exceeds inflow, and the excess fluid remains in the interstitial spaces between the tissue cells. This fluid, the *lymph*, slowly drains into tiny lymph vessesl or lymph capillaries. It is now accepted that this flow is not an entirely passive drainage. The small lymph vessels undergo continual rhythmic contractions whose pumping action causes a slight negative pressure to develop, and as long as the amount of fluid filtering through the capillary wall remains slight, the lymphatic system is capable of removing all the excess fluid (Guyton 1976).

The smaller lymph vessels gradually merge with other lymph vessels and eventually join into larger lymph ducts. These, in turn, empty the lymph into the larger veins, returning the fluid to the

FIGURE 4.15 The capillary wall is semipermeable, and the blood pressure forces fluid out by ultrafiltration (top). The plasma proteins remain within the capillary and oppose the ultrafiltration process. The blood pressure decreases along the length of the capillary, and when the blood pressure falls below the colloidal osmotic pressure exerted by the proteins, these cause an osmotic flow of fluid back into the capillary (bottom). The wall is freely permeable to salts and other small-molecular-size solutes, and the fluid movements therefore take place as if these were not present.

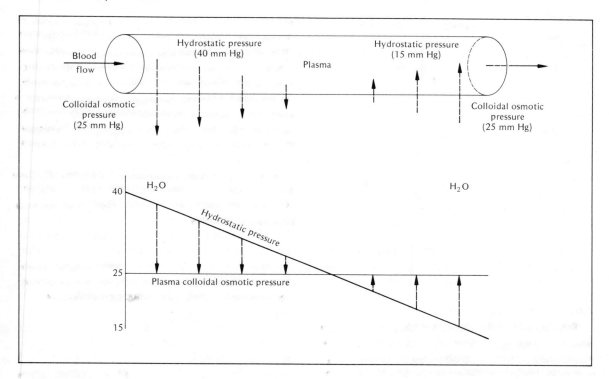

blood system. The lymphatic system can be considered part of the vascular system, but there is no direct circulation of the lymph as such; it is, rather, a drainage system that returns to the blood excess fluid lost from the capillary system.

If plasma proteins were absent, any hydrostatic pressure in the capillary in excess of that in the surrounding tissue spaces would cause fluid to filter out, and no force would be available to cause fluid to reenter. The plasma proteins are therefore essential to retain the blood fluid within the vascular system.

Plasma colloidal osmotic pressure is higher in mammals than in other vertebrates (Table 4.5). Surprisingly, birds have a rather low colloidal pressure, although their arterial blood pressure usually far exceeds the common values for mammals. Some birds have blood pressures above 200 or even 250 mm Hg (27 to 33 kPa), against about 100 mm Hg (13 kPa) commonly found in mammals. These differences are not well understood. It seems either that birds must have a greater resistance to blood flow at the level of the arterioles so that the capillary pressure is not very high, or that the avian capillary wall has properties that differ from those of mammalian capillaries. A higher resistance in the arterioles seems meaningless for why should birds then have the high arterial pressure to begin with?

Other vertebrates – reptiles, amphibians, and fish – also have relatively low colloidal osmotic

TABLE 4.5 Colloidal osmotic pressure of vertebrate blood plasma (1 mm Hg = 0.13 kPa). [Data from Altman and Dittmer 1971]

Animal	Pressure (mm Hg)
Cow	21
Sheep	22
Dog	20
Cat	24
Chicken	11
Dove	8.1
Alligator	9.9
Turtle	6.4
Frog	5.1
Toad	9.8
Codfish	8.3
Flounder	8.5

pressure. This is consistent with the fact that these groups in general have lower arterial blood pressures than mammals.

Circulation during exercise

During muscular activity the demand for oxygen is increased and thus the amount of oxygen the heart must deliver to the tissues. Two avenues are open to meet this increased demand: increasing the volume of blood pumped by the heart (the cardiac output) or increasing the amount of oxygen delivered by each volume of blood. Arterial blood is already fully saturated and cannot take up more oxygen, but venous blood normally still contains more than half the oxygen content of arterial blood. Increasing the oxygen extraction from the blood is an obvious way of obtaining more oxygen from each volume of blood.

Let us first examine oxygen extraction. The total muscle mass of a lean man, which makes up nearly half his weight, uses about 50 ml O_2 per minute. This oxygen is supplied by a blood flow of about 1 liter (i.e., the oxygen content of arterial blood, 200 ml O_2 per liter blood, is reduced to 150

ml O_2 per liter in venous blood). Because a quarter of the oxygen in arterial blood is removed, we say that the oxygen extraction is 25%. In a normal man during heavy exercise the blood flow to the muscles may be 20 liters per minute (in well-trained athletes, even higher), and the oxygen extraction in the muscles increases to 80 or 90%; in other words, the venous blood coming from heavily working muscles has very little oxygen left in it (Folkow and Neil 1971).

The second avenue available for increasing the delivery of oxygen is increased cardiac output. This can be achieved by increasing both heart rate and stroke volume. Because of the interest in man, both for medical reasons and in relation to athletic performance, much more information is available for man than for other animals. At rest the human heart beats at a rate of about 70 per minute, with a stroke volume of about 70 ml (from each side), giving a total cardiac output of about 5 liters per minute. In heavy exercise, the cardiac output can readily increase fivefold or more (if the oxygen extraction is tripled, this corresponds to a 15-fold increase in oxygen delivery). Most of the increase in cardiac output is attributable to increased pulse rate, which may rise to 200 strokes per minute, but there is also an increase in stroke volume, which may exceed 100 ml.

The distribution of the increased cardiac output to the muscles and to the remainder of the body in rest and exercise is shown in Figure 4.16. The blood flow to the muscles of an athlete in top condition may increase as much as 25- or 30-fold, while there is a slight decrease in circulation to the remainder of the body. In the athlete the oxygen consumption of the muscles may be increased as much as 100-fold; this is possible only because of an approximately threefold increase in the oxygen extraction.

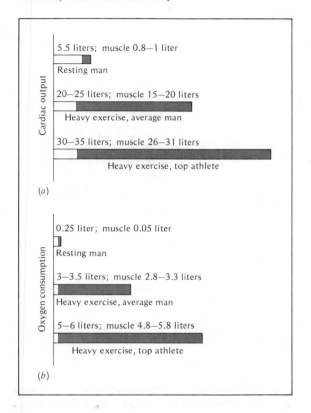

FIGURE 4.16 Distribution of total blood flow (cardiac minute volume) (*a*) and of oxygen consumption (*b*) between the muscles (shaded bars) and all other parts of the body (unshaded bars). Data for resting man, heavy exercise in a normal man, and heavy exercise in a top athlete. [Folkow and Neil 1971]

5.5 liters; muscle 0.8—1 liter
Resting man

20—25 liters; muscle 15—20 liters
Heavy exercise, average man

30—35 liters; muscle 26—31 liters
Heavy exercise, top athlete

(*a*)

Cardiac output

0.25 liter; muscle 0.05 liter
Resting man

3—3.5 liters; muscle 2.8—3.3 liters
Heavy exercise, average man

5—6 liters; muscle 4.8—5.8 liters
Heavy exercise, top athlete

(*b*)

Oxygen consumption

INVERTEBRATE CIRCULATION

Many invertebrates have well-developed circulatory systems; examples are annelids, echinoderms, arthropods, and molluscs. Most of these have open systems. Cephalopods (octopus and squid) have closed circulations, although the other classes of molluscs have open systems. It is worth noting that insects, which are highly organized and complex animals and can sustain exceptionally high metabolic activity, have open circulatory systems.

Annelids

Some annelids have well-organized circulatory systems and their blood often contains respiratory pigments dissolved in the plasma. Hemoglobin is the most common pigment, but some annelids use chlorocruorin or hemerythrin.

For the earthworm, *Lumbricus*, the general body surface serves as a respiratory organ and is supplied with a dense capillary network (Figure 4.17). The circulatory system has two longitudinal vessels: a dorsal vessel in which the blood is pumped forward, and a ventral vessel in which it flows in the opposite direction. The nerve cord of many annelids, including the earthworm, is supplied through additional longitudinal vessels, and some of the oxygenated blood from the integument is routed directly to this important organ. The closed system of blood vessels thus has a distinctive advantage in controlling precisely where the blood will flow.

An annelid does not have one single distinct heart, although several of the blood vessels have dilations that are contractile. The dorsal blood vessel is the most important vessel, in which peristaltic waves propel blood forward. In addition, blood vessels that on each side connect the main dorsal with the main ventral longitudinal vessel are contractile and serve as accessory hearts.

Echinoderms

Echinoderms (e.g., starfish, sea urchins, and sea cucumbers) have three fluid systems: the *coelomic system*, the blood or *hemal system*, and the *water vascular system*. The last functions primarily in locomotion; it is a hydraulic system used in the movement of the tube feet and is filled with a fluid similar to sea water.

An echinoderm has a large coelom between the body wall and the digestive tract, filled with a coe-

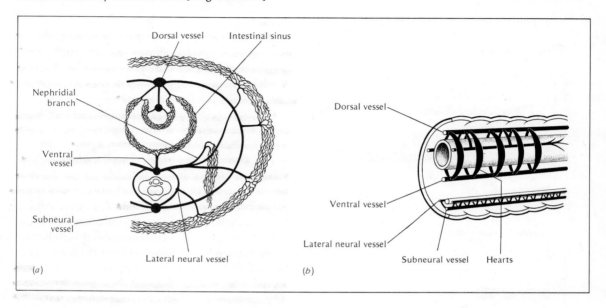

FIGURE 4.17 The circulation in the earthworm. The dorsal vessel is contractile and pumps blood into the larger vessels. The integument serves in gas exchange and is heavily vascularized. Special vessels from the integument supply the ventral nerve cord with oxygen-rich blood directly from the skin. [Meglitsch 1972]

lomic fluid that seems to be important in transporting nutrients between the digestive tract and other parts of the body (Farmanfarmaian and Phillips 1962).

The hemal or blood vascular system of the echinoderm contains a separate fluid that, in some sea cucumbers (holothurians), contains hemoglobin. The hemoglobin from the sea cucumber *Cucumaria* has a rather high oxygen affinity, with half saturation pressures between 5 and 10 mm Hg (depending on temperature) and a striking absence of a Bohr effect between pH 7.5 and 6.5 (Manwell 1959). These characteristics are apparently well suited to the often oxygen-deficient mud environment in which sea cucumbers live.

The relationship between the hemal system and the coelomic system is not well understood, and the role of the hemal system in respiration and oxygen supply to the animal is not clear (Martin and Johansen 1965). The role of the water vascular system is primarily in locomotion and is better understood.

Molluscs

With the exception of octopus and squid, molluscs have open circulatory systems. The blood of many molluscs contains hemocyanin, and a few have hemoglobin. The oxygen-carrying capacity of the blood tends to be correlated with the size of the animal and in particular, with its activity.

Molluscs, in general, have a well-developed heart, and the heartbeat is adjusted to meet the physiological demands for oxygen. An increased venous return of blood to the heart causes an increase in both amplitude and frequency of the heartbeat. The mollusc heart continues to beat if it is separated from the nervous system; it has an inherent rhythmicity with a pacemaker. However, the heart is also under the influence of neurosecretions that modify its beat. Acetylcholine inhibits the heart, and serotonin acts as an excitatory substance.

In clams, active movement of the foot is based on the use of blood as a hydraulic fluid. Large blood sinuses in the foot are controlled by valves that help in adjusting both the size and the movements of the foot. This is particularly important in burrowing species, which use the foot in moving through the bottom material.

Cephalopods (octopus and squid) have highly developed closed circulatory systems with distinct arteries, veins, and capillary networks. This is related to the high organization and activity of these animals and to the importance of the blood in respiratory exchange in the gills as well as in the function of the kidney.

The fact that cephalopods have a closed circulatory system means that the blood remains confined within the blood vessels. Therefore, the blood remains distinctly separate from the interstitial fluid. The blood makes up about 6% of the body weight, and the interstitial fluid, about 15%. These magnitudes are strikingly similar to those of the corresponding fluid volumes in vertebrate animals. The open system of a noncephalopod mollusc does not separate blood from interstitial fluid, and blood flows freely throughout the extracellular space; the blood volume is therefore quite large and may constitute as much as 50% of the wet body weight (without the shell) (Martin et al. 1958).

Insects

An insect usually has a major blood vessel running along the dorsal side. The posterior part of this vessel functions as a "heart" and is provided with a series of valved openings through which blood can enter. The anterior part of the blood vessel, which could be called the "aorta," is contractile and may have peristaltic waves propelling the blood forward. This main blood vessel branches

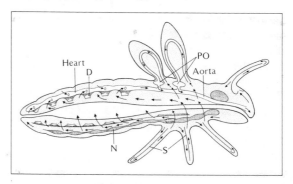

FIGURE 4.18 The well-developed circulatory system in an insect has a main dorsal vessel that carries blood from the heart. Although the circulation is open, the blood follows certain channels because of the presence of longitudinal membranes, especially in the legs. Arrows indicate course of circulation. D, dorsal diaphragm with muscles; N, nerve cord; PO, pulsatile organs; S, septa dividing appendages. [Wigglesworth 1972]

and continues to the head, where it ends. The branches from the aorta supply most of the body, where blood flows freely among the tissues and slowly percolates back toward the heart.

In many insects the blood that flows among the tissues is given some direction by longitudinal membranes. In the antennae and the limbs, the blood may enter on one side and leave on the other. In this way the blood is led into certain pathways, although it does not run in discrete blood vessels (Figure 4.18).

Because the blood system is open, the blood pressure of an insect can barely exceed the tissue pressure. Insect blood plays no particular role in respiration; its main functions are to carry nutrients and metabolites around in the body and to provide a transport system for hormones, which in insects are of great importance in physiological coordination (growth, molting, etc.). In some insects the circulation of blood is also essential in the distribution of heat and, in particular, in the regulation of heat loss in highly active flying insects (Heinrich 1971).

Although the heart and the dorsal vessel are the principal components of the circulatory system of insects, many also have accessory pumping organs or hearts. These are particularly important for maintaining circulation in the appendages (wings, legs, and antennae). The flow of blood in the wings is aided by pulsating organs within the wings themselves, located in the channels that return the blood to the body (Thomsen 1938). Blood flow into the antennae is aided by contractile organs located at their bases. These function by aspiration of blood from blood sinuses in the head. Many insects also have leg hearts that aid in circulating blood to these important appendages. The accessory hearts of an insect and their contraction are independent of the dorsal heart: They may beat at different rates; one may stop while others continue to beat and none is dependent on the dorsal heart.

The filling of the dorsal heart of an insect takes place by suction. A set of muscles (the *aliform muscles*) radiate from the heart; when they contract, the expansion of the heart pulls blood through valved openings (*ostia*) into the lumen of the heart. The subsequent contraction of the heart propels blood forward, in the direction of the head, and the valves prevent the blood from flowing out through the ostia (Figure 4.19).

Arachnids

The circulatory system of spiders and scorpions is similar to that of insects, but it may have a greater role in respiration than is usual in insects. Hemocyanin has been found in the blood of some scorpions, and both scorpions and spiders have definite respiratory organs that are perfused by blood. This should not be surprising, for arachnids lack the tracheal system that is characteristic of insects.

The arachnid heart is located dorsally in the abdomen. It is filled through valved openings (ostia) and empties into arteries. In spiders, distinct arteries lead to the legs, where a relatively high blood pressure is of importance in locomotion. Spider

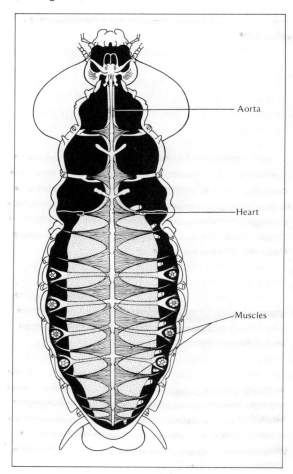

FIGURE 4.19 The insect heart consists of a contractile portion of the main dorsal blood vessel. Filling of the heart with blood is achieved by suction, provided by muscles that pull externally on the walls of the heart. This figure shows the dissection of a cockroach (*Blaberus*). [Nutting 1951]

Aorta

Heart

Muscles

legs lack extensor muscles, and blood is used as a hydraulic fluid for leg extension (see Chapter 11).

Crustaceans

The circulatory system of crustaceans is extremely variable. Small crustaceans have poorly developed circulatory systems, often without any heart; large crustaceans, in particular decapods (lobster, crabs, and crayfish), have well-developed

FIGURE 4.20 Circulation in the lobster (a). In large crustaceans the oxygenated blood from the gills is conveyed directly to the heart, which pumps it on to the tissues (b). In fish the heart is located differently, before the gills (see Figure 4.3). [Meglitsch 1972]

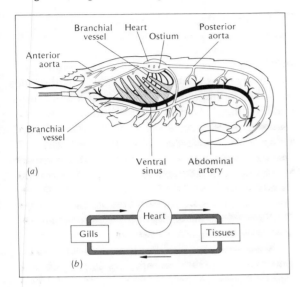

(a)

(b)

circulatory systems and blood with respiratory pigments (hemocyanin).

The crustacean circulatory system is based on the same morphological structures as in other arthropods, but it differs from insect circulation in one important physiological aspect: The crustaceans have gills and, therefore, also well-defined circulation to the gills.

The dorsal heart of a large crustacean lies in a pericardial sinus from which blood enters the heart through valved ostia. From the heart there is usually a main artery running in the anterior and another in the posterior direction. As the arteries branch, the blood leaves the vessels and flows between the tissues to a system of ventral sinuses. From these the blood flows into the gills and then in discrete vessels back to the heart.

As a result of this arrangement the decapod heart is directly supplied with oxygenated blood, which is then pumped to the tissues (Figure 4.20). This is the opposite of the arrangement in fish; in a fish the heart receives deoxygenated venous blood, which is then pumped to the gills and on to the tissues. The fish heart itself, however, is supplied with oxygenated blood through a branch of the gill circulation from which oxygenated blood reaches the heart muscle directly.

BLOOD COAGULATION AND HEMOSTASIS

Several mechanisms help prevent loss of blood from ruptured blood vessels. Severe blood loss leads to a decrease in blood pressure and thus reduces the flow of blood from the damaged area. Damaged blood vessels constrict and thereby decrease blood flow. The most important mechanism, however, is the closing of blood vessels at the site of injury by a plug consisting of coagulated protein and blood cells. Such a plug or clot is important in the complete arrest of bleeding from minor injuries, but if major blood vessels have been ruptured, it does not suffice.

The clotting or *coagulation* mechanism has been well studied in mammals, particularly man, because blood clotting is of great medical importance. To be effective, a clotting mechanism must act rapidly; yet blood must not clot within the vascular system. Blood must therefore have the inherent ability to clot, and the clotting mechanism should be ready to be turned on when needed. On the other hand, this mechanism is like sitting on a bomb, it must not go off inadvertently.

In vertebrates the blood clot consists of the protein *fibrin*, an insoluble fibrous protein formed from *fibrinogen*, a soluble protein present in normal plasma in an amount of about 0.3%. The transformation of fibrinogen to fibrin is catalyzed by the enzyme *thrombin*, and the reason blood does not clot in the vascular system is that thrombin is absent from the circulating blood. Thrombin, however, can be formed rapidly because its

RED BLOOD CELL A human red blood cell caught in the meshwork of fibrin in a blood clot. The cell, whose diameter is about 6μm, has the biconcave shape characteristic of mammalian red blood cells. [Courtesy of Emil Bernstein, the Gillette Company Research Institute; from *Science,* 1971, vol. 173, No. 3993, cover picture]

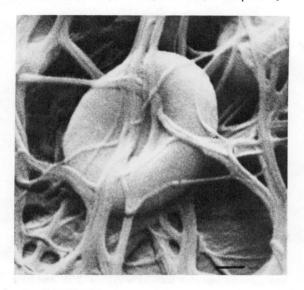

The many steps in the clotting mechanism may seem an unnecessary complication which indeed has been a severe hindrance in the clarification of what takes place. Biologically, the importance of the many steps seems to be explained by the fact that the clotting mechanism works as a biochemical amplifier (MacFarlane 1964). Clotting is normally initiated when blood contacts "foreign" surfaces or damaged tissues. This initiates a series of enzymatic steps in which the enzyme formed in the first step serves as a catalyst or activator for the next step, and so on. The series of steps thus forms an enzyme cascade, which ends with the final formation of the clot in which the soluble fibrinogen is changed to the insoluble fibrin.

The enzymatic amplification allows clotting to take place rapidly, yet provides a considerable safety margin to prevent spontaneous coagulation within the vascular system. The analogy to an electronic amplification system is obvious. If we want an amplification system with a low noise level, we use several steps with low amplification in each step, rather than a single step with a high amplification. This minimizes the chance that random noise in the system may set off the final step and ensures an adequate margin of safety.

A hemostatic mechanism is as essential for most invertebrates as for vertebrates. The fact that many have open circulatory systems makes the situation more difficult, for in such a system the contraction of blood vessels is of no help. On the other hand, open systems always have lower blood pressures, and this decreases the tendency for loss of large volumes of fluid.

The two distinct hemostatic mechanisms of vertebrates – clotting of blood and local vasoconstriction – have their counterparts in invertebrates. The simplest invertebrate mechanism is the agglutination of blood corpuscles without participation

precursor, *prothrombin*, is already present in the plasma. What is necessary to initiate coagulation is the formation of thrombin from prothrombin. This is only the final step in a complex sequence of biochemical events that has been slowly unraveled in studies of human patients with various deficiencies in the clotting mechanism (e.g., hemophilia). A total of 12 clotting factors have been identified, numbered I through XIII (factor VI is a term no longer used). A few of the final steps of the clotting mechanism are shown in the following diagram.

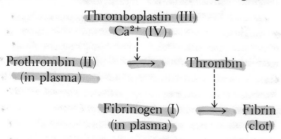

of plasma proteins (Grégoire and Tagnon 1962). Agglutination is followed by the formation of a cellular meshwork that shrinks and helps close a wound. This is often helped by the contraction of the muscles of the body wall, thus aiding in wound closure.

A true clotting, induced by enzymatic changes of unstable blood proteins, has been described for many arthropods, especially crustaceans. The clotting mechanism of invertebrates, where it occurs, is biochemically distinctly different from the vertebrate mechanism. For example, vertebrate clotting is inhibited by heparin, a mucopolysaccharide that can be isolated from mammalian liver. Heparin has no effect on the clotting system of the horseshoe crab *Limulus* (an arachnid, not a crab), and virtually no effect on crustacean blood (Needham 1970).

Knowledge of the clotting mechanisms of invertebrates is very incomplete, but present information indicates that such mechanisms must have evolved independently many times in the course of evolution.

In the four preceding chapters we have discussed respiratory gases and their transport. We shall now turn to questions concerning the supply of energy and other related matters.

REFERENCES

Altman, P. L., and Dittmer, D. S. (eds.) (1971) *Biological Handbooks: Respiration and Circulation.* Bethesda: Federation of American Societies for Experimental Biology. 930 pp.

Bond, C. F., and Gilbert, P. W. (1958) Comparative study of blood volume in representative aquatic and nonaquatic birds. *Am. J. Physiol.* 194:519–521.

Burton, A. C. (1972) *Physiology and Biophysics of the Circulation: An Introductory Text.* 2nd ed. Chicago: Year Book. 217 pp.

DeLong, K. T. (1962) Quantitative analysis of blood circulation through the frog heart. *Science* 138:693–694.

Farmanfarmaian, A., and Phillips, J. H. (1962) Digestion, storage, and translocation of nutrients in the purple sea urchin (*Strongylocentrotus purpuratus*). *Biol. Bull.* 123:105–120.

Folkow, B., and Neil, E. (1971) *Circulation.* New York: Oxford University Press. 593 pp.

Foxon, G. E. H. (1955) Problems of the double circulation in vertebrates. *Biol. Rev.* 30:196–228.

Gordon, M. S. (1972) *Animal Physiology: Principles and Adaptations*, 2nd ed. New York: Macmillan. 592 pp.

Grégoire, C., and Tagnon, H. J. (1962) Blood coagulation. *Comp. Biochem.* (B) 4:435–482.

Guyton, A. C. (1976) Symposium: Interstitial fluid pressure and dynamics of lymph formation. *Fed. Proc.* 35:1861–1885.

Heinrich, B. (1971) Temperature regulation of the sphinx moth, *Manduca sexta*. 2. Regulation of heat loss by control of blood circulation. *J. Exp. Biol.* 84:153–166.

Hopper, J., Jr., Tabor, H., and Winkler, A. W. (1944) Simultaneous measurements of the blood volume in man and dog by means of Evans blue dye, T1824, and by means of carbon monoxide. 1. Normal subjects. *J. Clin. Invest.* 23:628–635.

Johansen, K. (1960) Circulation in the hagfish, *Myxine glutinosa* L. *Biol. Bull.* 118:289–295.

Johansen, K. (1962) Double circulation in the amphibian *Amphiuma tridactylum. Nature, Lond.* 194:991–992.

Johansen, K. (1968) Air-breathing fishes. *Sci. Am.* 219:102–111.

Johansen, K. (1970) Air breathing in fishes. In *Fish Physiology*, vol. 4 (W. S. Hoar and D. J. Randall, eds.), pp. 361–411. New York: Academic Press.

Johansen, K., Lenfant, C., and Hanson, D. (1968) Cardiovascular dynamics in the lungfishes. *Z. Vergl. Physiol.* 59:157–186.

Johansen, K., and Martin, A. W. (1965) Comparative

aspects of cardiovascular function in vertebrates. In *Handbook of Physiology*, sect. 2, *Circulation*, vol. 3 (W. F. Hamilton and P. Dow, eds.), pp. 2583–2641. Washington, D.C.: American Physiological Society.

Krijgsman, B. J. (1956) Contractile and pacemaker mechanisms of the heart of tunicates. *Biol. Rev.* 31:288–312.

Lasiewski, R. C., and Calder, W. A., Jr. (1971) A preliminary allometric analysis of respiration variables in resting birds. *Respir. Physiol.* 11:152–166.

Lasiewski, R. C., Weathers, W. W., and Bernstein, M. H. (1967) Physiological responses of the giant hummingbird, *Patagona gigas. Comp. Biochem. Physiol.* 23:797–813.

MacFarlane, R. G. (1964) An enzyme cascade in the blood clotting mechanism, and its function as a biochemical amplifier. *Nature, Lond.* 202:498–499.

Manwell, C. (1959) Oxygen equilibrium of *Cucumaria miniata* hemoglobin and the absence of the Bohr effect. *J. Cell. Comp. Physiol.* 53:75–84.

Martin, A. W., Harrison, F. M., Huston, M. J., and Stewart, D. M. (1958) The blood volumes of some representative molluscs. *J. Exp. Biol.* 35:260–279.

Martin, A. W., and Johansen, K. (1965) Adaptations of the circulation in invertebrate animals. In *Handbook of Physiology*, sect. 2, *Circulation*, vol. 3 (W. F. Hamilton and P. Dow, eds.), pp. 2545–2581. Washington, D.C.: American Physiological Society.

Meglitsch, P. A. (1972) *Invertebrate Zoology*, 2nd ed. New York: Oxford University Press. 834 pp.

Melbin, J., and Noordergraaf, A. (1971) Elastic deformation in orthotropic vessels: Theoretical and experimental results. *Circulation Res.* 29:680–692.

Morrison, P., Ryser, F. A., and Dawe, A. A. (1959) Studies on the physiology of the masked shrew *Sorex cinereus. Physiol. Zool.* 32:256–271.

Needham, A. E. (1970) Haemostatic mechanisms in the invertebrata. *Symposia of the Zoological Society of London*, No. 27 (R. G. MacFarlane, ed.), pp. 19–44. London: Academic Press.

Nutting, W. L. (1951) A comparative anatomical study of the heart and accessory structures of the orthopteroid insects. *J. Morphol.* 89:501–597.

Randall, D. J. (1970) The circulatory system. In *Fish Physiology*, vol. 4 (W. S. Hoar and D. J. Randall, eds.), pp. 133–172. New York: Academic Press.

Satchell, G. H. (1970) A functional appraisal of the fish heart. *Fed. Proc.* 29:1120–1123.

Skalak, R., and Branemark, P. I. (1969) Deformation of red blood cells in capillaries. *Science* 164:717–719.

Snyder, G. K. (1973) Erythrocyte evolution: The significance of the Fåhraeus-Lindqvist phenomenon. *Respir. Physiol.* 19:271–278.

Stahl, W. R. (1965) Organ weights in primates and other mammals. *Science* 150:1039–1042.

Stahl, W. R. (1967) Scaling of respiratory variables in mammals. *J. Appl. Physiol.* 22:453–460.

Starling, E. H. (1896) On the absorption of fluids from the connective tissue spaces. *J. Physiol., Lond.* 19:312–326.

Studier, E. H., and Howell, D. J. (1969) Heart rate of female big brown bats in flight. *J. Mammal.* 50:842–845.

Thomsen, E. (1938) Ueber den Kreislauf im Flügel der Musciden, mit besonderer Berücksichtigung der akzessorischen pulsierenden Organe. *Z. Morphol. Oekol. Tiere* 34:416–438.

Thorson, T. B. (1958) Measurement of the fluid compartments of four species of marine Chondrichthyes. *Physiol. Zool.* 31:16–23.

Thorson, T. B. (1959) Partitioning of body water in sea lamprey. *Science* 130:99–100.

Thorson, T. B. (1961) The partitioning of body water in Osteichthyes: Phylogenetic and ecological implications in aquatic vertebrates. *Biol. Bull.* 120:238–254.

Thorson, T. B. (1964) The partitioning of body water in Amphibia. *Physiol. Zool.* 37:395–399.

Thorson, T. B. (1968) Body fluid partitioning in Reptilia. *Copeia* 1968 (3):592–601.

Van Citters, R. L., Kemper, W. S., and Franklin, D. L. (1968) Blood flow and pressure in the giraffe carotid artery. *Comp. Biochem. Physiol.* 24:1035–1042.

White, F. N. (1956) Circulation in the reptilian heart (*Caiman sclerops*). *Anat. Rec.* 125:417–431.

White, F. N. (1959) Circulation in the reptilian heart (*Squamata*). *Anat. Rec.* 135:129–134.

Wigglesworth, V. B. (1972) *The Principles of Insect Physiology*, 7th ed. London: Chapman & Hall. 827 pp.

ADDITIONAL READING

Altman, P. L., and Dittmer, D. S. (eds.) (1971) *Biological Handbooks: Respiration and Circulation*. Bethesda: Federation of American Societies for Experimental Biology. 930 pp.

Burton, A. C. (1972) *Physiology and Biophysics of the Circulation: An Introductory Text*, 2nd ed. Chicago: Year Book. 217 pp.

Caro, C. G., Pedley, T. J., Schroter, R. C., and Seed, W. A. (1978) *The Mechanics of Circulation*. Oxford: Oxford University Press. 540 pp.

Chien, S., Usami, S., Dellenback, R. J., and Bryant, C. A. (1971) Comparative hemorheology: Hematological implications of species differences in blood viscosity. *Biorheology* 8:35–57.

Folkow, B., and Neil, E. (1971) *Circulation*. New York: Oxford University Press. 593 pp.

Hamilton, W. F., and Dow, P. (eds.) (1962, 1963, 1965) *Handbook of Physiology*, sect. 2, *Circulation*, vol. 1, pp. 1–758 (1962); vol. 2, pp. 759–1786 (1963); vol. 3, pp. 1787–2765 (1965).

MacFarlane, R. G. (ed.) (1970) *The Haemostatic Mechanism in Man and Other Animals*. Symposium of the Zoological Society of London, No. 27. London: Academic Press. 248 pp.

Martin, A. W. (1974) Circulation in invertebrates. *Annu. Rev. Physiol.* 36:171–186.

Merrill, E. W. (1969) Rheology of blood. *Physiol. Rev.* 49:863–888.

Taylor, M. G. (1973) Hemodynamics. *Annu. Rev. Physiol.* 35:87–116.

PART TWO

FOOD AND ENERGY

5

Food, fuel, and energy

Animals need food (1) for energy to keep alive and carry on their functions and (2) for building and maintaining their cellular and metabolic machinery.

Most plants use the energy of sunlight and carbon dioxide from the atmosphere to synthesize sugars and, indirectly, all the complicated compounds that constitute a plant. All animals use chemical compounds to supply energy as well as building materials; they must obtain these from outside sources, either directly from plants by eating them or by eating other organic material. Ultimately the chemical energy and organic compounds animals need are derived from plants and thus indirectly from the energy of sunlight.

The acquisition and ingestion of food are referred to as *feeding*. Virtually all food, whether of plant or animal origin, consists of highly complex compounds that cannot be incorporated by the organism or used for fuel without being broken down to simpler compounds. In fact, these complex substances usually cannot even be taken into or absorbed by the organism without such breakdown, called *digestion*.

A variety of chemical compounds can be used to provide energy, but in addition animals have specific needs for compounds they cannot synthesize: amino acids, vitamins, certain salts, and so on. Both the need for food to provide energy and the need for specific food components belong to the subject of *nutrition*.

In this chapter we shall discuss food and food intake, beginning with the mechanism of feeding, proceeding to the processes of digestion that make the food compounds available to the organism, and then briefly specifying some needs with regard to the quality of the food and specific nutritional requirements. Finally, we shall discuss how plants may defend themselves by producing substances that make them *inedible* to many animals.

TABLE 5.1 Methods used by various animals to obtain food, classified by character of the food.

Type of food	Feeding method	Animals using method
Small particles	Formation of digestive vacuoles	Amebas, radiolarians
	Use of cilia	Ciliates, sponges, bivalves, tadpoles
	Formation of mucus traps	Gastropods, tunicates
	Use of tentacles	Sea cucumbers
	Use of setae, filtering	Small crustaceans (e.g., daphnia), herring, baleen whales, flamingos, petrels
Large particles or masses	Ingestion of inactive masses	Detritus feeders, earthworms
	Scraping, chewing, boring	Sea urchins, snails, insects, vertebrates
	Capture and swallowing of prey	Coelenterates, fish, snakes, birds, bats
Fluids or soft tissues	Sucking plant sap, nectar	Aphids, bees, hummingbirds
	Ingestion of blood	Leeches, ticks, insects, vampire bats
	Sucking of milk or milklike secretion	Young mammals, young birds
	External digestion	Spiders
	Uptake through body surface	Parasites, tapeworms
Dissolved organic material	Uptake from dilute solutions	Aquatic invertebrates (probably of secondary importance)
Symbiotic supply of nutrients	Action of intracellular symbiotic algae	Paramecia, sponges, corals, hydras, flatworms, clams

FEEDING

Food is obtained by a diversity of mechanical means, and these determine the nature of the food a given animal can obtain and utilize. Major feeding mechanisms and examples of animals that use the different means to obtain food are listed in Table 5.1.

Small particles

Microscopic algae and bacteria can be taken directly into a cell (e.g., into the digestive vacuole of an ameba). A variety of organisms, even some quite large ones, catch microscopic organisms with the aid of cilia. A few animals, notably tunicates and some gastropods, use a sheet of mucus that traps fine suspended particles; the mucus sheet is then ingested and the food organisms digested.

Tentacular feeding is used by sea cucumbers, which live burrowed in the mud with their tentacles extended above the mud surface. The tentacles entangle fine particles and at intervals are retracted into the animal's mouth, where digestible material is removed.

Various filtering devices are common in a wide variety of small planktonic crustaceans such as copepods and in amphipods, sponges, and bivalves. Quite a few vertebrates, ranging from fish to mammals, also use filtering to obtain food, but their filters usually strain off organisms larger than the microscopic food commonly utilized by invertebrates.

Many pelagic fish are plankton eaters. Herring and mackerel have gill rakers structured so that they function as a sieve that catches plankton, predominantly small crustaceans. Some of the largest sharks, the basking sharks and whale sharks, feed exclusively on plankton, which is strained from the water that enters the mouth and flows over the gills. It has been estimated that a large basking shark, in 1 hour, strains the plankton from 2000 tons of water (Matthews and Parker 1950).

A few marine birds also feed on plankton. One of the petrels, the whale bird or prion, has a series of lamellae extending along the edge of the upper beak. It feeds by straining crustaceans from the surface water in a way similar to what baleen whales do on a much larger scale. The flamingo, also, is a plankton eater, using its beak to strain small organisms from the water.

The baleen whale is entirely specialized for feeding on plankton. Its filtering apparatus consists of a series of horny plates attached to the upper jaw and hanging down on both sides. As the whale swims, water flows over and between the plates, and plankton is caught in the hairlike edges of the plates. The plankton-eating whales include the largest living animal, the blue whale, which may reach a weight of over 100 tons. It is interesting that both the largest whales and the largest sharks are plankton feeders rather than carnivores and thus avoid any extra links in the food chain.

Large food masses

Methods for dealing with large particles or masses of food include a tremendous variety of mechanisms and structures. Some animals ingest the medium in which they live and digest the organic material it contains, whether small living organisms or dead organic matter. Such animals frequently pass large amounts of relatively inert material through their digestive tracts and utilize whatever organic material it contains.

Many animals use mechanical methods such as chewing and scraping to obtain their food, which is frequently of plant origin. The range of these animals includes such a variety that to list them would be pointless. We need only think of the large numbers of insects and other invertebrates, as well as the many vertebrates, that are herbivores. Some carnivores are also included in this group. They capture prey and tear, shred, or chew it before swallowing.

Some carnivores capture and swallow their prey whole. Relatively few invertebrates feed this way, but many vertebrates do, including representatives of all classes: fish, amphibians, reptiles, birds, and mammals.

The mouth and oral cavity may have various kinds of jaws or teeth that serve in the mechanical breakdown of food into smaller particles (*mastication*), making it more accessible to the digestive enzymes.

In some animals the mechanical breakdown of food occurs in a special compartment with thick muscular walls, the *gizzard*, where a grinding action takes place. An earthworm, for example, has a gizzard for such grinding, called *trituration*. A bird, which has no teeth, also has a gizzard and often swallows small stones that help in the trituration of food. This is particularly important in a seed-eating bird, for its food is hard and would, in the absence of grinding, take a long time to soften in the digestive tract. The gizzard may be preceded by a simple storage chamber, a *crop*, in which the moisturizing or softening begins, but without any important grinding action.

At the beginning of the digestive tract there is usually a special larger compartment, a *stomach*, which serves as a storage organ and in which digestive action also takes place. In all vertebrates, with a few exceptions, the stomach has a major role in the digestion of proteins, and the most important enzyme in the stomach is the protein-digesting pepsin, which acts best in highly acidic solution. The stomach of a vertebrate also has special cells that secrete acid (hydrochloric acid).

The stomach is followed by a hollow tube, the *intestine*, in which further digestion takes place and which is particularly important for the absorption of the products of digestion. These products are mainly simple sugars, glycerol and fatty acids, and amino acids. The length of the intestine varies a great deal, and it can usually be subdivided into several parts. The posterior portion is of little importance in digestion and serves mainly in absorption. Periodically the remaining material is expelled as feces through the anus.

Fluids

Many liquids are of value as foods. Animals feeding on liquids are often highly specialized and adapted to their food sources, and in some cases there is a mutual benefit to the feeding animal and the provider. For example, insects obtain nectar from flowers and in turn provide pollination by moving from flower to flower. In many cases, fluid feeders are parasitic; this is true of aphids, which suck plant sap, and of animals such as leeches, mosquitoes, and ticks that suck the blood of other animals.

All mammals begin their lives as fluid feeders when for a period they live exclusively on the milk produced by their mothers. It is less well known that some birds feed their young on a milklike secretion. In the pigeon this secretion is formed in the crop, it is known as *crop milk*, and is regurgitated to feed the nestlings. Curiously, the formation of crop milk is stimulated by the same hormone, prolactin, that in mammals stimulates the mammary glands to produce milk. The biological advantage of feeding the young this way is that it allows the parents to be opportunistic in their own feeding and frees them from the need to find special kinds of food (e.g., insects) for the young. It also buffers the young against fluctuations and shortages in the food supply.

This advantage is particularly striking in the emperor penguin, which can feed its young on "milk"

TABLE 5.2 Composition of "milk" from pigeons and emperor penguins, as percentage of total dry matter. The composition of rabbit milk is given for comparison. [Prévost and Vilter 1962]

Constituent	Pigeon	Penguin	Rabbit
Protein	57.4	59.3	50.6
Lipid	34.2	28.3	34.3
Carbohydrate	0	7.8	6.4
Ash	6.5	4.6	8.4
Total	98.1	100.0	99.7

secreted by the esophagus (Prévost and Vilter 1962). The emperor penguin reproduces in midwinter. When the Antarctic winter approaches in March, the birds leave the sea and walk over the ice to the rookeries, where the female leaves a single egg with her mate to be incubated for over 2 months while he stands on the ice in the cold and dark of winter. In the meantime, the female walks back to the sea to feed and returns later to the fasting male to feed the young. If she is delayed, however, the still-fasting male begins to feed "milk" to the young, which is kept alive and even may gain in weight. In protein and fat content pigeon and penguin milk (Table 5.2) is similar to mammalian milk (although the milk of many mammals has a much higher carbohydrate content).

Spiders provide a special case of fluid feeding. A spider's prey is often as large or larger than its own body, usually covered with hard chitin, and not easily torn apart. Spiders have solved the problem of obtaining food from their insect prey by piercing it with their hollow jaws and pumping digestive juices into its body. These liquefy the tissues and the spider then sucks the prey empty.

Some parasites (e.g., tapeworms) live in highly nutritive media and have completely lost their digestive tracts. Because their hosts provide the digestive system, the parasites lack both digestive tract and the digestive enzymes that are otherwise necessary before food can be absorbed, and all their nutrients are taken up directly through the body surface.

FLUID-FEEDING INSECT. Rove beetles (Staphylinidae) are predators that feed on various invertebrates. They pierce their prey with sharp fangs (mandibles) and pump it full of enzymes that digest the prey from inside. The beetles then suck the liquefied food through a strainer composed of hair at the entrance to the mouth [Courtesy of John F. Lawrence, Museum of Comparative Zoology, Harvard University]

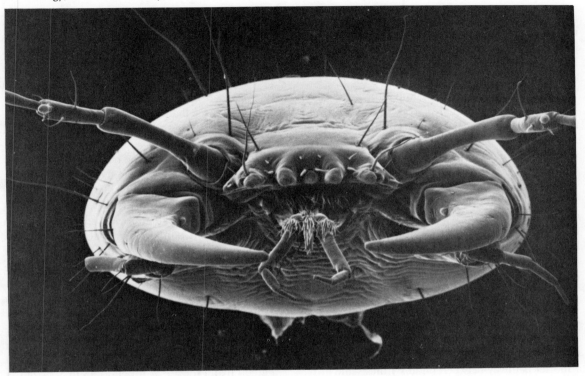

Dissolved organic material

Whether organic matter dissolved in the water can be utilized by aquatic animals is a question that remains controversial. It was suggested early in this century by the German biologist Pütter that dissolved organic matter in the sea, although present in a highly dilute state, could be used directly by aquatic animals. Pütter even suggested that such dissolved organic matter was more important than plankton, for plankton organisms are scattered so far apart in the water that often their capture is not feasible.

Whether the uptake of dissolved organic material is important has been difficult to study experimentally because of the extremely dilute solutions

to be examined. This changed with the introduction of radioisotopes to label organic compounds, and it became possible to demonstrate clearly that many aquatic invertebrates indeed can absorb both glucose and amino acids from exceedingly dilute solutions (Stephens 1962, 1964).

In such experiments it is not difficult to show that some uptake has taken place. This is beyond doubt; the difficulty is to show that the amount of a compound taken up exceeds the amount given off in the same period of time. The uptake must exceed the loss to be of net value to the animal. A comparison of simultaneous uptake and release of free amino acids in sea water has been made by Johannes et al. (1969). In experiments with a single species, a turbellarian, they showed that the release of free amino acids to the water exceeded by about fourfold the simultaneous uptake of ^{14}C-labeled amino acid. Another question that remains open is whether bacteria play a role in the uptake. The bacteria could absorb organic solutes and later be ingested by the animal. Bacteria can effectively remove glucose or acetate from solutions containing only 1 to 10 μg per liter; for common freshwater algae this is impossible (Wright and Hobbie 1966).

Although present evidence is uncertain, there is no strong indication that any aquatic invertebrate depends primarily on the uptake of organic materials from the highly dilute solution in which it lives. The elaborate mechanical feeding mechanisms animals possess also speak against any major role of dilute dissolved organic substances.

Symbiotic supply of nutrients

Some invertebrate animals contain within their cells symbiotic algae. These algae are of importance in supplying nutrients to the host, providing a shorter pathway between photosynthesis and animal than is possible in any other way. The algae, in turn, utilize ammonia given off by the animal host for synthesis of protein. Such symbiotic relationships between animal and plant occur among protozoans, sponges, corals, hydras, flatworms, and clams.

In this relationship the animal obtains organic compounds (primarily carbohydrates) from the living cells of the plant, and not from the destruction and digestion of the plant cells. In this respect the symbiotic supply of nutrients differs in principle from all the feeding methods discussed above, which always involve destruction of part or the whole of the organism that provides the food.

The importance of the symbiosis has been demonstrated in hydras by comparing animals that carry algae (green hydras) with animals that do not. The presence of symbiotic algae greatly prolongs the survival of hydras under conditions of starvation, and if a limited amount of food is given, the green hydras grow faster. The oxygen consumption of the green hydras is lower in light than in the dark and lower than in animals without symbiotic algae (Muscatine and Lenhoff 1965; Pardy and Dieckmann 1975).

Two major types of algae occur in the host animals: green algae (Zoochlorellae) or brown algae (Zooxanthellae) (Table 5.3). If we include in the term carbohydrate not only sugars, but also derivatives such as sugar alcohols and glycerol, we can say that the bulk of material transferred is of carbohydrate nature. The soluble carbohydrate given off to the host animal is invariably different from the major intracellular carbohydrate of the algae. Zoochlorellae, for example, synthesize and retain in their cells sucrose, but they excrete maltose or glucose to the host animal.

The amount of organic material obtained in a symbiotic relationship is substantial. The algae in a green hydra release as much as 45 to 50% of the carbon fixed in photosynthesis to their host. Isolated algae release carbohydrate, mainly maltose, to the medium, and the amount depends on the pH of the solution. At pH 4.5 these algae release as much as 85% of their photosynthesized carbon, but in alkaline medium the release declines to a few percent. This opens the possibility that the host animal may be able to control the amount of carbohydrate released by the algae by adjusting its intracellular pH.

When the Zooxanthellae are isolated from their host and incubated in sea water, they release some carbohydrate, but the amount increases if homogenized host tissue is added. In the case of Zooxanthellae from *Tridacna* (the giant clam) a 16-fold increase in carbohydrate release was ob-

TABLE 5.3 Release of carbohydrate from symbiotic algae to host animal. [Smith et al. 1969]

Type of alga	Host phylum	Host species	Carbohydrate released to host
Zoochlorellae	Protozoa	*Paramecium bursaria*	Maltose
	Porifera	*Spongilla lacustris*	Glucose
	Coelenterata	*Chlorohydra viridissima* (wild type)	Maltose
		C. viridissima (mutant)	Glucose
	Platyhelminthes	*Convoluta roscoffensis*	Unknown
	Mollusca	*Placobranchus ianthobapsus*	Unknown
		Tridacna crispata	
Zooxanthellae	Coelenterata	*Pocillopora damicornis*	Glycerol
		Anthopleura elegantissima	Glycerol
		Zoanthus confertus	Glycerol
		Fungia scutaria	Glycerol
	Mollusca	*Tridacna crocea*	Glycerol

tained in the presence of homogenate of the host. This is not a universal phenomenon, however; Zoochlorellae from a hydra are unaffected by homogenates of their own host animal (Smith et al. 1969).

DIGESTION

Animal food is made up of organic material, most of which belongs to three major groups: proteins, fats, and carbohydrates. These three kinds of organic compounds completely dominate the makeup of virtually all plants and animals. Most have fairly large molecules. Even the smallest (sugars) have molecular weights of a few hundred; fats are larger; many proteins have molecular weights of 100 000 to several million; and starch and cellulose are of indeterminate size, being polymers of small carbohydrate units. Whether the food is used for fuel or for building and maintenance, the large molecules of the food are first broken down into simpler units; these are then absorbed and either incorporated into the body or metabolized to provide energy.

The main function of digestion is to break up the large and complex molecules in the food so that they become absorbable and available for use in the body. This breakdown is achieved in the digestive tract with the aid of enzymes.

Intracellular and extracellular digestion

Digestion can take place within cells or outside. In a unicellular animal digestion is usually, of necessity, inside the cell. A protozoan takes food into the digestive vacuole, and enzymes that aid in the digestion of carbohydrates, fats, and proteins are secreted into the vacuole. Similar intracellular digestion occurs in sponges and to some extent in coelenterates, ctenophores, and turbellarians. It is found also in a number of more complex animals in conjunction with extracellular digestion. For example, a clam often takes small food particles into the cells of the digestive gland, where they are digested intracellularly.

Some animals that feed on larger pieces of food, such as coelenterates, have partly extracellular and partly intracellular digestion. Digestion begins in the digestive cavity, the coelenteron; and fragments of partly digested food are then taken into the cells of the cavity wall where they are digested completely.

Extracellular digestion has one obvious advantage: It permits the ingestion of large pieces of food, whereas intracellular digestion is limited to particles small enough to be taken into the individual cells of the organism.

Extracellular digestion is usually associated with a well-developed digestive tract that allows secreted enzymes to act on the food material. The digestive tract may have one opening, as in coelenterates, brittlestars, and flatworms. In these animals, any undigested material is expelled through the same opening that served as a mouth. In more complex animals the digestive tract has two openings, a mouth and an anus. This permits an assembly-line type of digestion. Food is ingested through the mouth and is passed on and acted upon by a series of digestive enzymes; the soluble products of digestion are absorbed, and in the end undigested material is expelled through the anus without interference with food intake. In this way food intake can continue while digestion takes place, and the passage of food through the digestive tract can continue uninterrupted.

All coelenterates are carnivorous. They have capture devices, tentacles, that help them catch and paralyze prey. The tentacles have specialized cells, *nematocysts*, which, when suitable prey comes in contact with them, fire tiny hollow spines that penetrate the prey. A poison is ejected from the cell through the spine, and this paralyzes the prey. The tentacles then draw the prey into the coelenteron for digestion.

The mouth of a flatworm (planarian) leads into a gastrovascular cavity that is widely branched and ramifies throughout the body. This cavity, because of its distribution, not only serves in digestion, but at the same time can transport food to all parts of the body. The extensive branching also increases the total surface of the gastrovascular cavity, aiding the absorption of digested food. In planarians extracellular digestion aids in breaking down the food, but most of the food particles are taken up by the cells that line the cavity and are digested intracellularly.

Enzymatic digestion

Most food compounds are very large molecules (proteins), highly insoluble in water (fats), or large as well as insoluble (starch, cellulose). Before the elements of the food can be absorbed and utilized, they must be brought into soluble form and broken down into smaller component units.

Proteins, starches, and cellulose are polymers formed from simpler building blocks by removal of water; fats are esters of fatty acids and glycerol, again formed by removal of water. The breakdown of these food compounds into simpler components involves the uptake of water and is therefore called *hydrolysis*. Hydrolysis is a spontaneous reaction, and energy in the form of heat is released in the process. Spontaneous hydrolysis of food compounds proceeds at an immeasurably slow rate, but can be speeded up by catalysts. Catalysts produced by living organisms are called *enzymes*, and enzymes are essential in all digestive processes. Virtually all metabolic processes within the cells are also catalyzed by enzymes.

All enzymes are proteins; many have been isolated in pure form, and their action has been studied in great detail. Many enzymes have been named according to the substrates on which they act, with the attachment of the ending -*ase*. Digestive enzymes frequently act on a broad group of substrates, and many are best known by common or trivial names, although they possess more precise names according to internationally accepted nomenclature. Because the digestive enzymes from only a few organisms are known in sufficient detail, we shall discuss most of them under their familiar common names, rather than following the technical nomenclature.

After being described by the substrate it acts on or the reaction it catalyzes, the most important measure of an enzyme is its activity, that is, the rate at which the catalyzed reaction proceeds. Many factors influence the activity, and if the protein nature of the enzyme is changed, it usually loses its activity. High temperature, strong chemicals, other enzymes that split protein, and heavy metals that bind to the protein may all inactivate the enzymatic activity.

The activity of all enzymes is greatly influenced by the pH of the solution. Most enzymes show their highest activity in a narrow range of pH, called the optimum pH for the enzyme in question. Two well-known protein-digesting enzymes are *pepsin*, found in the vertebrate stomach, and *trypsin*, produced in the vertebrate pancreas.

Pepsin has a pH optimum in highly acidic solutions, about pH 2, and trypsin has its optimum in slightly alkaline solutions, about pH 8. The optimum pH for a given enzyme is not necessarily identical for all substrates on which it acts. Pepsin, for example, shows an optimum pH of 1.5 for digestion of egg albumin, 1.8 for casein, and 2.2 for hemoglobin (White et al, 1964). On both sides of the optimum pH the activity is decreased, and at a very different pH the activity is nil. Pepsin still has some activity at pH 4, and resumes full activity if the pH is returned to the optimum. At pH 8, however, pepsin is destroyed and permanently inactivated.

The rate of enzymatic reactions is greatly influenced by the temperature. An increased temperature speeds the rate. Because heat usually makes proteins coagulate, it also inactivates enzymes; at temperatures above 50 °C or so, most enzymatic action is completely destroyed. Therefore, a mod-

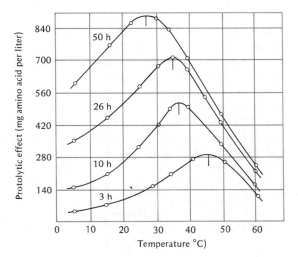

FIGURE 5.1 The effect of temperature on a protein-splitting enzyme (protease) from the ascidian *Halocynthia*. The enzyme effect seems to have a temperature optimum, but this optimum is lower the longer the duration of the experiment. This is explained in the text. [Berrill 1929]

erate increase in temperature gives an increased rate of reaction, but when the temperature increases further, thermal destruction of the enzyme catches up with the accelerated reaction rate. At some temperature we will observe a maximum, but because thermal destruction increases with time, we will find that the optimum temperature depends on the duration of the experiment. In a long-lasting experiment the temperature effect on the enzyme will have proceeded further, and the observed temperature optimum will therefore be lower; in a short experiment with the same enzyme the temperature optimum will be higher. Thus the temperature optimum is not a specific characteristic of an enzyme; it depends on the duration of an experiment (Figure 5.1).

Most enzymes are rapidly inactivated by temperatures above 45 or 50 °C. This is the temperature range that for most complex organisms forms the upper limit for life, except for some resting stages and certain thermophilic bacteria. A few enzymes, however, are heat-resistant; examples are the protein-splitting enzyme papain from the papaya fruit and a similar enzyme from pineapple. These enzymes are used as meat tenderizers because their proteolytic action continues at cooking temperatures as high as 70 or 80 °C, although at boiling temperature they are rapidly inactivated.

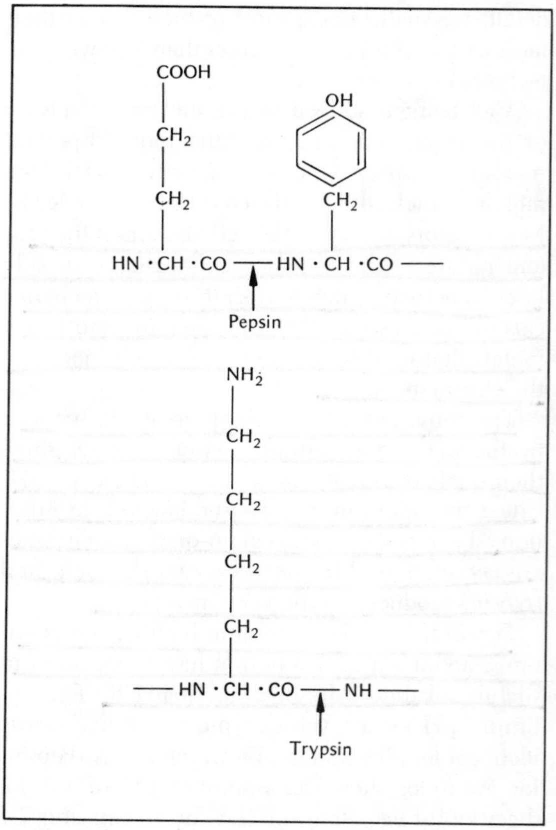

Protein digestion

Proteins are polymers consisting of amino acids bound together by peptide bonds. An amino acid is a relatively simple organic acid that on the carbon atom next to the acid group (carboxyl group, —COOH) carries an amino group (—NH$_2$). A peptide bond is formed by removal of water between a carboxyl and an amino group. There are about 20 common amino acids that in various proportions occur in virtually all living organisms. This means that when proteins are hydrolyzed into amino acids, these amino acids are the building blocks required for synthesis of nearly any specific protein.

The enzymes that digest proteins are divided into two groups, according to where on the protein molecule they act. *Exopeptidases* hydrolyze a terminal peptide bond in a long peptide chain; *endopeptidases* act in the interior of the peptide chain. The two best known endopeptidases in vertebrate digestion are pepsin and trypsin.

Pepsin is secreted by the stomach as an inactive precursor, *pepsinogen*. The stomach also secretes hydrochloric acid, which gives its contents a low pH. In an acid medium, below pH 6, pepsinogen is activated autocatalytically to form the active enzyme, pepsin. Pepsin specifically hydrolyzes peptide bonds between an amino acid that carries a phenyl group (tyrosine or phenylalanine) and a dicarboxylic acid (glutamic or aspartic acid). Pepsin therefore attacks only a few of the peptide bonds in a large protein and not others, and the result of pepsin digestion is a series of shorter chains or fragments that are not further broken down by pepsin.

FIGURE 5.2 The protein-digesting enzyme pepsin splits peptide bonds between a dicarboxylic and a cyclic amino acid. Trypsin acts on peptide bonds adjacent to an amino acid with two amino groups. Arrows show points of attack. See text for further details.

Trypsin is secreted by the pancreas in an inactive form, *trypsinogen*. It is activated in the intestine by the enzyme enterokinase, which is secreted by glands in the intestinal wall. Trypsinogen is also activated by active trypsin (i.e., trypsinogen is activated more and more rapidly as more trypsin is formed). This is known as *autocatalytic activation*.

Trypsin acts best at a slightly alkaline pH, between 7 and 9. It hydrolyzes peptide bonds that are adjacent to a basic amino acid (i.e., one that carries two amino groups, lysine or arginine) (Figure 5.2).

The protein fragments and peptide chains formed by the action of pepsin and trypsin are further digested with the aid of exopeptidases, which act only on terminal peptide bonds. Carboxypeptidase, secreted by the pancreas, hydrolyzes terminal peptide bonds next to a free carboxyl group, and aminopeptidase, secreted by the intestine, hydrolyzes peptide bonds next to a free amino group. Finally, dipeptidases hydrolyze the peptide bond of fragments that consist of only two amino acids.

Protein digestion in invertebrates, whether extracellular or intracellular, in principle proceeds as in vertebrates, except that invertebrates generally have no pepsinlike enzymes that act in acid solution. Otherwise, although the enzymes are not identical in their structure to the vertebrate enzymes, their action is similar.

Pepsinlike enzymes are not an exclusive characteristic of vertebrates, however. The maggot of the ordinary housefly (*Musca domestica*) secretes a protease that is active in the pH range of 1.5 to 3.5 with an optimum at 2.4. Whether this enzyme is of importance in the digestion of the maggot depends on whether a suitably acid medium can be produced, and this has not yet been completely clarified (Greenberg and Paretsky 1955; Lambremont et al. 1959).

Fat digestion

Digestion of fats is similar in vertebrates and invertebrates. Ordinary fats, whether of plant or animal origin, consist of esters between one molecule of glycerol (a trivalent alcohol) and three molecules of long-chain fatty acids. They are highly insoluble in water and as a result are not easily hydrolyzed by enzymes. The vertebrate pancreas secretes a fat-hydrolyzing enzyme, *lipase*, but to bring it into contact with the fat, the aid of a detergent is needed. The bile acids, secreted by the liver, serve this function. With the aid of the bile acids, and the mechanical movements of the intestine, the fats are emulsified, and as hydrolysis proceeds, the resulting fatty acids (also relatively insoluble) are kept in solution with the aid of the bile salts and thus can be absorbed. The glycerol (which is a normal intermediary in virtually all cell metabolism) is water-soluble and is easily absorbed and metabolized. Fats can, however, to some extent be absorbed by the intestinal epithelium without preceding hydrolysis; the tiny droplets formed when fats are emulsified with the aid of the bile are taken up directly by the epithelial cells.

Some fatty substances, among them waxes, are not hydrolyzed by ordinary lipases. Waxes are esters of one molecule of a long-chain fatty alcohol with one molecule of a fatty acid; if they could be hydrolyzed, the components would be metabolized with a high energy yield. What we usually think of as wax is beeswax, and with one exception, beeswax remains undigested by vertebrates and therefore has no nutritive value.*

Even though no vertebrates produce beeswax-digesting enzymes, beeswax is a main food item for a small group of birds, the South African honeyguides. These birds, related to the woodpeckers, are known for their habit of guiding the ratel (a badger) or a human being to the nest of wild bees. The bird attracts attention by chattering and noisy behavior, and when it is being followed, it leads the way to a bees' nest. It waits while the nest is plundered, and then eats the wax left by the predator. The honeyguide prefers to eat wax rather than honey and eats pure wax as well as honeycomb. The honeyguide can digest wax because its intestinal tract contains symbiotic bacteria that produce the necessary enzymes for attacking the wax. Wax is indigestible to virtually all other animals, ver-

---

*The "wax" often used in the household and for candle making is a hydrocarbon fraction of crude mineral oil. It is chemically inert, is attacked only by certain strains of bacteria, and is not known to be utilized by any animal.

tebrate and invertebrate alike. Only the larvae of the wax moth (*Galleria*), which is a common parasite in beehives, can live on beeswax (Friedmann and Kern 1956a,b).

The hypothesis that symbiotic bacteria are responsible for the wax digestion of the honeyguide is confirmed by transferring to domestic chicks pure cultures of bacteria obtained from the digestive tract of the honeyguide. Normal chicks are completely unable to digest wax, but if wax is fed together with cultures of the isolated microorganisms from the honeyguide, the chickens can digest and metabolize wax (Friedmann et al. 1957).

The digestion of wax by birds with the aid of symbiotic bacteria is a rare curiosity. In contrast, symbiotic digestion of cellulose, one of the commonest plant materials, is of immense importance to herbivorous animals, as we shall see later in this chapter.

Although waxes are insignificant as food for terrestrial animals, they are extremely important in the food chain of marine animals, side by side with ordinary fats and oils.

Waxes occur in a wide variety of marine organisms: molluscs, cephalopods, shrimps, sea anemones, corals, and many fish. The primary producers of waxes appear to be small planktonic crustaceans, especially copepods. In some of these, waxes may constitute as much as 70% of the animal's dry weight. The copepods feed on phytoplankton, which contains no wax. Diatoms and dinoflagellates accumulate oil globules, which are largely triglycerides. However, the fatty acids of the waxes of the copepods closely resemble characteristic fatty acids found in the phytoplankton, and it is reasonable to conclude that these are used directly for wax production by the copepods.

The role of waxes in the marine food chain is overwhelmingly more important than was imagined only a few years ago, for the crustacean plankton is the major step between the photosynthetic microscopic algae and the consumers in the sea. It has been estimated that because of this step, at least half of the earth's photosynthetic production is for a time converted to wax (Benson et al. 1972).

Fish that prey on copepods (e.g., herring, anchovies, sardines, and pilchards) have wax lipases in their digestive tracts (Sargent and Gatten 1976). The fatty alcohols are oxidized to fatty acids, which then enter into ordinary neutral fats as triglycerides. In some other fish the amount of wax lipases is much smaller, and this leaves open the question of how well they can digest waxes. Because waxes are found in the fats and oils of such a variety of marine animals, even in whales, it is difficult to say whether these waxes can be metabolized and serve as an energy reserve or are merely stored because they are ingested and not easily metabolized. This important question is the subject of intensive studies by marine biologists.

Carbohydrate digestion

There are no great differences between vertebrates and invertebrates in how carbohydrates are digested. Simple sugars, such as glucose and fructose, are absorbed without any change and directly utilized in ordinary metabolic pathways. Disaccharides, such as sucrose (of plant origin) or lactose (in milk), are broken down to monosaccharides before they are absorbed and can be utilized. The enzyme sucrase is secreted in the intestine, but it is absent from the cellular machinery of animals. Therefore, if sucrose is injected into the body of vertebrates, it is completely execreted again in the urine without any change.

Large numbers of plants store starch as a major energy reserve. Starch is a polymer of single glucose units. It is relatively insoluble, but it is hydrolyzed by the enzyme *amylase* (from Latin *amylum* = starch), secreted in the saliva of man (and some, but not all, other mammals) and in larger amounts by the pancreas.

Although starch is relatively insoluble, it is well digested by many animals. It is more easily attacked by amylase, however, if it first has been heated, and this is the major reason man prepares starchy foods by cooking (potatoes, bread, etc.). It is quite probable that in many animals that readily digest uncooked starch, bacteria aid in the initial breakdown of the starch, which is then more easily attacked by the amylase.

Symbiotic digestion of cellulose

The most important structural material of plants is cellulose, a glucose polymer that is extremely insoluble and refractory to chemical attack. Cellulose-digesting enzymes, *cellulases*, are absent from the digestive secretions of vertebrates, and yet many vertebrates digest cellulose and depend on it as their main energy source. True cellulases have been reported from the intestinal tracts of several invertebrates that feed on wood and similar plant products, but in many cases cellulose digestion is carried out by *symbiotic microorganisms* that live in the digestive tract of their host.

Invertebrate cellulose digestion

The question of whether a cellulase is produced by an animal or by symbiotic microorganisms is often difficult to decide with certainty. A few examples will illustrate the difficulties.

Snails. It has long been accepted that the garden snail, *Helix pomatia* digests cellulose and actually secretes a cellulase. However, careful examination of extracts of the snail's digestive gland and intestinal wall has failed to reveal any cellulase activity, although cellulase is readily demonstrated in the intestinal contents (Florkin and Lozet 1949). This seems to indicate a major role of intestinal microorganisms, but the fact that an extract of digestive organs does not reveal the enzyme does not necessarily exclude the possibility that the animal may produce the enzyme.

Shipworms. Another organism that has been accepted as able to digest cellulose is the shipworm (*Teredo*), actually a wood-boring bivalve mollusc. True cellulase has been found in the gut of the shipworm, and gut extracts liberate sugar from cellulose, although no cellulose-digesting bacteria or protozoans could be isolated from the digestive system (Greenfield and Lane 1953). In this case it could be argued that the apparent absence of cellulose-digesting microorganisms is no proof that these do not exist in the living animal. The extracts from the intestine that showed cellulase activity were made with pieces of gut from which the contents had not been removed, and the source of the enzyme activity therefore remained uncertain.

Silverfish. There are, however, reports of true cellulases obtained from some animals under conditions that seem to exclude any possibility of symbiotic microorganisms. The silverfish (*Ctenolepisma lineata*) digests cellulose and can survive on a diet of cellulose alone, although this is not satisfactory for prolonged feeding. The gut of the silverfish contains many microorganisms, but no cellulose-digesting species has been isolated. More importantly, bacteria-free silverfish have been obtained by washing the eggs in a solution of mercuric chloride and ethanol and raising the nymphs on sterilized rolled oats and vitamins under aseptic conditions. Such bacteria-free silverfish digest cellulose marked with ^{14}C and exhale ^{14}C-labeled carbon dioxide. Finally, a cellulase has been demonstrated in extracts of the midgut, and it therefore seems certain that silverfish do produce the cellulase (Lasker and Giese 1956).

Termites. In the case of termites, which live virtually exclusively on wood, the role of symbiotic organisms is beyond doubt. The intestinal tract of a wood-digesting termite is virtually crammed full of several kinds of flagellates and bacteria. Several of the flagellates have been isolated and kept in pure culture, and this has permitted a careful study of their role in cellulose digestion.

The symbiotic flagellates from termites are obligate anaerobic organisms (i.e., they can live only in the absence of free oxygen). This sensitivity to oxygen can be used to remove the flagellates from the host animals and thus to obtain protozoan-free termites. If termites are exposed to oxygen at 3.5 atm pressure, the flagellates are selectively killed in about half an hour; the termites survive unharmed (Cleveland 1925). Such defaunated termites cannot survive if they are fed as usual on wood. They still have bacteria in the intestine, and this indicates that the protozoan symbionts, rather than the bacteria, are responsible for cellulose digestion. When the termites are reinfected with the proper flagellates, they can again digest cellulose (Hungate 1955).

The hypothesis that the flagellates carry out the cellulose digestion is further supported by studies of these organisms in isolated culture. One such flagellate (*Trichomonas termopsidis*, from the termite *Termopsis*), which has been kept in culture for as long as 3 years, yields extracts that contain cellulase (Trager 1932). In fact, this flagellate cannot utilize any carbon source other than cellulose, so there is no doubt about its role in the cellulose digestion of termites (Trager 1934).

Although the cellulose from wood can give a substantial amount of energy, wood has a very low nitrogen content. Because termites thrive on wood, it has been suggested that the intestinal microbes could be nitrogen-fixing bacteria, and thus be of aid in the protein supply. This hypothesis seems plausible. Many earlier attempts were un-

successful at demonstrating nitrogen fixation in the overall nitrogen balance of termites (Hungate 1941), but it is now clear that nitrogen fixation does take place with the aid of microorganisms (Benemann 1973; Breznak et al. 1973). Nitrogen fixation may, in fact, be more widespread among wood-boring animals than was formerly expected. In several species of shipworms (*Teredo* and others) nitrogen fixation is associated with bacteria in the gut, and these bacteria have been isolated and shown to be capable of nitrogen fixation (Carpenter and Culliney 1975).

Vertebrate cellulose digestion

Many mammals are herbivores, and most of them live on diets that make cellulose digestion essential. The ruminants, which include some of our most important domestic meat- and milk-producing animals (cattle, sheep, and goats) have specialized digestive tracts that are highly adapted to symbiotic cellulose digestion. However, many nonruminant mammals also depend on symbiotic microorganisms for cellulose digestion, although their anatomical adaptations differ from those of the true ruminants.

Ruminants. The stomach of a ruminant consists of several compartments (Figure 5.3), or, to be more precise, the true digestive stomach is preceded by several large compartments, the first and largest called the *rumen*. The rumen serves as a large fermentation vat in which the food, mixed with saliva, undergoes heavy fermentation. Both bacteria and protozoans are found in the rumen in large numbers. These microorganisms break down cellulose and make it available for further digestion. The fermentation products (mostly acetic, propionic, and butyric acids) are absorbed and utilized; carbon dioxide and methane (CH_4) formed in the fermentation process are permitted to escape by belching (eructation).

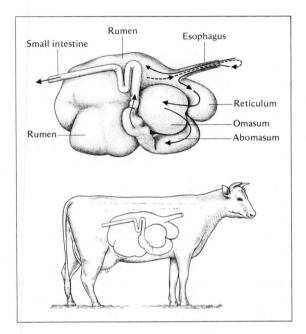

Rumination, or chewing the cud, is regurgitation and remastication of undigested fibrous material, which is then swallowed again. This process, rumination, has given the ruminant group its name. As the food reenters the rumen, it undergoes further fermentation. Broken-down food particles gradually pass to the other parts of the stomach, where they are subjected to the usual digestive juices in the fourth stomach (which corresponds to the digestive stomach of other mammals).

The fermentation products from the rumen are a major source of energy. It has been estimated that the energy available from the organic acids produced in the rumen makes up 70% of the total energy requirement of cattle. The methane, on the other hand, is completely lost to the ruminant, and the loss is appreciable. A cow fed 5 kg of hay each day gave off 191 liters of methane each day; the energy value of this methane was more than 10% of the daily digested food energy (Kleiber 1961).

The fermentation products are mostly short-chain organic acids, and the copious amounts of saliva secreted by ruminants serve to buffer these fermentation products in the rumen. The saliva of a ruminant is little more than a dilute solution of sodium bicarbonate, which serves both as a buffer and as a proper fermentation medium for the microorganisms.

The amount of saliva secreted by the ruminant is impressive. The total secretion of saliva per day has been estimated at 6 to 16 liters in sheep and goats, and at 100 to 190 liters in cattle (Bailey 1961; Blair-West et al. 1967). Because sheep and goats weigh perhaps 40 kg, and cattle some 500 kg, the daily production of saliva may reach roughly one-third of the body weight. As two-thirds of the body weight is water, about half of the total body water passes through the salivary glands (and the rumen) each day.

The protozoans of the rumen include ciliates that superficially resemble free-living forms such as paramecia. They occur in numbers of several hundred thousand per milliliter of rumen contents. Many of the rumen organisms have been cultured in the laboratory, and extracts from pure cultures show cellulase activity (Hungate 1942). The ciliates are obligate anaerobic organisms and must meet their energy requirements through fermentation processes. This gives a relatively low energy yield for the microorganisms, but it means that the fermentation products become available to the host animal, which in turn uses them in oxidative metabolism. Because symbiotic cellulose digestion is the only way cellulose becomes available to mammals, the relatively small loss to the symbiotic microorganisms is quantitatively unimportant. What is important is that the rumen microorganisms contribute in several other ways to the nutrition of the host.

The rumen microorganisms can synthesize protein from inorganic nitrogen compounds, such as ammonium salts (McDonald 1952). It is particularly useful that urea, which is normally an excretory product eliminated in the urine, can be added to the feed of ruminants and increase the protein synthesis. This has been used in the dairy industry, for urea can be synthesized cheaply, and it is therefore less expensive to supplement the diet of milk cows with urea than to use more expensive high-protein feed.

The rumen contents of a cow may weigh as much as 100 kg, with a total weight of protozoans of about 2 kg, and these again contain 150 g protozoan protein. The propagation time for these protozoans makes it possible to estimate that 69% of the protozoan population passes into the omasum each day, and that this contributes a protein supply of over 100 g per day (Hungate 1942).

The microbial protein synthesis in the rumen is of special importance when the animal is fed on low-grade feed. It has been found that a camel fed a nearly protein-free diet (inferior hay and dates) excretes virtually no urea in the urine. Urea continues to be formed in metabolism, but instead of being excreted, this "waste" product reenters the rumen, in part through the rumen wall and in part with the saliva. In the rumen the urea is hydrolyzed to carbon dioxide and ammonia, the latter being used for resynthesis of protein. In this way a camel on low-grade feed can recycle much of the small quantity of protein nitrogen it has available (Schmidt-Nielsen et al. 1957).

Similar reutilization of urea nitrogen in animals fed low-protein diets has been observed in sheep (Houpt 1959), and under certain conditions the rabbit (a nonruminant) can utilize urea to a signifi-

cant extent in its nitrogen metabolism (Houpt 1963).

If inorganic sulfate is added to the diet of a ruminant, the microbial synthesis of protein is improved and, particularly important, the sulfate is incorporated in the essential amino acids cysteine and methionine (Block et al. 1951). Thus, the rumen microorganisms contribute both to protein synthesis and to the quality of the protein. Because the microbes in the rumen can synthesize all the essential amino acids, ruminants are nutritionally independent of these, and the quality of the protein they receive in their feed is of minor importance.

Another nutritional advantage of ruminant digestion is that some important vitamins are synthesized by the rumen microorganisms. This applies to several of the vitamin B group; in particular, the natural supply of vitamin B_{12} for ruminants is obtained entirely from microorganisms.

Nonruminant animals. Cellulose digestion in many nonruminant herbivorous mammals is also aided by microorganisms. Cellulose-containing feed is usually bulky, and fermentation is relatively slow and takes time. Much space is required, and the part of the digestive tract used for fermentation is therefore of considerable size. In some animals the stomach is large and has several compartments, and the digestion has obvious similarities to ruminant digestion. In others the major fermentation of cellulose takes place in a large diverticulum from the lower intestine, the *cecum.*

Multiple-compartment stomachs are found not only in some nonruminant ungulates, but in very far removed animals such as the sloth (Denis et al. 1967) and the langur monkey (Bauchop and Martucci 1968) (Figure 5.4). Even among marsupials there are animals that have a rumen-like stomach. One is the rabbit-sized quokka, which weighs 2 to 5 kg. Its large stomach, which harbors microorga-

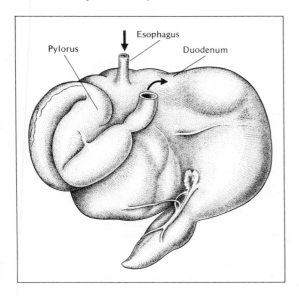

FIGURE 5.4 The stomach of the sloth (*Bradypus tridactylus*) is complex and highly reminiscent of the stomach of a ruminant. [Grassé 1955]

Pylorus

Esophagus

Duodenum

nisms that participate in cellulose digestion (Moir et al. 1956), may contain nearly 0.5 kg of moist material. For an animal weighing about 3 kg, this equals 15% of the body weight, an amount similar to that which is often found in ruminants.

Microbial fermentation in the cecum has considerable similarity to fermentation in the rumen, but the rumen has three definite advantages over the cecum. One is that rumen fermentation takes place in the anterior portion of the gastrointestinal tract, so that the products of digestion can pass through the long intestine for further digestion and absorption.

The second advantage of the ruminant system is that the mechanical breakdown of the food can be carried much further, for coarse and undigested particles can be regurgitated and masticated over and over again. This difference is clearly visible if we compare the fecal material of cattle (ruminants) and horses (nonruminants). Horse feces contain coarse fragments of the food intact; cow feces are a well-ground-up, smooth mass with few large visible fragments.

The third, and perhaps greatest, advantage of the ruminant system with microbial fermentation in the anterior portion of the gastrointestinal tract is the opportunity to recycle the urea nitrogen that otherwise would be excreted and lost to the organism. As we saw earlier, this is important for ruminants eating low-grade feed. The ruminant-like large marsupials – kangaroos and wallabies – in which microbial fermentation also takes place anterior to the digestive stomach, utilize the same mechanism. When the nitrogen content in the vegetation declines at the beginning of the dry season, wallabies begin to recycle urea and continue to do so throughout the prolonged dry season, thus gaining a relative independence of the low quality of the available feed (Kinnear and Main 1975).

An improvement in the digestion of plant materials through fermentation is not restricted to mammals. For example, the willow ptarmigan in Alaska subsists during several winter months on a diet of only willow buds and twigs. Most gallinaceous birds have two large ceca suitable for cellulose fermentation. In the willow ptarmigan the fermentation products are ethanol and acetic, propionic, butyric, and lactic acids in various proportions. The contribution of cecal fermentation to the basal energy requirement is up to 30%, a substantial contribution indeed (McBee and West 1969).

Coprophagy. The disadvantage of locating cellulose fermentation in the posterior part of the intestinal tract can be circumvented in an interesting way. Many rodents as well as rabbits and hares form a special kind of feces from the contents of the cecum, and these are reingested so that the food passes through the entire digestive tract a second time. There are, in fact, two kinds of feces: the well-known ordinary, firm, dark rodent fecal pellets, and softer, larger and lighter feces that are not dropped by the animal but are eaten directly from the anus. This latter type is kept separate from the ordinary droppings in the rectum, and reingestion permits a more complete digestion and utilization of the food.

Coprophagy (from the Greek *copros* = excrement and *phagein* = to eat) is common in rodents and is of great nutritional importance. For example, if coprophagy is prevented, rats require supplementary dietary sources of vitamin K and biotin. In addition, deficiencies of other vitamins develop more rapidly. What is more interesting is that, even if the rats are provided with a variety of dietary supplements, their growth rate is still depressed by about 15 to 25%. Even if the rats are given access to eating normal rat feces, normal growth is not attained by rats prevented from coprophagy (Barnes et al. 1963).

In rabbits prevention of coprophagy leads to a decrease in the ability to digest food, as well as a decrease in protein utilization and nitrogen retention. When coprophagy is permitted again, there is a corresponding increase in the ability to digest cellulose (Thacker and Brandt 1955).

The special soft feces that a rabbit reingests originate in the cecum. When ingested, these feces are not masticated and mixed with other food in the stomach; they tend to lodge separately in the fundus of the stomach (Figure 5.5). The soft feces are covered by a membrane, and they continue to ferment in the stomach for many hours, one of the fermentation products being lactic acid (Griffiths and Davies 1963). In this way the fundus of the stomach serves as a fermentation chamber, analogous to the rumen of sheep and cattle, and thus provides essential nutritional advantages to the animal.

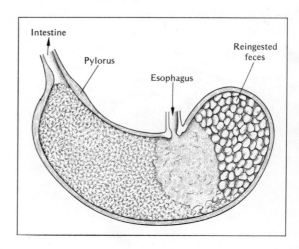

FIGURE 5.5 In the stomach of the rabbit ingested food is located in the pyloric part (left), which contains digestive glands. Reingested fecal pellets are located in the large fundus (right), where they remain separate from the food material while fermentation continues. [Harder 1949; Grassé 1955]

Intestine

Pylorus

Esophagus

Reingested feces

NUTRITION

The mechanical properties of the food must correspond to the animal's feeding apparatus, and the chemical nature of the food must be suited to the available digestive mechanisms. Once these requirements have been met, the next question is: Is the food adequate in regard to the needs of the organism? Nutrition deals specifically with this question.

The problem can conveniently be divided into two parts: the quantity of food and the quality of food. The organism needs (1) energy for external activity and internal maintenance and (2) a supply of specific substances for building and upkeep. These specific substances include amino acids, vitamins and certain other essential nutrients, and various minerals and other elements, some of them in such minute quantities that it has been difficult to establish whether they are really needed.

The following discussion will first deal with the need for energy and energy-yielding components of the diet. We will then proceed to specific substances needed by the organisms. Finally, we will mention some less desirable compounds that may occur in food and their biological importance.

Energy supply, fuel

The bulk of organic material used for food consists of proteins, carbohydrates, and fats. The main digestion products of these compounds are amino acids (from proteins), various simple sugars (present in the food or derived from starch digestion), short-chain fatty acids (primarily from cellulose fermentation), and long-chain fatty acids (from fat digestion).

The oxidation of these digestion products yields virtually all the chemical energy needed by animal organisms. All are quite simple organic products, consisting mostly of carbon, hydrogen, and oxygen in various proportions, except that the amino acids also contain nitrogen (in the amino group, $-NH_2$). If amino acids are used to supply chemical energy, the amino group is first removed by deamination, and the remaining relatively small fragments (various short-chain organic acids) enter the usual pathways of intermediary metabolism.

These various organic compounds are, within wide limits, interchangeable in energy metabolism, although there are certain limitations. For example, the human brain needs carbohydrate (glucose), but most other organs can use fatty acids, which constitute the main fuel for mammalian muscle metabolism. In contrast, the flight muscles of the fruit fly, *Drosophila*, require carbohydrate, and when the supply is exhausted, the fruit fly cannot fly, although stored fat is used for other energy-requiring processes (Wigglesworth 1949). Not all insects need glucose for flight, however; the migratory locust uses primarily fat for its long migratory flights (Weis-Fogh 1952).

For an animal to remain in a steady state, the total energy expenditure must be covered by an equal intake of food. If the amount of food is insufficient, the remaining energy requirement is covered by the consumption of body substance, primarily stored fat. If the food intake exceeds the energy used, most animals store the surplus as fat, irrespective of the nature of the food. This is the reason domestic livestock, such as beef and pigs, can be fattened by feeding on grain or corn, which contain mostly starch. Some animals, however, store primarily carbohydrate in the form of glycogen. This is important, for glycogen can provide energy in the absence of oxygen (Chapter 6).

Regulation of food intake

In many animals intake of food is remarkably well adjusted to energy expenditure (Mayer and Thomas 1967). If the energy needs are increased because of physical activity, the food intake is adjusted accordingly. The regulatory mechanisms for the food intake of mammals are located in the hypothalamic area of the brain, but specific peripheral influences play a major role in the momentary level of hunger and food intake. One of the important influences is the blood sugar level. It has been demonstrated that hunger pangs coincide with contractions of the stomach and that these occur especially when the level of blood glucose decreases. Also, the degree of filling of the stomach is of obvious importance in inducing a feeling of satiation and termination of food intake.

These factors can be of only minor importance in the regulation of long-term food intake and maintenance of body weight. Increasing the bulk or volume of the food by adding non-nutritive material to the food has a considerable effect on both gastric filling and motility, but the animal nevertheless rapidly adjusts by increasing food intake. Rats fed a calorically dilute diet after the first day

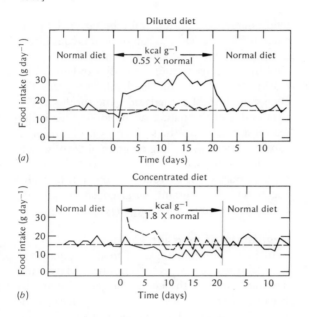

FIGURE 5.6 Changes in amount of food eaten per day by adult rats when the energy content per unit weight of food was (a) reduced and (b) increased. Solid line, weight of food consumed daily; broken line, energy content of food eaten, expressed as equivalent weight of initial diet (1 kilocalorie = 4.184 kilojoule). [Hervey 1969]

maintain their caloric intake by ingesting larger amounts. In rats fed a calorically concentrated diet, the caloric intake remains too high for several days, but then decreases to the basic level (Figure 5.6).

The exact mechanism for this control is not fully understood, but it is known that the hypothalamic region of the brain is important. If specific areas in the hypothalamus are destroyed, the food intake is greatly altered and may be either reduced or increased. Carefully placed hypothalamic lesions may lead to such tremendous increases in food intake that an animal eats continuously and grows so fat it is unable to move.

Regulation of food intake is not limited to higher vertebrates. Goldfish, which continue to grow throughout their lifetimes (although at a decreasing rate), have been used in similar experiments with caloric dilution of their diet. If per-

mitted to control their food intake, goldfish respond to caloric dilution by increased food intake. Also, a decrease in the water temperature from 25 °C to 15 °C causes a decrease in food intake by one-half to one-third. Goldfish therefore seem to eat for calories, for a 10 °C drop in temperature causes roughly a two- to three-fold decrease in metabolic rate and energy requirement (Rozin and Mayer 1961).

Specific nutritional needs

A diet that furnishes adequate fuel, measured in terms of energy units, can be utterly inadequate in regard to raw materials for growth and building and maintaining the cellular and metabolic machinery.

For most animals the energy-requiring processes can be fully satisfied by glucose alone, but glucose contributes nothing toward other needs. There are three major categories of such needs: proteins and amino acids, vitamins and related compounds, and minerals and trace elements.

Proteins and amino acids

When an animal grows, protein is continually synthesized and added to the organism. In the adult organism, however, the amount of protein remains much the same throughout life. It might therefore seem that once an organism has reached its adult size, dietary protein would be less important. This is not true; an inadequate supply of protein leads to serious malnutrition.

Why is this so? It would appear that, once proteins have been incorporated into an organ (a muscle, for example), they would remain as part of the permanent structure. It was, therefore, a great surprise when the biochemist Rudolf Schoenheimer found that the body proteins are constantly broken down and resynthesized. Schoenheimer introduced in the diet of adult rats amino acids labeled with the heavy nitrogen isotope ^{15}N, and these amino acids were invariably found incorporated in the tissue proteins. Thus, it was shown that cell components once regarded as stable structures have a high turnover rate of proteins (Schoenheimer 1964).

Not all proteins have the same turnover rate. In man, for example, the half life of serum proteins is approximately 10 days, but connective tissue has a low rate of protein turnover. Once these latter structural proteins are formed, they remain relatively stable compared with proteins of metabolically active tissues such as blood, liver, muscle, and so on.

New protein is synthesized from single amino acids, not from partially degraded fragments of other protein molecules. An adult man synthesizes and degrades about 400 g protein per day, and this far exceeds his total protein intake, which on an average diet may be between 100 and 150 g per day. Thus, the synthesis utilizes not only amino acids from the food, but also, in part, those available from protein breakdown in the body.

Each organism contains a large number of different proteins as cell components, enzymes, and so on. Furthermore, every animal or plant species has proteins that are specific for that species, and even individuals within a species may have characteristically different proteins. For example, the blood groups of man are related to the different structures of some of the blood proteins.

Only about 20 different amino acids occur commonly in proteins, but because a protein molecule contains from less than 100 to many thousand amino acids, there is virtually no limit to the number of different proteins that can be built from the 20 basic building blocks.

The amino acids needed by the body are usually supplied by the protein of the food, but they can be

TABLE 5.4 Classification of the amino acids with respect to dietary needs of rats.

Essential	Nonessential
Lysine	Glycine
Tryptophan	Alanine
Histidine	Serine
Phenylalanine	Norleucine
Leucine	Aspartic acid
Isoleucine	Glutamic acid
Threonine	Proline
Methionine	Citrulline
Valine	Tyrosine
Arginine[a]	Cystine

[a]Arginine can be synthesized, but not at a sufficiently rapid rate to meet the demands of normal growth.

replaced by chemically purified amino acids in suitable proportions. However, it is not necessary to supply every one of the 20 different amino acids. Some can be formed in the body from other amino acids, but others cannot be synthesized by the organism and must be supplied in the diet. These latter are called *essential amino acids*.

The terms *essential* and *nonessential* describe the dietary requirements for amino acids, but say nothing about how important these amino acids are. The nonessential amino acids are just as important as the others as structural components of body proteins. In fact, they are probably so important that the dietary supply cannot be trusted, and the organism cannot afford to be without a mechanism for their synthesis.

Of the 20 or so common amino acids, 10 are essential to rats (Table 5.4) (Rose 1938). The exclusion of any one of them (except arginine) produces profound nutritional failure. The absence of arginine gives a decreased rate of growth, but arginine can be synthesized, although not at a rate sufficient for optimal growth. Man and other mammals have the same requirements, except that histidine is not needed by adult humans although it seems to be required by all other higher animals tested (Rose 1949). It is particularly striking that fish, many insects, and even protozoans have practically overlapping needs for essential amino acids (Tarr 1958; Taylor and Medici, 1966).

Plants, in contrast to animals, do synthesize protein from inorganic nitrogen (ammonia salts or nitrate), and some microbes can even use molecular nitrogen. Such *nitrogen-fixing bacteria* are found in the root nodules of clover and other leguminous plants, which therefore are particularly important in increasing the protein yield from agriculture.

We have already mentioned that termites and other wood-boring insects use nitrogen-fixing bacteria in the gut to supplement the extremely low protein content of the wood they eat. Probably many more animals, themselves unable to utilize molecular nitrogen, take advantage of nitrogen-fixing microbes. A case in point is the high rate of biological productivity of coral reefs that has puzzled marine biologists for years. Tropical water masses characteristically have very low levels of dissolved nutrients, and especially the low level of fixed nitrogen is often the major limiting factor in phytoplankton production. It appears that, as the ocean water flows over shallow coral reefs, there is a marked increase in its nitrogen content, caused mainly by the nitrogen-fixing blue-green alga *Calothrix*. Without this aid to the supply of nitrogen, the high productivity of the reef community would be difficult to explain (Webb et al. 1975).

Ultimately, all animal proteins are of plant origin. This holds even for the protein synthesized from ammonia or urea in the rumen of a ruminant, for without microbes this synthesis does not take place. Microbial protein synthesis in the rumen has an interesting consequence; it makes the ruminant animal nutritionally independent of essential amino acids, for these are synthesized by the microorganisms. Ruminants therefore have no dietary need for the essential amino acids, although metabolically they are indispensable (Loosli et al. 1949; White et al. 1964).

Accessory foodstuffs: vitamins

Early in this century it was discovered that a diet of carbohydrates, fats, and proteins is insufficient in regard to a number of minor components that have become known as *vitamins*. The quantities needed are small, in the range of milligrams or micrograms. For this reason these compounds are also known as *accessory foodstuffs*. The initial discovery of a vitamin was associated with the human disease beri-beri, which is caused by the absence of thiamine (vitamin B_1) in the diet. Originally, the vitamins were given letters of the alphabet, but as their chemical structure and exact biochemical roles have been clarified, it has become increasingly common to use their chemical names instead. Detailed lists of the vitamins and their metabolic roles can be found in any adequate textbook of nutrition and will therefore be omitted here.

The vitamin requirements are not the same for all animals. For example, ascorbic acid (vitamin C) is needed by man, but not by other mammals (except some primates, bats, and the guinea pig). Other mammals synthesize ascorbic acid and do not need a dietary source, but fish do require dietary vitamin C (Hegsted 1971).

A striking difference in synthesizing ability exists in regard to cholesterol, which vertebrates readily synthesize. For man, cholesterol is, in fact, considered undesirable because it is a major factor in the development of arteriosclerosis. Insects, however, cannot synthesize cholesterol, and it must therefore be supplied in the food; for insects, cholesterol is a vitamin (Horning 1958). All insects so far studied are incapable of synthesizing cholesterol from nonsteroid precursors and need a dietary source of sterols. The same applies to crustaceans, although the cholesterol concentration in crustacean tissues is among the highest recorded for any animal. Several other invertebrates need a dietary source of sterols; among protozoans some require sterols and others do not.

Symbiotic organisms are often important sources of vitamins. For example, vitamin K, which is necessary for normal blood coagulation in vertebrates, is synthesized by bacteria in the digestive tracts of mammals. Its physiological role was discovered in birds fed a cholesterol-free diet for the purpose of studying cholesterol synthesis. The birds developed severe bleedings, which were traced to a vitamin deficiency. Only then was it established that this vitamin is needed also by mammals. Its discovery, however, depended on the accidental discovery of its importance in birds, which require it in their food.

Symbiotic synthesis of vitamins is particularly important for ruminants, which appear not to need several of the vitamins in the B group. Again, it is the microbial synthesis that makes these animals independent of a dietary source (McElroy and Goss 1940).

Information about the vitamin requirements of lower vertebrates and invertebrates is inadequate. Studies are difficult, for they require the use of diets composed of purified components, and many animals will not feed on such diets in the laboratory. Broadly based nutritional studies have therefore mostly been confined to protozoans, insects, and higher vertebrates. Also, to establish nutritional requirements with certainty, it is necessary to know whether symbiotic microorganisms contribute to meet the vitamin needs. This can be done with the aid of germ-free animals, but the procedures then become very complicated and costly.

TABLE 5.5 Approximate elementary composition of the body of a 70-kg man.

Element	Weight (g)	Percent body weight
Oxygen	45 500	65.0
Carbon	12 600	18.0
Hydrogen	7 000	10.0
Nitrogen	2 100	3.0
Calcium	1 350	1.93
Phosphorus	785	1.12
Potassium	245	0.35
Sulfur	175	0.25
Sodium	105	0.15
Chlorine	105	0.15
Magnesium	35	0.05
Total	70 000	100.00

Minerals and trace elements

How many elements are essential to life? The answer to this question is known for no more than a few laboratory animals. Of the over 90 naturally occurring elements, only 26 are known to be necessary for rats and/or chicks.

The four commonest elements – oxygen, carbon, hydrogen, and nitrogen – make up 96% of the total body weight of a mammal (Table 5.5). The seven next most abundant elements make up very close to the remaining 4%; they are calcium, phosphorous, potassium, sulfur, sodium, chlorine, and magnesium, in that order. Fifteen additional elements are required, but their total combined amount in the mammalian body is less than 0.01% of the body weight.

Water and the organic building blocks of the body – proteins, nucleic acids, fats, and carbohydrates – contain most of the oxygen, hydrogen, carbon, and nitrogen, plus part of the sulfur and phosphorus. The mineral compounds of the skeleton together with dissolved ions make up the remaining few percent. The quantitatively most important ions in animals are sodium, potassium, chloride, and bicarbonate. The maintenance of normal concentrations of sodium and chloride ions is necessary for the osmotic balance of animals, as will be discussed in Chapter 9. Although sodium is usually the major cation in blood and extracellular fluids, potassium is the dominant cation inside the cells, where sodium is correspondingly lower in concentration. Chloride is usually higher in the extracellular and lower in the intracellular fluid. Bicarbonate ion is intimately connected with respiration and carbon dioxide transport and is of major importance in the regulation of the pH of the organism, as described in previous chapters.

Calcium, magnesium, phosphate, and sulfate are also important ions; all are necessary for many physiological processes. Any great change in their concentration is deleterious or fatal. For example, calcium ions are necessary for normal cell function, for nerve conduction, for muscle contraction, and for blood coagulation. If the calcium concentration in the blood of a mammal falls to about half its normal value (about 5 mEq per liter) the result is severe or fatal tetanic cramps.

Calcium is also an important constituent of skeletons and many other rigid mechanical structures. Vertebrate bones and teeth consist mainly of calcium phosphate (hydroxyapatite). Invertebrate mineralized structures, such as clam shells and coral, are mainly calcium carbonates.

Not all structural substances in animals are mineralized. The exoskeleton of an insect consists of the organic material chitin. In crustaceans the exoskeleton may be entirely organic, but in many larger forms (such as crabs and lobsters), it is encrusted with calcium carbonate. Invertebrate skeletons frequently contain some magnesium carbonate in addition to the calcium carbonate, and often minor amounts of calcium sulfate as well. A peculiar exception is the skeletal material of the radiolarian *Acantharia* (a protozoan), which con-

sists entirely of strontium sulfate (Odum 1951, 1957).

The supporting structures of many plants, in addition to cellulose, often contain amorphous silicon oxide (silica). Of animals, some sponges have skeletons composed of spicules of silica (SiO_2).

Trace elements

Let us now consider the 15 elements that make up less than 0.01% of the body of a mammal. These elements occur in such small amounts that, although early investigators could demonstrate their presence in the body, analytical methods were not sufficiently accurate for a precise analysis. These elements therefore were mentioned as occurring in "trace" amounts, and they became known as *trace elements*. With better analytical methods, and in particular with the aid of radioisotopes, it has become possible not only to determine these elements more accurately, but also to clarify their physiological and biochemical roles.

The trace elements can be divided into three groups: (1) those known to be essential, (2) those that have metabolic effects but have not been shown to be indispensable, and (3) those that occur widely in living organisms but seem to be only incidental contaminants. Some trace elements are needed in such small amounts that it is difficult to devise a diet completely free of them, and it is probable that as research progresses, some elements of the second and perhaps even of the third group will be moved to the first group.

The mere occurrence of a trace element gives no information about its function, and the chemical demonstration of its presence is therefore rather uninformative, for virtually every element in the periodic table can be found in some minute quantity.

An understanding of the role of trace elements is

FIGURE 5.7 Periodic system of the naturally occurring elements. Elements in bold type are known to be essential for animals.

1a	2a	3a	4a	5a	6a	7a	8a			1b	2b	3b	4b	5b	6b	7b	8b
1 H																	2 He
3 Li	4 Be											5 B	6 C	7 N	8 O	9 F	10 Ne
11 Na	12 Mg				Transition elements							13 Al	14 Si	15 P	16 S	17 Cl	18 Ar
19 K	20 Ca	21 Sc	22 Ti	23 V	24 Cr	25 Mn	26 Fe	27 Co	28 Ni	29 Cu	30 Zn	31 Ga	32 Ge	33 As	34 Se	35 Br	36 Kr
37 Rb	38 Sr	39 Y	40 Zr	41 Nb	42 Mo	43 Tc	44 Ru	45 Rh	46 Pd	47 Ag	48 Cd	49 In	50 Sn	51 Sb	52 Te	53 I	54 X
55 Cs	56 Ba	57-71 ΣLa	72 Hf	73 Ta	74 W	75 Re	76 Os	77 Ir	78 Pt	79 Au	80 Hg	81 Tl	82 Pb	83 Bi	84 Po	85 At	86 Rn
87 Fr	88 Ra	89 Ac	90 Th	91 Pa	92 U												

obtained mainly in two ways: (1) feeding of synthetic diets composed of highly purified components and designed to lack the element under study, and (2) the study of disease, in both man and animals, which are believed to result from a deficiency of a trace element. Such deficiency diseases are often associated with exceptionally low concentrations of certain elements in the soil in specific geographical areas. The first method has the disadvantage that it is difficult to eliminate every trace of a given element from the diet, and a highly purified diet also readily becomes deficient in other respects. For example, should a diet be inadequate with regard to some vitamin as well as the trace element under study, the animal will not show the expected improvement if the trace element alone is added. Therefore, progress in the study of trace elements was difficult before vitamin research was sufficiently advanced to design adequate test diets.

Iron [26].* The presence of iron in living animals has been known for so long that usually iron is not considered a trace element. Nevertheless, it is essential. It is grouped in the periodic system among the "transition elements" (see Figure 5.7), which encompass a total of nine known trace elements. Another six trace elements are located in the periodic table in close proximity to those nonmetallic elements that are the most abundant in living organisms.

Iron is a constituent of hemoglobin and of several intracellular enzyme systems, notably the cytochromes. The total amount of iron in the organism is not very great, in an adult human it is about 4 g. Of this amount 70% is found in hemoglobin, 3.2% in myoglobin, 0.1% in cytochromes, 0.1% in catalase, and the remainder in storage

* The number in brackets after the name of an element is the atomic number of that element in the periodic table.

compounds mostly in the liver (O'Dell and Campbell 1971). The dietary requirement for an adult mammal is very small, for iron from the breakdown of hemoglobin is stored in the liver and used again for hemoglobin synthesis. Growing animals need more iron, and adult females need to replace that which is lost in reproductive processes (growth of fetus, menstruation).

Cobalt [27]. Cobalt is unique among the trace elements because the cobalt ion as such is not known to be needed by any animal organism. All higher animals, however, need cobalt in the specific form of vitamin B_{12}, which is required for blood formation. Deficiency leads to severe anemia. In ruminants vitamin B_{12} is formed in the rumen by the microbial flora, provided a sufficient amount of cobalt is present in the diet.

The first conclusive evidence that cobalt is an essential trace element came from Australia, where a serious disease of cattle and sheep caused great economic losses. Addition of small amounts of cobalt prevents this disease. The use of cobalt in fertilizers is expensive because fairly large amounts are needed. Instead, it has been found practical to make each sheep swallow a small cobalt-containing ceramic ball that remains in the rumen and slowly releases cobalt over several years.

Nickel [28]. Nickel is adjacent to cobalt in the periodic table and is thought to be essential. Although nickel is known to influence the growth of feathers in chickens, it is not certain that the element is essential for life. The problem is that small amounts of nickel are nearly always associated with iron and therefore extremely difficult to eliminate completely from a diet.

Copper [29]. Copper occurs widely in soil, but in certain areas the concentration is so low that both plants and animals suffer a deficiency. One of the most conspicuous symptoms is anemia. Copper is not part of the hemoglobin molecule, but copper deficiency prevents the formation of hemoglobin. The exact mechanism of this effect is not known.

Copper is also a constituent of a dozen or more enzymes, one of the most important being cytochrome oxidase, which contains both iron and copper.

Zinc [30]. Zinc is a constituent of many important enzymes, including carbonic anhydrase and several peptidases. Deficiency symptoms can be produced by feeding a purified diet.

Vanadium [23]. Vanadium exerts a pronounced growth-promoting effect on rats kept in an all-plastic isolated environment on a highly purified amino acid diet. The amounts needed, around $1\,\mu g$ per day for a rat, are within the range of concentrations found naturally in food and in tissues of higher animals (Schwarz and Milne 1971).

It is peculiar that some species of tunicates accumulate very high concentrations of vanadium in their blood, although other tunicates have no unusual concentrations of this element. The function of this vanadium enrichment, if any, remains uncertain.

Chromium [24]. Chromium is a fairly recent addition to the list of essential trace elements. It is needed for insulin to exert its maximal effect in promoting glucose uptake by the tissues (Mertz 1969). So far there is no evidence that chromium is a component of any enzyme system or serves as an enzyme activator.

Molybdenum [42]. Molybdenum is required by all nitrogen-fixing microorganisms, by higher plants, and by animals. Molybdenum is a constituent of the enzyme xanthine oxidase and thus has a role in the oxidation of purine to uric acid.

Manganese [25]. Manganese is widespread in soil and is found in all living plants and animals. It was established rather late that manganese is an essential dietary component. Deficiency symptoms include abnormal development of the bones, and the element is also needed for the function of ovaries and testes. Its biochemical effect has become known only recently. It functions as an activator of certain enzyme systems, but the connection with the deficiency symptoms is not entirely clear.

Silicon [14]. It has long been known that silicon, which next to oxygen is the most abundant element in the earth's crust, is essential for plants. Certain knowledge that it is essential for growth and normal development in animals is, however, more recent (Carlisle 1972). Silicon-deficient chicks develop bone and skull abnormalities, apparently due to an effect on the collagen matrix of the bone (Carlisle 1977).

Tin [50]. Located in the same group of the periodic table as carbon and silicon, tin is among the most recent additions to the list of essential trace elements. Inorganic tin salts are necessary for normal growth of rats kept on highly purified diets with other known trace elements added (Schwarz et al. 1970).

Arsenic [33]. This element probably is essential for rats and, by implication, for other animals. When fed on diets with extremely low arsenic content, rats develop rough hairy coats and grow more slowly than control animals. They also show other disturbances, with enlarged spleens and low hematocrit values. For some unknown reason males are consistently more affected than females (Nielsen et al. 1975).

Selenium [34]. Selenium has, because of its toxicity, been of greater practical importance in animal nutrition than as an essential element. In some areas selenium is found in the soil in such high concentrations that plants become toxic to livestock, and grazing animals then may suffer acute selenium poisoning. Selenium, however, is also an essential trace element; deficiency symptoms include liver necrosis and muscular dystrophy. The biochemical role of selenium seems to be related to the function of vitamin E, which is necessary for growth and normal reproduction in rats.

Fluorine [9]. Until recently fluorine was not classed as essential, although some of its physiological effects were well known. This changed, however, when fluorine was found to be essential for the normal growth of rats (Schwarz and Milne 1972).

It has long been known that fluorine is present in small amounts in teeth and bone, where the fluoride ion is incorporated into the crystal lattice of calcium phosphate (apatite), the main mineral component. Whether fluorine is essential in this role is doubtful, but it makes the teeth more resistant to dental caries. The addition of fluorine to drinking water has therefore been recommended as a public health measure to reduce the incidence of tooth decay. The effectiveness is beyond doubt, and the amounts added are so small that there are no known toxic effects.

In some geographical areas, however, fluorine is present in the natural water supply in such high concentrations that it causes abnormalities of the tooth enamel, known as "mottled enamel." Cosmetically, mottled enamel is undesirable, but mottled teeth are particularly resistant to decay. Even higher intake of fluorine leads to deformation of bones and serious poisoning; these have occurred near industrial plants and smelteries where fluorine is discharged to the atmosphere.

Iodine [53]. The physiological significance of iodine relates to its role as a constituent of the thyroid hormones. These accelerate the cellular metabolism and are necessary for the control of metabolic rate and for normal growth of vertebrates. It has long been known that the incidence of human goiter (an enlargement of the thyroid gland) is concentrated in geographical areas where the iodine content of the soil is particularly low. This enlargement of the thyroid appears to be a response to iodine deficiency, and the addition of small amounts of iodine to the food (usually by adding a minute quantity of sodium iodide to ordinary table salt) has virtually eliminated this rather serious public health problem.

Iodine-deficient regions are mostly inland and mountainous areas far from the ocean. Sea water contains a small amount of iodine, and spray that enters the atmosphere from wave action is precipitated with rain and meets the needs of both animals and man, even several hundred kilometers from the coast.

Other trace elements. Many additional elements (e.g., lithium, rubidium, beryllium, strontium, aluminum, boron, germanium, bromine) are widely present in living material and may have functional significance, although definite evidence is lacking. Some are highly toxic, but even so may be needed in trace amounts. Furthermore, at present there is no reason to believe that all animals have exactly the same requirements for trace elements.

NOXIOUS COMPOUNDS AND CHEMICAL DEFENSE

One way to avoid being eaten is to make oneself inedible. This simple principle is widely used by both plants and animals. Plants are the primary producers in the food chain, and because they cannot run away, they remain in place, ready to be eaten.

Making oneself inedible with the aid of toxic compounds is a much used defense, and an amazing number of plants are poisonous to animals. Their "predators," in turn, have developed mechanisms to circumvent or counteract these defenses,

so that hardly any plant is completely without enemies (Whittaker and Feeny 1971). An example of such adaptation is the cabbage butterfly, whose larva feeds on a number of cruciferous plants (cabbage, mustard, etc.). These plants contain mustard oils that are toxic or lethal to other butterflies, but not only are tolerated by the cabbage butterfly larva, but even act as attractants to lure the adult butterfly to deposit its eggs on these plants.

Animals likewise use venoms and toxins for defense; for some (e.g., snakes, spiders, scorpions), the venom serves both for defense and as an attack mechanism for catching prey.

Some animals have not only circumvented the chemical defenses of plants, but have even exploited the toxic principles of the plants on which they feed to make themselves distasteful or inedible to their predators.

Plant poisons

Various plant compounds affect almost every physiological system. They may act on the skin, on the digestive system, on the blood, on various parts of the nervous system, on the metabolism, on the nutritional value of food, and on the hormonal and reproductive systems of animals. The variety of compounds and effects makes them difficult to classify but they can be placed in some major groups, in part because of chemical similarities, and in part because of similarities in effects. A few examples will illustrate the immense variety of compounds and actions.

Alkaloids

Alkaloids belong to a group of organic compounds that have little in common except that they all contain nitrogen. Among these are toxic compounds that have profound effects on the central nervous system of vertebrates (e.g., opium, which is a crude product of the opium poppy containing the medically useful alkaloids morphine and codeine). A widely used alkaloid is nicotine; other well-known and highly toxic alkaloids are strychnine and the arrow poison curare.

Glycosides

Glycosides are compounds that on hydrolysis yield one or more sugar molecules and a variety of other compounds. One of the most widely known is digitalis, an important drug in the treatment of certain afflictions of the heart. It is found in the foxglove and is lethal in overdose.

Several plants contain glycosides that upon hydrolysis yield cyanide (HCN). One of the best known is the glycoside amygdalin, which occurs in bitter almonds and in the kernels of apricots, prunes, and cherries. Apple seeds also contain small amounts of this highly toxic compound. For most of us this is unimportant, but it has been reported that a man who was fond of apple seeds as a delicacy saved a cupful of them and, after eating them all at one time, died from cyanide poisoning (Kingsbury 1964).

A major staple food in many tropical countries, cassava (manioc), releases so much cyanide that even the preparation of the tubers for eating involves a risk of fatal poisoning. The tubers are harvested and shredded with quantities of water, and after the starch granules have settled, the poisonous liquid is poured off. Among some Central American Indians, the process of preparing cassava is considered so dangerous that only older people are given this job, presumably because they are considered expendable.

It is interesting that cyanide smells strongly of bitter almonds to some persons, and to others is completely odorless. This difference is not the result of an inability to smell "bitter almond," for benzaldehyde, which to cyanide smellers smells very similar to cyanide, has the characteristic "bit-ter almond" smell to cyanide nonsmellers. Cyanide nonsmellers, who are quite common, should be careful if they use this dangerous substance, for their sense of smell will not give them any warning.

Oils and resins

Oils and resins of various kinds are widely distributed in plants and have many different effects. One of the popularly known such resins is the active principle in marijuana, cannabis, which through its effect on the central nervous system distorts the conscious mind and causes hallucinations.

Other oils and resins are irritants to the skin or act on the gastrointestinal tract, causing vomiting, diarrhea, and a variety of other symptoms. Several compounds act as irritants to the nose and mouth and thus produce their deterrent effects most promptly.

Oxalic acid

Oxalic acid is found in a large number of plants from widely different families. It owes its acute toxicity to the fact that it precipitates calcium in the form of insoluble calcium oxalate. In acute poisoning the calcium concentration in the blood is reduced, causing nervous symptoms, muscle cramps, and acute damage to the kidneys.

One plant that is important because of its oxalate content is *Halogeton*, common on the ranges of the western United States. It is known as locoweed and over 30% of the dry weight of the plant may be soluble oxalate. It is highly toxic to grazing livestock and has killed virtually thousands of sheep.

Two plants commonly eaten by man, spinach and rhubarb, belong to the same family, the Chenopodiaceae. Neither is usually eaten in sufficient quantities to cause acute poisoning. However, the leaves of rhubarb, which have occasionally been used as a vegetable, contain more oxalic acid than the stems and have caused several deaths. If large amounts of spinach are eaten daily, the precipitation of calcium oxalate in the intestine will reduce the calcium available for absorption and may over a period of time lead to calcium deficiency.

Enzyme inhibitors

Some plants, including several of the pea family, contain substances that inhibit the action of trypsin, the important proteolytic enzyme of the pancreas. Such trypsin inhibitors are found in soybeans, lima beans, peanuts, and a number of plants from other families.

Most of the trypsin inhibitors are heat-sensitive, and because humans mostly eat cooked food, they are of minor importance in human nutrition. However, trypsin inhibitors reduce the protein utilization of animals that eat raw or untreated soybeans and related products, and are therefore of importance in animal husbandry. The deleterious effects can be reduced or eliminated by heat treatment (Rackis 1965).

Compounds with hormonal effects

Cabbage, mustard, radishes, and a number of other plants of the family Cruciferae owe their sharp taste to mustard oils, which are irritants and may even cause fatal poisoning. In addition, many contain substances that suppress the function of the thyroid gland and cause an enlargement of the gland, known as goiter. When laboratory animals are fed diets containing large amounts of cabbage leaves or turnips, they develop goiter.

The antithyroid substances are chemically related to drugs such as thiourea that are used in the treatment of thyroid hyperactivity. It is well known that some persons find the compound phenyl-thiourea extremely bitter; others find it tasteless. The nontasting condition is inherited as a recessive gene. One of the antithyroid substances from turnips and cabbage is similarly tasteless to some and bitter to others, with tasters and nontasters matching the tasters and nontasters of phenylthiourea (Boyd 1950). This suggests a possible physiological role for the gene responsible for the tasting ability.

Compounds that affect animal reproduction

One of the defense strategies used by plants is to interfere with the reproduction of their enemies through compounds that foul up the animals' endocrine systems. Both insects and vertebrates suffer the effects.

One way is to interfere with the normal development of insects, which as larvae go through a number of molts before they change into adult form. A hormone, the juvenile hormone, is essential in directing larval development, but the change into the mature adult form depends on the *absence* of this hormone. Several American evergreen trees produce substances that act as juvenile hormone (see Chapter 13), and the effect is simple: The insects do not develop into the mature form and thus are prevented from reproducing.

Many synthetic compounds with juvenile hormone effects have been developed, some of them far more powerful than the naturally occurring hormones. Under certain conditions they can be used as potent insecticides with minimal effects on other animals, including domestic animals, which are insensitive to insect hormones. One disadvantage of these insecticides is that they are effective only during the brief period when the immature insect develops into the adult form.

A broader effect is produced by plant substances that have anti-juvenile hormone activity (Bowers et al. 1976). They are effective over a much longer time and upset normal development as well as a number of other processes regulated by juvenile hormone. The action is to stop or depress the production of juvenile hormone; this not only leads to precocious metamorphosis, but also interferes with induction of diapause, development of eggs, production of sex attractants, and so on. The substances promise to be of great interest as more information is obtained about their physiological effects.

Vertebrate reproduction is also affected by plant compounds that act on the endocrine system. Several plants that commonly are eaten by grazing animals, especially plants of the pea family, contain biologically active compounds with estrogenic effects that mimic the normal activity of animal estrogens. These estrogens cause reproductive disturbances in sheep and cattle and may be a factor in controlling wild rodent populations.

The harmful effects of plant estrogens have been extensively studied in Australia, where they are important in sheep production. Animals that eat subterranean clover show declined fertility, caused by inability to conceive, and the number of pregnant ewes therefore decreases drastically (Shutt 1976).

Plant estrogens are not only important to animal husbandry, they have a much broader role in nature. In the arid regions of California the annual plants produce phytoestrogens in dry years. When the leaves are eaten by quail, these compounds inhibit reproduction and prevent the production of young that would have inadequate food. In wet years, the plants grow vigorously and estrogen concentration is low, and in such years the quail reproduce prolifically. The abundant crop of plants and seeds then assures the survival of the enlarged population. The close relationship between animal reproduction and plant productivity has long been known, and the mechanism is, at least in this instance, a well-adjusted natural biological control of population size (Leopold et al. 1976).

Animal use of plant poisons

The common milkweed, *Asclepias*, and other plants that belong to the same family produce cardiac glycosides. In addition to affecting the heart of the animal that ingests these toxic plants, they also affect the nerve center in the brain that controls vomiting. When an animal eats this plant, vomiting usually occurs before a lethal amount of the poison has been absorbed, and the effect is limited to a very unpleasant gastric upset.

In spite of their toxicity, milkweeds are the exclusive food of the larva of a group of insects that includes the monarch butterfly. In nature various insect-eating vertebrates avoid these butterflies and their larvae for the simple reason that they have accumulated the poisonous substances from the milkweed. If a bird eats only one of the strikingly colored monarch butterflies, it will be violently ill and never try again. The butterflies have thus developed the ability not only to feed on the poisonous plants, but also to use the poisons as protection against their own predators (Brower et al. 1967).

Such use of toxic substances obtained from the feed is widespread among insects, but is not restricted to this class. Some marine animals feed on various coelenterates that otherwise are well protected by their stinging cells, or nematocysts. One coelenterate that is painfully known to man is the Portuguese man-of-war, *Physalia*, which may even be lethal to swimmers who come in close contact with them. Certain sea slugs (nudibranch snails), which feed on the Portuguese man-of-war, ingest the tissues that contain nematocysts and digest them without triggering the release of the poison. The nematocysts are instead incorporated in the tissues of the snail and serve for the defense

BOMBARDIER BEETLE. These insects defend themselves with a chemical spray, ejected at a temperature
of 100 °C. The spray is aimed directly at the attacker
(e.g., an ant biting at one of the legs). This photograph
shows a bombardier beetle kept in place for photo-
graphy by a metal wire cemented to its back. The spray
is directed toward the "attacker," a pair of forceps
used to hold one of the legs. Bombardier beetles occur
in all major faunal regions of the world, except
Oceania. The species shown here, *Stenaptinus insignis*,
is from Kenya and is about 20 mm long. [Courtesy of
Thomas Eisner and D. Aneshansley, Cornell University]

Animal use of their own poisons

Animals also use a wide variety of venoms and
toxins that they produce themselves. Well-known
examples are snakes, scorpions, spiders, bees, and
wasps. More unusual types of defense include the
hydrogen cyanide produced by millipedes when
disturbed (Eisner et al. 1963) and, even more dramatic, the chemical spray ejected at a temperature
of 100 °C by bombardier beetles (Aneshansley et
al. 1969).

The bombardier beetles produce their defensive
spray from a pair of glands that open at the tip of
the abdomen. The spray is visible as a fine mist,
and the discharge occurs with an audible detona-
tion. The secreting gland is a two-chambered apparatus. The larger, inner compartment contains a
solution of hydroquinones and hydrogen peroxide;
the outer compartment contains an enzyme mixture of catalases and peroxidases. When the beetle
discharges, it squeezes fluid from the inner reser-

voir into the outer chamber. An instantaneous explosive reaction takes place as the hydrogen peroxide is decomposed by the enzymes and the hydroquinones are oxidized with a tremendous release of heat. Under the pressure of the evolved gas, the mixture is ejected as a fine spray that has a strongly repellent effect on predators. The beetle can spray repeatedly and in quick succession and can aim the abdominal tip in virtually any direction from which it is attacked.

In summarizing this section we can conclude that an overwhelming number of animals and plants use chemical defense mechanism, either producing their own poisons, or taking advantage of compounds produced by others. In fact, probably only a minority of organisms is not protected in some such way against at least some potential enemies.

REFERENCES

Aneshansley, D. J., Eisner, T., Widom, J. M., and Widom, B. (1969) Biochemistry at 100 °C: Explosive secretory discharge of bombardier beetles (*Brachinus*). *Science* 165:61–63.

Bailey, C. B. (1961) Saliva secretion and its relation to feeding in cattle. 3. The rate of secretion of mixed saliva in the cow during eating, with an estimate of the magnitude of the total daily secretion of mixed saliva. *Br. J. Nutr.* 15:443–451.

Barnes, R. H., Fiala, G., and Kwong, E. (1963) Decreased growth rate resulting from prevention of coprophagy. *Fed. Proc.* 22:125–133.

Bauchop, T., and Martucci, R. W. (1968) Ruminant-like digestion of the langur monkey. *Science* 161:698–700.

Benemann, J. R. (1973) Nitrogen fixation in termites. *Science* 181:164–165.

Benson, A. A., Lee, R. F., and Nevenzel, J. C. (1972) Wax esters: Major marine metabolic energy sources. *Biochem. Soc.* (*Lond.*) *Symp.*, no. 35, pp. 175–187.

Berrill, N. J. (1929) Digestion in ascidians and the influence of temperature. *J. Exp. Biol.* 6:275–292.

Blair-West, J. R., Coghlan, J. P., Denton, D. A., and Wright, R. D. (1967) Effect of endocrines on salivary glands. In *Handbook of Physiology*, sect. 6, *Alimentary Canal*, vol. II, *Secretion* (C.F. Code, ed.), pp. 633–664. Washington, D.C.: American Physiological Society.

Block, R. J., Stekol, J. A., and Loosli, J. K. (1951) Synthesis of sulfur amino acids from inorganic sulfate by ruminants. 2. Synthesis of cystine and methionine from sodium sulfate by the goat and by the microorganisms of the rumen of the ewe. *Arch. Biochem. Biophys.* 33:353–363.

Bowers, W. S., Ohta, T., Cleere, J. S., and Marsella, P. A. (1976) Discovery of insect anti-juvenile hormones in plants. *Science* 193:542–547.

Boyd, W. C. (1950) Taste reactions to antithyroid substances. *Science* 112:153.

Breznak, J. A., Brill, W. J., Mertins, J. W., and Coppel, H. C. (1973) Nitrogen fixation in termites. *Nature, Lond.* 244:577–580.

Brower, L. P., Brower, J. V. Z., and Corvino, J. M. (1967) Plant poisons in a terrestrial food chain. *Proc. Natl. Acad. Sci. U.S.A.* 57:893–898.

Carlisle, E. M. (1972) Silicon: An essential element for the chick. *Science* 178:619–621.

Carlisle, E. M. (1977) A silicon requirement for normal skull formation. *Fed. Proc.* 36:1123.

Carpenter, E. J., and Culliney, J. L. (1975) Nitrogen fixation in marine shipworms. *Science* 187:551–552.

Cleveland, L. R. (1925) Toxicity of oxygen for protozoa in vivo and in vitro: Animals defaunated without injury. *Biol. Bull.* 48:455–468.

Denis, G., Jeuniaux, Ch., Gerebtzoff, M. A., and Goffart, M. (1967) La digestion stomacale chez un paresseux: L'unau *Choloepus hoffmanni* Peters. *Ann. Soc. R. Zool. Belg.* 97:9–29.

Eisner, T., Eisner, H. E., Hurst, J. J., Kafatos, F. C., and Meinwald, J. (1963) Cyanogenic glandular apparatus of a millipede. *Science* 139:1218–1220.

Florkin, M., and Lozet, F. (1949) Origine bactérienne de la cellulase du contenu intestinal de l'escargot. *Arch. Int. Physiol.* 57:201–207.

Friedmann, H., and Kern, J. (1956a) The problem of

cerophagy or wax-eating in the honey-guides. *Q. Rev. Biol. 31*:19–30.

Friedmann, H., and Kern, J. (1956b) *Micrococcus cerolyticus*, nov. sp., an aerobic lipolytic organism isolated from the African honey-guide. *Can. J. Microbiol. 2*:515–517.

Friedmann, H., Kern, J., and Hurst, J. H. (1957) The domestic chick: A substitute for the honey-guide as a symbiont with cerolytic microorganisms. *Am. Nat. 91*:321–325.

Grassé, P.-P. (1955) *Traité de Zoologie: Anatomie, Systématique, Biologie*, vol. 17, fasc. 2; *Mammifères*, pp. 1173–2300. Paris: Masson.

Greenberg, B., and Paretsky, D. (1955) Proteolytic enzymes in the house fly, *Musca domestica* (L.). *Ann. Entomol. Soc. Am. 48*:46–50.

Greenfield, L. J., and Lane, C. E. (1953) Cellulose digestion in *Teredo. J. Biol. Chem. 204*:669–672.

Griffiths, M., and Davies, D. (1963) The role of the soft pellets in the production of lactic acid in the rabbit stomach. *J. Nutr. 80*:171–180.

Harder, W. (1949) Zur Morphologie und Physiologie des Blinddarmes der Nagetiere. *Verh. Dtsch. Zool. 2*:95–109.

Hegsted, D. M. (1971) Fish require dietary vitamin C. *Nutr. Rev. 29*:207–210.

Hervey, G. R. (1969) Regulation of energy balance. *Nature, Lond. 222*:629–631.

Horning, M. G. (1958) The sterol requirements of insects and of protozoa. In *Cholesterol: Chemistry, Biochemistry, and Pathology* (R. P. Cook, ed.), pp. 445–455. New York: Academic Press.

Houpt, T. R. (1959) Utilization of blood urea in ruminants. *Am. J. Physiol. 197*:115–120.

Houpt, T. R. (1963) Urea utilization by rabbits fed a low-protein ration. *Am. J. Physiol. 205*:1144–1150.

Hungate, R. E. (1941) Experiments on the nitrogen economy of termites. *Ann. Entomol. Soc. Am. 34*:467–489.

Hungate, R. E. (1942) The culture of *Eudiplodinium neglectum*, with experiments on the digestion of cellulose. *Biol. Bull. 83*:303–319.

Hungate, R. E. (1955) Mutualistic intestinal protozoa. In *Biochemistry and Physiology of Protozoa*, vol. II.

(S. H. Hutner and A. Lwoff, eds.), pp. 159–199. New York: Academic Press.

Johannes, R. E., Coward, S. J., and Webb, K. L. (1969) Are dissolved amino acids an energy source for marine invertebrates? *Comp. Biochem. Physiol. 29*:283–288.

Kingsbury, J. M. (1964) *Poisonous Plants of the United States and Canada*. Englewood Cliffs, N.J.: Prentice-Hall. 626 pp.

Kinnear, J. E., and Main, A. R. (1975) The recycling of urea nitrogen by the wild tammar wallaby (*Macropus eugenii*), a "ruminant-like" marsupial. *Comp. Biochem. Physiol. 51A*:793–810.

Kleiber, M. (1961) *The Fire of Life. An Introduction to Animal Energetics*. New York: Wiley. 454 pp.

Lambremont, E. N., Fisk, F. W., and Ashrafi, S. (1959) Pepsin-like enzyme in larvae of stable flies. *Science 129*:1484–1485.

Lasker, R., and Giese, A. C. (1956) Cellulose digestion by the silverfish *Ctenolepisma lineata. J. Exp. Biol. 33*:542–553.

Leopold, A. S., Erwin, M., Oh, J., and Browning, B. (1976) Phytoestrogens: Adverse effects on reproduction in California quail. *Science 191*:98–100.

Loosli, J. K., Williams, H. H., Thomas, W. E., Ferris, F. H., and Maynard, L. A. (1949) Synthesis of amino acids in the rumen. *Science 110*:144–145.

Matthews, L. H., and Parker, H. W. (1950) Notes on the anatomy and biology of the basking shark (*Cetorhinus maximus* (Gunner)). *Proc. Zool. Soc. Lond. 120*:535–576.

Mayer, J., and Thomas, D. W. (1967) Regulation of food intake and obesity. *Science 156*:335–337.

McBee, R. H., and West, G. C. (1969) Cecal fermentation in the willow ptarmigan. *Condor 71*:54–58.

McDonald, I. W. (1952). The role of ammonia in ruminal digestion of protein. *Biochem. J. 51*:86–90.

McElroy, L. W., and Goss, H. (1940) A quantitative study of vitamins in the rumen contents of sheep and cows fed vitamin-low diets. 1. Riboflavin and vitamin K. *J. Nutr. 20*:527–540.

Mertz, W. (1969) Chromium occurrence and function in biological systems. *Physiol. Rev. 49*:163–239.

Moir, R. J., Somers, M., and Waring, H. (1956) Studies on marsupial nutrition. 1. Ruminant-like digestion in a herbivorous marsupial (*Setonix brachyurus* Quoy and Gaimard). *Aust. J. Biol. Sci.* 9:293–304.

Muscatine, L., and Lenhoff, H. M. (1965) Symbiosis of hydra and algae. 2. Effects of limited food and starvation on growth of symbiotic and aposymbiotic hydra. *Biol. Bull.* 129:316–328.

Nielsen, F. H., Givand, S. H., and Myron, D. R. (1975) Evidence of a possible requirement for arsenic by the rat. *Fed. Proc.* 34:923.

O'Dell, B. L., and Campbell, B. J. (1971) Trace elements: Metabolism and metabolic function. *Comp. Biochem.* 21:179–266.

Odum, H. T. (1951) Notes on the strontium content of sea water, celestite radiolaria, and strontianite snail shells. *Science* 114:211–213.

Odum, H. T. (1957) Biogeochemical deposition of strontium. *Univ. Texas Publ. Inst. Mar. Sci.* 4:38–114.

Pardy, R. L., and Dieckmann, C. (1975) Oxygen consumption in the symbiotic hydra *Hydra viridis*. *J. Exp. Zool.* 194:373–378.

Prévost, J., and Vilter, V. (1962) Histologie de la sécrétion oesophagienne du Manchot Empereur. In *Proceedings, 13th International Ornithological Congress*, Vol. 2. pp. 1085–1094. Baton Rouge, La.: American Ornithologists' Union.

Rackis, J. J. (1965) Physiological properties of soybean trypsin inhibitors and their relationship to pancreatic hypertrophy and growth inhibition of rats. *Fed. Proc.* 24:1488–1500.

Rose, W. C. (1938) The nutritive significance of the amino acids. *Physiol. Rev.* 18:109–136.

Rose, W. C. (1949) Amino acid requirements of man. *Fed. Proc.* 8:546–552.

Rozin, P., and Mayer, J. (1961) Regulation of food intake in the goldfish. *Am. J. Physiol.* 201:968–974.

Sargent, J. R., and Gatten, R. R. (1976) The distribution and metabolism of wax esters in marine invertebrates. *Biochem. Soc. (Lond.) Trans.* 4:431–433.

Schmidt-Nielsen, B., Schmidt-Nielsen, K., Houpt, T. R., and Jarnum, S. A. (1957) Urea excretion in the camel. *Am. J. Physiol.* 188:477–484.

Schoenheimer, R. (1964) *The Dynamic State of Body Constituents*. New York: Hafner. 78 pp.

Schwarz, K., and Milne, D. B. (1971) Growth effects of vanadium in the rat. *Science* 174:426–428.

Schwarz, K., and Milne, D. B. (1972) Fluorine requirement for growth in the rat. *Bioinorg. Chem.* 1:331–338.

Schwarz, K., Milne, D. B., and Vinyard, E. (1970) Growth effects of tin compounds in rats maintained in a trace-element-controlled environment. *Biochem. Biophys. Res. Commun.* 40:22–29.

Shutt, D. A. (1976) The effects of plant oestrogens on animal production. *Endeavour* 35:110–113.

Smith, D., Muscatine, L., and Lewis, D. (1969) Carbohydrate movement from autotrophs to heterotrophs in parasitic and mutualistic symbiosis. *Biol. Rev.* 44:17–90.

Stephens, G. C. (1962) Uptake of organic material by aquatic invertebrates. 2. Uptake of glucose by the solitary coral, *Fungia scutaria*. *Biol. Bull.* 123:648–659.

Stephens, G. C. (1964) Uptake of organic material by aquatic invertebrates. 3. Uptake of glycine by brackish-water annelids. *Biol. Bull.* 126:150–162.

Tarr, H. L. A. (1958) Biochemistry of fishes. *Annu. Rev. Biochem.* 27:223–244.

Taylor, M. W., and Medici, J. C. (1966) Amino acid requirements of grain beetles. *J. Nutr.* 88:176–180.

Thacker, E. J., and Brandt, C. S. (1955) Coprophagy in the rabbit. *J. Nutr.* 55:375–385.

Thompson, T. E., and Bennett, I. (1969) Physalia nematocysts: Utilized by mollusks for defense. *Science* 166:1532–1533.

Trager, W. (1932) A cellulase from the symbiotic intestinal flagellates of termites and of the roach, *Cryptocercus punctulatus*. *Biochem. J.* 26:1762–1771.

Trager, W. (1934) The cultivation of a cellulose-digesting flagellate, *Trichomonas termopsidis*, and of certain other termite protozoa. *Biol. Bull.* 66:182–190.

Webb, K. L., DuPaul, W. D., Wiebe, W., Sottile, W., and Johannes, R. E. (1975) Enewetak (Eniwetok) atoll: Aspects of the nitrogen cycle on a coral reef. *Limnol. Oceanogr.* 20:198–210.

Weis-Fogh, T. (1952) Fat combustion and metabolic rate of flying locusts (*Schistocerca gregaria* Forskål). *Philos. Trans. R. Soc. Lond. B.* 237:1–36.

White, A., Handler, P., and Smith, E. L. (1964) *Prin-*

ciples of Biochemistry, 3rd ed. New York: McGraw-Hill. 1106 pp.

Whittaker, R. H., and Feeny, P. P. (1971) Allelochemics: Chemical interactions between species. Science 171:757–770.

Wigglesworth, V. B. (1949) The utilization of reserve substances in Drosophila during flight. J. Exp. Biol. 26:150–163.

Wright, R. T., and Hobbie, J. E. (1966) Use of glucose and acetate by bacteria and algae in aquatic ecosystems. Ecology, 47:447–464.

ADDITIONAL READING

Bücherl, W., Buckley, E. E., and Deulofeu, V. (eds.) (1968, 1971) Venomous Animals and Their Venoms, vol. 1, 707 pp. (1968); vol. 2, 687 pp. (1971); vol. 3, 537 pp. (1971). New York: Academic Press.

Code, C. F., and Heidel, W. (eds.) (1967–1968) Handbook of Physiology, sect. 6, Alimentary Canal, vols. I–V, pp. 1–2896. Washington, D.C.: American Physiological Society.

Crawford, M. A. (ed.) (1968) Comparative Nutrition of Wild Animals. Symposium, Zoological Society of London, no. 21. London: Academic Press. 429 pp.

Eisner, T. (1970) Chemical defense against predation in arthropods. In Chemical Ecology (E. Sondheimer and J. Simeone, eds.), pp. 157–217. New York: Academic Press.

Fenchel, T. M., and B. B. Jørgensen (1976) Detritus food chains of aquatic ecosystems: The role of bacteria. Adv. Microbl. Ecol. 1:1–58.

Florkin, M., and Stotz, E. H. (eds.) (1971) Comprehensive Biochemistry, vol. 21, Metabolism of Vitamins and Trace Elements. Amsterdam: Elsevier. 297 pp.

Goreau, T. F., Goreau, N. I., and Yonge, C. M. (1973) On the utilization of photosynthetic products from zooxanthellae and of a dissolved amino acid in Tridacna maxima f. elongata (Mollusca: Bivalvia). J. Zool. Lond. 169:417–454.

Halver, J. E. (ed.) (1972) Fish Nutrition. New York: Academic Press. 714 pp.

Jørgensen, C. B. (1975) Comparative physiology of suspension feeding. Annu. Rev. Physiol. 37:57–79.

Labov, J. B. (1977) Phytoestrogens and mammalian reproduction. Comp. Biochem. Physiol. 57A:3–9.

Liener, I. E. (1969) Toxic Constituents of Plant Foodstuffs. New York: Academic Press. 500 pp.

McDonald, P., Edwards, R. A., and Greenhalgh, J. F. D. (1973) Animal Nutrition, 2nd ed. New York: Hafner. 487 pp.

Stephens, G. C. (1975) Uptake of naturally occurring primary amines by marine annelids. Biol. Bull. 149:397–407.

Underwood, E. J. (1977) Trace Elements in Human and Animal Nutrition, 4th ed. New York: Academic Press. 545 pp.

CHAPTER SIX

Energy metabolism

The preceding chapter dealt with food and feeding; this chapter will be concerned with the use of food to provide energy. Animals need chemical energy to carry out their various functions, and their overall use of chemical energy is often referred to as their *energy metabolism.*[*]

Animals obtain energy mostly through the oxidation of foodstuffs, and their consumption of oxygen can therefore be used as a measure of their energy metabolism. Much of this chapter will be concerned with the rate of oxygen consumption, which we often take to mean the rate of energy metabolism. However, chemical energy can also be obtained in other ways, without the use of oxygen, and oxygen consumption is therefore not always a measure of energy metabolism. For example, some animals can live in the complete absence of free oxygen; they still utilize chemical energy for their energy needs, although the metabolic pathways are different. We say that such animals meet their energy needs through *anaerobic metabolism*. This situation is normal for quite a few animals that live in oxygen-poor environments and/or tolerate prolonged exposure to lack of oxygen. They still need energy, which they obtain without utilizing oxidation processes.

The energy-requiring processes and reactions in the living organism use a common source of energy, *adenosine triphosphate* (ATP). This ubiquitous compound, through the hydrolysis of an "energy-rich" phosphate bond, seems to be the immediate source of chemical energy for processes

---

[*]The word *metabolism* is used in several different ways. For example, in the phrase iron metabolism, the term applies to the total physiological role of iron – its intake, absorption, storage, synthesis of hemoglobin and cytochromes, excretion, etc. The term intermediary metabolism has come to mean a description of all the biochemical reactions and interconversions that take place in the organism. Thus the term metabolism is poorly defined. In this chapter it will mean energy metabolism.

such as muscle contraction, ciliary movement, firefly luminescence, discharge of electric fish, cellular transport processes, all sorts of synthetic reactions, and so on. ATP is formed at the various energy-yielding steps in the oxidation of foodstuffs, and also in anaerobic energy-yielding processes, but in smaller amounts. ATP is a universal intermediate in the flow of the chemical energy of the food to the energy-requiring processes of metabolism in both aerobic and anaerobic organisms.

METABOLIC RATE

Metabolic rate refers to the energy metabolism per unit time. It can, in principle, be determined in three different ways.

The first is by calculating the difference between the energy value of all food taken in and the energy value of all excreta (primarily feces and urine). This method assumes that there is no change in the composition of the organism and therefore cannot be used in growing organisms or in organisms that have an increase or a decrease in storage of fat or other material. The method is technically cumbersome and is accurate only if carried out over a sufficiently long period of observation to assure that the organism has not undergone significant changes in size and composition.

The second method of determining metabolic rate is from the total heat production of the organism. This method should give information about all fuel used, and in principle it is the most accurate method. In practice, determinations are made with the organism inside a calorimeter. This can yield very accurate results, but technically is a complex procedure. Such items as heating of ingested food and vaporization of water must be entered into the total heat account. Also, if the organism performs any external work that does not appear as heat, this work must be added to the account. Assume, for example, that a man located inside a large calorimeter lifts a ton of bricks from the floor onto a table. The increase in potential energy of the bricks is derived from the energy metabolism of the man and could be released as heat if the bricks were permitted to slide down again.

The third measure that can be used to determine the metabolic rate is the amount of oxygen used in oxidation processes, provided information is available about which substances have been oxidized (and there is no anaerobic metabolism).

The determination of oxygen consumption is technically easy and is so commonly used for estimation of metabolic rates that the two terms are often used interchangeably. This is obviously not correct; for example, a fully anaerobic organism has zero oxygen consumption but certainly does not have a zero metabolic rate!

The reason oxygen can be used as a practical measure of metabolic rate is that the amount of heat produced for each liter of oxygen used in metabolism remains nearly constant, irrespective of whether fat, carbohydrate, or protein is oxidized (Table 6.1, column c). The highest figure (5.0 kcal per liter O_2 for carbohydrate metabolism) and the lowest (4.5 kcal per liter O_2 for protein) differ by only 10%, and it has become customary to use an average value of 4.8 kcal per liter O_2 as a measure of the metabolic rate.* The largest error resulting

---

* The traditional unit for heat, the calorie (cal), is not a part of the International System of Units (the SI System). However, it is in universal use in physiological tables, books, and other publications, and it is impractical not to retain it in this book. The energy unit of the SI System is the joule (J), and the calorie is defined as 1 cal = 4.184 J (1 kcal = 4.184 kJ). Further, 1 cal is the amount of heat required to heat 1 g water from 14.5 °C to 15.5 °C, and 1 kcal = 1000 cal. Often, especially in elementary books and popular writings, the calorie and kilocalorie are confused. At times, a kilocalorie is called a "large calorie" and is written "Cal." Because the calorie is a precisely defined unit, it is meaningless to refer to "large" and "small" calories. For additional information about the use of symbols and units, see Appendix A.

TABLE 6.1 Heat produced and oxygen consumed in the metabolism of common foodstuffs. The values for protein depend on whether the metabolic end product is urea or uric acid. The ratio between carbon dioxide formed and oxygen used is known as the respiratory quotient (RQ). [Data based on Lusk 1931; King 1957]

Food	(a) kcal g^{-1}	(b) liter O_2 g^{-1}	(c) kcal per liter O_2	(d) $RQ = \dfrac{CO_2 \text{ formed}}{O_2 \text{ used}}$
Carbohydrate	4.2	0.84	5.0	1.00
Fat	9.4	2.0	4.7	0.71
Protein (urea)	4.3	0.96	4.5	0.81
Protein (uric acid)	4.25	0.97	4.4	0.74

from the use of this mean figure would be 6%, but if the metabolic fuel is a mixture of the common foodstuffs (carbohydrate, fat, and protein), the error is usually insignificant. Most determinations of metabolic rates are not very precise anyway, not because the available apparatus is inherently inaccurate, but because the amounts of animal activity and other physiological functions tend to vary a great deal.

Although 1 liter of oxygen gives similar amounts of energy for the three major foodstuffs, each gram of foodstuff gives a very different amount of energy. The energy derived from the oxidation of 1 g fat is more than twice as high as from the oxidation of 1 g carbohydrate or protein (Table 6.1, column a).

The high energy value of fat is widely known; the biologically most important consequence is that fat stores energy with a smaller addition of weight and bulk than would otherwise be possible. It may be of interest that the energy value of alcohol is also quite high, nearly 7 kcal g^{-1}. This is one reason persons who wish to lose weight by dieting usually avoid alcohol; another reason is that alcohol readily weakens the resolution to restrict food intake and thus contributes further to the lack of success in controlling energy intake.

The energy value given for protein, 4.3 kcal g^{-1}, differs from the combustion value determined for proteins in the chemical laboratory, which is usually given as 5.3 kcal g^{-1}. The reason is that protein is not completely oxidized in the body. Mammals excrete the nitrogen as urea, $CO(NH_2)_2$, an organic compound whose combustion value accounts for the difference. This difference must be considered if the energy balance of an animal is calculated from combustion values determined by employing a bomb calorimeter.

The amount of oxygen needed to oxidize 1 g fat is more than twice the amount needed for oxidation of carbohydrate or protein (Table 6.1, column b). In combination with the high energy yield from fat, the result is that the energy per liter oxygen is approximately the same for fat as for carbohydrate and protein (see above).

The last column in Table 6.1 shows the ratio between the carbon dioxide formed in metabolism and the oxygen used. This ratio, known as the *respiratory quotient* (RQ) or *respiratory exchange ratio* (R), is an important concept in metabolic physiology. The RQ gives information about the fuel used in metabolism. Usually the RQ is between 0.7 and 1.0. An RQ near 0.7 suggests primarily fat metabolism; an RQ near 1.0 suggests primarily carbohydrate metabolism. For an intermediate RQ, it is more difficult to say what foodstuffs have been metabolized; it could be protein, a mixture of fat and carbohydrate, or a mixture of all three. It is possible, however, to determine the amount of protein metabolized from the nitrogen excretion (urea in mammals, uric acid in birds and reptiles), and from this knowledge we can calculate that fraction of the oxygen consumption and carbon dioxide production that is attributable to protein metabolism. The remainder of the gas exchange is attributable to fat and carbohydrate metabolism, and the amount of each can then be calculated. From determinations of oxygen consumption, carbon dioxide elimination, and nitrogen excretion, it is thus possible to calculate the separate amounts of protein, fat, and carbohydrate metabolized in a given period of time.

Could carbon dioxide production be used for determinations of metabolic rate equally as well as oxygen consumption? In theory the answer is yes, but there are difficulties that make it less practical and in fact far less accurate. There are two main reasons: One is that the body contains a large pool of carbon dioxide; the other has to do with the caloric equivalent of 1 liter of carbon dioxide.

The large pool of carbon dioxide that is always present in the body changes easily. For example, hyperventilation of the lungs causes large amounts of carbon dioxide to leave in the exhaled air. This carbon dioxide, of course, is no measure of the metabolic rate. Hyperventilation is followed by a period of reduced ventilation during which the carbon dioxide again builds up to its normal level, and during this period the reduced amount of carbon dioxide given off again is no measure of metabolic rate. Other circumstances also influence the pool of carbon dioxide. In heavy exercise, for example, lactic acid is formed in the muscles as a normal metabolic product. As the lactic acid enters the blood, it has the same effect as pouring acid on a bicarbonate solution: Carbon dioxide is driven out of the blood and released in the lung in excessive amounts. Again, the amount of carbon dioxide is not a measure of the metabolic rate.

The other reason carbon dioxide is less suitable than oxygen for measuring metabolic rates is that the different foodstuffs give rather different amounts of energy for each liter of carbon dioxide produced. Carbohydrate gives 5.0 kcal per liter

CO_2 formed; fat gives 6.7 kcal per liter CO_2. The value for fat is therefore one-third higher, and unless we know whether fat or carbohydrate is metabolized, the error can be substantial.

For these reasons the carbon dioxide production is not a suitable measure of metabolic rate.

ENERGY STORAGE: FAT AND GLYCOGEN

For most adult animals food intake and energy expenditure remain approximately equal. If energy expenditure exceeds food intake, the excess energy is covered by the consumption of body substance, primarily fat. If food intake exceeds energy used, the surplus is stored as fat, irrespective of the composition of the food. For example, pigs are fattened on feed consisting mostly of carbohydrate (grains). During fattening, if large amounts of carbohydrate are changed into fat, the RQ will exceed 1.0. This is because fats contain relatively less oxygen than do carbohydrates (i.e., some of the "excess" oxygen from the carbohydrate can be used in metabolism, thus reducing the respiratory oxygen uptake and increasing the respiratory carbon dioxide oxygen ratio.

Because fat yields more than twice as much energy as carbohydrate, it is better suited for energy storage. However, there are exceptions to this rule. For animals that do not move about, such as oysters and clams, weight is of minor consequence, and glycogen is used for storage. Also, many intestinal parasites, such as the roundworm Ascaris, store glycogen. In these animals glycogen is probably a more suitable storage substance than fat, for they are frequently exposed to conditions of low or zero oxygen, and in the absence of oxygen, glycogen can yield energy by breakdown to lactic acid. Therefore, as bivalves frequently close their shells for long periods and intestinal parasites live in environments virtually devoid of oxygen, it is advantageous for these animals to store glycogen rather than fat. Sessile animals and parasites have little need for weight economy, and from this viewpoint the use of glycogen instead of fat is no disadvantage.

For animals that move about, however, weight economy is of great importance. It is particularly important for migrating birds, which may fly nonstop for more than 1000 km. To have enough fuel, migrating birds may carry as much as 40 to 50% of their body weight as fat; if the fuel were heavier, the weight would be excessive, and long-distance migration would be impossible because much of the available energy would merely go into carrying the heavier body.

Storage of glycogen involves much more weight, not only because of the lower energy content of carbohydrate compared with fat, but also because glycogen is deposited in the cells with a considerable amount of water. It has been estimated that the deposition of glycogen in the cells of liver and muscle is accompanied by about 3 g water for each gram of glycogen stored. Some investigations even indicate that the amount of water may be as high as 4 or 5 g per gram glycogen (Olsson and Saltin 1968).

The simplest way of expressing the weight of fuel relative to the energy value is to compute the weight per energy unit, or the *isocaloric weight*. For the fuels we have discussed, the isocaloric weights, in grams of fuel per kilocalorie, are as follows:

Fat	0.11
Protein	0.23
Starch	0.24
Glycogen + H_2O	~1.0

Expressed in this way, it is immediately apparent that, with the extra weight of water, glycogen is some 10 times as heavy to carry as the same energy in fat.

In spite of this weight difference, glycogen is an important storage form for energy. Its advantage is twofold: It can provide fuel for carbohydrate metabolism very quickly, whereas the mobilization of fat is slow; and, perhaps more importantly, glycogen can provide energy under anoxic conditions. This is common during heavy muscular exercise when the blood does not deliver sufficient oxygen to meet demands. For long-term storage of large amounts of energy, however, glycogen is unsuited, and fat is the preferred substance.

The fact that water is stored with the glycogen probably explains a puzzling observation made by many persons when they begin reducing diets. During the first few days they lose weight rapidly, often far in excess of what can be attributed to the loss of fat alone. In this initial period the glycogen stores become depleted, and the excess water is excreted. Much of the weight loss is therefore only loss of water. After a few days, even if food intake is less than what is metabolized, the glycogen stores are replenished. When metabolism turns to the use of stored fat, there is no longer a rapid decrease in body weight; on the contrary, body weight may increase again, although the person remains in negative calorie balance and continues to lose fat. These unexpected changes in weight, which are rarely adequately explained, cause many dieters to be overoptimistic during the first days and then rapidly become discouraged when they lose weight less quickly or even gain weight despite their continued sacrifice of food intake.

EFFECT OF OXYGEN CONCENTRATION ON METABOLIC RATE

It is often assumed that oxygen consumption (or metabolic rate) within wide limits is independent of the oxygen concentration. For example, if we replace the ordinary air with pure oxygen, a mam-

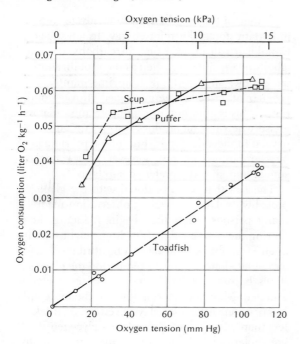

FIGURE 6.1 The oxygen consumption of fish changes with the oxygen tension in the water. For two fish, scup and puffer, the oxygen consumption decreases only moderately until the oxygen tension is quite low; for the third species, the toadfish, the oxygen consumption changes in proportion to the change in oxygen throughout the range. [Hall 1929]

Oxygen tension (kPa)

Scup

Puffer

Toadfish

Oxygen tension (mm Hg)

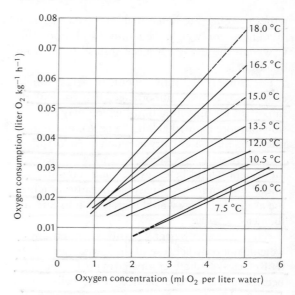

FIGURE 6.2 The oxygen uptake of lobsters is higher at high temperatures, but at any given temperature it varies linearly with the oxygen concentration in the water. [Thomas 1954]

18.0 °C
16.5 °C
15.0 °C
13.5 °C
12.0 °C
10.5 °C
6.0 °C
7.5 °C

Oxygen concentration (ml O_2 per liter water)

mal continues to consume oxygen at the same rate, although oxygen is present in nearly five times the usual concentration. The reverse is also true; if we reduce the oxygen concentration to half, which can be done by reducing the total atmospheric pressure to half (equivalent to about 6000 m altitude), the rate of oxygen consumption likewise remains virtually unaffected.

This independence of oxygen concentration is not universal, the situation may be just the opposite. Figure 6.1 shows how the oxygen uptake of some fish depends on the oxygen concentration in the water. For two fish (scup and puffer) the oxygen uptake decreases by about one-third as oxygen in the water is reduced, but at very low oxygen concentration (less than about 15 mm Hg) both species are unable to survive. The toadfish is

different: Its oxygen consumption decreases linearly with a decrease in available oxygen, giving a straight-line relationship. The toadfish can survive long periods in oxygen-free water, and in this situation its metabolic processes must be completely anaerobic with zero oxygen uptake.

Some invertebrates (e.g., the common lobster) respond in a similar way to changing oxygen (Figure 6.2). At high temperatures the oxygen consumption is higher than at low temperature, but at any one temperature the oxygen uptake is linearly related to the available oxygen.*

* In this book amounts of oxygen are given in volumes (i.e., volume of dry gas at 0 °C and 760 mm Hg pressure). The SI unit for amount of a substance is the mole (symbol: mol), and an amount of oxygen should be given not in liters or milliliters, but in moles and millimoles (mmol). For example, 1 liter O_2 at 0 °C and 760 mm Hg pressure = 44.64 mmol. We are presently in a transition period with increasing use of SI units, but many traditional units are more familiar and will persist for some time to come. Whenever practical, a change to SI units is preferred (see Appendix A).

FIGURE 6.3 At low oxygen concentrations the oxygen consumption of goldfish varies linearly with the oxygen content of the water, but at higher oxygen tension it is indepndent of the oxygen content. [Fry and Hart 1948]

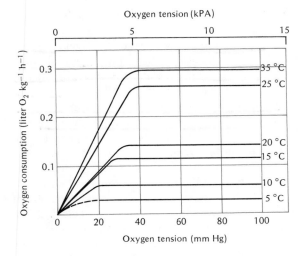

If the oxygen uptake is studied over a wider temperature range in an animal such as the goldfish, we obtain the relationship shown in Figure 6.3. At relatively high oxygen tension the oxygen uptake of goldfish is independent of the available oxygen, but at lower oxygen tensions there is a linear relationship. The inflection point, which indicates where the oxygen uptake changes from concentration dependence to independence, is lower at low temperature than at high temperature. Thus, in Figure 6.3 the inflection point changes from less than 20 mm O_2 at 5 °C to about 40 mm O_2 at 35 °C.

If we return to Figures 6.1 and 6.2, we can see that the straight-line relationships observed for toadfish and for lobster may indicate that we are looking at only the lower part of curves of the same general shape as the goldfish curves. If this is so, we can expect that the curves for toadfish and lobster at some unknown higher oxygen concentration would level off and become independent of further increases in oxygen.

Even mammals, whose metabolic rates show little dependence on available oxygen, may be more dependent on oxygen concentration than we usually assume. The oxygen consumption of isolated muscles of rats and rabbits is reversibly depressed at low oxygen concentrations, without alteration of other cell functions (Whalen 1966). This is of importance to our evaluation of the observation that diving animals, after a dive, do not increase their oxygen consumption enough to equal the oxygen debt they are expected to have acquired during the time of the dive (see later in this chapter). In the swimming animal the muscle mass is the major oxygen-consuming tissue, and an oxygen-dependent metabolic rate in this tissue would yield exactly the result observed in many diving animals. Perhaps an oxygen-independent metabolism as we usually think of it is a special case and oxygen dependence is the general rule.

Acclimation to low oxygen levels

Many fast-swimming, active fish are quite sensitive to low oxygen in the water. This is true of most salmonids – for example, the speckled trout (*Salvelinus fontinalis*). For this fish the lethal level of oxygen in the water depends on previous exposure to low but tolerable oxygen levels. If the fish has been kept in relatively oxygen-poor water, it becomes more tolerant to low levels of oxygen (Figure 6.4). This improved tolerance is probably attributable to an enhancement of the ability to extract oxygen from the water. If there was a true increase in the ability of the tissues to tolerate anoxic conditions, a fish acclimated to low levels of oxygen would be expected to survive longer than a nonacclimated fish when placed in completely anaerobic conditions. There is no evidence of such a difference (Shepard 1955). The improved oxygen extraction could result from increased volume of water pumped over the gills, increased ability to remove oxygen from the water flowing over the

OXYGEN CONCENTRATION AND METABOLIC RATE

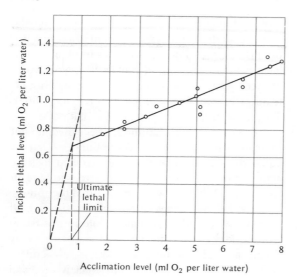

gills, improved oxygen transport capacity in the blood, improved function of the heart, or any combination of these.

Anaerobic metabolism

Animals that tolerate prolonged exposure to zero oxygen must of necessity obtain their metabolic energy from nonoxidative reactions. This holds true for many intestinal parasites, for animals that live in the oxygen-free mud of lakes and ponds, for bivalves that remain closed during long periods, and for others.

One common reaction available for anaerobic metabolism is the breakdown of carbohydrate into lactic acid. For example, 1 mol glucose can anaerobically be broken down to 2 mol lactic acid ($C_6H_{12}O_6 \rightarrow 2\ C_3H_6O_3$). This process, known as glycolysis, occurs commonly in vertebrate muscle when the energy demands are high and exceed the available oxygen. The process, which has a free energy yield (ΔF) of 50 kcal per mole glucose, makes available only about 7% of the energy yield from complete oxidation of glucose (691 kcal per mol).

Most anaerobic glycolysis depends on glycogen, rather than glucose, as the substrate. For glycogen the molar energy is 7 kcal higher than for glucose, and lactate formation yields 57 kcal. Thus, the energy available from glycogen in anaerobic glycolysis is slightly higher than that from glucose. Because ATP is the common energy source for energy-requiring metabolic processes, we can consider, instead of the free energy yields, the amounts of ATP formed in aerobic and anaerobic metabolism, respectively. Glycolysis (formation of 2 mol lactic acid from 1 mol glucose) yields 2 mol ATP. (If glycogen is the substrate, the yield is 3 mol ATP per 2 mol lactic acid.) In contrast, the complete oxidation of 1 mol glucose to carbon dioxide and water yields 36 mol ATP (6 in the formation of 2 mol pyruvic acid, and 30 in the complete oxidation of the pyruvic acid via the Krebs cycle).

If an animal is totally anaerobic, the yield of ATP from glycolysis is small, but an intestinal parasite probably doesn't care much about economy. However, animals that use glycolysis only intermittently can be expected to use lactic acid as a substrate for further oxidation (in the Krebs cycle) and thus in the end gain the full energy value of the original carbohydrate substrate.

The tolerance to anoxic conditions sometimes is amazing. It has been reported that the crucian carp (*Carassius carassius*) can live for 5½ months under the ice of a frozen lake when the water under the ice is oxygen-free because of fermentation of dead plant material and hydrogen sulfide is present (Blazka 1958). (Hydrogen sulfide is poisonous because it binds to and inactivates cytochrome oxidase, but because no oxygen is present to permit oxidative metabolism, this does not poison the fish.) These carp produce practically no lactic acid in the tissues, and other metabolic processes must therefore take place. It was suggested that in these carp the formation of fatty acids is the end result of metabolism, represented as an increase in the total amount of fat in the fish during the long period of anaerobic conditions. However, several other reactions are available for anaerobic metabolism, and several pathways are well known and established for a variety of organisms that can tolerate long periods or permanent absence of free oxygen (Hochachka and Mustafa 1972).

The oxygen minimum layer

Most of the world's oceans contain, at intermediate depth, zones where the oxygen content is very low. There is a large such area in the northeastern tropical Pacific, which is known as the *oxygen minimum layer*. Off the coast of central and southern Mexico the layer extends from approximately 100 to 900 m in depth, and the dissolved oxygen content is generally less than 0.1 cm^3 per liter sea water, with relatively little variation throughout the year (Douglas et al. 1976). Nevertheless, a surprisingly abundant population of animals lives in the oxygen-poor water.

Some of the animals, particularly the fish, make diurnal migrations to the surface, ascending at dusk and returning to depth before sunrise, in vertical movements that amount to 300 m or more. Paradoxically, the fish present in the low oxygen region often have gas-filled swimbladders that contain a high percentage of oxygen. During the day, when the fish are at depth, it seems likely that they must obtain energy by anaerobic pathways. Their blood has a low oxygen capacity and their hemoglobin such a strikingly low affinity for oxygen that no more than 1% of the hemoglobin would be oxygenated while the fish are at depth (Douglas et al. 1976).

Small crustaceans that live in the oxygen minimum layer are very effective at removing oxygen from the water. A study of 28 species from the oxygen minimum layer of southern California found that all the species are able to take up oxygen at least to the lowest P_{O_2} they encounter in nature, in some cases as low as 4 mm Hg. The exceptions are two parasites and one animal that makes daily vertical migrations to oxygen-rich water (Childress 1975).

At the present time, it appears that the existence of a fairly rich fauna in the oxygen minimum layer depends on the utilization of very low oxygen concentrations and, therefore, largely on aerobic metabolism. However, the exploration of the interesting biological problems of the oxygen minimum layer is just beginning, and it is not yet possible to say whether the fauna depends mostly on unique abilities to utilize low oxygen concentrations or on tolerance to long periods of anaerobic metabolism.

The latter option is not unthinkable, for several anaerobic biochemical mechanisms are potentially available. We have become conditioned to thinking of anaerobic glycolysis and lactic acid formation as the typical pathway for anaerobic metabolism because we know it is used in higher vertebrates. However, glycolysis is a fundamentally inefficient metabolic strategy, which makes available only a few percent of the energy obtainable from complete oxidation of the substrate. Many other pathways are available for production of ATP under extended periods of, or permanent, anoxia. The large free amino acid pool of many invertebrates is an important potential source of energy that represents astonishing energetic advantages over lactic acid formation (Hochachka et al. 1973).

PROBLEMS OF DIVING MAMMALS AND BIRDS

All aquatic mammals and birds have retained lungs and breathe air, but successful aquatic life requires a number of modifications in the usual terrestrial pattern. The most obvious one is that lengthy dives require effective utilization of a lim-

TABLE 6.2 Most of the major orders of mammals have a few aquatic representatives, and some are exclusively aquatic.

Mammalian order	Aquatic representative
Monotremata	Platypus (*Ornithorhynchus*)
Marsupialia	None
Insectivora	Water shrew
Chiroptera	None
Primates	None
Rodentia	Beaver, muskrat
Lagomorpha	None
Cetacea	Whales (all)
Carnivora	
Fissipedia	Sea otter
Pinnipedia	Seals (all)
Sirenia	Sea cows, manatee, dugong (all)
Ungulata	
Perissodactyla	Tapir
Artiodactyla	Hippopotamus

TABLE 6.3 Many bird families are almost completely aquatic (except for their reproductive activities); others are primarily or exclusively nonaquatic.

Primarily aquatic	Primarily nonaquatic
Penguins	Ratites (ostrich, rhea, etc.)
Loons	Hawks
Grebes	Fowls and pheasants
Petrels and albatrosses	Pigeons
Pelicans	Parrots
Herons	Owls
Ducks	Nightjars
Waders	Hummingbirds
Gulls	Woodpeckers
Auks	Passeriformes (finches, starlings, crows, etc.)

ited oxygen supply, but other functional modifications contribute importantly.

Adaptation to an aquatic mode of life is found in all major orders of mammals, except the Chiroptera (bats), Lagomorpha (rabbits and hares), primates, and marsupials (Table 6.2). A list of aquatic and nonaquatic birds shows that the successful ones are phylogenetically quite diverse (Table 6.3). However, the physiological adaptations these animals employ to cope with the special demands of their environment are similar. The facts that all divers use common mechanisms, and that to a lesser degree such mechanisms are present in nondiving vertebrates, suggest that the successful divers have taken advantage of and developed already existing mechanisms.

The problems that must be solved by diving animals can readily be explained by reference to the problems encountered by a man who dives, either with or without equipment. Some of man's difficulties have no direct bearing on diving animals; others are directly applicable to them. All air-breathing animals that dive must cope with the problem of oxygen supply, but those that make fairly deep dives encounter additional problems that fall into four main groups:

1. Bends or diver's disease
2. Oxygen toxicity
3. Narcotic effect of gases
4. Direct effects of high pressure

The bends

This dangerous syndrome is known under several names: diver's disease, caisson disease, and aeroembolism. It occurs when a human diver returns to the surface after spending a prolonged time at considerable depth, below 20 m or so, and is increasingly severe with increasing depth and diving time. It is caused by gas bubbles in the tissues and the bloodstream, formed in the same way that bubbles form in a bottle of soda water when the cap is removed. In both cases bubbles appear when the pressure is reduced over a liquid that is saturated with gas at a higher pressure. The bends may also occur during rapid ascent in aircraft (balloons or airplanes) with nonpressurized cabins; in these cases the danger of bends occurs when the pressure is reduced to about 0.5 atm (above 6000 m).

In soda water the dissolved gas is carbon dioxide. The gas causing the bends is always nitrogen (unless the diver has been breathing an artificial gas mixture that contains other inert gases such as helium). Even when the diver is at great depth, the carbon dioxide tension does not increase much above the physiologically normal of about 40 mm Hg, and to cause the bends, supersaturation with a gas at some 2 atm would be necessary. Oxygen does not cause the bends because it is rapidly used up by the tissues, and, as we shall see later, oxygen at a pressure of more than 1 atm is so toxic that higher concentrations must be avoided.

The pressure in water increases by about 1 atm for each 10 m increase in depth. To keep an ordinary diving suit inflated with air at 10 m depth requires 1 atm pressure above normal atmospheric pressure, or a total air pressure of 2 atm. If a diver at 10 m depth is supplied with ordinary air, the air he breathes has 0.4 atm oxygen pressure and 1.6 atm nitrogen pressure. The blood dissolves nitrogen at the higher pressure, and gradually all the tissue water becomes equilibrated with nitrogen at this pressure. Furthermore, the body fat, which in a normal lean male constitutes about 15% of body weight, dissolves about five times as much nitrogen per unit weight as does water. Our standard 70-kg man contains nearly 50 liters water and 10 kg fat; the total amount of nitrogen dissolved in the fat, therefore, about equals the amount dissolved in the body water.

It takes considerable time before a man at a given depth reaches full saturation with nitrogen. During short dives this is an advantage, but after a long dive, when the tissues are nearly saturated with nitrogen at high pressure, the danger increases. When the diver returns to the surface, several hours are required to eliminate the nitrogen,

for it is only slowly carried from the tissues, dissolved in the blood, and released in the lungs.

The danger of bubble formation is greater when gas is dissolved at a higher pressure. Increased movements, muscular effort, and increased circulation tend to increase bubble formation, analogous to the effect of shaking a beer can or a soda-water bottle before opening. Bubbles often form in the joints, which causes considerable pain. If bubbles occur in the bloodstream, they block the finer blood vessels, and when this happens in the central nervous system it is particularly dangerous and may cause rapid death. The only possible treatment for the bends is to increase the pressure quickly so that the bubbles redissolve. This can be done either by returning the diver to the depth from which he ascended or by transferring him to a pressure chamber where the air can be recompressed to the desired level.

The way to prevent the bends from developing is to return a diver slowly to the surface by stages. It is reasonably safe to let a diver ascend to a point where the pressure is half that at which he was working, and keep him there until a considerable amount of the nitrogen has been eliminated, say 20 to 30 minutes. It is then feasible to let him ascend part of the way again and continue the process of staging. Staging tables that give precise schedules for the fastest possible safe ascent from a given depth are widely used.

If a diver uses a tank of compressed air, he must breathe air at the pressure of the surrounding water or his chest would be collapsed by the water pressure. The scuba diver is therefore in the same danger of developing the bends as the diver using the ordinary diving suit.

The danger of bends can to some extent be reduced by using gases other than nitrogen. One widely used gas is helium, which has the advantages of being less soluble than nitrogen in water and fat and of diffusing several times faster. The faster diffusion reduces the time necessary for staging. On the other hand, it also speeds up the saturation of the tissues with helium, and for short dives this counteracts some of the advantages.

Our discussion so far has dealt only with divers supplied with air under pressure. A skin diver without equipment who repeatedly descends to some considerable depth is also in danger of developing bends. Each time he fills his lungs and descends, the water pressure on his chest compresses the rib cage so that the air pressure in the lungs increases. If he descends to 20 m, the air pressure in the lung will be 3 atm. Although one or a few dives to 20 m depth will not be of any significance, many repeated dives are risky.

A well-documented case of a skin diver developing the bends occurred when a Danish physician practiced underwater escape techniques in a 20-m-deep training tank. He used the so-called bottom drop technique, in which the diver after a deep breath pushes himself downward from the edge of the tank. After descending 2 or 3 m the chest is compressed so much that he continues to drop to the bottom with increasing speed. After 5 hours of spending about 2 minutes at the bottom during each of 60 dives, he developed pain in the joints, breathing difficulties, blurred vision, and abdominal pains. He was placed in a recompression chamber at 6 atm pressure, whereupon the symptoms rapidly disappeared. He was then staged to 1 atm pressure, according to the U.S. Navy staging tables, which required 19 hours and 57 minutes. This treatment was completely successful, but many other divers have not been so lucky (Paulev 1965).

If a man develops the bends from performing repeated breath-hold dives, how can seals and whales, which dive over and over again, avoid this danger? Often they come to the surface for seconds only, and they may dive to considerable depths.

The best diver among the seals is the Antarctic Weddell seal, which voluntarily dives to below 600 m and stays under water for as long as 43 minutes (Kooyman 1966). The deepest diving whale is the sperm whale (*Physeter catadon*), for which the known record is 1134 m. This curious record was documented because a trans-Atlantic cable was broken at this depth, and when the cable was taken up for repair, a dead sperm whale had its lower jaw entangled in the cable. The way the jaw was caught showed that the whale had been swimming along the bottom when it hit the cable and had tried to free itself.

Why can seals and whales tolerate deep dives and rapid ascent without developing the bends? There are three possibilities: (1) they are tolerant to bubbles, (2) they have mechanisms to avoid bubble formation in spite of supersaturation, or (3) there is no supersaturation. The information we have, although sparse, points to the third possibility as the likely answer.

The most important difference between seals and whales on the one hand and a man diving with equipment on the other is that the animals do not receive a continuous supply of air. However, as in man, many repeated dives with air-filled lungs should eventually produce the bends. In this regard it is significant that a seal exhales at the beginning of each dive, rather than diving with filled lungs as a man usually does. Whether a whale also exhales before diving is uncertain, but it has an exceptionally large and wide trachea. This may be an adaptation to enable the whale to ventilate its lungs rapidly when it briefly surfaces to breathe.

Equally or more important, the lungs are small relative to the large volume of the upper airways

(Scholander 1940). As a whale dives, the increasing water pressure compresses the lungs and forces air into the large trachea, which is rigidly reinforced with circular bone rings. If the volume of the noncompressible trachea is one-tenth the lung volume, the lung will be completely collapsed and all the air forced into the trachea at 100 m depth. With no air in the lung, nitrogen cannot enter the blood, as it does in man. Another reason nitrogen does not enter the blood is that the blood flow to the lung is minimal during the dive; thus, even if the lung contains some air before the animal reaches the depth where the lung is collapsed, little nitrogen enters the lung.

Oxygen toxicity

Pure oxygen at 1 atm pressure is harmful to most warm-blooded animals. A man can safely breathe 100% O_2 for 12 hours, but after 24 hours he feels substernal distress and increasing irritation of the lungs. Rats kept in pure oxygen die with symptoms of severe lung irritation after a few days. If the oxygen pressure is higher than 1 atm, nervous symptoms develop before the general irritation of the respiratory organs sets in. Continued breathing of oxygen at 2 atm pressure causes convulsions, and exercise reduces the tolerance. A pressure of 3 atm oxygen can be tolerated by man at most for a few hours, and longer exposure must be avoided.

The importance of oxygen toxicity to diving is obvious. If a diver descends to 40 m and is supplied with compressed air, he breathes at a total pressure of 5 atm. Because one-fifth of the air is oxygen, the partial pressure of oxygen is 1 atm (i.e., near the toxic limit). Dives with compressed air at even greater depths are, of course, increasingly harmful as the oxygen partial pressure exceeds 1 atm.

The only practical way for divers working at 40 m or more to avoid oxygen toxicity is to decrease the oxygen content of the air they breathe. A diver working at 40 m (5 atm) could be supplied with a gas containing 10% O_2 and 90% N_2. He would then breathe oxygen at a partial pressure of 0.5 atm, which is completely safe. At greater depth the oxygen content of the gas mixture should be further reduced, and to reduce the danger of bends during the decompression, nitrogen may be partly or completely replaced by other gases such as helium.

Narcotic effects of inert gases

The depth to which a man can dive safely is also limited by the direct physiological effects of inert gases. Nitrogen at several atmospheres pressure has a narcotic effect, similar to that of nitrous oxide (laughing gas), but nitrogen requires higher pressures.

The narcotic effect takes some time to develop, but at a depth of about 100 m it is so severe that nitrogen-oxygen mixtures cannot be used. It is therefore necessary to use other gases, helium again being favored. Hydrogen is another possibility, although much more hazardous to handle because of the danger of explosions. Eventually, helium under high pressure also has serious physiological side effects, thus limiting the depth to which it is possible to dive, even for the well-equipped man, unless he is within a completely rigid structure, such as a reinforced rigid diving suit or a bathyscaphe, in which a low pressure can be maintained.

Diving animals are not subject to the narcotic effects of inert gases (or to oxygen toxicity) for the simple reason that they do not breathe a continuous supply of air during the dive.

Oxygen supply during diving

For a man who swims under water without equipment, the most immediate need is to obtain

TABLE 6.4 Possible solutions for extending underwater time for diving animals. Solutions in brackets are not widely used or are improbable or physiologically impossible.

Increased oxygen storage
  [Increased lung size]
  Blood: higher blood volume and higher hemoglobin content
  Tissues: higher muscle hemoglobin
  [Unknown oxygen storage]

Use of anaerobic processes
  Lactic acid formation
  [Unknown hydrogen acceptors]

Decreased oxygen consumption
  Decreased metabolic rate

Aquatic respiration
  Cutaneous respiration (frog, sea snakes)
  Esophageal or rectal respiration (some turtles)
  [Breathing water (only tried experimentally)]

TABLE 6.5 Oxygen stores in seal and man. The figures are approximate (e.g., blood oxygen content will be lower than listed because a large part of the blood is located in the venous system and not fully saturated with oxygen). [Data for seal from Scholander 1940]

Location of oxygen	Amount of oxygen (ml)
Seal (30 kg)	
Air in lungs (350 ml, 16% O_2)	55
Blood (4.5 liters, 25 ml O_2 per 100 ml)	1125
Muscle (6 kg, 4.5 ml O_2 per 100 g)	270
Tissue water (20 liters, 5 ml O_2 per liter)	100
Total	1550
Milliliter O_2 per kilogram body weight	52
Man (70 kg)	
Air in lungs (4.5 liters, 16% O_2)	720
Blood (5 liters, 20 ml O_2 per 100 ml)	1000
Muscle (16 kg, 1.5 ml O_2 per 100 g)	240
Tissue water (40 liters, 5 ml O_2 per liter)	200
Total	2160
Milliliter O_2 per kilogram body weight	31

oxygen. He usually fills his lungs to increase the amount of air he carries with him, postponing the moment he must surface to breathe again. There are, however, many theoretical possibilities for extending diving time (Table 6.4). Some of these are physiologically not feasible or are not known to be used by animals. Most diving animals have adopted a combination of several possible methods, often resulting in diving performances that are striking when measured against human standards.

Oxygen storage

Increasing the amount of oxygen in the lung seems a simple solution to the problem of carrying more oxygen during the dive. However, diving animals have no greater lung volumes than other mammals, and many of the best divers have rather small lung volumes. A moment's thought helps to explain why this is so. First, a large amount of air in the lung may permit more nitrogen to invade the body fluids and thus increase the hazards of the bends. Furthermore, if the lungs are particularly large, an animal will have difficulty in initially

becoming submerged. This last argument may be a minor matter, however, for once an animal with large lungs is successfully submerged, compression of the lungs decreases the buoyancy and there would be no difficulty in remaining at depth.

Let us examine what other oxygen supplies are available to a seal at the beginning of a dive and how long this oxygen may last. Such an account has been given by the Norwegian physiologist, P. F. Scholander, who compared the oxygen stores in a small seal and a man (Table 6.5).

The most striking difference is the small amount of oxygen the seal carries in its lungs. Instead of filling the lungs at the beginning of a dive, seals seem to exhale and begin the dive with a minimal volume of air in the lungs. This is the reason for the very small amount of oxygen in the lungs of the seal compared with man, who fills his lung with air when he dives.

The amount of oxygen in the blood is a very different matter. Although the seal weighs less than half as much as the man, its total blood volume is almost as great. Furthermore, seal blood has a higher oxygen-carrying capacity than human blood, and as a result the total amount of oxygen in the seal's blood exceeds that in man's.

A high oxygen-carrying capacity of the blood is characteristic of good divers. Although the oxygen capacity of human blood (20 ml O_2 per 100 ml blood) is in the higher range among terrestrial mammals, seals and whales have oxygen capacities ranging up to between 30 and 40 ml per 100 ml blood. There is a limit to how much the oxygen capacity can be increased, however. The red blood cells are already about as densely packed with hemoglobin as possible, and if the red cells make up more than about 60% of the total blood volume the blood cannot flow freely; it becomes so viscous that the heart must work excessively to pump the blood through the vascular system.

Increasing the blood volume is one way of increasing the total amount of hemoglobin in the body without increasing blood viscosity. This solution is common among successful divers, but again, there are limits to how much blood it is feasible to have in the body.

The second largest amount of oxygen in the seal is stored in the muscle mass. The red color of mammalian muscle is caused by muscle hemoglobin (myoglobin). It is located within the muscle cell, and therefore remains even after all blood has been drained out. The hemoglobin content of muscle is much lower than of blood, but because of the large muscle mass of the body, about 20% of the body weight,* the muscle hemoglobin carries a substantial amount of oxygen. Seal and whale muscle has a much higher hemoglobin content

* This figure is low because of the large amount of fat in the seal's body. The muscles make up a much larger fraction of the fat-free body mass.

than muscle of terrestrial mammals (their meat is visibly darker than other mammalian muscle), and as a result the 30-kg seal has in its muscles about the same amount of stored oxygen as the 70-kg man.

All body water contains some dissolved oxygen. The amount depends on the water content of the body, which is similar in man and seal, but the total amount is small. Some oxygen is dissolved in the body fat, but it is unimportant as an oxygen store because fat is poorly vascularized and probably excluded from circulation during diving.

Adding up all the oxygen we can account for in Table 6.5 shows that a seal may have, per kilogram body weight, substantially more oxygen than a man, although not quite twice as much.

The stored oxygen, if used up at the normal metabolic rate of the seal at rest, would last at most 5 minutes. However, the seal is able to dive for at least three times as long. The total oxygen store in man would be sufficient to supply his metabolic rate at rest for perhaps 4 minutes, but this exceeds the maximum diving time for man (about 2 minutes, except for unusual record performances). The reason is that a man cannot tolerate full depletion of all free oxygen in the body, and the maximum diving time is therefore shorter than that calculated from the total oxygen stores. Why, then, can a seal dive for much longer?

We cannot a priori exclude some unknown oxygen store in the seal, but on the basis of present knowledge, it seems most unlikely. What we know about diving mammals indicates that the performance of even the best divers can be fully accounted for without invoking new and so far unknown mechanisms.

Circulation changes

One of the most characteristic physiological responses observed in a diving seal is that the heart

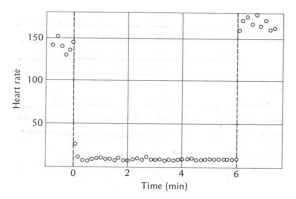

FIGURE 6.5 When a harbor seal is submerged, its heart rate drops precipitously from about 140 to less than 10 beats per minute. Beginning and end of dive are indicated by vertical broken lines. [Elsner 1965]

rate at the beginning of submersion decreases precipitously to one-tenth the normal value or even less (Figure 6.5). This sudden drop in heart rate must be caused by a nerve reflex, for if it were a response to a gradual depletion of oxygen it would develop more slowly.

In spite of the decrease in heart rate, the arterial pressure is maintained. This can happen only if the peripheral resistance in the vascular system is increased through vasoconstriction. With the reduction in cardiac output, only the most vital organs receive blood; these are the central nervous system and the heart itself. There is a complete redistribution of blood flow, and muscles as well as abdominal organs receive little or no blood during the dive. In a way this makes the seal into a much smaller animal: The oxygen is reserved for a small part of the body – the organs for which oxygen is most essential. The oxygen present in the circulating blood at the beginning of the dive is therefore used up only slowly, and the duration of a dive approximately coincides with the time needed to deplete fully the oxygen in the blood.

More accurate information about the distribution of blood flow during a dive has been obtained by injection of a radioactive isotope of rubidium, which rapidly exchanges with potassium inside the cells and thus gives an estimate of the amount of blood reaching an organ. Figure 6.6 shows the result of an experiment on a duck. During the dive the circulation is increased especially to head, heart, and eyes, and the greatest decrease is for the kidneys, which in the nondiving state receive a large fraction of the total blood supply (Johansen 1964).

Most muscles, whether used vigorously or not, receive virtually no blood during the dive. The work of the muscles is based entirely on anaerobic processes, which lead to the formation of lactic acid in the muscles. Because there is no blood flow to the muscles, this lactic acid accumulates in the muscle during the dive, and lactate concentration in the blood remains low. Immediately after the dive, when circulation to the muscles is resumed, the blood lactic acid concentration shoots up. (The ability of seal muscle to work anaerobically is not unique; lactic acid is normally formed in the muscles of man during strenuous exercise or athletic performances.)

The muscle hemoglobin in a diving seal provides some oxygen, but this oxygen is depleted early in the dive within a period much shorter than the normal duration of a dive. The amount of oxygen stored in the muscle hemoglobin is much too small to explain the long duration of a dive; the circulatory adjustments and the ability of muscle to work anaerobically are the major factors.

Slowing of the heartbeat during diving has been recorded in a large number of mammals (e.g., seals, whales, beavers, muskrats, hippopotamuses, ducks). In some of these animals the heart slows more gradually than is shown in Figure 6.5, but a substantial drop seems to occur in all diving mammals.

Nevertheless, we cannot consider diving bradycardia as an obligatory physiological reaction, for

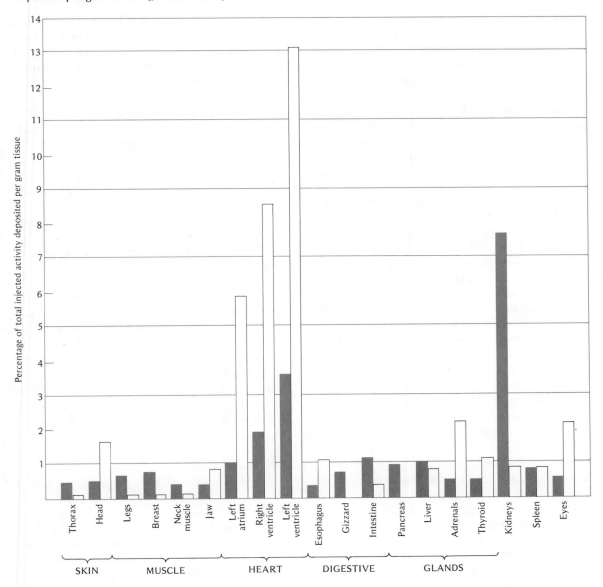

FIGURE 6.6 The circulation to various organs of a diving duck. Black bars indicate values during normal breathing of air; cross-hatched bars show values obtained during submersion. The circulation was determined with the aid of radioactive rubidium; the height of a bar indicates the percentage of the total injected amount deposited per gram tissue. [Johansen 1964]

177

many observations made during forced submersion of the experimental animal need reevaluation in the light of newer information.

Experimental observations on the alligator indicate that submersion causes a drop in the heart rate from more than 40 to 3 beats per minute. However, if the signals are transmitted from the animal to the recording apparatus by radio telemetry, we can obtain information from completely undisturbed animals, free of restraints and human presence. In telemetric studies of the caiman (*Caiman crocodilus*) the animals showed typical diving bradycardia when handled, with drops from 28 to 5 beats per minute. However, in the absence of humans, the heart rate of resting animals gradually dropped over a period of several hours to 10 to 12 beats per minute. When such animals submerged voluntarily, their heart rates dropped by only 1 or 2 beats per minute, or not at all. If the investigator entered the room, on the other hand, the animals dived and their heart rates dropped to as little as 5 beats per minute (Gaunt and Gans 1969). These results have been confirmed on alligators (*Alligator mississippiensis*) roaming free in a lake; voluntary short dives did not result in any reduction in heart rate (Smith et al. 1974). It is therefore important to make observations on animals under conditions that assure a minimum of external disturbance.

Interestingly, a reverse "diving" bradycardia has been observed in fish. When a fish is taken out of water, it cannot obtain enough oxygen, and its heart rate immediately decreases. This has been observed in both teleosts and elasmobranchs. An air-breathing fish, such as the mud skipper (*Periophthalmus*) (see Chapter 2), differs from fully aquatic fish in that its heart rate decreases when it escapes from air into its water-filled burrow. In this respect the mud skipper reacts as a terrestrial air breather which performs a dive (Gordon et al. 1969).

A record of respiration and blood gases in a diving seal is presented in Figure 6.7. The top graph shows that the blood oxygen gradually is depleted during the dive, while the carbon dioxide correspondingly increases. The blood lactic acid changes only a little during the dive, but after the dive a wave of lactic acid appears in the blood as soon as air breathing is resumed. This lactic acid, which signifies an oxygen debt, is afterward gradually removed from the blood. Part of the lactic acid is resynthesized into glycogen by the liver and muscles, and a smaller fraction is completely oxidized to carbon dioxide and water.

The center graph (Figure 6.7*b*) shows the ventilation volume, which during the dive is zero, followed by increased ventilation of the lung immediately after the dive.

The lower graph (Figure 6.7*c*) shows the corresponding respiratory oxygen uptake, which, immediately after the dive, shows a sharp increase. Most of the time carbon dioxide elimination is of the same magnitude as oxygen uptake, although a little lower. But immediately after the dive, carbon dioxide output is greatly increased because the surge of lactic acid in the blood drives out carbon dioxide from the bicarbonate in the blood, as vinegar poured on baking soda drives off the carbon dioxide.

Decreased metabolic rate.

Figure 6.7*c* shows that the oxygen debt acquired during the dive is followed by an increased oxygen uptake. The expected oxygen debt is represented by the shaded area (which indicates the normal metabolic rate of the seal). The increased oxygen uptake after the dive indicates that the oxygen debt acquired during the dive is being paid off, and the after-dive increase should equal the debt. If the area under the after-dive oxygen uptake curve is compared with the oxygen debt, however, the

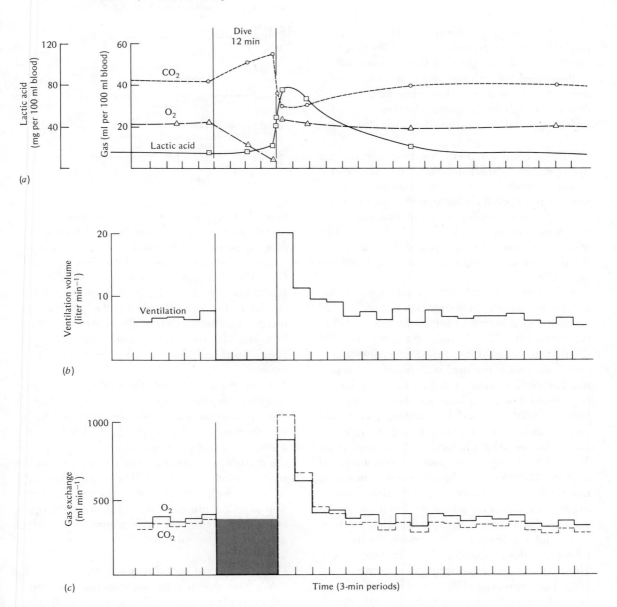

FIGURE 6.7 Record of respiratory exchange in a 29-kg seal during a 12-minute experimental dive: *a*, blood concentrations of oxygen, carbon dioxide, and lactic acid; *b*, ventilation volume; *c*, respiratory oxygen uptake and carbon dioxide release. [Scholander 1940]

179

amount repaid seems smaller. This could be caused by a long extension in time of the processes used to repay the oxygen debt that cannot easily be distinguished from normal small variations in resting metabolic rate. Or, there may be a real decrease in metabolic rate during the dive, so that the true oxygen debt is smaller than suggested by the graph.

Further comparisons of the after-dive oxygen uptake and the expected oxygen debt show that a reduced metabolic rate is the probable explanation. In the duck the observed after-dive increase in oxygen consumption is about one-quarter or one-third of the expected oxygen debt; this can be observed in consecutively repeated dives, although the duck shows no signs of an accumulating oxygen debt.

A decrease in metabolism to values below the normal resting rates is unexpected, for we are accustomed to thinking of metabolic rate as being very constant. This is not necessarily so. We should remember that during the dive many organs are poorly perfused with blood or not perfused at all. The kidneys, which normally have a high blood flow and a high oxygen consumption, have virtually no blood flow during a dive, and there is an immediate cessation of glomerular filtration and of urine production (Murdaugh et al. 1961). Other organs are also virtually excluded from circulation during diving, and we can except that their activity practically ceases, thus decreasing the amount of oxygen needed to repay an oxygen needed to repay an oxygen debt after the dive.

It is not easy to measure accurately the *true* metabolic rate during the dive of a mammal or a bird. However, aquatic turtles are highly tolerant to prolonged submersion and decreased oxygen and can be used for studies for the metabolic rate during diving. Oxygen consumption is obviously not a measure of metabolic rate under anaerobic condi-

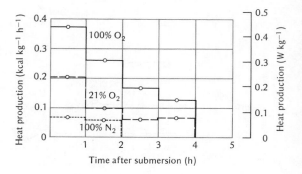

FIGURE 6.8 A diving turtle has a true metabolic rate which is indicated by its heat production. During the dive, the metabolic rate of a turtle that has been breathing air gradually decreases to a level of about 0.06 kcal per kilogram body weight per hour, the same level as when the turtle has been breathing pure nitrogen before the dive. [Jackson and Schmidt-Nielsen 1966]

tions, but heat production accurately reflects the metabolic rate. In a diving turtle the metabolic rate, measured as heat production, gradually decreases, as shown by the dashed curve in Figure 6.8. After 2 hours' submersion, metabolic heat production has reached a minimum value that is similar to the level obtained if the turtle is submerged after breathing nitrogen. This heat production indicates the level of totally anaerobic processes sustained in the complete absence of oxygen.

If, on the other hand, the turtle is given pure oxygen to breathe before a dive, its heat production during submersion is initially very high and gradually decreases as the oxygen is depleted. In the turtle, then, metabolic rate is clearly dependent on the available oxygen and gradually decreases as the oxygen supply is depleted. These results are fully in accord with the fact that the oxygen consumption of resting muscle is not constant, but depends on blood flow (i.e., the oxygen supply) (Whalen et al. 1973).

Cutaneous and rectal respiration

We have already seen how important cutaneous respiration is in frogs and how lungless salamanders obtain all their oxygen through the skin (Chapter 2). There is no reason to believe that a diving bird or mammal uses a similar route to cover any substantial part of its need for oxygen. For one thing, these are large animals, and their relative surface area is therefore small. Further, they are warm-blooded animals and their need for oxygen is high. Finally, the nature of their skin is very different from the thin, moist, highly vascularized integument of amphibians.

Diving reptiles may be a different matter. Aquatic turtles have a thick and probably impermeable carapace, but they may use the oral mucosa as an aid in gas exchange when they remain submerged for long periods of time. Some turtles take water into their rectum, but how important this is for their gas exchange is uncertain; it is more probable that they use the water to adjust their buoyancy, as a submarine takes water into its ballast tanks.

The marine turtles are, of course, excellent swimmers and divers. Nevertheless, it was quite a surprise when it was reported that green turtles (*Chelonia mydas*) can overwinter, submerged and in a dormant state, at the bottom of the Gulf of California (Felger et al. 1976). The turtles are found at a depth of 10 to 15 m or more and may remain immobile for several months. Once fishermen have discovered the existence of these turtles and how easily they can be located with modern gear, the inevitable result will be overfishing and rapid depletion of these unique animals. Whether the green turtle, while submerged, depends on anaerobic processes or can obtain enough oxygen for its low needs during dormancy is a subject that should be studied.

Sea snakes (*Pelamis platurus*), which prey on small fish and dive to depths of perhaps 20 m, remain for long periods of time floating at the surface, but with the head submerged while looking for prey. In this animal up to one-third of the oxygen uptake can take place through the skin, and over 90% of the carbon dioxide leaves via this route. This not only prolongs the time the snake can remain immobile with the head submerged, but also is useful when the snake subdues and swallows large prey, a process that may require up to 20 minutes and prevents air respiration (Graham 1974).

Liquid breathing

What would happen if a terrestrial animal attempted to breathe water? We know that for man this is impossible, he drowns. We also know that 1 liter of water contains only a small fraction of the amount of oxygen contained in 1 liter of air. It is therefore reasonable to ask what would happen if we were to increase the amount of oxygen dissolved in water, which can be achieved by equilibrating the water with oxygen under high pressure.

This has been tried. If water is equilibrated with pure oxygen at 8 atm pressure, it will dissolve about 200 ml O_2 per liter (at 37 °C); in other words, 1 liter of this water will contain about the same amount of oxygen as 1 liter of ordinary air. Another measure must also be taken. Instead of using water to breathe, we must use a balanced salt solution similar to blood plasma. If just water were used, rapid osmotic uptake of water into the blood in the lung would occur and cause hemolysis of the red corpuscles and serious loss of salts. In fact, the actual cause of death in fresh-water drowning is usually this effect of inhaled water. For this reason, artificial respiration is often ineffective in saving victims of fresh-water drowning and more successful in cases of salt-water drowning.

Complete submersion in oxygen-supersaturated water has been tried with success with both mice and dogs, and the animals have survived for several hours. Mice introduced into a pressure chamber and submerged in water supersaturated with oxygen first try to reach the surface. But if they are kept under, they inhale the oxygenated liquid and instead of drowning continue to breathe the oxygenated saline solution.

A major drawback to liquid breathing is that water is approximately 50 times more viscous than air, and the work of breathing is correspondingly increased. Another problem is that the normal surfactants that line the lungs are washed away during liquid breathing; this causes no difficulty during the experiment, but after return to air breathing the lungs tend to collapse.

A more important limitation on liquid breathing is caused by the need to eliminate carbon dioxide at the same rate as oxygen is taken up. If the P_{CO_2} of blood is to be maintained at the normal 40 mm Hg, several times as much liquid must be moved in and out of the lungs. Animals used in experiments with liquid breathing show this quite clearly. Although the arterial blood is fully saturated with oxygen, the carbon dioxide concentration increases greatly. This problem will be very difficult to overcome if an aqueous solution is used (Kylstra 1968).

Another approach to liquid breathing involves the use of certain synthetic liquids, such as fluorocarbons, instead of water. Oxygen is extremely soluble in some of these organic liquids and this alleviates the problem of saturation at high oxygen pressure. The solubility of carbon dioxide, however, is not as great, and this will cause difficulties in the practical application of fluorocarbons. Such liquids have been tried with animals, which have survived several hours' breathing without complications. For man, however, the scheme does not seem promising for practical applications.

Liquid breathing has one potential advantage that under certain very limited circumstances could be of value. Because only liquid and no gas is inhaled, a carrier gas such as nitrogen or helium is not needed. Theoretically, this could allow liquid breathing at very high pressures without supersaturating the body fluids with an inert gas, and the danger of bends could therefore in theory be avoided. The practical use of this possible advantage should be very limited.

Effects of high pressure

Human divers can operate fairly routinely down to about 200 m depth (20 atm), provided they use special gas mixtures such as helium − oxygen. Experimental dives are being carried out down to about 600 m, and this range may be extended by the use of other gas mixtures, including hydrogen. We have discussed the long periods of decompression necessary to avoid the bends and the narcotic and other neurological effects of gases that are inert at moderate pressures. In addition, it is possible that high pressure per se may have deleterious effects, unrelated to the effects of the gases. For an organism breathing air, there is no simple way to distinguish direct pressure effects from the effects of the gases, for the two variables cannot be varied independently. The question will therefore not be simple to resolve.

How important is the problem of high pressures for diving seals and whales, which, as we have seen, may attain depths exceeding 600 m (Weddell seal) and 1100 m (sperm whale)? Based on what we know, these animals do not suffer from bends because their tissues do not become heavily supersaturated with nitrogen. However, chemists and biochemists are well aware that chemical reactions and the structure of complex molecules are greatly influenced by pressure, and it was earlier believed that the immense pressures in the deeper parts of the oceans precluded all animal life. As exploration of the ocean bottom has proceeded, animal life has been found at even the greatest depths, in excess of 10 000 m (1000 atm).

If shallow-water animals are exposed to high pressures under otherwise normal conditions of temperature and oxygen supply, responses vary greatly from species to species. Many are stimulated to increased oxygen consumption, and often increased activity, by pressures of 50 or 100 atm. Higher pressures, of a few hundred atmospheres, often are inhibitory and may cause death (MacDonald 1975). It is quite conceivable that seals and whales are unaffected by hydrostatic pressures per se, for the deepest recorded dives are exceptional and probably rare events. When human divers operate close to the present limit, around 60 atm, it is possible that direct pressure effects may become important.

Many of the basic effects of pressure on biological systems can be referred to the effects of pressure on protein structure and to the effect on the ionization of weak acids. Increases in pressure tend to favor the dissociation of weak acids and bases, and pressure also affects the velocity constants of chemical reactions.

The biologically most important effect on proteins is probably on their tertiary structure – the way in which a protein molecule folds upon itself, with weak bonds and other weak interactions providing stabilization. Pressure has a considerable effect on the tertiary structure, which in turn is responsible for many of the biologically important characteristics of protein molecules, enzymes, contractile proteins of muscle, structural proteins in membranes, and so on (Somero and Hochachka 1976).

In the intact organism these effects are likely to be expressed as derangements in nerve transmission and the nervous system, muscle contraction, oxygen transport in the blood, membrane function and ion transport mechanisms. The complex biochemical processes of gene expression and transcription through the synthesis of messenger RNA are extremely pressure-sensitive, and embryonic development will therefore be seriously affected.

Organisms that live at abyssal depth in the oceans are physiologically adapted to function normally at the high pressures at which they normally live. The functional adaptations of their component proteins, enzymes, and other systems are currently the subject of intensive studies. The results begin to indicate which types of biochemical adaptations are prerequisite for life at these high pressures (Hochachka 1975).

METABOLIC RATE AND BODY SIZE

Mammals

Among the diving mammals we discussed before, the large whales can remain submerged for a couple of hours or so, but a very small diver, such as the water shrew, makes dives that rarely exceed half a minute. On the whole, the larger a diving mammal, the longer lasting are the dives it can perform. Why this difference? The answer is that the rate of oxygen consumption per gram body mass is much higher in the small mammal than in the larger mammal. Let us now compare small and large animals in this regard.

The rates of oxygen consumption of a variety of mammals are listed in Table 6.6. The largest animal, the elephant, is a million times larger than the smallest, the shrew, and its total oxygen consumption obviously must be much higher. However, we do not get a fair assessment of the two animals by comparing their total oxygen consumption. If instead we calculate the rate of oxygen consumption per unit body mass, the specific oxygen consumption* (last column), we get a striking pic-

*The word *specific* used before the name of a physical quantity means "divided by mass." Thus, *specific oxygen consumption* means "oxygen consumption per unit mass."

TABLE 6.6 Observed rates of oxygen consumption in mammals of various body sizes.

Animal	Body mass, M_b (kg)	Total O_2 consumption, $\dot{V}_{O_2}$ (liter O_2 h^{-1})	O_2 consumption per kilogram, $\dot{V}_{O_2}/M_b$ (liter O_2 kg^{-1} h^{-1})
Shrew[a]	0.004 8	0.035 5	7.40
Harvest mouse[b]	0.009 0	0.022 5	2.50
Kangaroo mouse[c]	0.015 2	0.027 3	1.80
Mouse[d]	0.025	0.041	1.65
Ground squirrel[e]	0.096	0.09	1.03
Rat[d]	0.290	0.25	0.87
Cat[d]	2.5	1.70	0.68
Dog[d]	11.7	3.87	0.33
Sheep[d]	42.7	9.59	0.22
Man[d]	70	14.76	0.21
Horse[d]	650	71.10	0.11
Elephant[d]	3 833	268.00	0.07

[a] Hawkins et al. (1960). [b] Pearson (1960). [c] Bartholomew and MacMillen (1961). [d] Brody (1945). [e] Hudson (1962).

ture of the relationship between body size and oxygen consumption.

We see that the rate of oxygen consumption per gram decreases consistently with increasing body size. This is even more apparent if the data are plotted graphically (Fig. 6.9). In this graph the abscissa has a logarithmic scale; otherwise all values for the medium-size to small animals would be crowded together at the left part of the graph. The ordinate, on the other hand, has a linear scale, and we can see that 1 g of shrew tissue consumes oxygen at a rate some 100-fold as great as 1 g elephant tissue. This tremendous increase in oxygen consumption of the small animal necessitates that the oxygen supply, and hence the blood flow, to 1 g tissue be about 100 times greater in the shrew than in the elephant. Other physiological variables – heart function, respiration, food intake, and so on – must be similarly affected.

If we plot the data from Table 6.6 on a graph with a logarithmic scale on both coordinates, the points fall more or less along a single straight line (Fig. 6.10). This regression line represents the generalization that the oxygen consumption of mammals per unit body mass decreases regularly with increasing body size; it also gives a quantitative expression of the magnitude of the decrease.

We can use this graph to read off the "expected" oxygen consumption for a mammal of any given size. For example, a typical mammal of 1 kg size can be expected to have an oxygen consumption of about 0.7 liter O_2 per hour. If we find that a mink which weighs 1 kg has a rate of oxygen consumption twice as high as indicated for a 1-kg mammal on the graph, we say that the mink has a higher oxygen consumption than expected for a typical mammal of its size.

Instead of using the graph, the regression line can be represented by the equation:

$$\dot{V}_{O_2}/M_b = 0.676 \times M_b^{-0.25}$$

in which $\dot{V}_{O_2}/M_b$ is the specific oxygen consumption in liters O_2 per kilogram body mass per hour (liter O_2 kg^{-1} h^{-1}) and M_b is body mass in kilograms. This general equation is based on numerous observations on a much greater variety of mammals than is listed in Table 6.6 and ranging

FIGURE 6.9 Observed rates of specific oxygen consumption of various mammals. The oxygen consumption per unit body mass increases rapidly with decreasing body size. Note that the abscissa has a logarithmic scale and the ordinate an arithmetic scale.

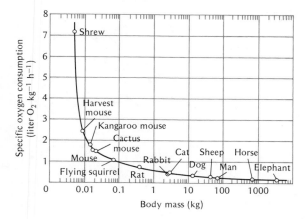

FIGURE 6.10 The rates of oxygen consumption of various mammals, when calculated per unit body mass and plotted against body mass on logarithmic coordinates, tend to fall along a straight regression line with a slope of −0.25. [Data from Table 6.6]

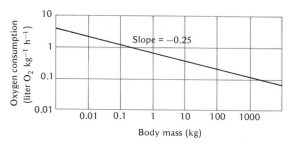

in size from a few grams to several tons. Observations on a single species may deviate more or less from the line, which only represents the best fitting straight line calculated from all available data. Various authors have arrived at slightly different numerical constants, (e.g., an exponent or slope of −0.27), but these differences are too small to be statistically valid (Kleiber 1961).

It is easier to use the above equation for calculations if it is written in the logarithmic form:

$$\log \dot{V}_{O_2}/M_b = \log 0.676 - 0.25\, M_b$$

This equation is of the general form $y = b + ax$, which represents a straight line with the slope a. In the metabolic equation above the slope is negative ($a = -0.25$) (see also Appendix C).

If we return to Table 6.6 and consider the oxygen consumption of the entire animal (column 2) we find, of course, that a whole elephant consumes more oxygen than a mouse. If the data are plotted on logarithmic coordinates, we obtain a straight line with a positive slope of 0.75. The equation for the regression line is:

$$\dot{V}_{O_2} = 0.676\, M_b^{0.75}$$

This equation can be derived from the preceding equation by multiplying both sides by body mass ($M_b^{1.0}$). For arithmetic calculations, it is again easier to use the equation in its logarithmic form, $\log \dot{V}_{O_2} = \log 0.676 + 0.75 \log M_b$, unless an electronic calculator is available.

Data that relate oxygen consumption to body size have been compiled for all sorts of animals, both vertebrate and invertebrate. Oxygen consumption is mostly lower for the cold-blooded than for the warm-blooded vertebrates, but the data again tend to fall on straight lines with slopes of about 0.75 (Figure 6.11).* Many invertebrate animals also have rates of oxygen consumption that fall on the same or similar regression lines, although there are some exceptions to this general

---

*Although the metabolic rate of an organism is most often measured as the rate of oxygen consumption, many authors recalculate their observations to heat production, most frequently by equating 1 liter oxygen (STP) to 4.8 kcal. The metabolic rates shown in Figures 6.11, 6.12, and 6.13 carry on the left ordinate the notation in kilocalories per hour and on the right ordinate the corresponding scale in liters of oxygen per hour.

The SI unit for energy (in this case heat) is the joule (J), and 1 cal = 4.184 J. A heat production of 1 kcal h⁻¹= 1.1622 J s⁻¹ or (because 1 J s⁻¹= 1 W) 1.1622 W. A metabolic rate of 1 liter O_2 per hour, if we assume 4.8 kcal per liter O_2, equals 5.5787 W. For further use of conversion factors, see Appendix A.

FIGURE 6.11 The rates of oxygen consumption for a wide variety of organisms when plotted against body mass (log coordinates) tend to fall along regression lines with a slope of 0.75. Note that each division on the coordinates signifies a 1000-fold change. [Hemmingsen 1960]

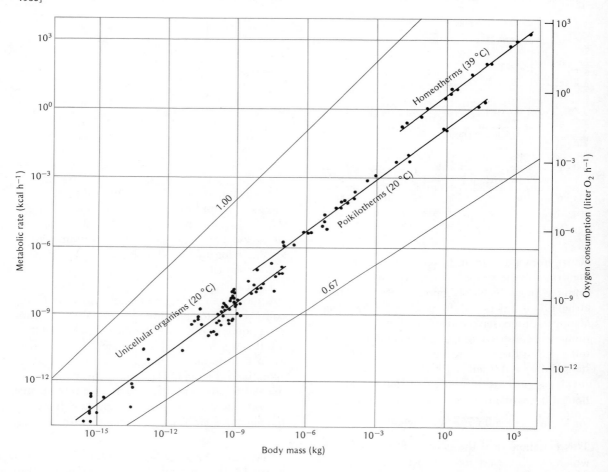

rule. For example, some insects, pulmonate snails, and a few other groups have oxygen consumption rates that fall on regression lines with a slope closer to 1.0. A slope of 1.0 means that the rate of oxygen consumption is directly proportional to the body mass, (i.e., an animal twice as big consumes twice as much oxygen, etc.)

The oxygen consumption rates of microorganisms also fall on lines with a similar slope, and even some trees show the same relationship

between oxygen consumption and size (Hemmingsen 1960). The fact that oxygen consumption bears the same relationship to size in so many different organisms suggests that the phenomenon represents a general biological rule.

Several physiologists who have worked with these problems have attempted to reach a rational explanation for the regular relationship between oxygen consumption, or metabolic rate, and body size.

BODY SIZE This high-speed photograph of a shrew, attacking a cockroach thrown to it for food, gives a good impression of the relative sizes of one of the smallest mammals and a moderately large insect. [Courtesy of Peter Morrison, University of Alaska; from *Physiological Zoology*, 32:263, 1959, © University of Chicago Press]

Nearly 100 years ago the German physiologist Max Rubner examined the metabolic rates of dogs of various sizes. He found that smaller dogs had a higher metabolic rate per unit body mass than larger dogs, a finding entirely in accord with the above discussion. Rubner had the clever idea that this relationship might be attributable to the fact that a smaller animal has a larger body surface relative to the body mass than a large animal. Because small and large dogs have the same body temperature, in order to keep warm they must produce metabolic heat in relation to their heat loss. Small dogs, because of their larger relative surface, must therefore produce a greater amount of heat per unit body mass. Rubner then calculated the heat production per square meter body surface and found that it was approximately 1000 kcal per 1 m² body surface per day, in both large and small dogs. He therefore thought that his theory had been confirmed and that metabolic rate was determined by surface area and the need to keep warm, a conclusion that became known as *Rubner's surface rule*.

Unfortunately, the slopes of the regression lines in Figure 6.11 cannot be explained by the need to compensate for heat loss, for temperature regulation is not a problem for fish or crabs or beech trees, which show the same relationship between metabolic rate and body size as mammals do. Furthermore, if the metabolic rate were really proportional to surface, the slopes of the regression lines

in Figure 6.11 should be 0.67.* Actually the slopes are closer to 0.75. The fact that cold-blooded vertebrates as well as many invertebrates (and at least some plants) have regression lines with the same slope excludes the possibility that temperature regulation is a primary cause of the regularity of the regression lines.

The regular relationship between metabolic rate and body size, and the value of 0.75 for the slope of the regression lines, are not easy to explain. We can say, however, that it would be virtually impossible to design mammals of widely different body sizes that would follow a metabolic regression line with a slope of 1.0, (i.e., with metabolic rates directly proportional to body mass). Kleiber (1961) has calculated that if a steer were designed with the weight-specific metabolic rate of a mouse, to dissipate heat at the rate it is produced, the steer's surface temperature would have to be well above the boiling point. Conversely, if a mouse were designed with the weight-specific metabolic rate of a steer, to keep warm it would need as insulation a fur at least 20 cm thick.

It is important to realize that many physiological processes, not only heat loss, are functions of the surface area. In fact, to design a workable organism it would be necessary to include a careful consideration of surface areas. A large number of physiological processes are surface-related: The uptake of oxygen in the lungs or in the gills depends on the area of these organs; the diffusion of oxygen from the blood to the tissues takes place across the capil-

---

* If two bodies of different size are geometrically similar, their surface areas will be related as the square of a corresponding linear dimension, and their volumes as the cube of the linear dimension. Their areas will therefore be related as their volumes raised to the power 2/3, or 0.67. A few minutes spent in relating edges, surfaces, and volumes of cubes of different sizes will clarify this rule, which applies to any geometrically similar bodies.

FIGURE 6.12 The resting oxygen consumption (metabolic rate) of birds increases with body size. The slope shown here indicates that for a nonpasserine bird of any given size, oxygen consumption is similar to that of a mammal of the same size. Passerine birds as a group have higher metabolic rates. [Lasiewski and Dawson 1967]

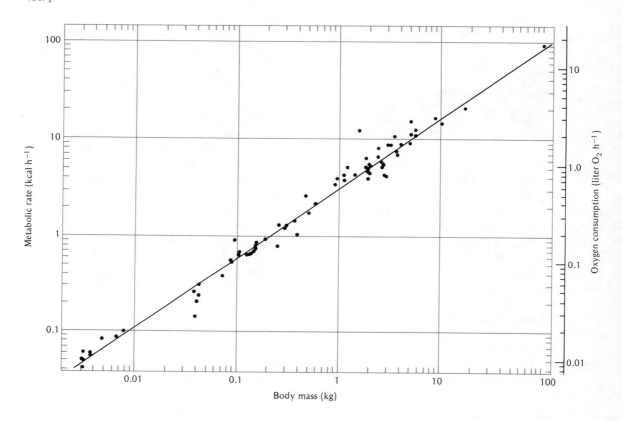

lary walls, again a surface function; food uptake in the intestine depends on the surface area of the intestine; and so on. In fact, all cells have surfaces through which nutrients and oxygen must enter the cell and metabolic products leave. The cells maintain a different ionic composition from the extracellular fluid, and the difference must again be maintained by surface- or area-related processes. It is therefore easy to understand that overall metabolism cannot be independent of surface considerations; it is more difficult to explain why it deviates in such a regular fashion that the slope of the regression lines we have discussed often is 0.75 or very close to this value.

Birds

We have discussed the metabolic rates of mammals, and it would be interesting to know whether the other major group of warm-blooded vertebrates, the birds, have similar metabolic rates. Passerine birds (sparrows, finches, crows, etc.) mostly have somewhat higher metabolic rates than nonpasserine birds, and we will therefore divide the birds into two separate groups. Metabolic rates have been compiled for 58 species of nonpasserine birds, ranging in size from a 0.003-kg hummingbird to the 100-kg ostrich (Figure 6.12). The equation for the regression line corresponds to

$$\dot{V}_{O_2} = 0.679 \times M_b^{0.723}$$

This equation is similar to that for mammals, $\dot{V}_{O_2} = 0.676 \times M_b^{0.75}$ ($\dot{V}_{O_2}$ in liter O_2 per hour and $\dot{M}_b$ in kilograms). This similarity between mammals and birds means that a mammal and a nonpasserine bird of the same body size are likely to have nearly the same metabolic rates.

The available data for 36 species of passerine birds, ranging in size from 0.006 to 0.866 kg, give the equation $\dot{V}_{O_2} = 1.11 \times M_b^{0.724}$. The slope of the regression line for passerine and nonpasserine birds thus is the same, but the metabolic rate of a passerine bird can be expected to be higher than that of a nonpasserine bird of the same size by an amount of about 65% (the difference between 1.11 and 0.679). Of course, any individual species may deviate more or less from the regression lines, which represent the best fit for all available data (Lasiewski and Dawson 1967).

Marsupials and monotremes

The body temperatures of most marsupials (about 35 °C) are somewhat lower than those of eutherian mammals (about 38 °C). This fact has been widely quoted and is often taken to mean that marsupials are physiologically intermediate between the "lower" monotremes (which have even lower body temperatures, about 30 °C) and the "higher" eutherian mammals. There is no a priori reason to regard a lower body temperature as physiologically inferior, for marsupials regulate their body temperature effectively over a wide range of external conditions. If a low body temperature indicated "lower" in the sense of "inferior," birds

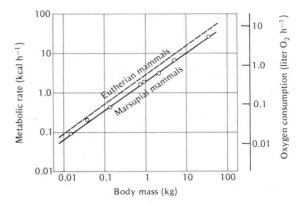

FIGURE 6.13 The resting metabolic rate of marsupials is consistently somewhat lower than that of eutherian mammals. The increase with increasing body size, however, follows regression lines of identical slope. [Dawson and Hulbert 1970]

would be "superior" to mammals because they have higher body temperatures (mostly about 40 to 42 °C).

These differences in body temperature make it interesting to compare the energy metabolism of the various groups and see whether marsupials as a whole adhere to a pattern similar to that of other warm-blooded vertebrates. A study of Australian marsupials, ranging in size from 0.009 to 54 kg, showed that their metabolic rates varied with body size in a way similar to eutherian mammals, but at an overall somewhat lower level (Dawson and Hulbert 1970).

Expressed with the same units used for eutherian mammals, the equation for marsupials is $V_{O_2} = 0.409 \times M_b^{0.75}$. The data fall on a line with the same slope as the general equation for eutherian mammals ($V_{O_2} = 0.676 \times M_b^{0.75}$), as shown in Figure 6.13. The consistently lower metabolism of the marsupials is evident. It is interesting that this lower metabolism of marsupials is associated with a 3 °C lower body temperature. However, it is not possible to say whether the lower body temperature is a result of the lower metabolism or vice versa.

If we estimate what the metabolism of marsupials would be in the event their body temperature were raised to 38 °C, we find that their "corrected" metabolic rate would be similar to that of eutherian mammals. This calculation,* carried out for the three major groups of mammals and for birds shows that monotremes, marsupials, eutherian mammals, and nonpasserine birds all have similar temperature-corrected metabolic rates. Passerine birds, however, fall distinctly outside the common range; their metabolic rates are, on the whole, more than 50% higher, even if recalculated to a lower body temperature than they normally have (Dawson and Hulbert 1970).

BODY SIZE AND PROBLEMS OF SCALING

In the preceding discussions we have related metabolic rates to body size, and we have derived general equations that represent this relationship. These equations express the consequences of a change in size, or in scale, for the metabolic rates of animals.

The study of the consequences of a change in size or scale is an important aspect of engineering science – so important that scaling is considered a separate discipline. The engineer meets problems of scaling every day when he builds taller buildings, longer bridges, and bigger ships, or changes from a small-scale working model to a full-size product.

In physiology, we define *scaling* as the structural and functional consequences of a change in size or in scale among similarly organized animals. As we examine available information, we find a multiplicity of functional variables that are related to body size. We have already discussed some of the

constraints on the rate of metabolism that prevail for a mammal of a given size; many other constraints exist on how an animal of a given size must be designed in order to function well. For example, it is familiar to all of us that the bones of a large animal, such as an elephant, are proportionately heavier than those of a small mammal, say a mouse. The reason is simple, the bones must be able to support the weight of an animal, which increases with the third power of the linear dimensions. To support the increased weight, the diameter of the bones of the large animal must therefore be given a disproportionate increase.

Other structural and functional variables are also scaled in relation to body size. For example, although the heart of a horse weighs far more than that of a mouse, in both species the heart is about 0.6% of the total body mass (M_b). If we express this in a general equation, it would be:

$$\text{heart weight} = 0.006\, M_b^{1.0}.$$

This equation, which is based on the measurement of hundreds of heart weights of different mammals, shows that the weight of the mammalian heart is proportional to the body weight (exponent of 1.0). There is, of course, still room for a certain margin of variation, but the equation gives the general trend.

Similarly, the lung volume of mammals is scaled in simple proportion to body size, as was shown in Figure 2.6 on page 32. The regression line in this graph has a slope of 1.02, which means that in all mammals the lungs constitute very nearly the same proportion of the body. As a general rule, then, all mammalian lungs are scaled similarly and have a volume of 63 ml per kilogram body weight (again with a certain margin of variation).

If we examine the surface area of the mammalian lung, rather than the volume, we find that this

---

* For information on how to calculate the effect of temperature change on metabolic rate, see Chapter 7.

FIGURE 6.14 The total surface area of the mammalian lung is proportional to the rate of oxygen consumption, as evident from the slope of 1.0 of the regression line that relates these two variables, [Tenny and Remmers 1963]

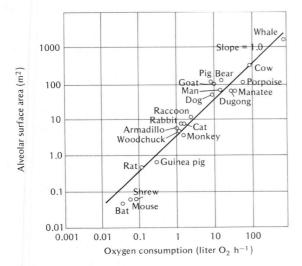

TABLE 6.7 Relationship for mammals between physiological variables and body mass (M_b in kilograms).[a] [Data selected from Adolph 1949; Drorbaugh 1960; Stahl 1967]

O_2 consumption (liter h^{-1})	=	$0.696 \times M_b^{0.75}$
O_2 consumption per kg (liter h^{-1} kg^{-1})	=	$0.696 \times M_b^{-0.25}$
Lung ventilation rate (liter h^{-1})	=	$20.0 \times M_b^{0.75}$
Lung volume (liter)	=	$0.063 \times M_b^{1.02}$
Tidal volume (liter)	=	$0.0062 \times M_b^{1.01}$
Blood volume (liter)	=	$0.055 \times M_b^{0.99}$
Heart weight (kg)	=	$0.0058 \times M_b^{0.99}$
Respiration frequency (min^{-1})	=	$53.5 \times M_b^{-0.26}$
Heart rate (min^{-1})	=	$241 \times M_b^{-0.25}$

[a] If equations listed in this text are compared with similar equations given elsewhere, it is necessary to pay close attention to the units used. In this chapter the units are consistently liters O_2, hours, and kilograms. Recalculations between liters and milliliters and between hours, minutes, and seconds are a matter of simple arithmetic. However, if the body mass is expressed in grams instead of kilograms, the conversion is more complex. As an example, consider the equation for metabolic rate $\dot{V}_{O_2} = 3.8\, M_b^{0.75}$ with the units milliliters, hours, and grams. If we convert the equation for use with kilograms, what will the coefficient 3.8 become? To insert kilograms, we must *divide* the gram mass (M_b) by 1000, but because the number 1000 is afterward raised to the 0.75 power, we must *multiply* the coefficient 3.8 by the same number ($1000^{0.75}$ or 177.83). This gives the equation $\dot{V}_{O_2} = 3.8 \times 177.83 \times M_b^{0.75} = 676\, M_b^{0.75}$ (ml, h, kg), or $0.676\, M_b^{0.75}$ (liter, h, kg). These last units are those used in all equations in this chapter.

variable is directly related to the oxygen consumption of the animal, rather than to body size (Figure 6.14). This does indeed make sense, for oxygen consumption is, as we have seen, related to body weight to the 0.75 power, and the surface area of the lung, or diffusion area for oxygen uptake, should therefore be scaled the same way, also related to the body weight to the 0.75 power.

Table 6.7 shows a number of physiological variables scaled to body size by similar equations. As we saw above, lung volume is directly related to body size. The tidal volume at rest is also directly related to body size (Table 6.7, line 5). If we divide the equation for tidal volume by the equation for lung volume, we obtain an expression for the tidal volume relative to lung volume. The body size exponents are so close that the residual exponent of 0.01 obtained in this division is insignificant compared with the inherent inaccuracy of the measurements used to establish the original equations. We thus obtain the value 0.0062 : 0.063, or very nearly 0.1. This dimensionless number predicts

that mammals in general will have a tidal volume at rest of one-tenth their lung volume, irrespective of their body size. Although there may be deviations from this general pattern, the statement has a predictive value that is useful. We can consider a rat or a dog and predict their expected physiological parameters, and we can then examine our measurements to see how they fit the general pattern or deviate from it.

When we continue to examine the equations in Table 6.7 we can see other interesting relationships of predictive value. The blood volume of mammals, for example, is universally 5.5% of the body mass, irrespective of body size (the exponent 0.99 can be considered as insignificantly different from 1.0). As we saw before, heart weight (the size of the circulatory pump) is also proportional to body mass. Because the blood supplies oxygen to the tissues at the rate required by metabolism, which is scaled to the 0.75 power of body size, the rate of the pump's beat (heart rate) must be adjusted accordingly, and we do indeed find this to be the case. The heart rate of the small animal is very high, and decreases with body size as indicated by the negative slope of -0.25 of the regression line (Table 6.7, last line). This shows that heart rate, and thus cardiac output, is indeed scaled to pump blood at the rate required to supply the oxygen needed for the metabolic rate.

The equations listed in Table 6.7 relate to variables that are either directly proportional to body size (exponent $= 1.0$) or are scaled to the metabolic rate (exponent $= 0.75$). We cannot assume that all anatomical and physiological variables are scaled in one of these two ways; indeed, some differ substantially. For example, for the size of kidney and liver the equations are:

kidney mass $= 0.021 \times M_b^{0.85}$
liver mass $= 0.082 \times M_b^{0.87}$

Both these organs are metabolically very important but they are not scaled in proportion to body mass; they are relatively smaller in the larger animal. On the other hand, they are not scaled as metabolic rate either, for their size increases with an exponent that exceeds the 0.75 power for metabolic rate. We can therefore conclude that the scaling of animals, their organs, and their physiological functions is not always a simple function of body size, and that more complex considerations enter into what is necessary to form an integrated and well-functioning organism.

ENERGY COST OF LOCOMOTION

Swimming, running, and flying require more energy than sitting still, but how do the different kinds of locomotion compare? Walking and running are more familiar to us than swimming and flying, and of more immediate interest, and we therefore know more about these ways of moving. Also, man is our best experimental animal, for he is highly cooperative and at times even highly motivated. This is of great help in studies of maximum performance, for when an animal refuses to run faster or longer, is it because he will not or cannot? Dogs have also been extensively studied, for they are often cooperative and easy to work with – at times they even seem to be highly motivated. Flying is a different matter, for man has minimal experience in flying under his own power, and he is a clumsy and ineffective swimmer compared with fish and whales.

When we compare the various kinds of locomotion, we should realize that the greatest differences between moving in water, on land, and in the air are attributable to the different physical qualities of these media. The two most important differences are in the support of the animal body that the medium provides and in the different resistance to movement provided by water and by air.

Most swimming animals are in near-neutral buoyancy; their weight is fully supported by the surrounding water, and no effort goes into supporting the body. Running and flying animals are in a very different situation; they must support the full weight of their bodies. The running animal has solid support under its feet, but the flying animal must continuously support its weight against a fluid medium of low density and low viscosity.

The resistance of the medium is of great importance to the swimming animal, for water has a high viscosity and density. In contrast, running and flying animals have the advantage of moving through a medium of low viscosity and low density. The physical characteristics of the medium have profound effects on the structural adaptations animals show in their modes of locomotion. Aquatic animals have streamlined bodies and propel themselves with fins and tail or modifications of these structures. Birds have streamlined bodies and wings, which basically function on the same principles of fluid dynamics as the fish tail. The smaller the size of the animal, however, the less effective is streamlining in reducing drag, and small insects are not obviously streamlined.

Running animals use their extremities as levers for moving over the solid substratum. Air resistance is of little importance, and running animals are not particularly streamlined. Streamlining is much more important for birds, for they move much faster than running mammals, and air resistance increases approximately with the square of the speed. Most mammals are quadrupedal, but an assortment of oddities such as man and kangaroos are primarily bipedal.

Running

Let us first consider the oxygen consumption of an animal, for example, a rat, as it runs at various speeds (Figure 6.15). The oxygen consumption increases with running speed, and within a wide range of speeds the increase is linear. (At even higher speeds the line may curve upward and the cost of running increase out of proportion to the increase in speed – an experience familiar to a man who runs a high-speed dash.) If we extend the straight line back to the ordinate, corresponding to zero running speed, the intersect is higher than the resting oxygen consumption. Therefore, if we were to draw a complete curve, covering all running speeds from zero to the maximum, we would not

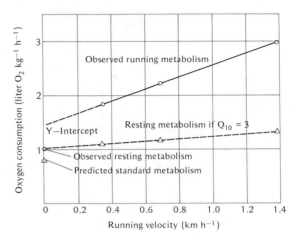

FIGURE 6.15 The oxygen consumption of running white rats increases linearly with increasing speed. [Taylor et al. 1970]

obtain a straight line. However, in the intermediate range of speeds, the curves for a wide variety of running mammals, including man, follow straight lines.

For a man, walking at moderate speed is less expensive than running, in terms of energy cost (Figure 6.16). With increasing speed, walking becomes increasingly more expensive, and where the curve for walking intersects the curve for running it is cheaper to change to running. Figure 6.16 also shows the well-known facts that walking or running uphill is energetically more expensive than moving on the level and that downhill locomotion is cheaper.

Let us now return to the running animal in Figure 6.15 and the straight-line relationship between cost and speed. The fact that the regression line does not extrapolate to the resting metabolic rate seems to be attributable to the metabolic cost of maintaining the posture of running. For example, the metabolic rate of a man who stands quietly instead of running falls close to the intercept on the ordinate. The simplest interpretation of this is to

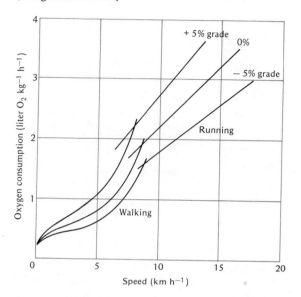

FIGURE 6.16 The energy expenditure of a man during walking and running, at three different grades. [Margaria et al. 1963]

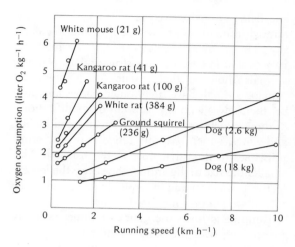

FIGURE 6.17 The oxygen consumption of a variety of running mammals increases linearly with speed. The increase per unit body weight is smaller the larger the body size of the animal. [Taylor et al. 1970]

assume that the difference between the intercept on the ordinate and resting metabolism results from the cost of maintaining the posture of locomotion and could therefore be called the postural effect. The linear increase in oxygen consumption above the intercept on the ordinate can therefore be used as a measure of the cost of increasing the speed, for this increase represents the excess that goes into moving at any speed. Thus, the metabolic cost of moving is given by the slope of the line, and as the line is straight, the cost (slope) remains constant and does not change with the speed of running.

If we determine oxygen consumption in relation to running speed for a number of mammals (Figure 6.17), we find that the smaller mammals have the steeper regression lines (i.e., the oxygen consumption increases more rapidly with increasing running speed).

A simple way of comparing the cost of running

for different animals is to calculate how much energy it takes to move one unit of body mass over one unit of distance: for example, to move 1 kg body mass over 1 km (given as liter O_2 kg^{-1} km^{-1}). If we now return to Figures 6.15 and 6.17, we see that the slopes of the regression lines give exactly this information. The slope (defined as an increment in the ordinate value over an increment in the corresponding abscissa value) is the increment in oxygen consumption (liter O_2 kg^{-1} h^{-1}) over the increment in speed (km h^{-1}). The units cancel out to give the slope in liter O_2 kg^{-1} km^{-1}, precisely the units used to define the cost of running at the beginning of this paragraph.

We can now compare directly the cost of running 1 km per kg body mass for the animals in Figure 6.17. The result is plotted in Figure 6.18. It is immediately apparent that the cost of running decreases quite regularly as the body size increases. The significance of this relationship is that, for a running mammal, it is metabolically relatively less expensive, and thus an advantage, to be of large body size.

The metabolic cost of locomotion can be seen

FIGURE 6.18 The cost of running for mammals of various body sizes. The net cost designates the cost of moving 1 kg body mass over a distance of 1 km, calculated from the increase in metabolism caused by running (and obtained from slopes of regression lines). The total cost includes the total metabolism while running and is therefore somewhat higher. [Schmidt-Nielsen 1972a]

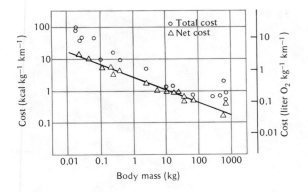

FIGURE 6.19 Oxygen consumption of a flying budgerigar (parakeet) in relation to flying speed. The lowest oxygen consumption occurs at about 35 km per hour, but the lowest cost of transport is at about 40 km per hour (see text). [Schmidt-Nielsen 1972b; data from Tucker 1968]

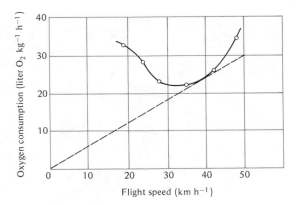

in a different light. The preceding discussion considered what it costs to run, above the cost of not running. However, an animal while it moves does in fact have to supply energy for its total metabolism. Thus, a sheep that grazes must eat enough food to cover it total metabolic rate, not merely the excess attributable to the act of locomotion per se. If we consider the total metabolic cost while moving, we obtain higher figures, for the nonmoving metabolic rate is added in.

Swimming and flying

We can examine the cost of swimming, or of flying, in terms similar to those we have used for running, and we again find that the cost of moving is related to the body size.

Although the oxygen consumption during running increases regularly and linearly with increasing speed (Figure 6.15), the situation is different for flying birds. Determinations on budgerigars (parakeets) have shown that there is an optimum speed of flight at which the bird's oxygen consumption is at a minimum (Figure 6.19). Flying slower or faster increases the oxygen consumption, so that the metabolic cost of flight relative to speed yields a U-shaped curve.

Although flying horizontally at 35 km h^{-1} gives

the lowest metabolic rate for the flying budgerigar, the speed that enables the bird to fly a certain distance most economically is higher, in the case of the budgerigar about 40 km h^{-1}. This is because the bird that flies faster covers the same distance in less time. The most economical flight speed can be calculated, but it can also be found directly from the graph by taking the tangent to the curve from the origin. The point of contact gives the lowest possible slope for the oxygen consumption relative to flight speed, and this is the cost of transport as previously defined (expressed in liter O_2 kg^{-1}). (The cost of transport derived by using the tangent to the curve in Figure 6.19 gives the total cost of transport, rather than the net transport as we calculated for the mammals. For birds this is relatively unimportant, because their metabolic rate during flying is about 10 times as high as the resting metabolic rate, and subtracting the latter makes only a minor difference in the result. For mammals the difference between net and total cost of locomotion is greater.)

Information about the cost of locomotion for swimming, flying, and running has been put together in Figure 6.20. It is interesting that for a given body size, flying is a far less expensive way to

OXYGEN CONSUMPTION DURING FLIGHT The budgerigar, or parakeet (*Melopsittacus undulatus*), photographed during flight in a wind tunnel. To determine the rate of oxygen consumption in flight, this trained bird is equipped with a plastic face mask that, through the attached tube, permits the collection of all exhaled air. [Courtesy of V. A. Tucker, Duke University]

move to a distant point than running. At first glance this is surprising, for we have the intuitive feeling that flying must require a great deal of effort just to stay up in the air. We know, however, that migrating birds fly nonstop for more than 1000 km, and it would be difficult to imagine a small mammal such as a mouse which could run that far without stopping to eat and drink.

Swimming fish use even less energy than flying birds. This may seem surprising, for if we consider the high viscosity of water, we would expect considerable resistance to moving a body through water. There are two factors to take into account, however. One is that fish on the whole do not move very fast, and their streamlined bodies are highly adapted to moving in a medium of relatively high density and viscosity. The other factor is that fish are in near-neutral buoyancy; the body is fully supported by the medium, and no effort goes into keeping from sinking.

For animals less well adapted to swimming the situation is different. If we calculate the cost of transportation for a swimming duck, it turns out to be similar to the cost of running on land for a mammal of the same size. For a man, swimming is even more expensive. Compared with a fish he moves through water with difficulty; his body shape and his appendages make him an ineffective swimmer, and the cost of moving through the water is 5 or 10 times as high as moving the same distance on land. He is, in fact, so unsuited to swimming that the cost of transportation puts a swimming man well above the range represented by the regression line for running in Figure 6.20.

EFFECT OF HIGH ALTITUDE

Most of what we know about the responses to high altitude and low air pressure comes from studies on humans. As described in Chapter 1, the percentage composition of the atmosphere is unchanged at high altitudes, but the decrease in barometric pressure has severe physiological effects.

Mountain climbers unaccustomed to high alti-

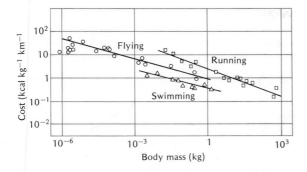

tude begin to feel decreased physical performance at some 3000 to 4000 m altitude, but after a few days they become acclimatized* and perform better. The highest known permanent human habitation is a mining camp in the central Andes in South America at an altitude of nearly 5000 m. Climbers on Mount Everest have shown that acclimatized humans can barely manage to ascend without supplementary oxygen to about 8600 m (28 200 ft.). It has been established on many mountain-climbing expeditions that acclimatization is progressive and increases with altitude, but at heights above 6000 m there is also an underlying process of deterioration which is more rapid and severe the greater the altitude. This eventually forces the climber to descend (Pugh 1964).

The symptoms of low oxygen in man is known as *mountain sickness*. The severity varies from person to person, but even at moderate altitude there is a feeling of weakness and disinclination to move about and perform work. This is followed by headache, nausea, sometimes vomiting, as well as a de-

---

* It is now agreed that the word *acclimatize* is used when an organism adjusts to a certain climate, say the seasonal change between summer and winter. For example, a fish caught in winter in cold water is acclimatized to cold water. The word *acclimate* is used when an organism adjusts to an artificially imposed condition. For example, a goldfish kept for two months in water at 10 °C is said to be acclimated to 10 °C.

terioration of mental functions, judgment, ability to perform simple arithmetic, and so on. Sleep and rest become difficult, which contributes to the problems.

Except for the danger of developing bends during rapid ascent in an unpressurized airplane or balloon, all the known symptoms of high altitude are related to the low partial pressure of oxygen and the inability to supply sufficient oxygen at the tissue level. Thus, breathing pure oxygen at an altitude of about 12 000 m, where the barometric pressure is about 150 mm, removes all symptoms of mountain sickness. At this altitude man could survive for no more than a few minutes if he were to breathe atmospheric air.

The adequate supply of oxygen to the tissues depends on a sequence of the following five steps:

1. Ventilation of the lungs
2. Diffusion from lung air to blood
3. Transport of oxygen in the blood
4. Diffusion of oxygen from blood to tissues
5. Tissue utilization of oxygen

Each of these steps is affected during acclimatization to high altitude.

The immediate response to an inadequate supply of oxygen is an increased ventilation of the lungs, due to increased tidal volume as well as frequency. This increased ventilation leads to an excessive loss of carbon dioxide from the blood, and thus causes a shift in the pH of the blood in the alkaline direction, which in turn reduces the normal stimulation of the respiratory center. The result is a decrease in ventilation and oxygen supply, and so on, leading to a cycling between increased and decreased respiration.

During acclimatization to altitude there is a change in the responses of the respiratory center and an adjustment of the bicarbonate concentration in the blood that permits the maintenance of acid–base balance and a normal blood pH at a lower P_{CO_2}. As these changes progress, the increased lung ventilation can be maintained with less difficulty.

The diffusion capacity of the lungs of the lowland natives remains unchanged at high altitudes. Highland natives, however, have larger lung volumes than lowland natives, and because pulmonary diffusing capacity is related to lung surface area, highland natives have greater pulmonary diffusing capacity. The difference results from an increased growth of the lungs during childhood. Similar changes occur in young rats that have been exposed to low pressure during growth. Thus, the difference in diffusion capacity seems related to a lifetime acclimatization.

Transport of oxygen in the blood is also affected by altitude acclimatization, in regard to both oxygen capacity and oxygen affinity of the blood. In humans the number of red blood cells is increased from a normal of about 5 million per cubic millimeter to as many as 8 million per cubic millimeter at altitudes about 4000 m. This change is found both in permanent residents and in acclimatized lowland natives when they reside at high altitudes.

The oxygen affinity of the blood was discussed in Chapter 3. Llamas and vicuñas, animals native to high altitude, have hemoglobin with a higher oxygen affinity than other mammals, and this facilitates the uptake of oxygen in the lung. Man, at altitude, on the other hand, displays a slight shift in the opposite direction with a decrease in the oxygen affinity of the blood. This is of no aid to the uptake of oxygen in the lung, but slightly facilitates the delivery of oxygen to the tissues.

Diffusion of oxygen from blood to tissues can be enhanced by decreasing the diffusion pathway and by increasing the myoglobin concentration in the tissues (facilitated diffusion). Both responses have been observed, with an increased number of capil-

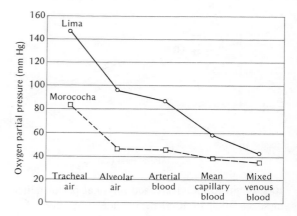

FIGURE 6.21 Mean oxygen partial pressures observed in native residents of Lima (sea level) and Morococha, Peru (4540 m altitude).[Hurtado 1964]

laries per unit volume of tissues (reducing the diffusion distance) and with an increased muscle myoglobin concentration (increasing the rate of diffusion).

Utilization of oxygen at the tissue level is the last step. Adaptations in the enzymatic pathways and their importance in altitude acclimatization is a more poorly understood subject.

The importance of these steps in the oxygen supply from air to tissue in humans is summarized in Figure 6.21. This graph compares native residents of Lima, which is at sea level, with native residents of Morococha, a mining town located at 4540 m altitude. The average barometric pressure at Morococha is 446 mm Hg, compared with the normal 760 mm at sea level.

The observations on the oxygen partial pressure, (P_{O_2}) in the two groups of permanent residents are summarized in Table 6.8.

The drop in P_{O_2} from tracheal air (essentially humidified outside air) to venous blood was more than twice as great at sea level as at altitude. Nevertheless, the P_{O_2} of mixed venous blood was only 7.3 mm Hg lower at altitude; this indicates that the oxygen supply to the tissues in the acclimatized na-

TABLE 6.8 Oxygen partial pressures (mm Hg) observed in native residents of Lima (sea level) and Morococha, Peru (4540 m). Mean of observations on eight healthy subjects in each group. Same data as in Figure 6.21. [Hurtado 1964]

	Tracheal air	Mixed venous blood	Drop
Lima	147.2	42.1	105.1
Morococha	83.4	34.8	48.6
Difference	63.8	7.3	56.5

tives was nearly as adequate as at sea level (Hurtado 1964).

Birds are conspicuously more tolerant of high altitude than mammals, and mountain climbers have often reported seeing birds flying overhead at altitudes where they themselves suffered severely from oxygen deficiency. There are several reasons for this difference. One is the greater effectiveness of the avian respiratory system, in which air moves through the lungs during both inhalation and exhalation. Bird blood has roughly the same oxygen affinity as mammalian blood, but the cross-current flow in the lungs permits a higher degree of oxygenation (Chapter 2). A further reason is the exceptional tolerance of birds to alkalosis; when excess carbon dioxide is lost from the blood, this is well tolerated and permits the Bohr shift to take effect, thus increasing the amount of oxygen taken up by the blood. A final factor is that birds, as opposed to mammals, maintain a normal blood flow to the brain at low blood P_{CO_2}; this relieves the birds from some of the most severe effects on the brain which are observed in mammals at altitude.

To summarize the preceding discussions, we can see the close connection between the energy metabolism of animals and not only their supply of food, but also the essential roles of oxygen and respiration. When we move on to temperature as another environmental variable, we will again find that we must consider many of the principles discussed in the preceding chapters and that there are close connections among the functions of all physiological systems.

REFERENCES

Adolph, E. F. (1949) Quantitative relations in the physiological constitutions of mammals. *Science* 109:579–585.

Bartholomew, G. A., and MacMillen, R. E. (1961) Oxygen consumption, estivation, and hibernation in the kangaroo mouse, *Microdipodops pallidus*. *Physiol. Zool.* 34:177–183.

Blazka, P. (1958) The anaerobic metabolism of fish. *Physiol. Zool.* 21:117–128.

Brody, S. (1945) *Bioenergetics and Growth: With Special Reference to the Efficiency Complex in Domestic Animals.* New York: Reinhold. 1023 pp. Reprinted (1964), Hafner, New York.

Childress, J. J. (1975) The respiratory rates of midwater crustaceans as a function of depth of occurrence and relation to the oxygen minimum layer off southern California. *Comp. Biochem. Physiol.* 50A:787–799.

Dawson, T. J., and Hulbert, A. J. (1970) Standard metabolism, body temperature, and surface areas of Australian marsupials. *Am. J. Physiol.* 218:1233–1238.

Douglas, E. L., Friedl, W. A., and Pickwell, G. V. (1976) Fishes in oxygen-minimum zones: Blood oxygenation characteristics. *Science* 191:957–959.

Drorbaugh, J. E. (1960) Pulmonary function in different animals. *J. Appl. Physiol.* 15:1069–1072.

Elsner, R. (1965) Heart rate response in forced versus trained experimental dives in pinnipeds. *Hvalrådets Skrifter*, no. 48, pp. 24–29.

Felger, R. S., Cliffton, K., and Regal, P. J. (1976) Winter dormancy in sea turtles: Independent discovery and exploitation in the Gulf of California by two local cultures. *Science* 191:283–285.

Fry, F. E. J., and Hart, J. S. (1948) The relation of temperature to oxygen consumption in the goldfish. *Biol. Bull.* 94:66–77.

Gaunt, A. S., and Gans, C. (1969) Diving bradycardia and withdrawal bradycardia in *Caiman crocodilus*. *Nature, Lond.* 223:207–208.

Gordon, M. S., Boëtius, I., Evans, D. H., McCarthy, R., and Oglesby, L. C. (1969) Aspects of the physiology of terrestrial life in amphibious fishes. 1. The mudskipper, *Periophthalmus sobrinus. J. Exp. Biol.* 50:141–149.

Graham, J. B. (1974) Aquatic respiration in the sea snake *Pelamis platurus. Respir. Physiol.* 21:1–7.

Hall, F. G. (1929) The influence of varying oxygen tensions upon the rate of oxygen consumption in marine fishes. *Am. J. Physiol.* 88:212–218.

Hawkins, A. E., Jewell, P. A., and Tomlinson, G. (1960) The metabolism of some British shrews. *Proc. Zool. Soc. Lond.* 135:99–103.

Hemmingsen, A. M. (1960) Energy metabolism as related to body size and respiratory surfaces, and its evolution. *Rep. Steno Hosp.* 9(2):1–110.

Hochachka, P. W. (ed). (1975) Pressure effects on biochemical systems of abyssal and midwater organisms: The 1973 Kona Expedition of the Alpha Helix. *Comp. Biochem. Physiol.* 52(1B):1–199.

Hochachka, P. W., Fields, J., and Mustafa, T. (1973) Animal life without oxygen: Basic biochemical mechanisms. *Am. Zool.* 13:543–555.

Hochachka, P. W., and Mustafa, T. (1972) Invertebrate facultative anaerobiosis. *Science* 178:1056–1060.

Hudson, J. W. (1962) The role of water in the biology of the antelope ground squirrel *Citellus leucurus. Univ. Calif. Publ. Zool.* 64:1–56.

Hurtado, A. (1964) Animals in high altitudes: Resident man. In Handbook of Physiology, sect. 4, *Adaptation to the Environment* (D. B. Dill, E. F. Adolph, and C. G. Wilber, eds.), pp. 843–860. Washington, D.C.: American Physiological Society.

Jackson, D. C., and Schmidt-Nielsen, K. (1966) Heat production during diving in the fresh water turtle, *Pseudemys scripta. J. Cell. Physiol.* 67:225–232.

Johansen, K. (1964) Regional distribution of circulating blood during submersion asphyxia in the duck. *Acta Physiol. Scand.* 62:1–9.

King, J. R. (1957) Comments on the theory of indirect calorimetry as applied to birds. *Northwest Sci.* 31:155–169.

Kleiber, M. (1961) *The Fire of Life: An Introduction to Animal Energetics.* New York: Wiley. 454 pp.

Kooyman, G. L. (1966) Maximum diving capacities of the Weddell seal, *Leptonychotes weddelli. Science* 151:1553–1554.

Kylstra, J. A. (1968) Experiments in water-breathing. *Sci. Am.* 219:66–74.

Lasiewski, R. C., and Dawson, W. R. (1967) A re-examination of the relation between standard metabolic rate and body weight in birds. *Condor* 69:13–23.

Lusk, G. (1931) *The Elements of the Science of Nutrition,* 4th ed. Philadelphia: Saunders. 844 pp.

MacDonald, A. G. (1975) *Physiological Aspects of Deep Sea Biology.* London: Cambridge University Press. 450 pp.

Margaria, R., Cerretelli, P., Aghemo, P., and Sassi, G. (1963) Energy cost of running. *J. Appl. Physiol.* 18:367–370.

Murdaugh, H. V., Jr., Schmidt-Nielsen, B., Wood, J. W., and Mitchell, W. L. (1961) Cessation of renal function during diving in the trained seal (*Phoca vitulina*). *J. Cell. Comp. Physiol.* 58:261–266.

Olsson, K. E., and Saltin, B. (1968) Variation in total body water with muscle glycogen changes in man. In *Biochemistry of Exercise,* Medicine and Sport Series, vol. III, pp. 159–162. Basel: Karger.

Paulev, P. (1965) Decompression sickness following repeated breath-hold dives. *J. Appl. Physiol.* 20:1028–1031.

Pearson, O. P. (1960) The oxygen consumption and bioenergetics of harvest mice. *Physiol. Zool.* 33:152–160.

Pugh, L. G. C. E. (1964) Animals in high altitudes: Man above 5,000 meters – mountain exploration. In *Handbook of Physiology,* sect. 4, *Adaptation to the Environment* (D. B. Dill, E. F. Adolph, and C. G. Wilber, eds.), pp. 861–868. Washington, D.C.: American Physiological Society.

Schmidt-Nielsen, K. (1972a) Locomotion: Energy cost of swimming, flying, and running. *Science* 177:222–228.

Schmidt-Nielsen, K. (1972b) *How Animals Work.* Cambridge: Cambridge University Press. 114 pp.

Scholander, P. F. (1940) Experimental investigations on the respiratory function in diving mammals and birds. *Hvalrådets Skrifter,* no. 22, pp. 1–131.

Shepard, M. P. (1955) Resistance and tolerance of

young speckled trout (*Salvelinus fontinalis*) to oxygen lack, with special reference to low oxygen acclimation. *J. Fish. Res. Bd. Can.* 12:387–446.

Smith, E. N., Allison, R. D., and Crowder, W. E. (1974) Bradycardia in a free-ranging American alligator. *Copeia* 1974:770–772.

Somero, G. N., and Hochachka, P. W. (1976) Biochemical adaptations to pressure. In *Adaptation to Environment: Essays on the Physiology of Marine Animals* (R. C. Newell, ed.), pp. 480–510. London: Butterworth.

Stahl, W. R. (1967) Scaling of respiratory variables in mammals. *J. Appl. Physiol.* 22:453–460.

Taylor, C. R., Schmidt-Nielsen, K., and Raab, J. L. (1970) Scaling of the energetic cost of running to body size in mammals. *Am. J. Physiol.* 219:1104–1107.

Tenney, S. M., and Remmers, J. E. (1963) Comparative quantitative morphology of the mammalian lung: Diffusing area. *Nature, Lond.* 197:54–56.

Thomas, H. J. (1954) The oxygen uptake of the lobster (*Homarus vulgaris* Edw.). *J. Exp. Biol.* 31:228–251.

Tucker, V. A. (1968) Respiratory exchange and evaporative water loss in the flying budgerigar. *J. Exp. Biol.* 48:67–87.

Whalen, W. J. (1966) Intracellular P_{O_2}: A limiting factor in cell respiration. *Am. J. Physiol.* 211:862–868.

Whalen, W. J., Buerk, D., and Thuning, C. A. (1973) Blood-flow-limited oxygen consumption in resting cat skeletal muscle. *Am. J. Physiol.* 224:763–768.

ADDITIONAL READING

Andersen, H. T. (ed.) (1969) *The Biology of Marine Mammals.* New York: Academic Press. 511 pp.

Bennett, P. B., and Elliott, D. H. (1975) *The Physiology and Medicine of Diving and Compressed Air Work.* London: Bailliere Tindall. 566 pp.

Brody, S. (1964) *Bioenergetics and Growth: With Special References to the Efficiency Complex in Domestic Animals,* reprint ed. New York: Hafner. 1023 pp.

Frisancho, A. R. (1975) Functional adaptation to high altitude hypoxia. *Science* 187:313–319.

Heath, D., and Williams, D. R. (1977) *Man at High Altitude.* Edinburgh: Churchill Livingston. 292 pp.

Heatwole, H., and Seymour, R. (1975) Diving physiology. In *The Biology of Sea Snakes* (W. A. Dunson, ed.), pp. 289–327. Baltimore: University Park Press.

Hemmingsen, A. M. (1960) Energy metabolism as related to body size and respiratory surfaces, and its evolution. *Rep. Steno Hosp.* 9(2):1–110.

Kleiber, M. (1961) *The Fire of Life: An Introduction to Animal Energetics.* New York: Wiley. 454 pp.

Kooyman, G. L. (1972) Deep diving behaviour and effects of pressure in reptiles, birds, and mammals. *Symp. Soc. Exp. Biol.*, no. 26, pp. 295–311.

Kylstra, J. A. (1975) Liquid breathing and artificial gills. In *The Physiology and Medicine of Diving and Compressed Air Work,* 2nd ed. (P. B. Bennett and D. H. Elliott, eds.), pp. 155–165. London: Bailliere Tindall.

Newell, R. C. (ed.) (1976) *Adaptation to Environment: Essays on the Physiology of Marine Animals.* London: Butterworth. 539 pp.

Schmidt-Nielsen, K. (1972) *How Animals Work.* Cambridge: Cambridge University Press. 114 pp.

Schmidt-Nielsen, K. (1975) Scaling in biology: The consequences of size. *J. Exp. Zool.* 194:287–308.

Sleigh, M. A., and MacDonald, A. G. (eds.) (1972) *The Effects of Pressure on Organisms.* Symposia, Society for Experimental Biology, no. 26. Cambridge: 516 pp.

PART THREE

TEMPERATURE

CHAPTER SEVEN

Temperature effects

In the preceding chapter we were concerned with the energy metabolism of living animals and the influence of such variables as oxygen concentration, body size, and activity. In this chapter we shall discuss the profound influence temperature has on living organisms and their metabolic processes.

Active animal life is limited to a narrow range of temperatures, from a few degrees below the freezing point of pure water (0 °C) to approximately + 50 °C.* Compared with cosmic temperatures, this is a very narrow range indeed. However, temperatures suitable for life can be found throughout the oceans and over most of the surface of the earth, for at least part of the year. At the lower extreme, in Arctic waters there are fish and numerous invertebrates living near −1.8 °C throughout the year. At the other extreme, in hot springs, there are a few animals living at temperatures of about 50 °C. Some primitive plants live at even higher temperatures, and thermophilic bacteria thrive near the boiling point of water.

Outside the temperature range that permits active life, many animals can survive in an inactive or torpid state. In fact, some animals can survive extremely low temperatures, such as that of liquid air (about − 190 °C) or even liquid helium (− 269 °C). The tolerance to high temperatures is more limited, although some animals when in a resting stage can be quite resistant, and bacterial spores may tolerate temperatures in excess of 120 °C.

Most animals, among them all aquatic invertebrates, have nearly the same body temperature as their surroundings. Birds and mammals, in contrast, generally maintain their body temperature nearly constant and independent of large variations

* We are concerned with the temperature of the organism itself not the surroundings. When a man goes out in subfreezing weather, his body temperature remains at the usual 37 °C.

205

in the temperature of their environment. However, quite a few other animals, both invertebrates and vertebrates, can at times maintain a substantial difference between their own temperatures and the surroundings.

There is no simple and easy way to classify these differences. Traditionally, birds and mammals have been called warm-blooded and the other animals cold-blooded. These terms are so well established that it is convenient to continue to use them. They are, however, rather inaccurate and may be quite misleading. A cold-blooded animal is not necessarily cold; a tropical fish or a desert lizard or an insect sitting in the sun may have a higher body temperature than a mammal. Furthermore, quite a few mammals and some birds undergo periods of torpor or hibernation during which their temperature may decrease to near the freezing point of water, with no harm to the animal. In that state it seems odd to call them warm-blooded.

The corresponding scientific terms, *poikilothermic* for cold-blooded and *homeothermic* for warm-blooded, are also imprecise. The word poikilothermic (Greek *poikilos* = changeable) refers to the fact that the temperature of a cold-blooded animal fluctuates with that of its surroundings. A fish has the temperature of the water it swims in and an earthworm that of the soil it crawls in. However, a deep-sea fish that spends its entire life in water that has barely measurable temperature fluctuations is truly an animal with a virtually constant body temperature. It might be logical to call such a fish homeothermic,* but this term is used specifically to refer to birds and mammals, animals that in fact have body temperatures that normally fluctuate by

*The terms *homothermic, homoiothermic,* and *homeothermic* are used interchangeably. They are derived from the Greek *homos* = like and *homoios* = similar. Homeothermic is an anglicized spelling form that is most commonly used.

several degrees and, in hibernation, may even drop to near 0 °C. Animals that at times have high and well-regulated body temperatures, but at other times are more like cold-blooded animals, are often called *heterothermic* (Greek *heteros* = different).

Homeothermic animals (birds and mammals) usually maintain a high body temperature and remain active in cold as well as in warm surroundings. Most poikilothermic animals become more and more inactive as the temperature decreases. There are some remarkable exceptions. For example, by exposing itself to the sun, a lizard can maintain its body temperature far above that of the surrounding air. To distinguish the lizard from birds and mammals, which maintain their high body temperature by metabolic heat production, we have the terms *ectothermic* and *endothermic*.

Endothermic animals are able to maintain a high body temperature by internal heat production; ectothermic animals depend on external heat sources, primarily solar radiation. These definitions also have their limitations; as we shall see later, a tuna fish may keep the temperature of its muscles 10 or 15 °C above that of the water. The heat is derived from muscular metabolism, but we do not consider the tuna an endothermic animal in the same sense as birds and mammals. Insects make this terminology even more difficult to apply. They may sit in the sun and warm up and at the same time produce additional heat by intense muscular contractions. Obviously, this case does not clearly fit one of the terms to the exclusion of the other. The same is true of some mammals and birds that, under appropriate circumstances, take advantage of solar heat to reduce internal heat production.

The choice of terminology is primarily a matter of convenience, and it may at times be difficult to find terms that are suitable in every situation. The question is not whether a certain terminology is right or wrong, but how useful it is for a given purpose. Nevertheless, the terms we use must always be accurately defined so that there is no uncertainty about their precise meaning.

PHYSIOLOGICAL EFFECTS
OF TEMPERATURE CHANGE

Temperature change has striking effects on many physiological processes. Within limits, a temperature increase accelerates most processes. For example, let us consider its effect on the rate of oxygen consumption, which is a convenient expression for the overall metabolic activity of an animal.

Within the temperature range an animal can tolerate, the rate of oxygen consumption is often found to increase in a fairly regular manner with increasing temperature. In general, a rise of 10 °C in temperature causes the rate of oxygen consumption to increase about twofold or threefold.

The increase in a rate caused by a 10 °C increase in temperature is called the Q_{10}. If the rate doubles, Q_{10} is 2; if the rate triples, Q_{10} is 3; and so on. This term is used not only for oxygen consumption but for all rate processes affected by temperature (see Appendix D.)

If an animal has a wide range of temperature tolerance, its rate of oxygen consumption may accelerate vastly as temperature increases. Thus, with a Q_{10} of 2, and a beginning temperature of 0 °C, the rate would double with a temperature increase to 10 °C; it would quadruple with an increase to 20 °C; and it would increase 8-fold with a temperature increase to 30 °C. With a Q_{10} of 3, the increases in oxygen consumption would be 3-fold, 9-fold, and 27-fold, respectively, for the same temperature intervals.

If we plot these figures on graph paper (Figure

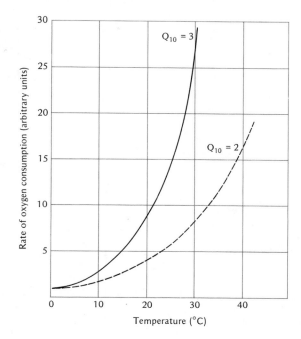

FIGURE 7.1 Many temperature-dependent rate processes proceed more and more rapidly as temperature increases. See text for further discussion.

7.1), we obtain rapidly rising curves, known as exponential curves because mathematically they are described by exponential functions,* of the general form:

$$y = b \cdot a^x$$

If R_2 and R_1 are the rates at two temperatures T_2 and T_1, and we use the customary symbol Q_{10}, the equation will be:

$$R_2 = R_1 \cdot Q_{10}^{\frac{T_2 - T_1}{10}}$$

Many rate processes, such as acceleration, radioactive decay, and growth curves, are described by exponential equations. Calculations that involve exponential equations can conveniently be carried out if we use the logarithmic form:

* See also Appendix C.

FIGURE 7.2 A temperature-dependent rate process of the kind shown in Figure 7.1 will, when plotted on a logarithmic ordinate, give a straight line. Note that the abscissa is linear and only the ordinate is logarithmic. For further explanation, see text.

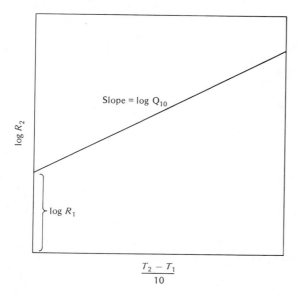

TABLE 7.1 Rates of oxygen consumption of the Colorado potato beetle (*Leptinotarsa decemlineata*) at various temperatures between 7 and 30 °C. Before the experiment the animals had been maintained at 8 °C. [Data from Marzusch 1952]

Temperature (°C)	O_2 consumed ($\mu l\ g^{-1}\ h^{-1}$)	Temperature interval (°C)	Q_{10}
7	61		
10	81	7–10	2.57
15	126	10–15	2.41
20	200	15–20	2.52
25	290	20–25	2.10
30	362	25–30	1.56

Q_{10} when the rates at two different temperatures have been observed, we can simply change the equation and use one of the two following forms:

$$\log Q_{10} = (\log R_2 - \log R_1) \cdot \frac{10}{T_2 - T_1}$$

or

$$Q_{10} = \left(\frac{R_2}{R_1}\right)^{\frac{10}{T_2 - T_1}}$$

We can then calculate the Q_{10} for the interval between T_1 and T_2. However, most often Q_{10} does not remain the same throughout the range of temperatures an animal can tolerate, and it is necessary to specify accurately the conditions under which the observations were made.

Let us now turn to a real example. The rates of oxygen consumption of the Colorado potato beetle between 7 and 30 °C are given in Table 7.1. Throughout this range the oxygen consumption increases with temperature, the Q_{10} for the entire range being 2.17. However, if we calculate the Q_{10} for each temperature interval for which observations were made, we find that up to 20 °C the Q_{10} remains rather constant at nearly 2.5, but at higher temperatures the Q_{10} falls off (Table 7.1, last column). If the same data are plotted in graph form,

$$\log y = \log b + x \cdot \log a$$

or

$$\log R_2 = \log R_1 + \log Q_{10} \cdot \frac{T_2 - T_1}{10}$$

We can now see that $\log R_2$ increases linearly with the temperature change $(T_2 - T_1)$. This means that if we plot log rate against temperature, we get a straight line. This is commonly done by using graph paper with a logarithmic ordinate and a linear abscissa (so-called semilog graph paper). A graph representing this form of the equation is given in Figure 7.2.

Obviously it is not necessary to determine two rates exactly 10 °C apart in order to calculate the Q_{10}. Any two temperatures can be used, provided they are sufficiently far apart to give reliable information about the temperature effect. To solve for

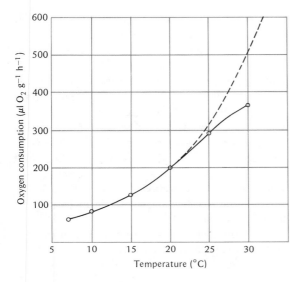

FIGURE 7.3 The rate of oxygen consumption of the Colorado potato beetle increases with temperature (solid line). The broken line shows the expected curve had the Q_{10} remained constant at 2.5. Same data as in Table 7.1.

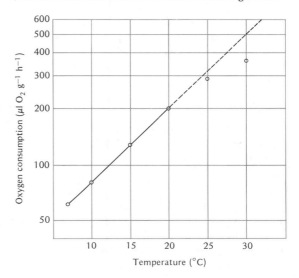

FIGURE 7.4 When plotted on a logarithmic ordinate, the rate of oxygen consumption of the Colorado potato beetle shows a linear dependence on temperature up to 20 °C. Above this temperature the rate does not increase as rapidly as expected from a constant Q_{10} of 2.5 (broken line). Same data as in Table 7.1 and Figure 7.3.

we can more clearly see that the observations, when the temperature exceeds about 20 °C, deviate from a regular exponential function (Figure 7.3).

It is a bit cumbersome to calculate the Q_{10} for each set of observations, and it is also inconvenient to construct an exponential curve on arithmetic coordinates for comparison with the observed data. If, instead, we plot the figures on a logarithmic ordinate, we obtain a straight line for that interval in which the temperature effect remains constant and the Q_{10} does not change (Figure 7.4). The straight line in this figure corresponds to a Q_{10} of 2.50, and we can see that the observations up to 20 °C follow this line completely. Above 20 °C, however, the rate of oxygen consumption deviates increasingly from the straight line (i.e., although there is still an increase in rate with temperature, the extent of the temperature effect is diminished). At even higher temperatures the drop in oxygen consump-

tion becomes more pronounced, and eventually the lethal temperature limit is reached.

EXTREME TEMPERATURES: LIMITS TO LIFE

Animals differ in the range of temperatures they can tolerate. Some have a very narrow tolerance range; others have a wide tolerance range. Furthermore, temperature tolerance may change with time, and a certain degree of adaptation is possible so that continued exposure to a temperature close to the limit of tolerance often extends the limit. Some organisms are more sensitive to extreme temperatures during certain periods of their lives, particularly during the early stages of development.

When discussing tolerance to extreme temperatures, we must distinguish between temperatures an organism can survive and those at which it can carry out its entire life cycle. We should also realize that we cannot determine an exact temperature that is lethal for a given organism because the duration of exposure is of great importance. A certain high temperature that can be tolerated for several minutes may be lethal if continued for several hours.

One further matter is important: We are concerned with the temperature of the organism itself, not that of its environment. For virtually all aquatic organisms the two are nearly equal, but for terrestrial organisms, as we have indicated, they may differ a great deal. A lizard sitting in the sun may, because of solar radiation, acquire a body temperature 10° or 20° above the surrounding air temperature. Among the so-called warm-blooded animals, mammals and birds, most can tolerate only a rather narrow range of body temperatures, and this must not be confused with the wide range of temperatures of the environment in which they live – from the Arctic to the hottest deserts.

Tolerance to high temperature

No animal is known to carry out its complete life cycle at a temperature over 50 °C. Some plants are more heat-tolerant than animals; for example, the unicellular blue-green alga *Synechococcus* is found in hot springs at temperatures as high as 73 to 75 °C. This seems to be the upper limit for photosynthetic life. Thermophilic bacteria are even more heat-tolerant and have been found to live and grow in the hot springs of Yellowstone National Park, where the temperature is about 92 °C (which is the boiling point of water at the altitude of Yellowstone) (Bott and Brock 1969; Brock 1970).

As indicated at the beginning of the chapter, an animal in a resting stage may be extremely tolerant to high temperatures. For example, a fly larva (*Polypedilum*, from Nigeria and Uganda) can tolerate dehydration, and in the dehydrated state it can survive a temperature of 102 °C for 1 minute and afterward grow and metamorphose successfully (Hinton 1960). Another example of extreme tolerance is provided by the eggs of a fresh-water crustacean (*Triops*, from Sudan); these eggs survive through winter and early summer in dry mud, where they may be exposed to temperatures up to 80 °C. In the laboratory they withstand temperatures within 1 °C of boiling. With the boiling point of water increased by subjecting the vessel to pressure, they have withstood 103 °C ± 1 °C for 16 hours, although they are killed within 15 minutes at 106 °C (Carlisle 1968).

From these examples it should be clear that the upper temperature limit for life cannot be accurately defined.

Determinations of lethal temperatures

When a group of animals is exposed to a temperature close to the limit of their tolerance, some may die and others survive. What is the exact *lethal temperature* for this species?

The lethal temperature is commonly defined as that temperature at which 50% of the animals die and 50% survive, often written as T_{L50}. To find this exact temperature by trial and error would require a great deal of experimentation before one experiment ends with exactly half the animals dead. Instead, after the approximate lethal temperature is established, the procedure indicated in Figure 7.5 is followed. Several groups (say, four) are exposed for the same time period (say, 2 hours) to a series of temperatures, and the percent survival at each temperature is plotted. The T_{L50} can now be readily found from the graph.

This procedure gives the lethal temperature only for the exposure time used in the experiment; shorter exposure gives higher survival and longer exposure usually lower survival. The effect of ex-

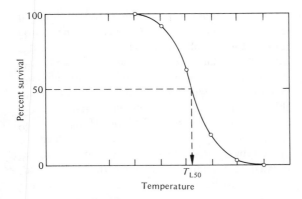

FIGURE 7.5 To determine the temperature that is lethal to 50% of a group of organisms (the T_{L50}), experiments are carried out at a range of temperatures, and after the data are plotted, the temperature for 50% survival is read from the graph.

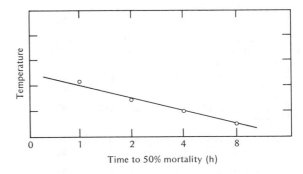

FIGURE 7.6 The effect of exposure time on temperature tolerance becomes evident if the T_{L50} is plotted against the duration of each experiment in which the T_{L50} was determined.

posure time can be determined by carrying out tests similar to that in Figure 7.5 for different periods of exposure – say 1, 2, 4, and 8 hours. If the various T_{L50} values obtained are plotted against exposure time, the points will often fall on a straight line on a log scale abscissa (Figure 7.6). This is the reason the time intervals in this example were chosen to correspond directly to a logarithmic scale. We now have a single graph that gives a quite complete picture of the temperature tolerance of the organism in question.

Lethal temperatures and cause of heat death

Although active animal life has an upper limit at about 50 °C, many animals die when they are exposed to much lower temperatures. This is particularly true of aquatic animals, and especially marine animals, which usually are not exposed to very high temperatures. Even in tropical seas the temperature is rarely as high as 30 °C, although it may be higher in small enclosed bays and pools.

For intertidal animals, however, the situation is different. At low tide, intertidal animals are exposed to warm air and to solar radiation, and their temperatures may increase appreciably. To some extent they remain cool by evaporation of water,

but while in air their restricted water reserves limit this way of cooling.

Intertidal snails show a clear correlation between heat resistance and their location in the tidal zone (Table 7.2). Those snails that are located high in the tidal zone, and are thus out of water and exposed for the longest time, have substantially higher heat resistance than those located in the lower part of the tidal zone, close to the low-water mark.

The most heat-tolerant fish of any are probably the desert pupfish that live in warm springs in California and Nevada. The species *Cyprinodon diabolis* lives in a spring known as Devil's Hole, in which the temperature (33.9 °C) probably has remained almost unchanged for at least 30 000 years (Brown and Feldmeth 1971). The upper lethal limit for this tiny fish, which as an adult weighs less than 200 mg, is about 43 °C – as high as is known for any species of fish.

In contrast, some Arctic and Antarctic animals have a surprisingly low heat tolerance. Antarctic fish of the genus *Trematomus* are particularly heat-sensitive and have an upper lethal temperature of about 6 °C. In nature these fish live in water with an average temperature of −1.9 °C that through the year varies by only about 0.1 °C (Somero and DeVries 1967).

TABLE 7.2 The heat resistance of snails found in the upper part of the tidal zone is almost 10 °C higher than the resistance of snails from near the low-water mark. The temperature given is the highest from which the snails can recover after 1 hour's exposure. [Fraenkel 1968]

Occurrence	Snail	Temperature (°C)
Spray zone, upper part of intertidal zone	Tectarius vilis	48.5
	Planaxis sulcatus	48
	Nodilittorina granularis	47
	Littorina brevicola	47
Shallow splash pools	Paesiella raepstorttiana	47
Middle part of intertidal zone	Nerita japonica	46
	Nerita albicilla	44
Lower part of intertidal zone in sheltered or shadowy places or under stones	Lunella coronata	43
	Drupa granulatus	42
	Purpura clavigerus	42
	Monodonta labis	42
Lower part of intertidal zone at water's edge	Tegula lischkei	39

The extremely low thermal tolerance of these animals is important to the understanding of the mechanism of heat death. Some factors that have been suggested as contributing to heat death are:

1. Denaturation of proteins, thermal coagulation
2. Thermal inactivation of enzymes at rates exceeding rates of formation
3. Inadequate oxygen supply
4. Different temperature effects (Q_{10}) on interdependent metabolic reactions
5. Temperature effects on membrane structure

Let us consider heat denaturation of proteins. It is true that many animals die at temperatures where thermal denaturation of protein takes place, for many proteins are denatured at temperatures above 45 to 55 °C. It is difficult to imagine, however, that thermal denaturation of any protein could take place at +6 °C, the lethal temperature for *Trematomus*.

Another suggestion for the cause of thermal death is thermal inactivation of certain particularly temperature-sensitive enzyme systems. Again, it is difficult to imagine enzymes so thermolabile that they become inactivated at +6 °C. Some enzyme systems from *Trematomus* that have been tested in this regard actually show increased activity up to about 30 °C (Somero and DeVries 1967).

The third possibility, that thermal death is caused by inadequate oxygen supply as the increased temperature accelerates the demand for oxygen, can be excluded in many cases. For example, supplying insects with pure oxygen instead of air does not enable them to survive higher temperatures. Likewise, the trout, a cold-water fish, dies at the same temperature in warm water even if the oxygen content of the water is increased several-fold by aeration with pure oxygen.

The fourth possibility is consistent with much of the information at hand. If various processes in the intermediary metabolism are influenced differently by temperature (have different Q_{10} values), it may lead to the depletion or accumulation of certain intermediary metabolic products. Let us consider the following scheme:

$$A \rightarrow B \rightarrow C \rightarrow D$$
$$\downarrow$$
$$E$$

If in this scheme the process $C \rightarrow D$ is accelerated by increased temperature more than the process $B \rightarrow C$, the intermediary C will be depleted at high temperature. If intermediary C is also needed for some other metabolic purpose, E, it will no longer be available in sufficient quantity. Different temperature sensitivities of the several hundred metabolic enzymes that participate in intermediary metabolism may therefore easily lead to a complete derangement of the normal biochemical balance of the organism.

Although the possibility of different thermal sensitivities of enzyme systems gives a plausible explanation of heat death under some circumstances (e.g., the Antarctic fish that die at +6 °C), it is not necessarily the only factor in heat death. Any of the other possibilities mentioned above may be contributory, and for organisms that have a very high lethal temperature, it is more likely that several of the listed factors play a role.

The last possibility, changes in membrane structure, actually covers a broad range of subjects. The generally accepted view is that a membrane consists of a lipid bilayer with a variety of functional proteins embedded or attached, and that any change in interactions among these can be expected to change the functional properties of the membrane. Temperature (i.e., changes in kinetic energy) has profound effects on such interactions by affecting higher orders of protein structure, protein – lipid interactions, lipid – lipid interactions, and so on – in short, molecular structures that depend on weak interactions that are easily changed by temperature. Such disturbances in the integrity of membrane function appear to be a primary factor in heat damage to organisms.

Tolerance to low temperature

The effects of low temperature are at least as perplexing as those of high temperature. Some organisms can tolerate extensive freezing, but most animals cannot. Some can survive submersion in liquid nitrogen at −196 °C or even liquid helium at −269 °C (about 4 K), but some animals are so sensitive to cold that they die at temperatures far above freezing. People who keep tropical fish in home aquaria know this. If the heater is off, the fish may be found dead on a cold morning. Guppies (*Lebistes reticulatus*), for example, that have been kept at 23 °C room temperature or above will die if they are cooled to about 10 °C (Pitkow 1960).

For this fish the evidence indicates that death is caused by cold depression of the respiratory center, followed by damage from anoxia, for an increase in the oxygen content of the water improves survival, and a decrease in the oxygen content increases susceptibility to cold.

Cold tolerance and freezing tolerance

Animals that live in temperate and cold regions withstand long periods of winter temperatures that are far below the freezing point. They can escape cold injury by two means: *supercooling* and *freezing tolerance*. The former is the lowering of the temperature of a fluid to below its freezing point without formation of ice; the latter is a tolerance to freezing of water and formation of ice in the body. Those animals that cannot tolerate ice formation in the body are called *freezing-susceptible*; those that survive freezing are *freezing-tolerant*.

How does supercooling occur? If water or aqueous solutions are cooled to below the freezing point, freezing does not necessarily take place.*

* The term *freezing point* can have several different meanings. In our use we shall define freezing point as the temperature at which a minute amount of ice is present in a liquid sample that is in thermodynamic equilibrium at that temperature.

Pure water has a freezing point of 0 °C, but an aqueous solution that contains dissolved salts or other solutes must be at a lower temperature for ice to form. Consider such a solution at a temperature somewhat below 0 °C with a few ice crystals present. If heat is added, some ice melts; and if heat is removed, some water turns to ice. In the latter case, the solute concentration in the remaining liquid increases, thus gradually lowering the temperature at which more ice can form. This process could be continued until the entire sample is frozen throughout, but this point is difficult to determine and is only rarely considered. The freezing point, as defined above, is most often determined by freezing the entire sample, observing it during slow heating, and reading the temperature when the last minute crystal of ice is about to disappear. A more correct designation for this point might be the *melting point*, a term not often used in this context.

Pure water can be cooled far below 0 °C without any ice formation. The probability that such supercooled water will freeze depends on three important variables: the temperature, the presence of nuclei for ice formation, and time. In the absence of foreign nucleating materials, pure water is readily supercooled to −20 °C before it freezes, and if extraordinary precautions are taken, water can be supercooled to near −40 °C. The moment an initial ice nucleus is formed, freezing progresses rapidly throughout the sample. The presence of certain solutes not only lowers the freezing point, but also influences the extent of supercooling that can take place before freezing occurs.

We should now consider the possibility that, although an animal is exposed to temperatures considerably below the freezing point of its body fluids, it may remain supercooled. Such supercooling is in fact of great importance to survival. During an occasional cold night, for example, it may be essential for freezing-susceptible animals that are unable to escape. Experiments with reptiles and various other vertebrates, whose body fluids have a freezing point of −0.6 °C, have shown that they can be supercooled to as low as −8 °C without freezing (Lowe et al. 1971).

Other animals survive in spite of extensive ice formation in their bodies. Midge larvae (*Chironomus*) can be frozen to −25 °C and thawed repeatedly without injury. It may seem difficult to decide whether these tiny animals are supercooled or frozen to ice. However, the actual amount of ice can be accurately determined without injury to the animal by an ingenious method that takes advantage of the change in volume as water freezes to ice. By determining the specific gravity of the frozen larva, it was found that at −5 °C 70% of its body water is frozen, and at −15 °C 90% is frozen. As no harm is done to the larva by these determinations, the same specimen in which the ice formation is known can be thawed and tested for survival (Scholander et al. 1953a).

Although many invertebrate animals at high latitudes are normally exposed to very low temperatures during winter, few are as exposed to as rapidly changing conditions as tidal organisms in the subpolar regions. In winter, the temperature of the tidal zone regularly alternates between freezing and thawing twice a day. As the water recedes, an animal in the tidal zone has most of its body water frozen to ice, and as the water rises again, it thaws out. Such a tidal animal may be exposed to air temperatures as low as −30 °C for 6 hours or more, and its internal temperature then approaches that of the air. At −30 °C more than 90% of the body water is frozen and the remaining fluid therefore contains solutes in extremely high concentrations. This means that the cells, in addition to losing water to the ice crystals, must be able to tolerate an exceptional increase in osmotic concentration.

Examination of tidal animals shows that they indeed freeze rather than become supercooled, and that there is a remarkable distortion of muscles and internal organs by the ice. The ice crystals, however, are generally located outside the cells, which are shrunken and possibly have no ice formed within them. Within a few seconds of thawing, the tissues again assume a normal appearance (Kanwisher 1959).

Many plants are also highly tolerant to freezing, and on cold winter days the water in the plants freezes without causing permanent damage. The tolerance changes with the seasons, so that a plant that in summer is killed at temperatures a few degrees below freezing may survive cooling to −196 °C in winter (Weiser 1970).

In addition to the degree of cooling, the dura-

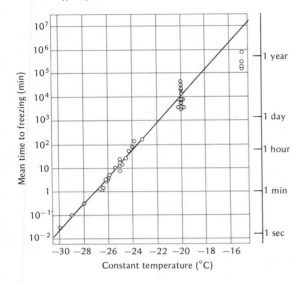

FIGURE 7.7 The relation between maintenance temperature and mean time to freezing for larvae of the wheat-stem sawfly. Each point shows the time required for 50% of a group of larvae to freeze. [Salt 1966]

tion of exposure to a given low temperature is important for determining whether freezing takes place. This time dependence is shown in Figure 7.7 for hibernating larvae of the wheat-stem sawfly, a plant-feeding wasp, *Cephus cinctus*. When these larvae are kept at constant subfreezing temperature, no freezing takes place at temperatures above −15 °C. As the temperature is lowered further, freezing takes place more and more rapidly so that at −30 °C, freezing occurs in about 1.2 seconds, whereas at −17 °C it takes more than a year for 50% of the larvae in a large sample to freeze.

It has long been known that *glycerol* protects red blood cells and mammalian spermatozoa from injury caused by freezing. Glycerol is widely used for this purpose, and samples of human or bull sperm can be kept frozen and remain viable for several years if glycerol is added in a suitable concentration before freezing. Without such treatment, freezing is lethal to spermatozoans.

Because of its well-known protective action, the natural occurrence of high concentrations of glycerol in some insects has been suggested as an explanation of their ability to survive low temperatures. Glycerol could influence the cold resistance of insects in two ways: (1) by its protective action against freezing damage, glycerol could help insects that do freeze to survive, and (2) by lowering the freezing point and increasing the degree of supercooling, glycerol could increase the probability that an insect will completely avoid ice formation.

In many insects the glycerol concentration increases before winter. In the parasitic wasp *Bracon cephi*, the concentration becomes as high as 5 molal (roughly 30%) and decreases again in the spring. This large amount of solute depresses the freezing point of the blood to −17.5 °C. The supercooling points are lowered even more than the freezing points, so that larvae of *Bracon* can be supercooled without ice formation to as low as −47.2 °C. The relationship between supercooling points and freezing points (actually melting points) in the *Bracon* larvae is shown in Figure 7.8. The solid line represents the calculated regression line for all observations; the broken line indicates a constant degree of supercooling of 29 °C below the actual freezing point. Nearly all the observations are located below this line, so that in these animals virtually every individual tested showed a degree of supercooling in excess of 30 °C, and the degree of observed supercooling actually increased with decreasing melting points.

Glycerol, however, is not unique in protecting insects against cold and freezing injury. Of 11 insect species collected in Alberta, Canada, 3 species were freezing-tolerant, although one of these contained less than 3% glycerol and another no glycerol at all. Eight other species, most of which contained more than 15% glycerol, were killed by

FIGURE 7.8 The relation between observed supercooling and melting points for larvae of a parasitic wasp, *Bracon cephi*. Solid line is the calculated regression line for all observations; broken line indicates a constant degree of supercooling of 29 °C. [Salt 1959]

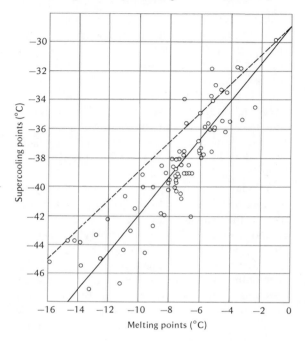

freezing. Evidently glycerol alone cannot protect insects against freezing injuries. There is also a considerable degree of adaptation taking place, for when insects are exposed to moderate cold, they afterward show increased tolerance to freezing temperatures and an increase in the degree of supercooling. This happens both in glycerol-forming species and in non-glycerol-forming species. Thus, the effect of glycerol does not fully explain cold tolerance; other solutes in the hemolymph must contribute (Sømme 1964, 1967).

Antifreeze in fish

Teleost fish have an osmotic concentration in their body fluids of about 300 to 400 mOsm; this corresponds to a freezing point of about −0.6 to −0.8 °C. Sea water in the polar regions often has a

temperature of about −1.8 °C, yet a variety of fish lives in these waters.* When fish swim in water at −1.8 °C, why don't they freeze? Do they have a lower freezing point than ordinary fish, or do they remain supercooled throughout life? It seems that both possibilities can be realized.

These problems have been studied in the Hebron Fjord in northern Labrador, where the temperature of the surface water in summer is several degrees above the freezing point, while the water at the bottom remains at −1.73 °C throughout the year. Several species of fish live in this fjord, both near the surface and at the bottom, and they provide excellent material for studies of the freezing problem.

In summer the blood of all the fish in the Hebron Fjord has approximately the same freezing point as the blood of teleost fish in general, about −0.8 °C (Figure 7.9). For the fish at the surface this causes no problem, but those from the deeper water evidently remain supercooled. If they are caught and removed to the laboratory, they can be supercooled to the temperature of the water in which they were caught (−1.73 °C) and will not freeze. If, however, they are touched with a piece of ice, ice formation rapidly spreads in their bodies and they die almost immediately. It can therefore be assumed that the bottom fish survive throughout their lifetimes in the supercooled state.

Invertebrate animals, of course, have no special problem. Marine invertebrates in general are in os-

---

* The salt content of sea water corresponds to a concentration of nearly 1.0 osmole per liter. The molar freezing point depression for water is 1.86 °C, and sea water therefore does not freeze until its temperature is this low. When ice begins to form, it floats at the surface, both in fresh water and sea water. Fresh water has a point of highest density at + 4 °C, and when ice forms at the surface the bottom water in a lake may still be at + 4 °C. Sea water, however, does not exhibit this peculiarity; the temperature of the polar seas, when there is ice at the surface, is about −1.8 °C throughout.

FIGURE 7.9 In summer, the surface fish in Hebron Fjord, Labrador, have no freezing problem. The fish that live deeper, however, where the water is at −1.73 °C, have a body-fluid freezing point of −1.0 °C and must remain supercooled. [Scholander et al. 1957]

FIGURE 7.10 The antifreeze substance from the Antarctic fish *Trematomus* is abnormally effective in preventing formation of ice in an aqueous solution. On a molar basis it is several hundred times as effective in this regard as expected. The abscissa gives the molal concentration of the substance; the ordinate gives the apparent osmolal concentration as determined from freezing point depression. [DeVries 1970]

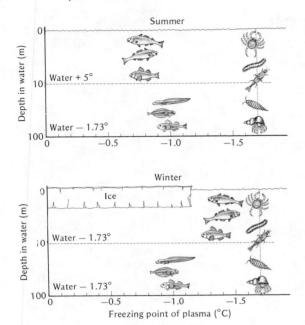

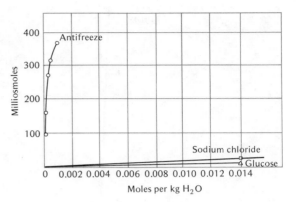

motic equilibrium with the sea water in which they live, and therefore avoid freezing by just remaining in the water.

In winter the bottom fish in Hebron Fjord have the same problem as in summer: Their freezing point is still about −0.8 °C. Because the water temperature remains unchanged at −1.73 °C, they must remain supercooled, for ice formation would kill them immediately.

The fish near the surface, however, come into contact with ice crystals in winter, and if they were only supercooled, they would have no means of avoiding freezing. Instead, in winter the freezing point of the surface fish is lowered to about that of the water in which they swim. These fish show only small increases in the major dissolved ions, sodium and chloride; the most important contribution to lowering the freezing point is made by an

antifreeze substance that at the time the study was done remained unidentified (Gordon et al. 1962).

The antifreeze in the blood of an Antarctic fish, *Trematomus*, has been isolated and its chemical nature studied in detail. Chemically, this antifreeze is a glycoprotein that occurs in the blood with three distinct molecular weights: 10 500, 17 000, and 21 500. At low concentrations (6 g per liter) this glycoprotein is more effective than sodium chloride in preventing formation of ice in water (Komatsu et al. 1970). This is amazing, for the glycoprotein molecule is several hundred times larger than the sodium chloride molecule. This means that if we dissolve equal weights of the glycoprotein and of NaCl in water, the molar concentration of the glycoprotein is several hundred times less. The freezing point depression is normally determined by the number of dissolved particles, and on a molar basis the glycoprotein is therefore abnormally effective in preventing ice formation.

Figure 7.10 shows the freezing point depression caused by various concentrations of glucose, sodium chloride, and the antifreeze. For both sodium chloride and glucose the freezing point depression increases linearly with the concentration, but sodium chloride, because of its dissociation

ICEFISH The blood of the Antarctic fish *Trematomus borchgrevinki* contains a glycoprotein that acts as an antifreeze substance. This permits the fish to swim in sea water at a temperature of −1.8 °C, although the osmotic pressure of its blood, in the absence of the antifreeze, is insufficient to prevent ice formation at this temperature. [Courtesy of A. L. DeVries, University of California, San Diego]

into sodium and chloride ions, has twice as high freezing point depression as glucose. In contrast, even a minute concentration of the antifreeze substance prevents the formation of ice in water about 200 to 500 times as effectively as sodium chloride.

Chemically the glycoproteins from *Trematomus* consist entirely of repeating units of the two amino acids alanine (23%) and threonine (16%), with a disaccharide derived from galactose attached to the threonine unit. This seems to be the first case where the entire structure of a naturally occurring glycoprotein has been completely clarified (DeVries et al. 1970; Shier et al. 1972).

There are at least two ways in which glycoproteins could act to depress the freezing point of aqueous solutions. One is that the water is structured by hydrogen bonding to the glycoprotein; another is that the glycoprotein inhibits the growth of ice crystals by being absorbed to their surfaces, thus hindering the addition of further water molecules to the crystal lattice. The latter possibility is strongly supported by the finding that once a solution of the glycoprotein has been frozen, the temperature at which it melts is much higher than the temperature to which it had to be cooled to get ice formation. The "melting point" thus is not depressed, as it would be if structuring of the liquid water were responsible. The finding is better explained by the model that the glycoproteins lower the point at which ice crystals will grow, apparently because the many −OH groups prevent the orientation of H_2O molecules at the surface of the crystal (DeVries 1974).

The concept that the growth of ice crystals is

inhibited agrees with information from a large number of Arctic and Antarctic fish. Those fish that live in deep water and near the bottom where they can never contact the ice have freezing points of about -0.8 °C, and thus are supercooled by about 1.0 °C. Shallow-water fish, which live in contact with sea ice at -1.8 °C, also have freezing points about -0.8 °C, but their blood plasma is much more resistant to freezing because ice propagation is inhibited. They have higher plasma protein concentrations than most other vertebrates, and the proteins impede the growth of ice crystals (Hargens 1972).

Macromolecular antifreezes have been found in a large number of fish species that belong to 11 families of distant phylogenetic relationships (Duman and DeVries 1975). It appears that the structure of the various antifreezes differs substantially, and therefore that the ability to produce a macromolecular antifreeze has evolved independently in the different families.

PHYSIOLOGICAL TEMPERATURE ADAPTATION

The limits of temperature tolerance for a given animal are not fixed. Exposure to a near-lethal temperature often leads to a certain degree of adaptation so that a previously lethal temperature becomes tolerable.

Frequently the range of thermal tolerance is different for the same species in summer and in winter. A winter animal often tolerates and even is active at a temperature so low that it is lethal to a summer animal; conversely, the winter animal is less tolerant to high temperature than a summer animal. Such changes in the temperature tolerance with climatic changes are called *acclimatization*. Similar effects can be simulated in laboratory experiments by keeping animals for some time at

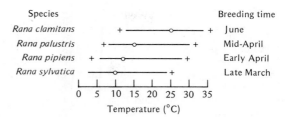

FIGURE 7.11 The relation between temperature tolerance and breeding time in four species of North American frogs. The horizontal lines show the temperature range for normal development, with the average water temperature at the time of egg-laying indicated by a black circle. The crosses indicate lethal temperatures. [Moore 1939]

given temperatures. To distinguish the adaptations or adjustments that take place in laboratory experiments from natural acclimatization, the response to experimental conditions is often described by the term *acclimation* (Hart 1952; Prosser 1973).

Acclimatization and acclimation can take place in response to many environmental factors, such as temperature, oxygen tension, nature of the food, moisture. Here we shall be concerned primarily with responses to temperature, a subject about which we have more information than is available about the responses to other environmental factors.

Geographical differences and seasonal adjustments

It is self-evident that many Arctic animals live, breathe, and move about at temperatures close to freezing and in a range that would completely inactivate or kill similar animals from the tropics. Arctic fish and crustaceans are active in ice-cold water and die when the temperature exceeds 10 to 20 °C. Comparable tropical forms die below 15 to 20 °C. There are obvious differences in the temperature range that can be tolerated by animals from different geographical areas (Scholander et al. 1953b).

Closely related animals, such as various species of frogs of the genus *Rana*, have a geographical distribution that is closely related to their temperature tolerance. Figure 7.11 shows that, of four spe-

TABLE 7.3 Comparison of breeding and development of four species of the frog *Rana*. [Moore 1939]

	R. sylvatica	R. pipiens	R. palustris	R. clamitans
Order of breeding	1	2	3	4
Water temperature at time of breeding (°C)	10	12	15	25
Northernmost record (°N)	67	60	55	50
Lower limiting embryonic temperature (°C)	2.5	6	7	11
Upper limiting embryonic temperature (°C)	24	28	30	35
Time between stages 3 and 20 at 18.5 °C (h)	87	116	126	138

cies of North American frogs, the species that has the northernmost geographical limit is also the most cold-tolerant.

The correlation between distribution and temperature tolerance extends to the temperature of the water at the time of breeding and to the lower as well as the upper limiting temperatures for embryonic development. Frogs that breed at relatively low environmental temperatures have low minimal and maximal temperatures for breeding and for embryonic development (Table 7.3).

The geographically most widely distributed frog of these four species is *R. pipiens*. The population of this species from the northern part of the range differs from southern populations in the same way that the northern species differs from southern species. Thus, *R. pipiens* exists in races that show physiological differences in their responses to temperature. This is clearly related to the unusually wide geographical range over which this species is distributed (Moore 1949).

We should now examine more closely some natural and experimentally induced changes in temperature tolerance. In large parts of the world the seasonal temperature fluctuations are appreciable, and during one season animals tolerate temperature extremes that in other seasons are fatal. For example, the bullhead, a catfish (*Ictalurus* [*Ameiurus*] *nebulosus*) has an upper lethal temperature of nearly 36 °C in summer, but in winter this fish will die if the temperature of the

water exceeds 28 °C (Figure 7.12). These changes in temperature tolerance occur in response to the natural climatic change between summer and winter and, therefore, present an example of natural acclimatization.

The lower lethal temperature also changes with the seasons. As an example, summer individuals of the beetle *Pterosticus brevicornis*, from Alaska, invariably die if they are frozen (which happens at −6.6 °C). In winter, however, they tolerate temperatures below −35 °C and complete freezing as well (Miller 1969). The fact that this beetle actually tolerates freezing is important, for it has been stated that no adult insect is able to tolerate freezing and that freezing tolerance is limited to larval and pupal forms. Obviously, this does not hold as a general rule.

We have now seen that there are phylogenetic differences among animals that are reflected in their geographical distribution and that there are seasonal changes in their thermal tolerance (acclimatization), changes that can be mimicked by laboratory exposure to appropriate thermal conditions (acclimation) while the phylogenetic limits remain unchanged.

Maximum range of tolerance

Since the limits of temperature tolerance are modified by the previous thermal history of an animal, how far can the limit be pushed by slow and gradual acclimation? Is there an ultimate limit

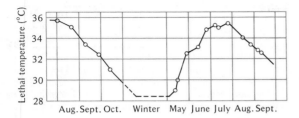

FIGURE 7.12 The bullhead, *Ictalurus* (*Ameiurus*) *nebulosus*, has an upper lethal temperature limit that changes with the seasons. Each point indicates the temperature that is fatal to 50% of the samples after an exposure of 12 hours. [Fry 1947]

FIGURE 7.13 The complete range of thermal tolerance of goldfish is shown by the solid line. If a goldfish is kept at some temperature on the diagonal broken line until it is fully acclimated, its tolerance to high temperature can be read on the upper solid line, and its tolerance to cold on the lower solid line. For example, a fish kept at 30 °C has an upper lethal limit of 38 °C and a lower lethal limit of 9 °C. [Fry et al. 1942]

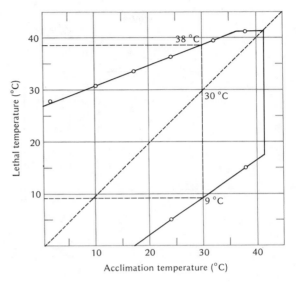

that cannot be exceeded? To answer this question groups of animals are kept at various constant temperatures for sufficient periods of time to ensure that they are fully acclimated at these temperatures. They can then be tested in order to establish the upper and lower limits of thermal tolerance, and when this has been done for a series of different temperatures, the results can be plotted in a single graph. Figure 7.13 is such a plot for the goldfish, showing its complete thermal tolerance.

The goldfish has an exceptionally wide temperature tolerance, and the area enclosed by the tolerance curve is therefore large. The tolerance curve for a fish with a narrower range, such as the chum salmon (*Oncorhynchus keta*), covers a much smaller area. Like other salmonids the chum salmon is a typical cold-water fish, and its upper lethal temperature cannot be extended beyond 24 °C. This is shown in Figure 7.14, which gives the thermal tolerance range for the chum salmon and for the brown bullhead, *Ictalurus nebulosus*, a fish with a large tolerance range, similar to the goldfish. The total area enclosed by the tolerance curve for the bullhead is more than twice as large as that for the chum salmon. To quantify the area, we can take as our unit of measurement a small square which corresponds to the division of 1 degree on each coordinate. The area enclosed by the bullhead curve is 1162 such squares, which properly have the dimension degrees squared, (°C)²; the curve for the chum salmon encloses 468 squares.

Most other fish whose thermal tolerance has been established fall between these two extremes. Both the bullhead and the chum salmon tolerate water at freezing temperatures, but many warm-water species have curves that do not extend down to 0 °C, but are closed at some higher temperature. However, the complete thermal tolerance is known for only a relatively small number of fish, and it seems likely that, on the whole, fresh-water fish may have a wider tolerance than marine fish.

The tolerance of fresh-water fish ranges from 450 (°C)² for salmon to 1220 for goldfish (Brett 1956). Of three salt-water fish that have been examined, the silverside (*Menidia menidia*) has the largest range, 715 (°C)²; a puffer (swellfish, *Spheroides maculatus*) has the smallest, 550 (°C)²; and the winter flounder (*Pseudopleuronectes americanus*) has a range intermediate between the other two (Hoff and Westman 1966).

For easy comparison these data are combined with information about other fish in Table 7.4.

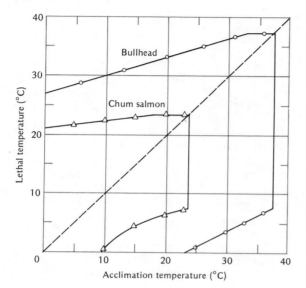

FIGURE 7.14 The ultimate thermal tolerance of the chum salmon, a cold-water fish, covers a much smaller range than the bullhead's. The latter has a wide temperature tolerance similar to that of the goldfish (Figure 7.13). [Brett 1956]

TABLE 7.4 Thermal tolerance of various fresh-water and salt-water fish, expressed in units of degrees squared (see text for explanation). The American lobster is included for comparison with the fish.

Fish	Thermal tolerance $(°C)^2$
Goldfish (*Carassius auratus*)	1220[a]
Bullhead (*Ictalurus nebulosus*)	1162[a]
Greenfish (*Girella nigricans*)	800[b]
Silverside (*Menidia menidia*)	715[c]
Winter flounder (*Pseudopleuronectes americanus*)	685[c]
Speckled trout (*Salvelinus fontinalis*)	625[d]
Puffer (*Spheroides maculatus*)	550[c]
Chum salmon (*Oncorhynchus keta*)	468[a]
Lobster (*Homarus americanus*)	830[e]

[a] Brett (1956).
[b] Doudoroff (cited in McLeese 1956).
[c] Hoff and Westman (1966).
[d] Fry et al. (1946).
[e] McLeese (1956).

The much greater thermal range of bullheads and goldfish is attributable to their better tolerance of high water temperatures, a fact obviously related to their habitat, which is shallow fresh water in which the temperature can rise more readily than in the cool streams where the salmon is at home. The American lobster is added at the bottom of the list as a representative of an economically important invertebrate; if the lobster were a fish it would be closer to the top of the list.

More information about the thermal tolerance of fish is much needed, for the limits to thermal adaptation will become increasingly important in connection with problems of thermal pollution from industrial activity, especially power generating plants and atomic reactors.

Rate of acclimation

The time needed for a fish to increase its ability to tolerate high temperatures is relatively short. It takes less than 24 hours for a bullhead to adjust

fully when it is moved from 20 °C to 28 °C (Figure 7.15). The process of temperature acclimation depends on an adequate supply of oxygen. Although the bullhead is tolerant to low oxygen and survives quite well, the process of thermal acclimation seems to be completely impeded in the absence of sufficient oxygen. In general, the acclimation goes faster the higher the temperature. Therefore it seems to be related to metabolic processes, but there is no clear understanding of the mechanisms responsible for the modification of the lethal limit, although changes in metabolic enzyme systems have an important role (Hochachka and Somero 1973).

If bullheads are given the opportunity to select their own preferred water temperature by presenting them with a temperature gradient, the water temperature they select depends on their previous thermal history (Table 7.5). Individuals acclimated to 7 °C initially select water at 16 °C, but they soon drift upward in their preference. After 10

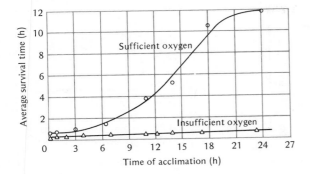

FIGURE 7.15 The thermal acclimation of the bullhead takes place within about 24 hours when the water contains adequate oxygen, but proceeds very slowly at low oxygen concentration. [Fry 1947]

TABLE 7.5 Brown bullheads (*Ictalurus nebulosus*) acclimated to different temperatures between 7 and 32 °C initially show different thermal preferences, but during 24 hours all approach the same final thermal preference of about 29 °C. [Crawshaw 1975]

Acclimation temperature (°C)	Preference temperature (°C)		
	Immediate	10 h	24 h
7	16	25	~ 29
15	21	25	~ 29
24	26	29	~ 29
32	31	30	~ 29

hours they prefer 25 °C, and after 24 hours they select the same range of 29 to 30 °C that the fish warm-acclimated at 32 °C also select after 24 hours.

When a fish is moved from a higher to a lower temperature, the loss in tolerance to high temperatures and the gain in resistance to low temperatures are slower. In a minnow (*Pimephales promelas*) the complete adaptation takes between 10 and 20 days when the fish is moved from 24 °C to 16 °C (Brett 1956).

We have now seen that the thermal tolerance of a fish depends on the thermal history of the individual and that oxygen plays a role in the speed with which adaptation to a new temperature regimen takes place. However, other factors are important, such as the age and size of the fish and the quality of the water. Of particular interest is the finding that goldfish, even if they are kept in the laboratory at completely constant conditions of temperature and diet, show a marked seasonal change in thermal tolerance, with a greater cold resistance in the winter. This difference has been ascribed to the photoperiod (Hoar 1956). In general, a long photoperiod, as in summer, increases resistance to heat, and a short photoperiod, as in winter, causes an increase in tolerance to cold (Roberts 1964).

Thermal acclimation and metabolic rate

Until now the discussion has been concerned with lethal temperatures and adaptation as a result of the thermal history of an animal. However, adaptation is not restricted to a change in lethal temperature limits; there are other compensating mechanisms as well, and these usually tend to counteract the acute effects of temperature change. Anybody who has caught fish in winter has noticed that the fish are not appreciably slower and more sluggish than in summer. Laboratory studies confirm the impression that fish, if given time, to a great extent can compensate for temperature changes by appropriate alterations in metabolic rate. This is true not only for fish, but for many other poikilothermic animals as well.

Let us assume that a fish, adapted to a winter temperature of 5 °C, has a temperature tolerance range of 0 to 25 °C and that the Q_{10} throughout this range is 2.0. The metabolic rate of this fish follows the straight line marked a in Figure 7.16. Assume that we acclimate this fish to a summer temperature of 25 °C and that its upper lethal limit is extended to 40 °C. If no metabolic compensation takes place, the fish will have a metabolic rate extending along the broken line marked a'. If, on the other hand, the fish compensates fully for the change in temperature, the summer fish after adaptation to 25 °C will have the metabolic rate it previously had at 5 °C. If the Q_{10} remains the same (2.0), the oxygen consumption of the fish will now follow the line marked b. Thus, the fish will have

FIGURE 7.16 Hypothetical change in metabolic rate (oxygen consumption, on logarithmic ordinate) in a fish acclimated to 5 °C water (*a*). If the fish compensates fully, acclimation to 25 °C water will make its metabolic rate follow the line marked *b*. If it does not compensate at all, its metabolic rate will follow the broken line marked *a'*.

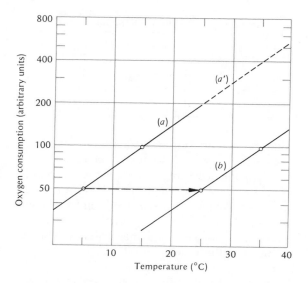

FIGURE 7.17 The compensation for a temperature change, from T_1 to T_2, may take any of several courses, as indicated by the various curves. For details, see text. [Precht et al. 1955]

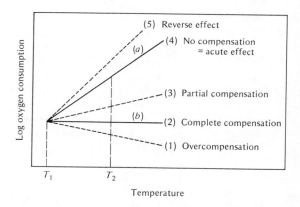

compensated fully for the change in temperature and will have returned to its previous metabolic rate.

In addition to the two possibilities just mentioned, no compensation or complete compensation, there are several other long-term types or patterns of adaptation. The various patterns can be generalized as indicated in Figure 7.17. If an animal shows no compensation and continues to consume oxygen at the same rate as caused by an acute effect of a temperature change, its metabolic rate will follow line *a* (the same as *a'* in Figure 7.16). Complete compensation, on the other hand, involves a return to the original oxygen consumption after the animal has had time to adapt to a new temperature, and the oxygen consumption will therefore follow line *b*, showing complete compensation. Any response between these two lines indicates partial compensation for the temperature change. The possibility also exists that the animal

may overcompensate, as indicated by line 1. A few cases have been reported when a temperature change has a reverse effect, as indicated by line 5, at least within part of the temperature range.

In reality, animals do not often follow these idealized patterns; their responses are frequently far more complex. As an example, we can examine the responses of two aquatic animals, the trout and the European crayfish, to a 10 °C drop in temperature, followed by a 10 °C increase in temperature after full adaptation has taken place (Figure 7.18).

An animal that shows little or no compensation is the beach flea (*Talorchestia megalophthalma*) from the area of Woods Hole, Massachusetts. The oxygen consumption of this animal in winter is the same as in summer, when measured at the same temperature. Thus there is virtually no acclimatization or compensation for the seasonal change in temperature. Accordingly, the winter animals have such a low metabolic rate that they remain inactive, and they can be found immobile in small burrows, hibernating beneath the sand. If they are disturbed in winter, they respond very sluggishly, but when warmed they soon become as active as in summer (Edwards and Irving 1943b).

FIGURE 7.18 Metabolic response to temperature change of crayfish and trout. Initially the animals were acclimated to 15 °C. When the water temperature was dropped by 10 °C, the trout rapidly achieved a new plateau; the crayfish reached a plateau more slowly. On raising the temperature to 15 °C again, the trout overcompensated before returning to the initial level; the crayfish only slowly responded to the temperature rise. [Bullock 1955]

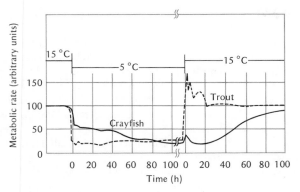

In contrast, another crustacean from the same habitat, the sand crab (*Emerita talpoida*), compensates for low winter temperatures by increasing its metabolic rate. At 3 °C, winter animals consume oxygen at a rate four times as high as that of summer animals at the same temperature. These animals can remain active in cold weather, but because the sandy beaches of Woods Hole are most inhospitable during winter, the animals move from the intertidal zone out to deeper water (Edwards and Irving 1943a). Thus the sand crab adjusts to the seasonal changes in temperature so that it can remain active at the low winter temperatures, instead of becoming inactive and hibernating. Its lethal temperature also changes with the seasons. Temperatures above 27 °C are lethal in winter, but for summer animals death occurs above 37 °C.

We have now discussed a variety of important physiological effects of temperature, not only the lethal effects of very high or very low temperatures, but also the way organisms react and adjust within the range that is normal to them. We shall next turn to the subject of how animals make themselves independent of the thermal whims and vagaries of their environment.

REFERENCES

Bott, T. L, and Brock, T. D. (1969) Bacterial growth rates above 90 °C in Yellowstone hot springs. *Science* 164:1411–1412.

Brett, J. R. (1956) Some principles in the thermal requirements of fishes. *Q. Rev. Biol.* 31:75–87.

Brock, T. D. (1970) High temperature systems. *Annu. Rev. Ecol. Syst.* 1:191–220.

Brown, J. H., and Feldmeth, C. R. (1971) Evolution in constant and fluctuating environments: Thermal tolerances of desert pupfish (*Cyprinodon*). *Evolution* 25:390–398.

Bullock, T. H. (1955) Compensation for temperature in the metabolism and activity of poikilotherms. *Biol. Rev.* 30:311–342.

Carlisle, D. B. (1968). *Triops* (Entomostraca) eggs killed only by boiling. *Science* 161:279–280.

Crawshaw, L. I. (1975) Attainment of the final thermal preferendum in brown bullheads acclimated to different temperatures. *Comp. Biochem. Physiol.* 52A:171–173.

DeVries, A. L. (1970) Freezing resistance in antarctic fishes. In *Antarctic Ecology*, vol. 1 (M. Holdgate, ed.), pp. 320–328. New York: Academic Press.

DeVries, A. L. (1974) Survival at freezing temperatures. In *Biochemical and Biophysical Perspectives in Marine Biology*, vol. 1 (D. C. Malins and J. R. Sargent, eds.), pp. 289–330. New York: Academic Press.

DeVries, A. L., Komatsu, S. K., and Feeney, R. E. (1970) Chemical and physical properties of freezing point-depressing glycoproteins from antarctic fishes. *J. Biol. Chem.* 245:2901–2908.

Duman, J. G., and DeVries, A. L. (1975) The role of macromolecular antifreezes in cold water fishes. *Comp. Biochem. Physiol.* 52A:193–199.

Edwards, G. A., and Irving, L. (1943a) The influence of temperature and season upon the oxygen consumption of the sand crab, *Emerita talpoida* Say. *J. Cell. Comp. Physiol.* 21:169–182.

Edwards, G. A., and Irving, L. (1943b) The influence of season and temperature upon the oxygen consumption of the beach flea, *Talorchestia megalophthalma*. *J. Cell. Comp. Physiol.* 21:183–189.

Fraenkel, G. (1968) The heat resistance of intertidal snails at Bimini, Bahamas; Ocean Springs, Mississippi; and Woods Hole, Massachusetts. *Physiol. Zool. 41:* 1–13.

Fry, F. E. J. (1947) Effects of the environment on animal activity. University of Toronto Studies, Biological Services, no. 55. *Publ. Ontario Fish. Res. Lab.* 68:1–62.

Fry, F. E. J., Brett, J. R., and Clawson, G. H. (1942) Lethal limits of temperature for young goldfish. *Rev. Can. Biol. 1:*50–56.

Fry, F. E. J., Hart, J. S., and Walker, K. F. (1946) Lethal temperature relations for a sample of young speckled trout (*Salvelinus fontinalis*). University of Toronto Studies, Biological Series, no. 54. *Publ. Ontario Fish. Res. Lab. 66:*1–35.

Gordon, M. S., Amdur, B. H., and Scholander, P. F. (1962) Freezing resistance in some northern fishes. *Biol. Bull. 122:*52–56.

Hargens, A. R. (1972) Freezing resistance in polar fishes. *Science 176:*184–186.

Hart, J. S. (1952) Geographic variations of some physiological and morphological characters in certain freshwater fish. University of Toronto Studies, Biological Series, no. 60 *Publ. Ontario Fish. Res. Lab. 72:*1–79.

Hinton, H. E. (1960) A fly larva that tolerates dehydration and temperatures of −270 °C to +102 °C. *Nature, Lond.* 188:336–337.

Hoar, W. S. (1956) Photoperiodism and thermal resistance of goldfish. *Nature, Lond.* 178:364–365.

Hochachka, P. W., and Somero, G. N. (1973) *Strategies of Biochemical Adaptation.* Philadelphia: Saunders. 358 pp.

Hoff, J. G., and Westman, J. R. (1966) The temperature tolerances of three species of marine fishes. *J. Mar. Res.* 24:131–140.

Kanwisher, J. (1959) Histology and metabolism of frozen intertidal animals. *Biol. Bull.* 116:258–264.

Komatsu, S. K., DeVries, A. L., and Feeney, R. E. (1970) Studies of the structure of freezing point-depressing glycoproteins from an antarctic fish. *J. Biol. Chem.* 245:2909–2913.

Lowe, C. H., Lardner, P. J., and Halpern, E. A. (1971) Supercooling in reptiles and other vertebrates. *Comp. Biochem. Physiol.* 39A:125–135.

Marzusch, K. (1952) Untersuchungen über die Temperaturabhängigkeit von Lebensprozessen bei Insekten unter besonderer Berücksichtigung winterschlafender Kartoffelkäfer. *Z. Vergl. Physiol.* 34:75–92.

McLeese, D. W. (1956) Effects of temperature, salinity, and oxygen on the survival of the American Lobster. *J. Fish. Res. Bd. Can.* 13:247–272.

Miller, L. K. (1969) Freezing tolerance in an adult insect. *Science 166:*105–106.

Moore, J. A. (1939) Temperature tolerance and rates of development in the eggs of Amphibia. *Ecology* 20:459–478.

Moore, J. A. (1949) Geographic variation of adaptive characters in *Rana pipiens* Schreber. *Evolution* 3:1–24.

Pitkow, R. B. (1960) Cold death in the guppy. *Biol. Bull.* 119:231–245.

Precht, H., Christophersen, J., and Hensel, H. (1955) *Temperatur und Leben.* Berlin: Springer-Verlag. 514 pp.

Prosser, C. L. (1973) *Comparative Animal Physiology*, 3rd ed. Philadelphia: Saunders. 966 pp.

Roberts, J. L. (1964) Metabolic responses of fresh-water sunfish to seasonal photoperiods and temperatures. *Helgol. Wiss. Meeresunter* 9:459–473.

Salt, R. W. (1959) Role of glycerol in the cold-hardening of *Bracon cephi* (Gahan). *Can. J. Zool.* 37:59–69.

Salt, R. W. (1966) Relation between time of freezing and temperature in supercooled larvae of *Cephus cinctus* Nort. *Can. J. Zool.* 44: 947–952.

Scholander, P. F., Flagg, W., Hock, R. J., and Irving, L. (1953a) Studies on the physiology of frozen plants and animals in the arctic. *J. Cell. Comp. Physiol.* 42: (Suppl. 1); 1–56.

Scholander, P. F., Flagg, W., Walters, V., and Irving, L. (1953b). Climatic adaptation in arctic and tropical poikilotherms. *Physiol. Zool.* 26:67–92.

Scholander, P. F., Van Dam, L., Kanwisher, J. W., Hammel, H. T., and Gordon, M. S. (1957) Supercooling and osmoregulation in arctic fish. *J. Cell. Comp. Physiol.* 49:5–24.

Shier, W. T., Lin, Y., and DeVries, A. L. (1972) Structure and mode of action of glycoproteins from an antarctic fish. *Biochim. Biophys. Acta* 263:406–413.

Somero, G. N., and DeVries, A. L. (1967) Temperature tolerance of some antarctic fishes. *Science* 156:257–258.

Sømme, L. (1964) Effects of glycerol on cold-hardiness in insects. *Can. J. Zool.* 42:87–101.

Sømme, L. (1967) The effect of temperature and anoxia on haemolymph composition and supercooling in three overwintering insects. *J. Insect Physiol.* 13:805–814.

Weiser, C. J. (1970) Cold resistance and injury in woody plants. *Science* 169:1269–1278.

ADDITIONAL READING*

Asahina, E. (1969) Frost resistance in insects. *Adv. Insect Physiol.* 6:1–49.

Hochachka, P. W., and Somero, G. N. (1973) *Strategies of Biochemical Adaptation. Philadelphia: Saunders.* 358 pp.

Precht, H., Christophersen, J., Hensel, H., and Larcher, W. (1973) *Temperature and Life.* New York: Springer-Verlag. 779 pp.

Somero, G. N., and Hochachka, P. W. (1976) Biochemical adaptations to temperature. In *Adaptation to Environment: Essays on the Physiology of Marine Animals* (R. C. Newell, ed.), pp. 125–190. London: Butterworth.

* See also reference list at end of Chapter 8.

CHAPTER EIGHT

Temperature regulation

Chapter 7 was concerned with the effect on physiological processes of the temperature of the organism, which for many animals is virtually the same as that of the surroundings. It would seem that great benefit could be derived from independence of this environmental variable, and in this chapter we shall be concerned with animals that keep their body temperature more or less constant, independent of the environment. What birds and mammals can achieve in this regard is much more effective than what other animals can do, and it is reasonable to treat them first. This will set the stage for a discussion of other animals that to some extent can control their temperature independently of the environment.

First, it is necessary to clarify what we mean by body temperature, by no means a simple concept. Second, for the body to maintain a given constant temperature heat transfer must be balanced so that the loss and the gain of heat are equal. To understand these two processes requires familiarity with the simple physics of heat transfer.

In a cold environment heat balance can be achieved by manipulating the heat loss and/or the heat gain (heat production). Most mammals and birds do this very successfully, but some mammals and a few birds appear to give up and permit their temperature to drop precipitously; they go into torpor or hibernation. But they have not fully abandoned temperature regulation; on the contrary, hibernation is a well-regulated and controlled physiological state.

In a hot environment the problems of maintaining the body temperature are reversed: The animal must keep the body temperature from rising and is often compelled to cool itself by evaporation of water.

Birds and mammals are so successful at regulating their body temperature that they live through most of their lives with body temperature fluctua-

TABLE 8.1 Heat production in the major organs of a man at rest (body weight, 65 kg; heat production, 1872 kcal per day = 78 kcal per hour = 90.65 W). The main internal organs weigh about 5 kg but account for 72% of the total heat production. [Aschoff et al. 1971]

Organ	Organ weight				Heat production		
	kg		% of body weight		kcal h^{-1}	% of total	
Kidneys	0.29		0.45		6.0	7.7	
Heart	0.29		0.45		8.4	10.7	
Lungs	0.60	5.03	0.9	7.7	3.4	4.4	72.4
Brain	1.35		2.1		12.5	16.0	
Splanchnic organs[a]	2.50		3.8		26.2	33.6	
Skin	5.00		7.8		1.5	1.9	
Muscle	27.00	59.97	41.5	92.3	12.2	15.7	27.6
Other	27.97		43.0		7.8	10.0	
Total	65.00		100.0		78.0	100.0	

[a] Abdominal organs, not including kidneys.

tions of no more than a few degrees. However, some other animals are also amazingly adept at keeping their temperatures different from that of the environment. This applies to terrestrial vertebrates (e.g., lizards); to many insects, particularly highly active ones; and amazingly, even to some fish, which maintain parts of their body at temperatures approaching those of "warm-blooded" animals.

BODY TEMPERATURE OF BIRDS AND MAMMALS

What is body temperature?

The heat produced by an animal must be transported to the surface before it can be transferred to the environment. Therefore, the surface of the organism must be at a lower temperature than the inner parts, for if the temperature were the same throughout, no heat could be transferred. The conclusion is that the temperature of an organism of necessity cannot be uniform throughout.

If we examine where in the mammalian body heat production takes place, we find that some parts produce more heat than others. In man the

organs of the chest and the abdomen, although they make up less than 6% of the body mass, produce 56% of the total heat (Table 8.1). If we include the brain, which in man is large and has a high heat production, we have accounted for 72%, or more than two-thirds, of the total heat production.

We can thus consider that the body consists of a core where most of the heat production takes place and a much larger shell that includes skin and muscles and produces only a small fraction of the total body heat. During exercise, the situation is different, for the total metabolic rate may increase 10-fold or more. Most of this increase occurs in the muscles (including the diaphragm and other respiratory muscles).

During exercise, then, for the internal temperature to remain constant, more than 10 times as much heat as was produced at rest must be transported to the surface of the organism.

Temperature distribution in the body

The inner, or core temperature remains reasonably constant, but this does not mean that the temperature throughout the core is uniform. Organs that have a high rate of heat production may be

the legs and arms. At 20 °C room temperature, the temperature gradients in the shell extend throughout the legs and arms, and the core temperature is restricted to the trunk and head. [Aschoff and Wever 1958]

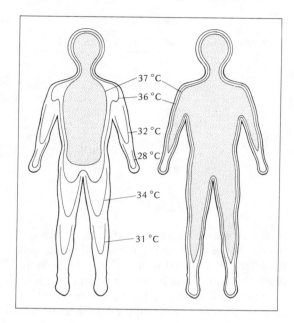

tissues, including a large part of the muscle mass, can take on temperatures considerably below the core temperature (Figure 8.1).

What is the mean temperature of the body? We can calculate a mean body temperature from multiple measurements at various sites, provided we have a weighting procedure to apply to the various measurements. However, mean body temperature calculated from multiple measurements is physiologically relatively meaningless. We have already seen that the core temperature does not represent the heat status of the whole body, that the temperature of the shell can vary widely, and that the depth in the body to which the shell temperature extends can change drastically. A change in the temperature of the shell means that the total heat content of the body changes, although the core temperature may remain constant. If a man moves from a room temperature of 35 °C to 20 °C, the drop in shell temperature may involve a heat loss of 200 kcal (> 800 kJ) from the shell. The magnitude of this much heat becomes clear when we realize that it corresponds roughly to 3 hours' resting metabolism of the man.

"Normal" body temperature of birds and mammals

Daily fluctuations in core temperature

The core temperature of man and of other mammals and birds undergoes regular daily fluctuations. Within 24 hours these fluctuations are usually between 1 and 2 °C. Diurnal animals show a temperature peak during the day and a minimum at night; nocturnal animals show the reverse pattern. However, these daily cycles are not caused directly by the alternating periods of activity and rest, for they continue even if the organism is at complete rest.

The daily body temperature pattern of many

warmer than others, but they are cooled by the blood (i.e., the venous blood that leaves these organs is warmer than the arterial blood). The temperature differences in the core may be as much as 0.5 °C from one site to another. We therefore cannot speak about a single core temperature, but for practical purposes the deep rectal temperature is often used as a representative measure.

The surface temperature of a man who is in heat balance is always lower than the core temperature. This means that the arterial blood which flows to the shell loses heat and returns as colder venous blood. This is, of course, how most of the heat produced in the core is brought to the surface or, in other words, how the core organs are cooled. Depending on the circumstances (external temperatures and the need for heat loss), the surface temperature varies a great deal. Also, the underlying

FIGURE 8.2 When the towhee (a finch, *Pipilo aberti*) is kept at a constant room temperature of 23 °C, its body temperature varies with the light cycle. When the lights come on at 0600 hours, the body temperature rises by nearly 3 °C, to drop again when the lights go off at 1800 hours. If the room temperature is reduced to 5 °C, the body temperature cycle is similar, but at a slightly higher level. [Dawson 1954]

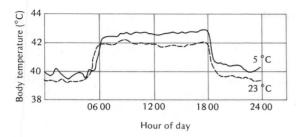

mammals and birds consistently follows the light cycle. When the towhee (*Pipilo aberti*) is kept in 12 hours of darkness and 12 hours of light, its body temperature follows a cycle synchronized with the light cycle. At a room temperature of 23 °C, the core temperature at night is about 39 °C and when the light comes on, it rapidly rises to nearly 42 °C (Figure 8.2). At a different room temperature (5 °C, for example) the core temperature follows the same day and night cycle, although it is, on the average, about 0.5 °C higher than the corresponding core temperature measured at 23 °C.

Similar temperature cycles have been recorded in a variety of other birds from several different families. Diurnal forms always have the highest temperature during the day and nocturnal species at night. One example of a nocturnal bird is the flightless kiwi (*Apteryx*) from New Zealand. Its core temperature is lower than in most other birds, with a daytime mean of 36.9 °C. At night, which is the normal period of activity for the kiwi, the temperature increases to a mean of 38.4 °C (i.e., at night the temperature is about 1.5 °C higher than in the daytime) (Farner 1956).

Both the temperature cycle and the corresponding cycle in metabolic rate can be reversed by reversing the periods of light and dark. This shows that the cycle is governed by the illumination (Dawson 1954). However, even if all variations in the light are removed and the animals are kept in continuous uniform light, the temperature variations continue on nearly the same timing. Because the regular cycle persists without any external cues, it must be inherent in the organism; it must be a truly endogenous cycle. The cycle is self-sustained in the sense that it continues for days, weeks, and even months, although the timing usually slowly drifts away from an exact 24-hour cycle. To make a measurement of body temperature really meaningful, it is therefore desirable to know details of activity, the time of day, and the usual temperature cycle of the animal.

Temperature differences among animal groups

If we disregard variations that amount to a couple of degrees, we see that the usual or "normal" core temperature is almost uniform within each of the major groups of warm-blooded vertebrates, but that there are characteristic differences among the groups (Table 8.2).

Allowing for the fact that it is difficult to establish what is normal body temperature for a given animal, and disregarding variations caused by external conditions and activity, we can as a rule of thumb say that most birds maintain their body temperature at 40 ± 2 °C, eutherian mammals at 38 ± 2 °C, marsupials at 36 ± 2 °C, and monotremes at 31 ± 2 °C. Small birds may have somewhat higher body temperatures than large birds (McNab 1966), but in mammals there is no clear relationship between body size and body temperature (Morrison and Ryser 1952). For marsupials the available information is insufficient to show whether there is any relationship between body size and body temperature.

The fact that the so-called "primitive" groups — insectivores, marsupials, and especially monotremes — consistently have low body temperatures raises some interesting evolutionary problems.

TABLE 8.2 Approximate normal and lethal core temperatures of some major groups of mammals and birds. The lethal temperatures are based on observations made under a wide variety of conditions. There is rather consistently an approximately 6 °C interval between the normal and the lethal temperatures for the same animal.

Animal	Approximate normal core temperature (°C)	Approximate lethal core temperature (°C)
Monotreme (echidna)	30–31[a]	37[a]
Marsupials	35–36[b]	40–41[e]
Insectivore (hedgehog)	34–36	41[f]
Man	37	43
Eutherian mammals	36–38[c]	42–44[g]
Bird (kiwi)	38[d]	
Birds, nonpasserine	39–40[b]	46[h]
Birds, passerine	40–41[b]	47[i,j]

[a] Schmidt-Nielsen et al. (1966).
[b] Dawson and Hulbert (1970).
[c] Morrison and Ryser (1952).
[d] Farner (1956).
[e] Robinson and Morrison (1957).
[f] Shkolnik and Schmidt-Nielsen (1976).
[g] Adolph (1947).
[h] Robinson and Lee (1946).
[i] Calder (1964).
[j] Dawson (1954).

These groups are considered to be very ancient and presumably have had more time than more recent groups to evolve toward a high body temperature, if this is "desirable" or advantageous. Have they remained in their more "primitive" stage because they have lacked the capacity to evolve in this direction? They have certainly been successful as witnessed by their ability to survive for so long.

The fact is that we do not fully understand the advantages of any given body temperature. In any event, it would be a mistake to interpret a low body temperature as a sign of "primitive" and thus inadequate temperature regulation. It has been said that the egg-laying echidna is halfway to being a cold-blooded animal and is unable to regulate its body temperature adequately. In fact, the echidna is an excellent temperature regulator and can maintain its core temperature over a wide range of ambient temperatures down to freezing or below,

although it has a poor tolerance to high temperature (Schmidt-Nielsen et al. 1966).

The approximate lethal body temperatures for the various groups of warm-blooded vertebrates are given in the last column of Table 8.2. It seems that the lethal temperature is regularly at roughly 6 °C above the normal core temperature. Thus, the echidna dies when its body temperature reaches 37 °C, which is a normal temperature for mammals and well below the normal temperature for birds. The margin of safety remains uniformly at about 6 °C from group to group, but we do not know why the lethal temperature is at such a constant level relative to the normal core temperature.

Temperature of cold-climate animals

Do Arctic birds and mammals maintain body temperatures within the same range as species from warmer climates?

To answer this question the body temperatures of a number of Alaskan birds and mammals were measured while the animals were exposed to a wide range of low air temperatures. The birds studied belonged to 30 different species and ranged in weight from 0.01 to 2 kg. In air temperatures from + 20 °C down to − 30 °C, their mean body temperature was 41.1 °C, which is within the normal range for birds from moderate or tropical climates.

Arctic mammals of 22 species, weighing from 0.1 to 1000 kg, were exposed to temperatures down to − 50 °C, and one species (the white fox) even to −80 °C. The individuals of all species maintained their body temperatures within normal mammalian limits. The mean for all the observed species was 38.6 °C, which is about 0.5 °C higher than the previously reported mean temperature for a large number of mammals from temperate regions, but because of the limited sample size, a difference of 0.5 °C is considered insignificant (Irving and Krog 1954).

We can therefore conclude that Arctic birds and mammals maintain body temperatures characteristic of their groups, although they live in some of the coldest areas on earth.

Animals in natural conditions

In contrast to the large amount of information on body temperature in captive and laboratory animals, there are few long-term studies on animals under natural conditions. The development of methods to measure temperature with transducers that transmit information by radio (telemetry) has made it possible to obtain recordings from unrestrained and completely undisturbed animals.

Surgical implantation of such a telemetric unit has enabled the monitoring of the temperature of a single animal for as long as 1 year. When this technique was used on sheep grazing under field conditions, the daily range of the deep body temperature was less than 1 °C regardless of sun, rain, or storm. The total temperature range observed during a whole year from summer to the most severe winter was 1.9 °C, from 37.9 to 39.8 °C (Bligh et al. 1965).

Fever

Fever is an increase in body temperature that usually is associated with bacterial or virus infections. Similar increases in body temperature can be produced by the injection of killed bacteria. These contain substances, *pyrogens*, that are the cause of the fever reaction. The organism behaves as if the set point of a thermostat has been increased by a few degrees and now regulates as if it wants to maintain the increased temperature.

Fever is definitely not a failure of temperature regulation. If a person with fever is challenged with cooling, increased heat production serves to maintain the higher temperature, and vice versa if an extra heat load is applied. The processes of temperature regulation respond as expected, except that the reactions revolve around the higher body temperature.

The concept that fever represents a well-regulated resetting of a "thermostat" is further supported by the following experiment. If a pyrogen is injected into a dog, the animal develops a fever. However, if heat is applied locally only to the hypothalamus, where the temperature regulation center is located, the fever in the body is suppressed. The body behaves as if the higher "set point" has already been reached. This indicates that the bacterial pyrogen does not somehow damage the regulatory mechanism in the hypothalamus, but that a certain hypothalamic temperature is "desirable," and when this has been attained, no further heating is called for (Andersen et al. 1961).

Both mammals and birds develop fever after becoming infected with bacteria, but it is unclear whether the fever is beneficial or harmful. An interesting attempt at settling this much argued question has been made by studying, not warm-blooded mammals or birds, but a cold-blooded animal, the lizard *Dipsosaurus dorsalis*. When *Dipsosaurus* is placed in an environment where there is a range of temperatures, it selects a preferred temperature where it tends to remain so that it maintains a body temperature of about 38.5 °C. If the lizard is injected with an appropriate bacterial suspension, it seeks a somewhat warmer environment and maintains a higher body temperature, a "fever" of approximately 2 °C.

Amphibians and fish show similar reactions in response to bacterial pyrogens (Reynolds et al 1976; Kluger 1977). They select a warmer environment and thus maintain higher preferred body temperatures, demonstrating that the "fever" response is present in a broad range of vertebrates and not only in the warm-blooded mammals and birds.

Let us return to the question of whether fever is beneficial or harmful. An answer has been sought by performing the following experiment on *Dipsosaurus*. Five groups of lizards were infected with bacteria (*Aeromonas hydrophila*) and placed in a neutral environment (38 °C), at lower temperatures (36 and 34 °C), and at higher temperatures (40 and 42 °C). There was a striking correlation between survival and temperature; all the animals at the lowest temperature died in less than 4 days; survival increased with temperature and was highest in the animals at 42 °C. This is approximately the same temperature that injected animals voluntarily chose if they were allowed to select their temperature (Kluger et al 1975). In these animals, then, a clear beneficial effect was associated with the higher body temperature.

The ability to maintain a constant body temperature, as birds and mammals do, and the ability to reset the temperature with full maintenance of regulation as occurs in fever, requires extraordinary capacities of the physiological mechanisms for heat exchange. Before we discuss these mechanisms we will need an elementary knowledge of the physics of heat transfer.

TEMPERATURE, HEAT, AND HEAT TRANSFER

In the preceding section we were concerned with temperature, and the concept of heat was mentioned only incidentally. It is important to understand the difference between these two physical quantities and to realize that the measurement of temperature does not necessarily give any information about heat.

Temperature is usually measured in degrees Celsius (°C), although in physical chemistry and thermodynamics we use absolute temperature expressed in kelvins (K).*

In biology heat is usually measured in calories, and 1 calorie (cal) is defined as the amount of heat needed to raise the temperature of 1 g water by 1 °C. (The calorie is not a part of the International System of Units (the SI System), but the term is so common that it will remain in use for some time to come. For conversion to SI units, 1 cal = 4.184 joules (J).)

To heat 1 g water from room temperature (25 °C) to the boiling point (100 °C) requires 75 cal of heat. To raise the temperature of 100 g water by 75 °C requires 7500 cal (7.5 kcal). In both cases the initial and final temperatures are the same, and temperature alone therefore gives no information about the amount of heat added. If we know the amount of water, however, we can calculate the amount of heat from the temperature change, for, by definition, the heat needed to raise the temperature of 1 g water by 1 °C is 1 cal. The amount of heat needed to warm 1 g of a substance by 1° is known as the *specific heat capacity* of that substance.† The specific heat capacity of water is 1.0 cal g^{-1} $°C^{-1}$, which compared with that of other substances is very high. The specific capacity of rubber is 0.5, of wood 0.4, and of most metals 0.1 or less. The specific heat capacity of air is 0.24 cal g^{-1} $°C^{-1}$, and because the density of air is 1.2 g per liter (at 20 °C), the heat capacity of 1 liter air is 0.3 cal.

The amount of heat needed to increase the temperature of the animal body is slightly less than the amount needed to heat the same weight of water.

* Absolute zero is at −273.15 °C. Celsius temperature (T_C) relates to absolute temperature (T_K) as follows: $T_K = T_C + 273.15$.

† It is recommended that the word *specific* before the name of a physical quantity be restricted to the meaning "divided by mass" (Council of the Royal Society 1975).

The mean specific heat capacity of the mammalian body is about 0.8. Thus, to increase the temperature of a 1000 g mammal by 1 °C requires about 800 cal. The exact value for the specific heat capacity of the animal body varies somewhat. For example, the specific heat capacity of mouse bodies was found to vary between 0.78 and 0.85 with a mean of 0.824 (Hart 1951). The bulk of the body is water, which has a specific heat capacity of 1.0, and the other components – proteins, bone, and fat – tend to reduce this value. Fat is particularly important, for its amount can vary within wide limits, and its specific heat capacity is only about 0.5. For many purposes it is sufficient to use a mean value of 0.8 for the specific heat capacity of the animal body, for in order to determine a change in heat content, we must know the mean body temperature, and as we have seen, this variable is very difficult to determine with accuracy.

Physics of heat transfer

For a body to maintain a constant temperature, there is one absolute requirement: Heat loss must exactly equal heat gain. For an animal to maintain a constant temperature, heat must be lost from the body at the same rate it is produced by metabolic activity. As we have seen, metabolic heat production can easily increase more than 10-fold with activity, and unless heat loss is increased in the same proportion, body temperature will rise rapidly. Furthermore, the conditions for heat loss vary tremendously with external factors such as air temperature and wind. An understanding of the physiological mechanisms involved in the regulation of heat production and heat loss requires an elementary knowledge of the physics of heat transfer.

Whenever physical materials are at different temperatures, heat flows from a region of higher temperature to one of lower temperature. This transfer of heat takes place by *conduction* and by *radiation*. A body cannot lose heat by conduction or radiation unless its environment, or some part of it, is at a lower temperature than the surface of the body. There is, however, a third way to remove heat: the *evaporation* of water. These three ways of heat transfer – conduction,* radiation, and evaporation – are the only means available for the removal of the heat produced in the metabolic activity of living organisms.

Conduction

Conduction of heat takes place between physical bodies that are in contact with each other, whether they are solids, liquids, or gases. Conduction of heat consists of a direct transfer of the kinetic energy of molecular motion, and it always occurs from a region of higher temperature to one of lower temperature.

Assume that we have a uniform conductor in which we keep one end warm and the other cold (Figure 8.3). The rate of heat transfer by conduction ($\dot{Q}$) can now be expressed as:

$$\dot{Q} = k\,A\,\frac{T_2 - T_1}{l}$$

where k is the thermal conductivity of the conductor, A is the area through which heat flows (normal to the direction of heat flow), and T_2 and T_1 are the temperatures at two points separated by the distance l. The fraction $(T_2 - T_1)/l$ is known as the *temperature gradient* and stands for the temperature difference per unit distance along the conductor.

This expression for heat flow in a conductor can

---

* Transfer of heat between a surface and a fluid (gas or liquid) in contact with it takes place by conduction. Mass movement in the fluid, termed *convection*, contributes to renewal of fluid in the boundary layer and thus complicates the conductive heat transfer (see later in this chapter).

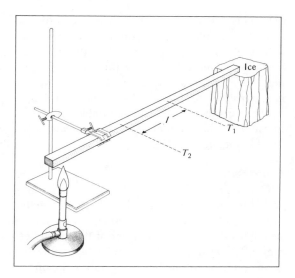

TABLE 8.3 Thermal conductivities (k) for a variety of common materials. [Hammel 1955; Hensel and Bock 1955; Weast 1969]

Material	k (cal s^{-1} cm^{-1} °C^{-1})
Silver	0.97
Copper	0.92
Aluminum	0.50
Steel	0.11
Glass	0.002 5
Soil, dry	0.000 8
Rubber	0.000 4
Wood	0.000 3
Water	0.001 4
Human tissue	0.001 1
Air	0.000 057
Animal fur	0.000 091

be stated in simple, intuitively obvious terms. Heat flow increases with the thermal conductivity (k) of the conducting material, with increasing cross-sectional area (A) of the conductor, and with an increased temperature difference between two points T_2 and T_1. Increasing the distance (l) between two given temperatures, T_2 and T_1 (if these remain unchanged), decreases the amount of heat flow.

The thermal conductivity coefficient (k), is an expression for how easily heat flows in a given material. Values for the thermal conductivity of some common materials are given in Table 8.3. As we know, metals are excellent conductors and have high conductivity coefficients. Glass and wood are poorer conductors; water and human tissues have slightly lower conductivities again, but of a similar order of magnitude. The similarity between the thermal conductivities of human tissue and water is, of course, attributable to the fact that most tissues consist of roughly two-thirds to three-quarters water. Air and animal fur have very low thermal

conductivities, which means that their insulation value (resistance to heat flow) is high. The main reason the thermal conductivity of fur is low is the large amount of air trapped between the hairs. Other materials that enclose a high proportion of air (e.g., felt, woolen fabrics, down) are also poor conductors or excellent insulators.

The simple equation for heat conduction given above unfortunately applies only when heat flows through a plane object such as a wall. Most animal surfaces are curved, and this makes the conduction equation considerably more complex. For practical purposes, we can consider the heat flow as dependent on temperature gradients and area, but if we wish to make a quantitatively satisfactory analysis of heat transfer in an animal, it is necessary to apply a more rigorous treatment of the physics of heat transfer.

The transfer of heat in fluids is almost invariably accelerated by the process of *convection*, which refers to mass movement of the fluid. Assume a cold fluid in contact with a warm solid surface. Heat flows into the fluid by conduction, and the fluid adjacent to the surface becomes warmer. If the fluid is in motion, the warm fluid adjacent to the solid surface is replaced by cold fluid, and the heat loss from the solid surface is therefore speeded up. Mass flow, or convection, in the fluid thus facilitates heat loss from the solid, although the transfer process between solid and fluid remains one of conduction.

Convection in a fluid may be caused by temperature differences or by external mechanical force. Heating or cooling of a fluid usually causes changes in its density, and this in turn causes mass flow. For example, if a warm solid surface is in contact with a cold fluid, the heated fluid expands and therefore rises, being replaced by cool fluid. In this case the mass flow, or convection, is caused by the temperature difference and is called *free* or *natural convection*. This term applies also if the wall is colder than the fluid and the fluid adjacent to the wall becomes denser and sinks.* Free convection, of course, can take place both in air and in water and contributes substantially to the rate of heat loss from living organisms.

Motion in the fluid can also be caused by external forces, such as wind, water currents, or an electric fan. Convection caused by external forces, as opposed to density changes, is referred to as *forced convection*.

Because convection depends on mass transfer in fluids, the process is governed by the rather complex laws of fluid dynamics, which include such variables as the viscosity and density of the fluids, in addition to their thermal conductivity. Convective heat loss does not depend only on the area of the exposed surface. Variables such as the curvature and the orientation of the surface give rise to

*Water has a higher density at +4 °C than at the freezing point. This is of great importance in bodies of fresh water, but in most physiological situations the anomalous density of water is of no significance. Sea water does not exhibit the anomalous density properties at near-freezing temperature (see also page 216).

rather complex mathematical expressions, which cause great difficulties in the analysis of the heat transfer from an animal. As we shall see below, however, we can find practical means for analyzing the heat transfer between an animal and the environment that circumvent the need for an exact analysis of the physical processes involved.

Radiation

Heat transfer by radiation takes place in the absence of direct contact between objects. All physical objects at a temperature above absolute zero emit electromagnetic radiation. The intensity and the wavelength of this radiation depend on the temperature of the radiating surface (and its emissivity, which we shall discuss below). All objects also receive radiation from their surroundings. Electromagnetic radiation passes freely through a vacuum, and for our purposes atmospheric air can be regarded as fully transparent to radiation.

The *intensity* of radiation from an object is proportional to the fourth power of the absolute temperature of the surface. This is expressed by the *Stefan-Boltzmann law* for heat radiation flux:

$$\dot{Q}_R \propto \sigma T^4$$

where T is the absolute temperature of the radiating surface (in kelvin, K), and σ is Stefan-Boltzmann's constant (1.376×10^{-12} cal s^{-1} cm^{-2} K^{-4} or 5.67×10^{-8} W m^{-2} K^{-4}). Because the amount of heat or energy radiated increases with the fourth power of the absolute temperature, the emission increases very rapidly indeed with the surface temperature.

The *wavelength* of the emitted radiation depends on the surface temperature, and the hotter the surface, the shorter is the emitted radiation. As the surface temperature of a heated object increases, the radiation therefore includes shorter and shorter wavelengths. The radiation from a

FIGURE 8.4 The thermal radiation from a body depends on its surface temperature, in regard to both the spectral distribution of the radiation and its intensity. The higher the surface temperature, the shorter is the wavelength and the higher is its intensity. This figure shows the spectral distribution of the thermal radiation from the sun (6000 K), a red-hot stove (1000 K), and the human body (300 K). [Hardy 1949]

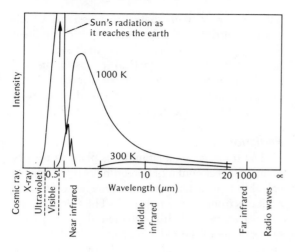

heated piece of iron just barely begins to include visible red light when its temperature is about 1000 K. If heated further, it emits shorter wavelengths (i.e., more visible radiation is included). Therefore, as the temperature increases, the visible color shifts from red to yellow to white. The sun's radiation, which has its peak in the visible part of the spectrum, corresponds to a surface temperature of about 6000 K and includes an appreciable amount of radiation in the near ultraviolet as well. Objects that are close to physiological temperatures emit most of the radiation in the middle infrared. For example, the infrared radiation from the living human skin ($T \approx 300$ K) has its peak at about 10 000 nm. As visible light is between 450 and 700 nm, the radiation from human skin includes no visible light. (The visible light from the skin that we perceive is, of course, only reflected light, and in darkness we see no light from the skin.)

Figure 8.4 shows how the wavelengths and intensity of emitted radiation change with the temperature of the radiating surface, as described in the preceding two paragraphs.

The next concept to consider is *emissivity*. The simplest way to approach this concept is to consider first the *absorptivity* of a surface for the radiation that falls on it. A black body, by definition, absorbs radiation completely in all wavelengths and reflects nothing. The absorptivity of a completely black body is therefore 100%. (Although we often think of "black" in relation to visible light, the physical concept applies to all wavelengths.)

In contrast to a black body, the surface that reflects all radiation is a perfect reflector, and its absorptivity is zero. This condition is approached by a highly polished metallic surface (e.g., a silver mirror). Because incident radiation is either absorbed or reflected, absorptivity and reflectivity *must* add to 1 (or 100%). If 30% of incident radiation is reflected, 70% is absorbed, and so on.

Usually, the absorptivity (and thus reflectivity) is different in different wavelengths of incident radiation. In visible light we recognize this as the color of objects; an object we perceive as yellow reflects mostly yellow light and absorbs other components of the visible light. In the middle infrared, which is of the greatest interest in connection with heat radiation at physiological temperatures, most surfaces are black bodies. The human skin, for example, absorbs virtually 100% of incoming infrared radiation, and thus is a black body in these wavelengths, irrespective of whether it is light or dark in the visible light.

The concept of surface *absorptivity* should now be clear, and we can return to the *emissivity*. The two are numerically equal, a fact that can be intuitively understood from the following. Consider an object suspended in a vacuum within a hollow sphere of uniform wall temperature. The object within the sphere receives radiation from the wall of the sphere, part of which is absorbed and part of which is reflected. Likewise, the object radiates to the spherical surface, and part of this radiation is absorbed while part is reflected. (The reflected portion, in turn, is either intercepted by the suspended object or reaches another point of the sphere.) When the system is in equilibrium, the object has attained the temperature of the surrounding spherical surface, and the absorbed and emitted radiations from the object are now precisely equal. If this were not so, the object would not be at the temperature of the sphere. This would be a physical impossibility, for otherwise we could tap energy from the system and have the makings of a *perpetuum mobile*. As a practical example, take a highly reflective metal coffee pot that has a reflectivity of 95% (i.e., an absorptivity of only 5%). Its emissivity, therefore, is also 5% (i.e., the polished coffee pot loses heat by radiation very slowly).

Let us next consider that the object within the sphere is nearly black and has an absorptivity of 99%. This object obviously will attain the same temperature as a highly reflective silver object similarly suspended, although the latter may have an absorptivity of only 1%. For each object at equilibrium the absorbed and the emitted radiation are precisely equal. The final temperature will be the same for both, but the highly reflective object will need longer to reach it.

To repeat, an object of high absorptivity also has high emissivity. A body that is a perfect "black body" in a given wavelength is a perfect emitter in the same wavelength.

Human skin, animal fur, and all sorts of other nonmetallic surfaces have high absorptivities in the middle infrared range of the spectrum, between 5000 and 10 000 nm. For all practical purposes human skin is black in this range, and there is no difference between heavily pigmented and unpigmented skin. The difference we perceive is in the visible region of the spectrum, but because no radiation is emitted in this range, the difference in pigmentation does not influence radiation from human skin.

As a consequence of the high absorptivity in the middle infrared, emissivity is also close to 100% in this region. Consequently, heat losses by radiation from pigmented and unpigmented skins are about equal; both radiate as virtually black bodies. The same is true of radiation from animal fur, which likewise is independent of the color in the visible part of the spectrum.

Failure to appreciate this simple physical fact has led to some ill-conceived speculation in regard to animal coloration. For example, it has been suggested that black-colored animals lose heat by radiation faster than white-colored ones. Because the emission of radiation in the infrared has no relation to visible coloration, a difference in emissivity in the infrared can be established only by direct measurement in this range of the spectrum. No such differences have been found.

Skin and fur color may, however, be important to the heat absorbed from solar radiation, which has its peak intensity in the visible range. About half of the energy carried in solar radiation falls within what we call the visible light (Figure 8.4), and it is important to the heat balance whether this light is absorbed or reflected. When exposed to direct solar radiation, dark-colored skin or fur absorbs more of the incident energy than light-colored skin or fur.

Net heat transfer by radiation. If two surfaces are in radiation exchange, each emits radiation according to *Stefan-Boltzmann's law*, and the net radiation transfer ($\dot{Q}_R$) between them is:

$$\dot{Q}_R = \sigma \, \epsilon_1 \, \epsilon_2 (T_2{}^4 - T_1{}^4) A$$

in which σ is Stefan-Boltzmann's constant, ϵ_1 and ϵ_2 are emissivities of the two surfaces, T_1 and T_2 are their absolute temperatures, and A is an expression for the effective radiating area.* If the environment is a uniform sphere, A is a simple expression of the integrated "visible" surface in the direction of radiation. If the environment is nonuniform, however, and in particular if it includes a point source of heat (such as the sun), the integration of the surfaces of exchange becomes more complex. In this regard, situations in nature are extremely complex, and to describe in exact terms the total heat transfer is very difficult. Nevertheless, once the elementary physics of radiation exchange is understood in principle, we can readily avoid some erroneous conclusions such as those relating to the role of surface pigmentation.

A *practical simplification.* Although radiation heat transfer changes with the fourth power of the absolute temperature, we can use a simplified expression, provided the temperature difference between the surfaces is not too great. Within a temperature range of about 20 °C, the error of not using the rigorous Stefan-Boltzmann equation can often be disregarded, and, as an approximation, we can regard the radiation heat exchange as being proportional to the difference in temperature between the two surfaces. For small temperature differences, the error is relatively insignificant, but it becomes increasingly important the greater the temperature difference.

We will now use this simplification and regard the radiation heat exchange as proportional to the temperature difference. The rate of heat loss from a warm-blooded animal in cool surroundings consists of conduction and radiation heat loss (for the moment disregarding evaporation). Because both can be considered proportional to the temperature difference $(T_2 - T_1)$, their sum will also be proportional to $(T_2 - T_1)$, or:

*This equation for net radiation transfer assumes that one or both surfaces has an emissivity close to or equal to unity.

$$\dot{Q} = C\,(T_2 - T_1)$$

in which all the constants that enter into the heat exchange equation have been combined to a simple proportionality factor, C.

We shall later return to the application of this simplified equation in the discussion of heat loss from warm-blooded animals in the cold.

Evaporation

The evaporation of water requires a great deal of heat. To transfer 1 g water at room temperature to water vapor at the same temperature requires 584 cal (2443 J). This is an amazingly large amount of heat, for when we consider that it takes 100 cal (418 J) to heat 1 g water from the freezing point to the boiling point, we see that it takes more than five times again as much heat to change the liquid water into water vapor at the same temperature.

The amount of heat required to achieve the phase change from liquid water to vapor is known as the *heat of vaporization* (H_v). The heat of vaporization changes slightly with the temperature at which the evaporation takes place: At 0 °C the H_v is 595 cal per gram water; at 22 °C it is 584 cal per gram; and at the boiling point, 100 °C, it is 539 cal per gram.

In physiology it is customary to use the figure 580 cal per gram water, which is an approximation of the value for vaporization of water at the skin temperature of a sweating man, about 35 °C.

The measurement of heat loss by vaporization of water has one great convenience: It suffices to know the amount of water that has been vaporized. When a man is exposed to hot surroundings, he cools himself by evaporation of sweat from the general body surface, but a dog evaporates most of the water from the respiratory tract by panting. The amount of heat transferred per gram water is, of course, the same in both cases (i.e., we do not have to know the exact location or the area of the evaporating surface).

The respiratory air of mammals and other air-breathing vertebrates is exhaled saturated with water vapor, and therefore there is normally a considerable evaporation of water from the respiratory tract, even in the absence of heat stress. This evaporation must, of course, be included in any consideration of the total heat balance of an animal, which is the subject we shall now discuss.

HEAT BALANCE

We have repeatedly emphasized that for the body temperature to remain constant, heat loss must equal heat gain. The body temperature does not always remain constant, however. Assume that heat loss does not quite equal the metabolic heat production, but is slightly lower. The body temperature inevitably rises. This means that part of the metabolic heat remains in the body instead of being lost, and the increase in body temperature thus represents a storage of heat. If the mean body temperature decreases, which happens when heat loss exceeds heat production, we can regard the excess heat loss as heat removed from storage. The amount of heat stored depends on the change in mean body temperature, the mass of the body, and the specific heat capacity of the tissues (which for mammals and birds usually is assumed to be 0.83).

The heat exchange between the body and the environment takes place by the three means described above: conduction (including convection), radiation, and evaporation. Usually each of these represents a heat loss from the body, but this is not always so. When the air temperature exceeds the body surface temperature, heat flow by conduction is to, not from, the body. When there is a strong radiation from external sources, the net radiation flux may also be toward the body. Evaporation is nearly always a negative entity, but under unusual circumstances it could be reversed; this happens, for example, when a cold body comes in contact with moist warm air.[*]

We can enter these variables into a simple equation:[†]

$$H_{tot} = \pm H_c \pm H_r \pm H_e \pm H_s$$

in which

H_{tot} = metabolic heat production
   (always positive)
H_c = conductive and convective heat exchange
   (+ for net loss)
H_r = net radiation heat exchange
   (+ for net loss)
H_e = evaporative heat loss (+ for net loss)
H_s = storage of heat in the body
   (+ for net heat gain by body)

---

[*] When a human with a skin temperature of, say, 35 °C, enters a Turkish steam bath in which the air is nearly saturated and above 40 °C, there is an immediate condensation of water on his skin. In this case the direction of evaporation heat exchange is the reverse of the usual. In the sauna bath, however, the air is usually dry, and the visible moisture seen on the skin shortly after entering the sauna is sweat.

[†] It is important to note that no work term appears in this equation. For heat balance, the equation is correct as it stands. However, if oxygen consumption is used to estimate metabolic rate and to calculate heat production, external work must be considered, for that fraction of the oxygen consumption that goes to perform external work does not appear as heat in the body. For example, a flying bird uses probably 20% of its metabolic rate to impart acceleration to the air through which it flies (ultimately this external work degenerates into heat in a trail of decaying air vortexes). For a flying bird, therefore, only 80% of the oxygen consumption represents heat released within the body, and only this part enters into the heat balance equation as metabolic heat (H_{tot}) to be dissipated through the means to the right of the equality sign. This also means that the caloric equivalent of oxygen (4.9 kcal formed per liter oxygen consumed) cannot be used to calculate the heat production of an animal performing external work.

The three components of heat exchange – conduction, radiation, and evaporation – depend on external factors, among which the most important single factor is temperature. It is obvious that heat losses increase when the external temperature falls. If, on the other hand, external temperature rises, the heat losses decrease, and if external temperature exceeds body surface temperature, both conduction and radiation heat exchange may be from the environment to the organism. The total heat gain is then the sum of heat gain from metabolism plus heat gain from the environment. This situation still permits the maintenance of a constant body temperature (storage = 0), provided evaporation is increased sufficiently to dissipate the entire heat load.

The physiological responses in cold and in heat differ in many respects, and it is convenient to treat the two conditions separately. We shall first deal with temperature regulation in cold surroundings, and afterward discuss temperature regulation in the heat.

Temperature regulation in the cold: keeping warm

To maintain a constant body temperature, an animal must satisfy the steady-state condition in which the rate of metabolic heat production $(\dot{H})$ equals the rate of heat loss $(\dot{Q})$. For the moment we will consider that the heat loss (in the cold) takes place only through conduction (including convection) and radiation, and that heat loss through evaporation can be disregarded.* To describe the heat loss we will use the simplified equation that relates heat loss $(\dot{Q})$ to ambient temperature as developed in the preceding section:

---

* At moderate to low temperature, evaporation takes place primarily from the respiratory tract, at a rate corresponding to a few percent of the metabolic heat production.

$$\dot{H} = \dot{Q} = C\,(T_b - T_a)$$

This equation simply says that the rate of metabolic heat production equals the rate of heat loss, which in turn is proportional to the temperature difference between the body and the environment $(T_b - T_a)$. The term C is a conductance term that will be discussed later.

What can an animal do to maintain the steady state? Among the terms in the above equation, ambient temperature (T_a) is one an animal can do little about, short of moving to a different environment. To adjust to an unfavorable T_a, an animal can adjust only the three remaining terms: heat production $(\dot{H})$, the conductance term (C), or body temperature (T_b). As we are concerned with the maintenance of the body temperature (T_b), we are left with adjustments to be made either in heat production $(\dot{H})$ or in the conductance term (C). (The third alternative, a change in body temperature, does occur in hibernation, a subject to be discussed later.)

Increase in heat production

Although heat production (metabolic rate) cannot be turned down below a certain minimum level, increased metabolic rate permits a wide range of adjustments. The major ways in which heat production is increased are through (1) muscular activity and exercise, (2) involuntary muscle contractions (shivering), and (3) so-called nonshivering thermogenesis. The last term refers to an increased metabolic rate that takes place without noticeable muscle contractions.

To clarify the role of increased heat production, let us consider a simple physical model of an animal (Figure 8.5). An insulated box is equipped with an electric coil that delivers heat at the rate $\dot{H}$. The temperature inside the box (T_b) is therefore higher than the ambient temperature (T_a). If we

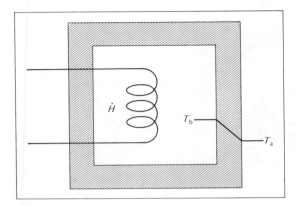

FIGURE 8.5 An insulated box is maintained at a given temperature (T_b) by a heater supplying heat at the rate $\dot{H}$. If the outside temperature (T_a) is lowered, T_b can be maintained constant by increasing the rate of heat input in proportion to the temperature difference ($T_b - T_a$). This is a model of thermoregulation in a mammal.

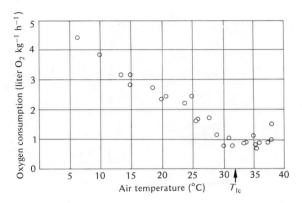

FIGURE 8.6 The oxygen consumption of the pigmy possum (*Cercaertus nanus*) at various ambient temperatures. Below a certain point, the lower critical temperature (T_{lc}), the oxygen consumption increases linearly with decreasing ambient temperature. [Bartholomew and Hudson 1962]

lower the ambient temperature (T_a), the temperature within the box can be maintained only by increasing the rate of heat input ($\dot{H}$). The increase in $\dot{H}$ must be proportional to the increase in temperature difference as long as the insulation (or conductance) in the wall remains unchanged.

This is precisely the situation for an animal exposed to cold. The lower the ambient temperature, the greater the increase in metabolic rate needed to stay warm. In Figure 8.6 we see that below a certain ambient temperature, called the *lower critical temperature* (T_{lc}), the metabolic rate increases linearly with decreasing temperature. Above the critical temperature, heat production, which cannot be reduced lower than the resting metabolic rate, remains constant as temperature is increased.

If we want to compare a variety of different animals, we can conveniently do so if we assign to the normal resting metabolic rate of each the value 100%. This has been done in Figure 8.7, where we can see that most tropical mammals have critical temperatures between +20 and +30 °C; below these temperatures their metabolic rates increase rapidly. Arctic mammals, on the other hand, have much lower critical temperatures, and a well-insulated animal such as an Arctic fox does not

increase its metabolic rate significantly until the air temperature is below −40 °C.

The slopes of the lines in Figure 8.7 can be regarded as an expression of the conductance term (C) in the equation we have been using.* As a rule, tropical mammals have high conductances, and Arctic mammals low conductances (high insulation values). As a consequence, a tropical mammal, because of its high conductance (low insulation), must increase its metabolism drastically for even a moderate temperature drop. For example, a monkey with a body temperature of 38 °C and a lower critical temperature of 28 °C must double its metabolic rate for a further 10° drop in the temperature, to 18 °C. An Arctic ground squirrel, on the other hand, which has a lower critical temperature at about 0 °C, will not need to double its metabolic rate until the temperature has dropped to nearly −40 °C. The Arctic fox, with its lower critical temperature at −40 °C, should be able to sustain the lowest temperatures measured in Arctic climates (−70 °C) with less than a 50% increase in its metabolic rate.

* When the data have been normalized by assigning to the resting heat production the value of 100%, the slope does not indicate conductance in absolute units.

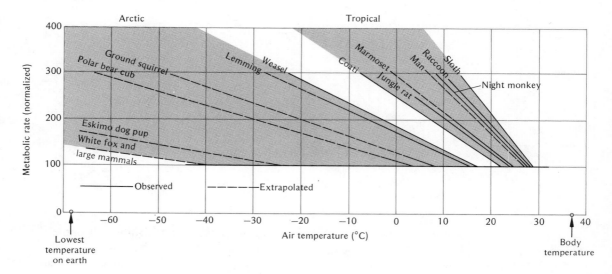

FIGURE 8.7 The metabolic rates of various mammals in relation to air temperature. The normal resting metabolic rate for each animal, in the absence of cold stress, is given the value 100%. Any increase at lower temperature is expressed in relation to this normalized value, making it possible to compare widely differing animals. [Scholander et al. 1950a]

The difference between Arctic and tropical animals is very clear in Figure 8.7. The width of the *thermoneutral zone** is much greater in the Arctic animal and the metabolic response to cold is less pronounced than in tropical mammals. The way the graph is drawn, however, is in some ways misleading. Assume that we examine two mammals of equal size and equal conductance value, but that one has a metabolic rate in the thermoneutral zone which is twice as high as the other. If, instead of normalizing the metabolic rate at 100%, we plot their actual metabolic rates (Figure 8.8), the lower critical temperature will be different for the two animals, but below this temperature, the two animals must expend the same number of calories to keep warm. It is therefore obvious that the location of the lower critical temperature (the width of the

thermoneutral zone) by itself is not sufficient information about how well adapted an animal is to cold conditions. Information about the level of the resting metabolic rate, or better, the conductance value, is also necessary to evaluate the relationship between energy metabolism and heat regulation.

Conduction, insulation, and fur thickness

Conductance, in the sense it has been used above, is a measure of the heat flow from the animal to the surroundings. The term includes the flow of heat from deeper parts of the body to the skin surface and from the skin surface through the fur to the environment. When conductance is low, the insulation value is high. In fact, insulation is the reciprocal value of conduction.*

---

* The *thermoneutral zone* is the temperature range within which the metabolic heat production is unaffected by a temperature change.

* *Thermal conductance* (heat flow per unit time per unit area per degree temperature difference) has the units $W\ m^{-2}\ °C^{-1}$ (or $cal\ s^{-1}\ cm^{-2}\ °C^{-1}$). *Insulation* is the reciprocal of conductance, and its units are therefore $W^{-1}\ m^2\ °C$ (or $cal^{-1}\ s\ cm^2\ °C$).

FIGURE 8.8 If two animals have the same body temperature and the same conductance, but have different resting metabolic rates, they will also have different lower critical temperatures (top). If their metabolic rates are normalized to 100%, this procedure will suggest, incorrectly, that they have different conductance values (bottom).

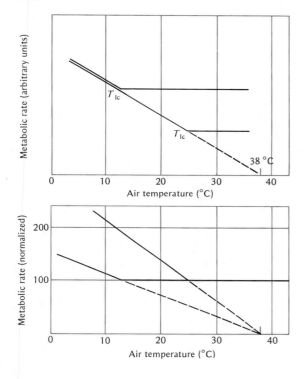

Our discussion is concerned with animals in the cold, where the fur is a major barrier to heat flow. The insulation values of the fur from various animals differ a great deal. Some data, plotted relative to the thickness of the fur, are shown in Figure 8.9. As expected, the insulation value increases with the thickness of the fur and reaches a maximum for some of the larger animals which have thick fur, such as the white fox. Among the smaller animals there is a clear correlation between fur thickness (and insulation) and the size of the animal. A small animal must have relatively short light fur or it could not move about. This is particularly true for the smallest mammals, small rodents and shrews. Because of their relatively

poor insulation, these animals must either take advantage of microclimates (e.g., by living in burrows) or hibernate to escape the problem of keeping warm.

The polar bear is interesting, for its open and coarse fur provides poor insulation relative to its thickness. More important, when polar bear skin is submerged in ice water, most of the insulation value of the fur is lost, and heat transfer is 20 to 25 times faster than in air. If the water is agitated (as it would be when the bear is swimming), the heat loss is increased even more, some 50 times. This is because water penetrates to the skin surface, dislodging all air from the fur. When in water, the polar bear is helped by subcutaneous blubber, which affords insulation (see later in this chapter).

The skin of the seal has a relatively thin fur, but a heavy layer of blubber affords considerable insulation. Therefore, the difference in insulation value for seal skin in air and in water is not very great. In air, seal skin with the blubber attached has only a slightly better insulation value than the skin of a lemming (i.e., 60 to 70 mm of mostly blubber insulates about as well as 20 mm of fur). When the seal skin is submerged in water, the insulation value is reduced, but not as drastically as for polar bear skin.

It is well known that the thickness of animal fur changes with the seasons and that winter fur is heavier and presumably affords better insulation than the thinner summer fur. The seasonal changes that have been measured are greatest in large animals and relatively insignificant in small rodents. The black bear, for example, in summer loses 52% of the insulation value of its winter fur (i.e., the summer pelt affords less than half the insulation of the winter pelt). The smallest seasonal change observed in a sub-Arctic mammal was a 12% reduction of the winter value in the muskrat (Hart 1956).

FIGURE 8.9 The insulation value of animal fur is related to fur thickness. Because small mammals of necessity carry relatively short fur, they have poor insulation compared with larger animals. Insulation values for fur in air are indicated by open circles; values for fur submerged in water, by triangles. Insulation values for fur (ordinate) are measured in air or water at 0 °C with the inside of the fur maintained at 37 °C. The sloping broken line represents the insulation values of cotton of various thicknesses. In this figure the insulation value of 4 cm of fur is about 1 W^{-1} m^2 °C, which equals 41 840 cal^{-1} s cm^2 °C. For 1 cm thickness of fur the insulation value will be one-fourth, or 10 460 cal^{-1} s cm^2 °C. The conductance value (the reciprocal of insulation) is therefore 0.000 096 cal s^{-1} cm^{-2} $°C^{-1}$, which is in good agreement with the value for fur listed in Table 8.3. [Scholander et al. 1950b]

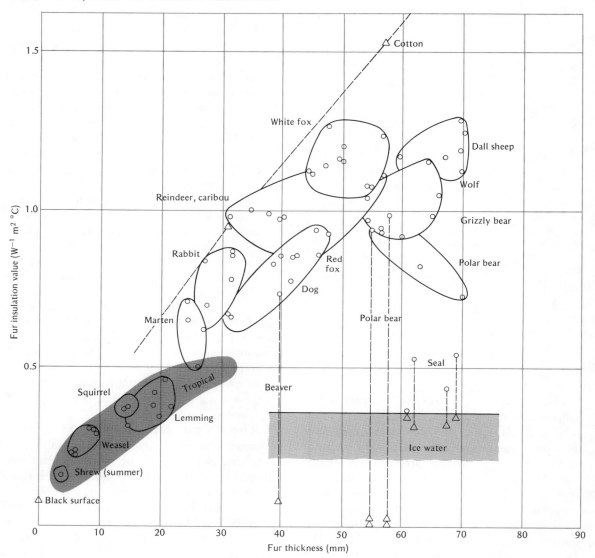

Conductance in birds

As we have seen, the conductance values for mammals in the cold require that the metabolic rates increase in proportion to the imposed cold stress. The straight lines describing the metabolic rates at low temperature extrapolate to body temperature, showing that mammals fit our assumed model (Figure 8.5).

FIGURE 8.10 The oxygen consumption of the pigeon and of the roadrunner increases at low temperature, but the regression line does not extrapolate to the body temperature of the animal as it commonly does in mammals. [Calder and Schmidt-Nielsen 1967]

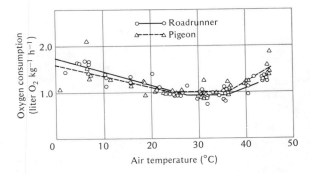

For birds the situation is not so simple. Some birds conform to the mammalian pattern with metabolic curves that extrapolate to body temperature. Other birds, however, deviate from this pattern. For example, the metabolic heat production of pigeons and roadrunners does not increase as much as expected at low ambient temperature (Figure 8.10). For these birds the metabolic line extrapolates not to the body temperature of about 40 °C, but to a much higher temperature, well above 50 °C.

There is only one possible explanation for such a slope: The bird does not adhere to the simple equation $\dot{H} = C\,(T_b - T_a)$, where C remains constant at low temperature. One way to explain the observed regression line is to assume that the conductance value (C) gradually decreases with falling ambient temperature (T_a). It is not well understood how the birds achieve this reduction in conductance below the lower critical temperature. The other way to obtain a change in slope, to permit body temperature to decrease, is not in accord with observed maintenance of deep body temperature.

How can the bird reduce conductance (increase insulation)? One possible way is to raise the feathers and draw the feet up into the feathers, thus making the body into a round feather ball. Another way is to allow the peripheral tissues to un-

dergo an appreciable temperature drop while the core temperature is maintained constant. A drop in shell temperature and/or an increase in shell thickness have several effects: The circulation to the shell is decreased; the heat production in the shell is decreased; the thickness of the shell is increased; and the volume of the core is reduced. Each of these changes contributes to a decrease in conductance and a smaller-than-expected increase in metabolic rate, and still permits the maintenance of constant core temperature (although in a smaller core volume than usual).

Huddling

Among the penguins of the Antarctic the most striking is the large emperor penguin (*Aptenodytes forsteri*), which lives under colder conditions than any other bird and has the unique characteristic of breeding during midwinter.

As winter approaches, emperor penguins leave the open water and walk on the sea ice for 50 or 100 km toward the permanent ice shelf to the rookeries. There the female lays a single egg, which is placed on the feet of the male, who remains at the rookery to incubate the egg while the female returns to the sea to feed.

While incubating, the male stands on the ice for over 2 months in air temperatures that may be as low as − 30 to − 40 °C with high wind velocities. The female usually returns to the rookery with a full stomach about the time the egg hatches; she feeds the chick while the male returns to the sea to feed.

Emperor penguins can feed only at sea, and to sustain himself during the long fast the male begins the period with large deposits of subcutaneous fat. The fast may exceed 100 days, and by the end the male may have lost up to 40% of his initial body weight.

How much energy is needed for the long walk to

HUDDLING PENGUINS Chicks of the emperor penguin photographed on the ice-covered Antarctic breeding grounds. The chicks huddle together and thus reduce the surface area exposed to cold air. This behavior leads to a substantial reduction in the metabolic cost of keeping warm in the frigid climate. The group in this photo contains about 50 birds; a few have momentarily raised their heads to watch the photographer. [Courtesy of Yvon LeMaho, University of Lyon, France]

and from the rookery, and how much is needed to keep warm in the Antarctic cold? And do penguins keep warm? These questions have recently been the subject of detailed studies (LeMaho et al. 1976; Pinshow et al. 1976).

The penguins do keep warm; their body temperature remains at the normal 38 °C during the entire breeding period. How much fuel do the birds need for this and how much for walking?

A large male penguin may weigh 35 kg when he leaves the sea and may have lost 15 kg at the end of the fast. The amount of energy needed for walking can be determined by training penguins to walk on a treadmill in the laboratory and determining their oxygen consumption. This shows that walking 200 km to the breeding grounds and back to the sea may require the use of about 1.5 kg of fat. Is the remainder sufficient to keep the bird warm for several months of starvation while he incubates the egg?

One might expect penguins to have physiological characteristics that distinguish them from other birds, such as a thermoneutral zone extending to extremely low temperatures (i.e., a very low thermal conductance). This is not the case. Penguins have a thermal conductance to be expected for any bird of the same body size, and their lower critical temperature is at −10 °C, far above the usual Antarctic winter temperatures. The metabolic rates of penguins under laboratory conditions at temperatures similar to those in nature suggest that perhaps 25 kg of fat would be needed to keep warm during the winter, and this by far exceeds the reserves even a large male carries.

The apparent contradiction is explained by the behavior of the penguins in the rookery, where they huddle together in large groups that may number several thousands. The metabolic rates of huddling penguins have not been determined, but weight loss, which presumably results from metabolism of fat, is a good indication. Single emperor penguins in the cold lose about 0.2 kg per day; huddling penguins lose only about half as much, about 0.1 kg. Huddling together obviously helps, and it is easy to understand why. Instead of being exposed to the cold on all sides, a major part of each bird's surface is in contact with neighboring penguins. When two bodies have the same temperature, there is no heat flow between them. Huddling together obviously is a prerequisite for survival and for the success of breeding. Why the emperor penguins choose to breed in winter is another question and remains unanswered.

Other animals also huddle together to keep warm, although they may be less spectacular than the penguins. Huddling reduces the exposed surface, thus reducing the cold stress and the metabolic requirements for heat production. This is of particular importance for newborn mammals and bird nestlings that live together as a litter: They can remain warm in the absence of the parent for longer periods, and the young animal's decreased dependence on the use of energy for heat production permits faster growth.

Insulation in aquatic mammals

Many seals and whales live and swim in the near-freezing water of the Arctic and Antarctic seas. Not only are the temperatures low, but water has a high thermal conductance and a high heat capacity, and the thermal loss to water is therefore much higher than to air of the same temperature. The heat conductivity coefficient for water is about 25 times as high as for air, but because convection, both free and forced, plays a major role, the cooling power of water may under some circumstances be even greater, perhaps 50 or even as high as 100 times as great as for air.

What can seals and whales do about their heat balance in water with such cooling power? Evidently they manage quite well, for both seals and whales are far more numerous in cold water than in the tropics. As there is nothing they can do about the water temperature (short of moving to warmer seas), they have a limited choice. They could either (1) live with a lowered body temperature, (2) increase the metabolic rate to compensate for the heat loss, or (3) increase the body insulation to reduce the heat loss.

With regard to body temperature, seals and whales are similar to other warm-blooded animals; their usual temperature is around 36 to 38 °C (Irving 1969). We must therefore look at the other possibilities for an explanation of their ability to live in ice water.

The metabolic rate has been measured in several species of seals and in some dolphins (porpoises) (but not in any of the large whales which are rather unmanageable as experimental animals). Most of these animals have resting metabolic rates about twice as high as would be expected from their body size alone (Irving 1969). In the harp seal (*Phoca groenlandica*), an Arctic species, the metabolic rate remained the same in water all the way down to the freezing point; in

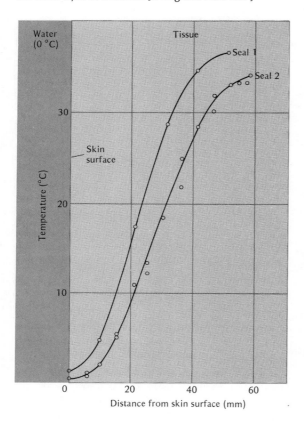

FIGURE 8.11 The temperature of the skin surface of a living seal immersed in ice water is nearly identical to that of the water. Most of the insulation is provided by the thick layer of blubber. [Irving and Hart 1957]

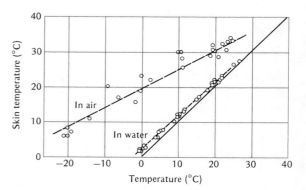

FIGURE 8.12 The skin surface temperature of seals in air and water. Solid line indicates equality of skin and environmental temperature. [Hart and Irving 1959]

other words, even the coldest water did not cause the heat loss to increase enough to require increased heat production. The lower critical temperature for the harp seal in water is thus below freezing and has not been determined (Irving and Hart 1957).

The third possibility, an effective insulation, is obviously the main solution to the problem. Both seals and whales have thick layers of subcutaneous blubber which afford the major insulation. Measurements of the skin temperature support this concept, for the temperature at the skin surface is

nearly identical to that of the water (Figure 8.11). If the surface temperature is nearly the same as that of the water, very little heat can be transferred to the water. The temperature gradient is sustained by the blubber, and at a depth of some 50 mm (roughly the thickness of the blubber), the temperature is nearly at body core temperature.

In Arctic land mammals the temperature gradients are different. The surface temperature of the body skin under the fur is regularly within a few degrees of core temperature (Irving and Krog 1955), and most of the insulation therefore resides outside the skin surface. The polar bear, whose fur loses most of its insulation value in water, also has a substantial layer of blubber that seems to be of major importance when the polar bear swims in ice water. Without the layer of blubber, the semiaquatic way of life would seem impossible (Øritsland 1970).

If seals and whales are so well insulated, how do they manage to avoid overheating when the water is warmer or when the metabolic rate is increased during fast swimming? If a seal is removed from water and placed in air, its skin temperature increases considerably (Figure 8.12). The higher skin temperature is necessary for dissipation of heat to the air, due to the reduced cooling power of air

SEAL BLUBBER This cross section of a frozen seal shows the thick layer of blubber. Of the total area in the photo, 58% is blubber and the remaining 42% is muscle, bone, and visceral organs. [Courtesy of P. F. Scholander, University of California, San Diego]

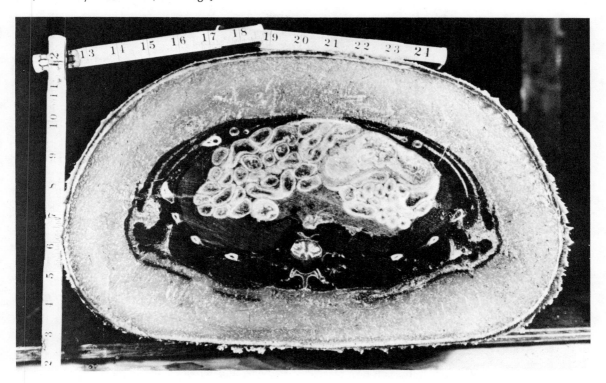

relative to water. The increase in skin temperature results from an increased blood flow through the blubber to the superficial layer of the skin, which is well supplied with blood vessels. This system of cutaneous blood vessels permits a precise regulation of the amount of heat that reaches the skin surface and thus is lost to the environment.

We can now conclude that the main difference in insulation between aquatic and terrestrial mammals is that the insulator of the aquatic mammal (the blubber) is located internally to the surface of heat dissipation. Therefore, the blood can bypass the insulator, and heat dissipation during heavy exertion or in warm water can be independent of the insulator. Terrestrial animals, in contrast, have the insulator located externally to the skin surface. Therefore, they cannot modulate the heat loss from the general skin surface to any great extent and must seek other avenues of heat loss when they need to dissipate excess heat to the environment (Figure 8.13).

Distribution of insulating material

We must remember that furred Arctic animals cannot afford to insulate all parts of the body equally well; they also need surfaces from which heat can be dissipated, especially during exertion. Because the main part of the body is well insulated for maximum heat retention, they need thinly covered skin areas on the feet, face, and other periph-

FIGURE 8.13 The insulation afforded by blubber can be bypassed when the need for heat dissipation increases. Fur, in contrast, is located outside the skin surface and its insulation value cannot be drastically changed by a bypass.

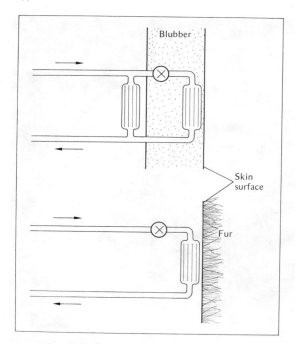

TABLE 8.4 Partitioning of surface areas with fur of different thicknesses in the South American guanaco (*Lama guanicoe*). About 40% of the area is covered by heavy fur, but almost 20% is nearly bare and permits extensive modulation of heat dissipation. [Morrison 1966]

	Depth of fur (mm)	Percent of total skin area
Nearly bare	1	19
Short hair	4	20
Medium length hair	15	20
Long hair	30	41

eral locations, from which heat loss can take place when the demand for heat dissipation increases.

Estimates of the fraction of the body surface area that is covered with fur of different thicknesses suggests that some animals have considerable flexibility in regulating their conductance by changes in posture. Estimates on a male guanaco are shown in Table 8.4 and in diagrammatic form in Figure 8.14 (Morrison 1966).

Because the thermal conductance through fur is inversely related to the thickness of the fur, the nearly bare areas can potentially transfer much more heat than the heavily furred areas. With the limbs extended and their surfaces fully exposed, the bare areas can serve as "heat windows" that can account for more than half the total heat transfer.

A guanaco lying or standing in a normal posi- tion would have the bare areas on the inside of the legs opposed, and this would reduce surface heat transfer by one-half. If the animal were to curl up to cover as much as possible of the remaining short-haired areas, surface conductance would be at a minimum: only one-fifth of that for the stretched out animal. Other animals have similar short-haired areas whose role in heat transfer can be modulated by changing their exposure and by changing the blood flow to these areas, thus giving a great deal of flexibility in the regulation of heat loss.

Heat exchangers

Seals and whales have flippers and flukes that lack blubber and are poorly insulated. These ap- pendages are well supplied with blood vessels and receive a rich blood supply. This means that these relatively thin structures with their large surfaces can lose substantial amounts of heat and aid in heat dissipation, but how is it possible to avoid ex- cessive heat loss from these structures when heat needs to be conserved? If the blood that returns to the body core from these appendages were at near- freezing temperature, the deeper parts of the body would rapidly be cooled.

Excessive heat loss from the blood in the flippers is prevented by the special structure of the blood vessels, which are arranged in such a way that they function as heat exchangers. In the whale flipper each artery is completely surrounded by veins (Fig-

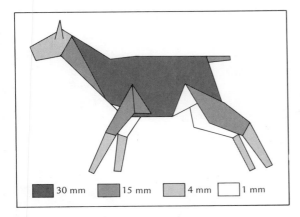

■ 30 mm   ■ 15 mm   ■ 4 mm   □ 1 mm

ure 8.15), and as warm arterial blood flows into the flipper, it is cooled by the cold venous blood that surrounds it on all sides. The arterial blood therefore reaches the periphery precooled and loses little heat to the water. The heat has been transferred to the venous blood, which thus is prewarmed before it reenters the body. If the heat exchange is efficient, the venous blood can reach nearly arterial temperatures and thus contribute virtually no cooling to the core. This type of heat exchanger is known as a *countercurrent heat exchanger* because the blood flows in opposite directions in the two streams (Scholander and Schevill 1955).

A diagram of a countercurrent heat exchanger may help explain how it works (Figure 8.16). Assume that a copper pipe carries water at 40 °C into a coil suspended in an ice bath, and that the water returns in a second pipe located in contact with the first so that heat is easily conducted between them. As the water leaves the coil, it is very nearly at the temperature of the ice bath, 0 °C. As this water flows adjacent to the warmer pipe leading into the coil, it picks up heat and thus cools the incoming water. After some time a steady state of temperatures is reached, which may be as indicated in the

diagram, depending on the conditions for heat exchange between the two pipes and their length.

For animals swimming in warm water, heat dissipation presents a greater problem than heat conservation. The anatomical arrangement of the heat exchanger is such that an increased blood supply and an increase in arterial blood pressure cause the diameter of the central artery to increase, and this in turn causes the surrounding veins to be compressed and collapse. The venous blood must now return in alternate veins, which are located closer to the surface of the flipper. Because the heat exchanger is now bypassed, the arterial blood loses heat to the water, and the venous blood returns to the core without being rewarmed, thus cooling the core. In this way the circulatory system of the flipper can function both in heat conservation and in heat dissipation.

It is interesting that such countercurrent heat exchangers are found in a number of other animals. For example, sea cows, which live in tropical and subtropical waters, have heat exchangers in their appendages. This might seem unnecessary, for they live in warm water, but sea cows move slowly and have relatively low metabolic rates for their size, and thus would have difficulties unless they could reduce the heat loss from the appendages.

Heat exchangers in the limbs are not restricted to aquatic animals. Even in the limbs of man some heat exchange takes place between the main arteries and the adjacent larger veins, located deep within the tissues. In cold surroundings most of the venous blood from the limbs returns in these deep veins, but in warm surroundings much of the venous blood returns in superficial veins under the skin and is thus diverted from heat exchange with the artery (Aschoff and Wever 1959).

In birds, heat exchange in the legs is very important, especially for birds that stand or swim in

FIGURE 8.15 In the flippers and fluke of a dolphin (porpoise) each artery is surrounded by several veins. This arrangement permits the venous blood to be warmed by heat transfer from the arterial blood before it reenters the body. [Schmidt-Nielsen 1970]

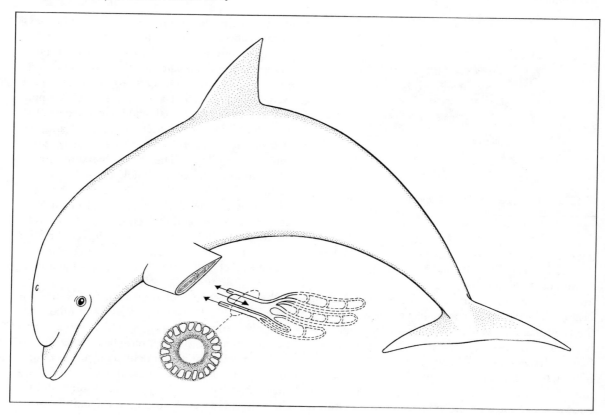

water. Unless the blood flowing to these thin-skinned peripheral surfaces went through a heat exchanger, the heat loss would be very great indeed. However, due to the heat exchanger that heat loss is minimal. A gull placed with its feet in ice water for 2 hours lost only 1.5% of its metabolic heat production from the feet, a quite insignificant heat loss (Scholander 1955).

Even some tropical animals have vascular heat exchangers in their limbs. In the sloth the artery to the foreleg splits into several dozen thin parallel arteries, which are intermingled with a similar number of veins. It may seem superfluous for a tropical animal to have such heat exchangers, but on a rainy and windy night the heat loss from an animal sleeping in the crown of a tree can be very great indeed. The value of the heat exchanger in the sloth has been clearly demonstrated by immersing a limb in ice water. The blood temperature below the heat exchanger decreased drastically, but in the part of the foreleg above the heat exchanger the temperature of the venous blood was nearly at core temperature (Scholander and Krog 1957).

It has been suggested that the vascular bundles in the forelegs of the sloth and the loris have an en-

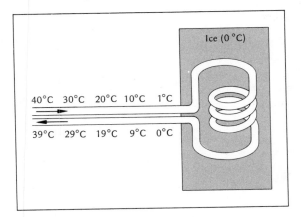

tirely different function: that they are related to the ability of these climbing animals to maintain a strong grip (Buettner-Janusch 1966; Suckling et al. 1969). However, the vascular structure is located in the upper part of the foreleg, and the muscles responsible for the grasp are in the lower part. The fact that the blood runs through a bundle of arteries that join again into a single vessel cannot have any imaginable influence on the muscles in the lower part of the foreleg. The blood still contains the same oxygen and nutrients, and the only difference these muscles could detect would be a slightly decreased blood pressure, which certainly could not improve the strength of the grip.

An animal in the cold has two opposed problems: to conserve heat and to keep peripheral tissues such as the feet from freezing. At ambient temperatures down to freezing, there is no conflict between the two; reduction in blood flow and cooling of the arterial blood in a heat exchanger help to minimize the heat loss.

However, when the temperature falls much below freezing, it is necessary to add enough heat to the peripheral tissues to keep them from freezing. The bare feet of the duck and other birds are suitable examples. As long as there is no danger of freezing, the heat loss from the feet is minimal, but as the temperature is lowered below 0 °C, the heat loss increases in proportion to the decrease in temperature (Kilgore and Schmidt-Nielsen 1975). The increase in heat drain is reflected in an increased metabolic heat production, appearing as a distinct inflection point in the metabolic curve (Figure 8.17).

Temperature regulation in the heat: keeping cool

We have seen that mammals and birds must increase their heat production in order to stay warm at ambient temperatures below a certain point, the lower critical temperature. We will now be concerned with what happens above this critical temperature.

Above the critical temperature the metabolic heat production (resting metabolic rate) remains constant through a range usually known as the *thermoneutral range* (see Figure 8.6). As before, for the body temperature to remain constant, metabolic heat must be lost at the same rate it is produced. If we return to the now familiar equation for heat balance, $\dot{H} = C(T_b - T_a)$, we see that if $\dot{H}$ and T_b remain constant and T_a changes, the conductance term (C) must also change.

The conductance term refers to the total heat flow from the organism, and it can be increased in several ways. One way is by increasing circulation to the skin so that heat from the core is moved more rapidly to the surface. Another way is by stretching out and exposing increased surface areas, especially naked or thinly furred areas.

Let us return to the basic equation for heat balance (page 241):

FIGURE 8.17 At low temperature, below 24 °C, the heat loss from the feet of a duck is a small fraction of the metabolic heat production. Below the freezing point the heat transferred to the feet (and lost to the water) increases in proportion to the drop in temperature. [Kilgore and Schmidt-Nielsen 1975]

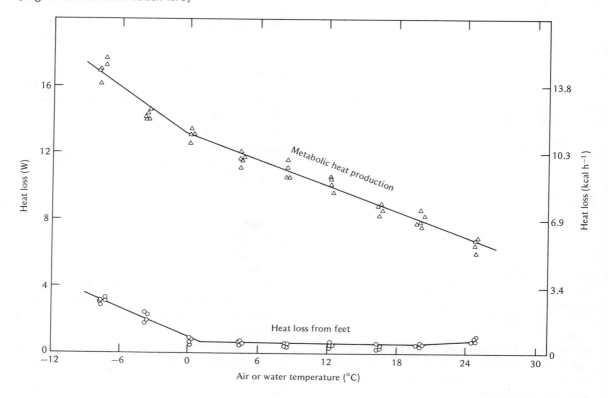

$$H_{tot} = \pm H_c \pm H_r \pm H_e \pm H_s$$

As the ambient temperature increases, the conditions for heat loss by conduction and convection (H_c), and radiation (H_r) become increasingly unfavorable. Because the metabolic heat production (H_{tot}) remains unchanged (or even increases slightly), the heat balance equation must increasingly emphasize the evaporation and storage terms (H_e and H_s).

In the discussions of heat regulation in the cold we disregarded heat removed by evaporation because it made up a fairly small fraction of the total heat exchange. At high temperature evaporation is

the key item in the heat balance. Assume that we have increased the temperature of the environment to equal the body temperature of an animal. If the temperature of the environment equals the body temperature, there can be no loss of heat by conduction, and the net radiation flux approaches zero. If we now wish to maintain a constant body temperature (i.e., keep the storage term H_s at zero), the consequence is that the entire metabolic heat production (H_{tot}) must be removed by evaporation of water.

This prediction has been tested on a number of mammals and birds. When they are kept in air that equals their body temperature, those that are able

FIGURE 8.18 The cooling by evaporation in jackrabbits, expressed in percent of the simultaneous metabolic heat production. Evaporation increases as the difference between body temperature (T_b) and air temperature (T_a) decreases. As this difference approaches zero (body and air temperatures equal), the entire heat production (100%) is dissipated by evaporation. [Dawson and Schmidt-Nielsen 1966]

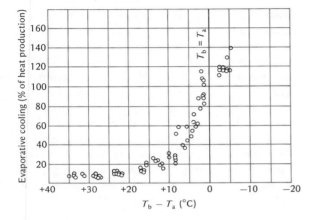

to keep their body temperature from rising (not all species can) evaporate an amount of water that is equivalent to their metabolic heat production. An example is given in Figure 8.18, which shows that the desert jackrabbit (actually a hare) behaves according to the prediction.

The ambient temperature can, of course, increase further and exceed the body temperature, as frequently happens in deserts. In this event the body receives heat from the environment by conduction from the hot air, by radiation from the heated ground surface, and in particular by radiation from the sun. Under these circumstances evaporation must be used to dissipate the sum of metabolic heat production and the heat gain from the environment. The possibility of storing heat is not a good solution, for as we saw before, animals have only limited tolerance to increased body temperature. Nevertheless, as we shall see later, even a moderate increase in body temperature can indeed be important for a desert animal.

To evaluate the importance of the heat load from the environment, consider how much water is used for evaporation (sweating) by a man in a hot desert. His total resting metabolic rate is about 70 kcal per hour, and this amount of heat, if dissipated entirely by evaporation, requires 120 ml, or 0.12 liter, water. We also know that a man exposed to the sun on a hot desert day may sweat at a rate of between 1 and 1.5 liters per hour. To make our calculations simple, let us say 1.32 liters per hour. If there is no change in his body temperature, and 0.12 liter is used to dissipate the metabolic heat, the excess of 1.2 liters used must be attributable to the heat gain from the environment. Thus the sum of conduction from the hot air, radiation from the hot ground, and radiation from the sun must require the evaporation of 1.2 liters of water; the heat load from the environment in this example is exactly 10 times as high as the metabolic heat production of the man at rest. If the body temperature were to increase by 1 °C, the storage of heat would be about 60 kcal, or the equivalent of 0.1 liter water evaporated – a fairly insignificant amount.

We now have a good idea of how important the environmental heat load can be when the temperature gradients are reversed, and it is worth considering how this influences animals. If a man were to continue evaporating at the indicated rate and had no water to drink, he would be near death from dehydration at the end of a single hot day in the desert. Yet the deserts of the world have a rich and varied animal life, although in most deserts there is no drinking water available. This requires further discussion.

The importance of body size

If we put a large block of ice outside in the hot sun, and a small piece next to it, the small piece will melt away long before the large block. If a big rock and a small pebble are placed on the ground in the sun, the pebble will be hot long before the big rock. The reason is that a small object has a much larger surface relative to its volume. If the

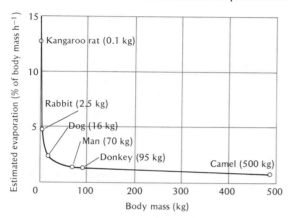

duction and convection, as well as radiation from ground and sun, are surface processes, and the total environmental heat load therefore is directly related to the surface area. In Chapter 6 we saw that the metabolic heat production of a mammal is not quite proportional to body surface, but sufficiently close that, as an approximation, we can assume this to be so. Therefore, the total heat load – the sum of metabolic and environmental heat gain – is roughly proportional to the body surface area. This puts the small animal, with its larger relative surface, in a far more unfavorable position with regard to heat load than a large animal.

If a man in the desert must sweat at a rate of one liter per hour (which equals about 0.60 liter per square meter body surface area per hour), we can use the surface relationship to estimate or predict how much water other animals should evaporate under similar desert conditions in order to dissipate the heat load. We thus obtain a theoretical curve that predicts the amount of water needed to keep cool (Figure 8.19). The amount is an exponential function of body mass, and on logarithmic coordinates the curve will be a straight line. However, Figure 8.19 is plotted on a linear ordinate in order to emphasize the exponential increase in water required if a small animal were to depend on evaporation to keep cool in the desert. Many small rodents weigh between 10 and 100 g and would have to evaporate water at a rate of 15 to 30% of their body weight per hour. As a water loss of between 10 and 20% is fatal for mammals, such rates of water loss are impossible. Obviously, because of their small size alone, desert rodents must evade the heat, which they do by retreating to their underground burrows during the day.

Large body size: the camel. Figure 8.19 shows that large animals, such as camels, derive a substantial advantage simply from being big. The ben-

big rock or block of ice is broken into many small pieces, numerous new surfaces, previously unexposed, are in contact with the warm air and receive solar radiation. Because of the many new exposed surfaces the total mass of rock heats up more rapidly and the crushed ice melts faster.

The relation between surface and volume of similar-shaped objects is simple. If a given cube is cut into smaller cubes with each side one-tenth of the larger cube, the total surface of all the small cubes is 10 times that of the original cube. If the linear dimension of the small cube is one-hundredth of the larger one, the aggregate surface is 100 times the original, and so on. This rule holds for any other similar-shaped objects. Thus, any small body has a surface area that, relative to its volume, increases as the linear dimension decreases. Small and large mammals have sufficiently similar shapes that their body surface area is a regular function of the body volume, and the same rule applies.

Let us return to the animal in hot surroundings. Its total heat gain consists of two components: heat gain from the environment and metabolic heat gain. The heat gain from the environment by con-

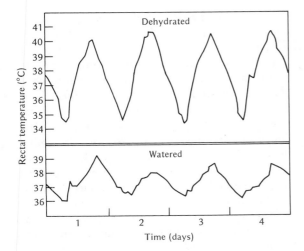

FIGURE 8.20 The daily temperature fluctuation in a well-watered camel is about 2 °C. When the camel is deprived of drinking water, the daily fluctuation may increase to as much as 7 °C. This has a great influence on the use of water for temperature regulation (see text). [Schmidt-Nielsen 1963]

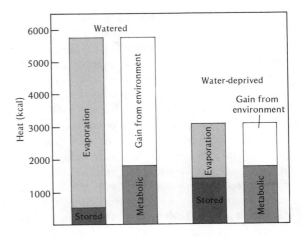

FIGURE 8.21 The change in body temperature of a camel deprived of drinking water greatly affects the heat gain from the environment and in turn the amount of water used for heat regulation (by evaporation). [Schmidt-Nielsen et al. 1957]

efit, however, does not increase much as the body size increases further. Theoretically, if a camel in other respects were like a man, its body size would make it evaporate water at about half the rate of a man. The camel, however, is not a man and does not use this much water for heat regulation. As we shall see, the camel uses a combination of several approaches that help reduce the heat gain from the environment and thus the use of water.

First consider what a camel can do about two of the variables in the heat balance equation, evaporation (H_e) and storage of heat (H_s). Storage of heat is reflected in an increase in body temperature. In a normal camel which is watered every day and is fully hydrated, the temperature varies by less than 2 °C, between about 36 and 38 °C. When the camel is deprived of drinking water, however, the daily temperature fluctuations become much greater (Figure 8.20) (Schmidt-Nielsen et al. 1957). The morning temperature may be as low as 34 °C, and the highest temperature in the late afternoon may be nearly 41 °C. This large increase in body temperature during the day constitutes a

storage of heat. For a camel that weighs 500 kg, the amount of heat stored by a 7 °C temperature rise corresponds to 2900 kcal of heat, which equals a saving of 5 liters of water. In the cool night, the stored heat can be unloaded by conduction and radiation without use of water.

The high body temperature during the day has a further advantage, beyond that of storing heat. When the body temperature is increased, the temperature gradient from the hot environment to the body (which determines the amount of heat gain from the environment) is reduced. This reduction in environmental heat gain is at least as important as heat storage with regard to water savings and is reflected in the reduced amount of water used. This is illustrated in Figure 8.21, which shows the heat balance of a small adult camel studied in the Sahara Desert. In the fully hydrated animal the evaporation during the 10 hottest hours of the day was 9.1 liters, corresponding to the dissipation of 5300 kcal heat. In the same camel when deprived of drinking water, the evaporation was reduced to 2.8 liters in 10 hours, corresponding to the dissipation of 1600 kcal heat. Thus, the use of water in the dehydrated camel was reduced to less than one-third that in the fully hydrated animal.

The main reason the camel in the desert uses substantially less water than a man (about 0.28 liter per hour in the camel compared with more than 1 liter per hour for a man who weighs about one-quarter as much) is the higher body temperature in the camel, which means both storage of heat and a reduced environmental heat gain. In contrast, man maintains his body temperature nearly constant at about 37 °C, and thus does not store appreciable amounts of heat. A man also maintains much steeper gradients between the environment and his body surface, for the mean skin surface temperature of a sweating man is about 35 °C.

One further important reason is that the camel has thick fur with a high insulation value. This imposes a heavy insulating layer between the body and the source of the heat and thus reduces the heat gain from the environment. The simplest way of testing the importance of this fur is to shear a camel. In one such experiment, the water expenditure of a camel, under otherwise comparable conditions, was increased by about 50% when the animal was shorn.

We have now seen some of the major reasons for a lower-than-predicted (from body size alone) water loss in the camel: (1) heat is stored because of an increase in body temperature; (2) the increase in body temperature decreases the heat flow from the environment; and (3) the fur is a substantial barrier to heat gain from the environment.

The camel has one further advantage over man: It can tolerate a greater degree of water depletion of the body. A man is close to the fatal limit when he has lost water amounting to between 10 and 12% of the body weight, but the camel can tolerate about twice as great a water depletion without apparent harm. As a result, the camel can go without

drinking for perhaps 6 to 8 days under desert conditions that would be fatal in a single day to a man without water.

In the end, however, a camel must drink to replenish its body water, and when water is available, it may drink more than a third of its body weight. The immense capacity for drinking has given rise to the legend that a camel, before a long desert journey, fills its water reservoirs; in reality, the camel, like other mammals, drinks to replenish lost water and to restore the normal water content of the body, and there is no evidence that it overdrinks in anticipation of future needs.

Small body size: the ground squirrel. We saw earlier that because of their small size and large relative surface, small rodents should be unable to remain above ground and active during the day. Seemingly, the small ground squirrels of the North American deserts defy this conclusion. These small rodents, which weigh from less than one hundred to a few hundred grams, are often seen outside their burrows in the daytime. They move around quickly, dashing from one place to another; they frequently disappear in a hole; but soon appear again, often within minutes. During the hottest midsummer days they are less active in the middle of the day, but are often seen during the morning and late afternoon hours.

Usually, there is no free water available in their environment. How is it possible for ground squirrels to remain active during the day? Apparently, heat storage plays a great role, but not on a diurnal cycle, as for the camel.

When they are outside their burrows on a hot day, the ground squirrels heat up very rapidly. They cannot tolerate any higher body temperature than other mammals and die if heated to 43 °C. However, a temperature of 42.4 °C can be tolerated without apparent ill effects. As an active ground squirrel has become heated outside, it returns to the relatively cool burrow, where it rapidly cools off, helped by its large relative surface. By moving in and out of the cool burrow the ground squirrel can rapidly pick up heat and rapidly unload it again and in this way avoid using water for evaporation. This permits it to move about and be active during the day.

Evaporation: sweating or panting

We know from our own experience that humans sweat to increase cooling by evaporation. Dogs, in contrast, have few sweat glands, and they cool primarily by panting – a very rapid, shallow breathing that increases evaporation from the upper respiratory tract. Some animals use a third method for increasing evaporation: They spread saliva over their fur and lick their limbs, thus achieving cooling by evaporation.

The amount of heat needed to evaporate 1 kg of water (580 kcal or 2426 kJ) is, of course, the same irrespective of the source of the water and where it is evaporated. Let us see if there are other differences that might make one method of evaporation more or less advantageous than another for a particular animal. We will therefore examine the characteristic features of each method.

In man, sweating carries virtually the entire burden of increased evaporation. Evaporation from the general body surface is an effective way of cooling, and in the absence of fur the water can evaporate readily. However, several furred mammals, in particular those of large body size, such as cattle, large antelopes, and camels, also depend mainly on sweating for evaporation. The camel in particular has a thick fur, but evaporation is not much impeded because in a desert atmosphere the air is very dry and the sweat evaporates so rapidly that some observers have reported that camels do not sweat at all. In contrast, a number of smaller ungulates – sheep, goats, and many small gazelles – evaporate mostly by panting. Those carnivores that have been studied also pant. The third method, salivation and licking, is common in a fairly large number of Australian marsupials, including the large kangaroos, and can also be found in some rodents, including the ordinary laboratory rat. The salivation-licking method is not very effective and seems to be used primarily as an emergency measure when the body temperature threatens to approach a lethal level.

In addition to being less effective than sweating and panting, salivation and licking are less widespread, and we shall not discuss the method further. Instead, we shall compare sweating and panting and see what possible advantages and disadvantages each method may have.

Birds, in contrast to mammals, have no sweat glands. They increase evaporation either by panting or by a rapid oscillation of the thin floor of the mouth and upper part of the throat, a mechanism known as *gular flutter*. Either mechanism seems to be an effective means of cooling, for birds may use one or the other or often a combination of both (Bartholomew et al. 1968).

An obvious difference between panting and sweating is that the panting animal provides its own air flow over the moist surfaces, thus facilitating evaporation. In this regard the sweating animal is not so well off. Another difference is that the sweat (at least in man) contains considerable amounts of salt, and a heavily sweating human may lose so much salt in the sweat that he becomes salt-deficient. This is the reason for the recommendation that we should increase the intake of sodium chloride when sweating is excessive in very hot situations. Panting animals, in contrast, do not lose any of the electrolytes secreted from glands in the nose and the mouth (unless the saliva actually drips to the ground) and are thus better off in this regard as well.

Panting has two obvious disadvantages. One is that increased ventilation easily causes an excessive loss of carbon dioxide from the lungs, which can result in severe alkalosis; the other is that increased ventilation requires muscular work, which in turn increases the heat production and thus adds to the heat load. The tendency to develop alkalosis can in part be counteracted by shifting to a more shallow respiration (smaller tidal volume) at an increased frequency, so that the increased ventilation takes place mostly in the dead space of the upper respiratory tract. Nevertheless, heavily panting animals regularly become severely alkalotic, and thus they do not utilize fully the possibility of restricting the ventilation to the dead space (Hales and Findlay 1968; Hales and Bligh 1969).

The increased work of breathing during panting would be a considerable disadvantage were it not for the interesting fact that the muscular work, and thus heat production, can be greatly reduced by taking advantage of the elastic properties of the respiratory system. When a dog begins to pant, its respiration tends to shift rather suddenly from a frequency of about 30 to 40 respirations per minute to a relatively constant high level of about 300 to 400. A dog subjected to a moderate heat load does not pant at intermediate frequencies; instead, it pants for brief periods at the high frequency, alternating with periods of normal slow respiration.

The meaning of this becomes clear when we realize that the entire respiratory system is elastic and has a natural frequency of oscillation, like other elastic bodies. That is, on inhalation, much of the muscular work goes into stretching elastic elements, which on exhalation bounce back again, like a tennis ball bouncing. To keep the respiratory system oscillating at its natural frequency (the *resonant frequency*) requires only a small muscular effort. As a consequence, the heat production of the respiratory muscles is small, adding only little to the heat load (Crawford 1962).

It has been estimated that if panting were to take place without the benefit of a resonant elastic system, the increased muscular effort of breathing at the high frequency of panting would generate more heat than the total heat that can be dissipated by panting. Apparently, birds can also pant at a resonant frequency, thus achieving the same benefits as mammals (Crawford and Kampe 1971).

Although we recognize some characteristic functional differences between panting and sweating, we do not have a clear picture of why one method or the other is used. Many animals use both methods, and this may indicate that both have valuable characteristics we do not yet fully appreciate.

The greatest advantage of panting may be that an animal under sudden heat stress, such as a fast-running African gazelle pursued by a predator, can reach a high body core temperature and yet keep the brain, the most heat-sensitive organ, at a lower temperature. This may, at first glance, seem impossible, for the brain is supplied with arterial blood at a high flow rate. It is achieved as follows (see Figure 8.22). In gazelles and other ungulates, most of the blood to the brain flows in the external carotid artery, which at the base of the skull divides into hundreds of small arteries, which then rejoin before passing into the brain. These small arteries lie in a large sinus of venous blood that comes from the walls of the nasal passages where it has been cooled. The blood that flows through these small arteries is therefore cooled before it enters the skull, and as a result the brain temperature may be 2 or 3 °C lower than the blood in the carotid artery and the body core.

Such temperature differences have been measured in the small East African gazelle, Thomson's

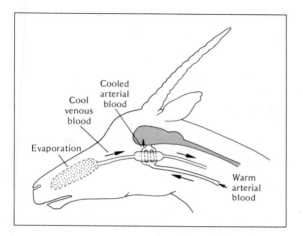

FIGURE 8.22 The brain of a gazelle can be kept at a lower temperature than the body core because the arterial blood, before it reaches the brain, passes in small arteries through a pool of cooler venous blood that drains from the nasal region, where evaporation takes place. [Taylor 1972]

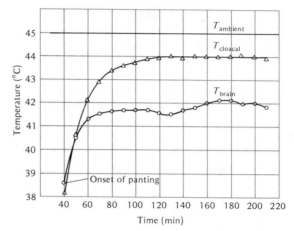

FIGURE 8.23 When the lizard *Sauromalus* is moved from 15 °C to 45 °C, cloacal and brain temperatures increase rapidly. As these temperatures reach about 41 °C, they separate and the brain remains about 2 °C below the cloacal temperature and 3 °C below the air temperature. [Crawford 1972]

gazelle, which weighs about 15 to 20 kg. When it ran for 5 minutes at a speed of 40 km per hour, a rapid buildup of body heat caused the arterial blood temperature to increase from a normal 39 °C to 44 °C. The brain temperature did not even reach 41 °C, which is a safe level (Taylor and Lyman 1972). It would be difficult to design a cooling system that can keep the entire body from heating up when an animal runs in hot surroundings at a speed that may require a 40-fold increase in metabolic rate, but by selective cooling of the blood to the brain the most serious hazard of overheating is avoided.

Such cooling of blood to the brain has also been observed in a number of domestic ungulates, and it may be a fairly common mechanism in animals that pant. However, it is unlikely that any such mechanism exists in man, for man does not pant and cooling takes place over the entire surface of the body.

It has been known for many years that some reptiles, when they are exposed to a severe heat stress, increase their respiration frequencies and breathe with open mouth, a situation reminiscent of panting. It has been difficult to establish with certainty that this increased respiration, which does increase respiratory evaporation, plays any major role in heat regulation. The amount of water evaporated does not suffice to keep the animals substantially cooler than the environment, and the mechanism therefore seems to be no more than a rudiment of panting found in birds and mammals. However, in view of the ability of mammals to cool the brain selectively, the panting of lizards can better be understood.

In the desert lizard, the chuckawalla (*Sauromalus obesus*), panting is of marginal importance in the animal's overall heat balance. When the ambient temperature is kept at 45 °C, the body is barely cooler (44.1 °C), but the brain remains at 42.3 °C, or nearly 3 °C below the ambient temperature. If a chuckawalla is moved from a room at 15 °C to 45 °C, the brain and cloacal temperatures initially increase rapidly (Figure 8.23), but as these temperatures reach 41 to 42 °C, the brain temperature stabilizes at this level, while the cloacal temperature continues to rise until it nearly reaches the ambient temperature.

In the chuckawalla the carotid arteries run very close to the surface of the pharynx, so close that they are visible through the open mouth of the animal. As the arteries pass right under the moist surfaces where evaporation takes place, the blood is cooled before it enters the brain. This is again an example of an animal that is able to maintain different parts of its body at different, well-regulated temperatures.

TORPOR AND HIBERNATION

Maintaining the body temperature in the cold at a cost of a several-fold increase in metabolic rate is expensive. Small animals have high metabolic rates to begin with, and a further increase may be too expensive when food is scarce or unavailable. The easy way out, and the only logical solution, is to give up the struggle to keep warm and let the body temperature drop. This not only eliminates the increased cost of keeping warm, but cold tissues use less fuel and the energy reserves last longer. This, in essence, is what hibernation is all about.*

Many mammals and a few birds hibernate regularly each winter. This means that the body temperature drops almost to the level of the surroundings; metabolic rate, heart rate, respiration, and many other functions are greatly reduced; and the animal is torpid and shows little response to external stimuli such as noise or being touched. With active life virtually suspended and the metabolic rate greatly reduced, the animal can survive a long winter. Before entering into a period of hiberna-

---

* The word hibernation (from Latin *hiberna* = winter) is also used to designate an overwintering, inactive stage of poikilothermic animals such as insects or snails.

tion, most hibernators also become very fat (i.e., they deposit large fuel reserves). Without the burden of keeping warm, the reserves can last for extended periods of unfavorable conditions.

Most animals that hibernate are small. This makes sense, for their high metabolic rates require a high food intake. Thus many rodents – hamsters, pocket mice, dormice – hibernate. Insect eaters at high latitudes can find little food in winter, and bats and insectivores (e.g., hedgehogs) could not possibly manage without hibernation. Hibernators are also found among Australian marsupials, such as the pigmy possum (Bartholomew and Hudson 1962). Hibernating birds include hummingbirds, the smallest of all birds, the insect-eating swifts, and some mouse birds (*Colius*, an African genus).

It is not easy to give a completely satisfactory definition of the term *hibernation*. In physiology the word refers to a torpid condition with a substantial drop in metabolic rate. Thus, bears may sleep during much of the winter, but most physiologists say that they are not true hibernators. Their body temperature drops only a few degrees, they show only a moderate drop in metabolic rate and other physiological functions, and the females often give birth to the cubs during the winter. In other words, the bear does not conform to what a physiologist customarily considers true hibernation.

Another term, *estivation* (Latin *aestas* = summer), refers to inactivity during the summer and is even less well defined. It may be applied to snails that become dormant and inactive in response to drought, or it may refer to ground squirrels that during the hottest months disappear into their burrows and remain inactive. The Columbia ground squirrel begins to estivate in the hot month of August, but then it remains inactive throughout the autumn and winter and does not appear again until the following May. Does it estivate or hibernate, and when does one condition change into the other? The fact is that no clear physiological distinction can be made between the two states. Furthermore, many hibernators (e.g., bats) may undergo daily periods of torpor with decreased body temperature and metabolic rate. Their physiological state is then quite similar to hibernation, although it lasts only for hours instead of weeks or months.

Because we lack precise definitions and there are no sharp lines between these different states, we will treat torpor and hibernation as one coherent physiological phenomenon.

It was earlier believed that hibernation and torpor result from a failure of temperature regulation in the cold and express some kind of "primitive" condition or poor physiological control. It is now quite clear, however, that hibernation is not attributable to inadequate temperature regulation; it is a well-regulated physiological state, and the superficial similarity between a hibernating mammal and a cold-blooded animal (say, a lizard or frog) is misleading. We shall find that hibernation can in no way be considered a physiological failure.

Body temperature and oxygen consumption

Bats are among the animals that may have daily periods of torpor and may also hibernate for longer periods. If exposed to low ambient temperature, bats may respond in one of two different characteristic ways. This is illustrated in Figure 8.24, which shows measured body temperatures of two North American bat species. At temperatures below about 30 °C these bats can be either torpid and have body temperatures within a degree or two of the air temperature, or they can be metabolically active with normal body temperatures between 32 and 36 °C. Even at air temperatures below 10 °C, these bats may be either active or torpid.

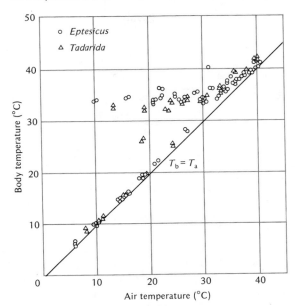

FIGURE 8.24 The body temperatures of two species of North American bats. At low air temperature these bats may be either active with normal body temperatures about 34 °C or torpid with body temperatures near the air temperature. [Herreid and Schmidt-Nielsen 1966]

The two bats are of similar size. The larger, the big brown bat (*Eptesicus fuscus*), which weighs about 16 g, ranges over most of the United States; and at least in the northern part of its range it hibernates. The smaller species, the Mexican free-tailed bat (*Tadarida mexicana*), weighs about 10 g; it inhabits the hot, dry southwestern United States and Mexico, and in winter it migrates southward and it not known to hibernate.

The response to low temperature of these two species is amazingly similar, although in nature one hibernates and the other does not. The oxygen consumption of those individuals that maintain a high body temperature at low air temperature shows the typical homeothermic pattern. As ambient temperature decreases, the cost of keeping warm is reflected in the increased oxygen consumption, which is at a minimum at about 35 °C (Figure 8.25). Those individuals that become tor-

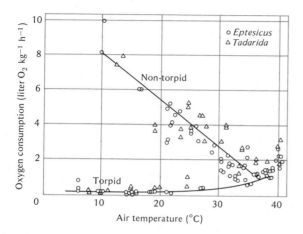

FIGURE 8.25 The oxygen consumption of active bats increases with decreasing air temperature; the oxygen consumption of torpid bats drops to a small fraction of the active rate (same bat species as in Figure 8.24). [Herreid and Schmidt-Nielsen 1966]

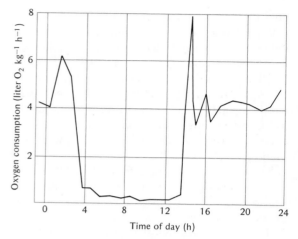

FIGURE 8.26 Oxygen consumption over a 24-hour period of a pocket mouse kept on a restricted food ration of 1.5 g seeds per day. During 9 hours the animal was torpid and had a very low rate of oxygen consumption followed by a peak as it returned to the active state. Air temperature was 15 °C. [Tucker 1965a]

pid, however, have very low metabolic rates. At 15 °C, for example, the difference in oxygen consumption between an active and a torpid bat is about 40-fold. The torpid bat thus uses fuel at a rate only one-fortieth the rate of the active bat, and its fat reserves would last 40 times as long.

The amount of energy that can be saved by going into torpor has been more carefully analyzed by Tucker (1965a, 1965b, 1966). The oxygen consumption during torpor is easily determined, but in addition, the energy cost required to rewarm the animal at the end of a period of torpor must be included in the cost. Tucker studied the California pocket mouse (*Perognathus californicus*) and found that for a short period of torpor the energy cost of reheating was a substantial part of the total cost.

The pocket mouse weighs about 20 g and readily becomes torpid at any ambient temperature between 15 and 32 °C. Below 15 °C the normal torpid state is disrupted and the animals usually cannot arouse on their own; in other words, if cooled below 15 °C they are unable to produce enough metabolic heat to get the rewarming cycle started.

Pocket mice enter torpor much more readily if

their food ration is restricted. With a gradual restriction of the food supply, longer and longer periods are spent in torpor. In this way a pocket mouse can maintain its body weight on a food ration that is about one-third of what it normally consumes when it remains active. A characteristic curve for the oxygen consumption of a pocket mouse on a restricted food ration is given in Figure 8.26. This particular individual received 1.5 g seed per day, and on this ration it was torpid for about 9 hours out of every 24. Although the food ration was less than half of what the animal would normally need if it remained active, it maintained a constant body weight.

When an animal goes into torpor, what happens first? Does heat loss first increase, the body temperature therefore drop, and the metabolic rate as a consequence decrease, thus accelerating the temperature drop? Or does the animal decrease its metabolic rate, the body temperature therefore begin to drop, the temperature drop causing a further decrease in heat production, and so on?

With information about heat production and heat loss during all stages of the torpor cycle, this question can be answered. The conclusion is that entry into torpor can result simply from a cessation of any thermoregulatory increase in metabolism at air temperatures below the lower critical temperature (about 32.5 °C in the pocket mouse). At any temperature below this point, a decrease in metabolic rate to the thermoneutral resting level alone will, without change in conductance, lead to a decrease in body temperature. In other words, there is no need for a special mechanism to increase the heat loss. As the body temperature begins to fall, metabolism decreases and the pocket mouse slides into torpor with a further drop in heat production and body temperature.

Arousal can take place at any temperature above 15 °C if the animal changes to maximum heat production at that temperature, which is about 10 to 15 times the minimum oxygen consumption at the same temperature. Arousal is thus an active process that requires a considerable expenditure of energy for a considerable period until the body temperature has reached normal.

If a pocket mouse goes into torpor at 15 °C and immediately arouses again, does it save any energy, or does rewarming cost too much? The decline in body temperature takes about 2 hours, and the total oxygen consumption during this period is 0.7 ml O_2 per gram. If arousal follows immediately, normal body temperature is reached in 0.9 hour at a cost of 5.8 ml O_2 per gram. Entering into torpor and then immediately arousing therefore costs a total of 6.5 ml O_2 per gram in 2.9 hours. The cost of maintaining normal body temperature for the same period of time at 15 °C ambient temperature would be 11.9 ml O_2 per gram. This period of torpor thus consumes only 55% of the energy required to keep warm for the same period of time (Tucker 1965b).

Remaining in torpor for longer periods, for example, 10 hours, is even more favorable for the

animal. We must now add to the previously calculated cost of entry and arousal the cost of 7.1 hours at 15 °C, which is 1.2 ml O_2 per gram, giving a total for a 10-hour cycle, including entry and arousal, of 7.7 ml O_2 per gram. This is less than 20% of the cost of maintaining normal body temperature for 10 hours, which at 15 °C would be over 40 ml O_2 per gram.

We can see that in all these cases the cost of arousal is the major part of the cost of the torpor cycle. For the 10-hour period of torpor the cost of arousal alone is 75% of the total energy expenditure for the period.

The widespread occurrence of torpidity as a response to unfavorable conditions suggests that this ability is a very fundamental trait. Animals that can enter torpor include the egg-laying echidna (the spiny anteater, *Tachyglossus aculeatus*), which is only remotely related to modern mammals. The echidna is normally an excellent temperature regulator in the cold and can maintain its normal body temperature at freezing temperatures (Schmidt-Nielsen et al. 1966). However, when it is without food and kept at 5 °C, it readily becomes torpid. The heart rate decreases from about 70 to 7 beats per minute, and the body temperature remains about +5.5 °C. In this state the oxygen consumption of the echidna is about 0.03 ml O_2 g^{-1} h^{-1}, roughly one-tenth its normal resting oxygen consumption (Augee and Ealey 1968).

Control mechanisms

Torpor and hibernation are under accurate physiological control. We have just seen how the duration of the daily torpor of a pocket mouse is adjusted to the food ration and the need for energy savings. Obviously, not only must entry into torpor be controlled, but the duration of the cycle and the arousal process must be accurately regulated.

In nature the beginning of the hibernation cycle

is usually associated with the time of the year, but it is not necessarily induced by low temperature or lack of adequate food. The yearly cycle of hibernation is influenced by the duration of the daily light cycle and is associated with endocrine cycles. For example, it is often impossible to induce even a good hibernator to begin a cycle of torpor during early summer and especially during the reproductive period.

The torpid animal with its low body temperature seems quite passive; it cannot perform coordinated movements, and it hardly responds to sensory stimuli. Superficially it resembles a cold-blooded animal that has been chilled. All of us have experienced how our fingers become numb when the hands are cold; this is because nerve conduction ceases at temperatures below some 10 to 15 °C, and we sense nothing. (The reason we can still move our cold fingers is that the appropriate muscles are located above the hand in the lower part of the arm.)

If the nerves of a hibernator became inoperative at lower temperatures, the nervous system could not remain coordinated at, say, 5 °C. Yet respiration and many other functions continue in a well-coordinated fashion although at a lower rate. If the ambient temperature decreases toward or below freezing, some hibernators die. Others, however, respond in one of two ways, leaving no doubt about the integrity of the central nervous system. Either they arouse and return to the fully active condition, or they resist the decrease in body temperature by a regulated increase in heat production, keeping the body temperature at some low level, say +5 °C.

Such a well-regulated heat production does indeed require a well-coordinated central nervous system. For example, the European hedgehog maintains its body temperature at +5 to +6 °C as the ambient temperature decreases to below freez-

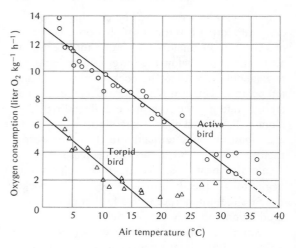

FIGURE 8.27 Oxygen consumption of the tropical hummingbird *Eulampis*. In the active *Eulampis* oxygen consumption increases linearly with decreasing temperature. In the torpid *Eulampis* oxygen consumption drops to a low level, but if the air temperature goes below 18 °C, the animal produces more heat and maintains its body temperature at 18 to 20 °C without arousing from torpor. [Hainsworth and Wolf 1970]

ing. This prevents freezing damage, and at the same time saves fuel, for it eliminates the need of going through an expensive complete arousal and the cost of then maintaining the high body temperature of the active state. For a hibernator that may be exposed to freezing temperatures several times during winter, such repeated arousals might prove too expensive; maintaining a low body temperature, just sufficient to keep from freezing, is far more economical.

A similar well-regulated state of torpor has been observed in the West Indian hummingbird, *Eulampis jugularis*, which like other hummingbirds readily becomes torpid. The temperature of *Eulampis*, when torpid, approaches air temperature, but if the air temperature falls below 18 °C, *Eulampis* resists a further fall and maintains its body temperature constant at 18 to 20 °C. Below this point the heat production must therefore be increased (Figure 8.27), and the increase must be linearly related to the drop in the ambient temperature.

Eulampis is particularly interesting because it is

a clear case of torpidity in a tropical warm-blooded animal, and this shows that torpidity is not restricted to animals from cold climates. Also, the well-regulated metabolic rate at two different levels of body temperature certainly shows that torpor in no way is a failure of the process of thermoregulation. Another interesting aspect is that the overall thermal conductance (the slope of the metabolic regression line) is the same in torpid and in normal nontorpid hummingbirds.

Arousal

The rewarming during arousal, as we have seen, is metabolically the most expensive part of the torpor cycle. The arousing animal displays violent shivering and muscle contractions and apparently uses fuel at a maximal rate. However, not only the muscles but also a particular kind of adipose tissue, known as *brown fat*, is important in the rewarming process.

In addition to the ordinary, or white, fat, many mammals have smaller or larger patches of brown fat, which differ from white fat both in color and in metabolic characteristics. Brown fat has a particularly high content of cytochrome and can consume oxygen at a high rate; white fat is metabolically rather inactive. Brown fat is particularly prominent in hibernators, and it has long been inferred that there is a direct connection between hibernation and the amount of brown fat. The precise connection remained obscure, however, until biochemical studies revealed that brown fat is capable of a high rate of oxygen consumption and thus heat production. During arousal the temperature of the brown fat, particularly in some large masses located between the shoulder blades, is among the highest in any part of the organism.

It is characteristic of an arousing hibernator that the temperature is not the same throughout the body; the rewarming is far from uniform. Figure

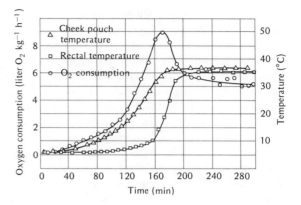

FIGURE 8.28 The temperature increase in a hamster arousing from hibernation proceeds more rapidly in the anterior part of the body (cheek pouch temperature) and more slowly in the posterior part (rectal temperature). [Lyman 1948]

8.28 shows temperatures recorded separately in the cheek pouch and in the rectum of an arousing hamster. The anterior part of the body, which contains vital organs such as heart and brain, warms much more rapidly than the posterior part. Rewarming of the heart is not only essential but must be an initial step, for the heart is needed to provide circulation and oxygen for all the other organs. The major masses of brown fat are also located in the anterior part of the body. The records show that not until the anterior portion of the body has reached near-normal temperatures does the posterior part enter into the reheating process. At this moment, when the entire body rapidly attains normal temperature, there is a peak in the oxygen consumption, which afterward declines and settles at a normal level.

The uneven temperature increase in the various parts of the body leads to an inevitable conclusion: During the early stages of arousal the blood flow is directed almost exclusively to the vital organs in the anterior part of the body, and not until these have been rewarmed does circulation increase substantially in the posterior parts as well.

The blood flow to the various organs can be followed by means of radioactive tracers, and such studies have shown that the skeletal muscles in the anterior part of arousing animals receive more than 16 times as much blood as in an awake, non-hibernating individual. This confirms that the muscles are highly involved in the increased heat production. During early arousal the muscles in the hind part of the animal receive only one-tenth as much blood. The brown fat, however, receives even more blood than the most active muscles, implicating this tissue in heat production during arousal. As could be expected, the digestive tract, in particular the small intestine, is among the tissues that receive the least blood during arousal (Johansen 1961).

It is perplexing that brown adipose tissue, which is so important for mammalian hibernators, appears to be absent from a number of birds that regularly hibernate, such as hummingbirds, swifts, and nighthawks (Johnston 1971).

BODY TEMPERATURE
IN "COLD-BLOODED" ANIMALS

Birds and mammals are traditionally known as warm-blooded, the term cold-blooded being used for all other animals. However poor the term is, it is well established and convenient, and we will use it, for no better single word is available that is both complete and accurate.

Most cold-blooded animals are more or less at the mercy of their environment in regard to the body temperatures they attain. However, some so-called "cold-blooded" animals can and indeed do stay warmer than the medium in which they live, whether it be air or water. What they can achieve in this regard is governed by simple physical principles, although some of the solutions seem quite ingenious.

Let us return to the heat balance equation on page 241 and rearrange the terms with only the term for heat storage (H_s) on the left side. On the right side are the total heat production (H_{tot}) and heat exchange by radiation (H_r), by conduction and convection (H_c), and by evaporation (H_e). If the body temperature is higher than the surrounding medium, the terms for conduction (H_c) and evaporation (H_e) will signify heat losses and be negative. The equation can therefore be written:

$$H_s = H_{tot} \pm H_r - H_c - H_e$$

If the goal is to increase the body temperature, we must see what the animal can do to maximize heat storage (H_s).

Aquatic animals

For an aquatic animal the situation is simple, for in water there is no evaporation and no significant radiation source (infrared radiation is rapidly absorbed in water). This leaves the equation: $H_s = H_{tot} - H_c$. Only two parameters need be manipulated in order to increase heat storage: Either total heat production must be increased, or conductive heat loss must be minimized.

Because of the high thermal conductivity and high heat capacity of water, a small animal loses heat rapidly and has no chance of attaining a body temperature very different from the medium. Even if it has a high level of heat production (metabolic rate) and could increase it further, the oxygen consumption must be increased.

Here is where the heat problem comes in. A high rate of oxygen uptake requires a large gill surface and blood for oxygen transport. As the blood flows through the gills, it is inevitably cooled to water temperature. The gill membrane, which must be thin enough to permit passage of oxygen, provides virtually no barrier to heat loss. The blood is therefore cooled to water temperature, and it is impossible for the animal to attain a high body

HEAT EXCHANGER Cross section of a 2-kg skipjack tuna (*Katsuwonis pelamis*) shows how this powerful swimming machine consists mostly of muscle (left). The red muscle, which is maintained at high temperature, appears nearly black in the photo. In the skipjack the main heat exchanger is located just below the vertebral column, almost exactly in dead center of the photo. A cross section of the vascular heat exchanger (right) shows a roughly equal number of arteries (smaller and thick-walled) interspersed with veins (larger and thin-walled). The diameter of the arteries is about 0.04 mm and of the veins 0.08 mm, and their length is about 10 mm. [Courtesy of E. D. Stevens, University of Guelph, Ontario]

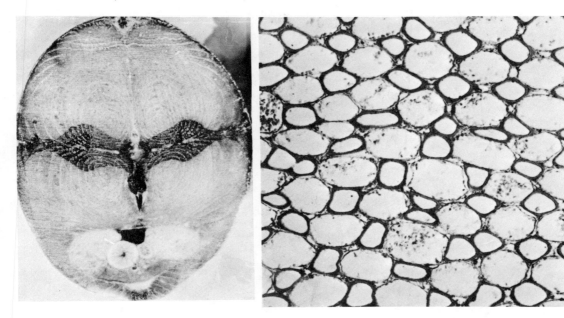

temperature unless a heat exchanger is placed between the gills and the tissues.

This solution is used by some large, fast-swimming fish (e.g., tunas and sharks) to achieve independent control of the temperature in limited parts of the body. They have heat exchangers that permit them to maintain high temperatures in their swimming muscles, independently of the water in which they swim.

The heat exchangers that supply the swimming muscles of the tuna are, in principle, similar to the countercurrent heat exchangers in the whale flipper, but anatomically they are somewhat differently arranged. In the ordinary fish the swimming muscles are supplied with blood from the large dorsal aorta that runs along the vertebral column and sends branches out to the periphery. In the tuna the pattern is different.

The blood vessels that supply the dark red muscles (which the tuna uses for steady, fast swimming) run along the side of the fish just under the skin. From these major vessels come many parallel fine blood vessels that form a slab in which arteries are densely interspersed between veins running in the opposite direction (Figure 8.29). This puts the cold end of the heat exchanger at the surface of the fish and the warm end deep in the muscles. The arterial blood from the gills is at water temperature, and as this blood runs in the fine arteries between the veins, it picks up heat from the venous blood that comes from the muscles. Heat exchange is facilitated by the small diameter of the vessels, about 0.1 mm. When the venous blood has reached the larger veins under the skin, it has lost its heat, which is returned to the muscle via the arterial blood. As a result, the tuna can maintain muscle temperatures as much as 14 °C warmer than the water in which it swims.

BODY TEMPERATURE IN "COLD-BLOODED" ANIMALS   **271**

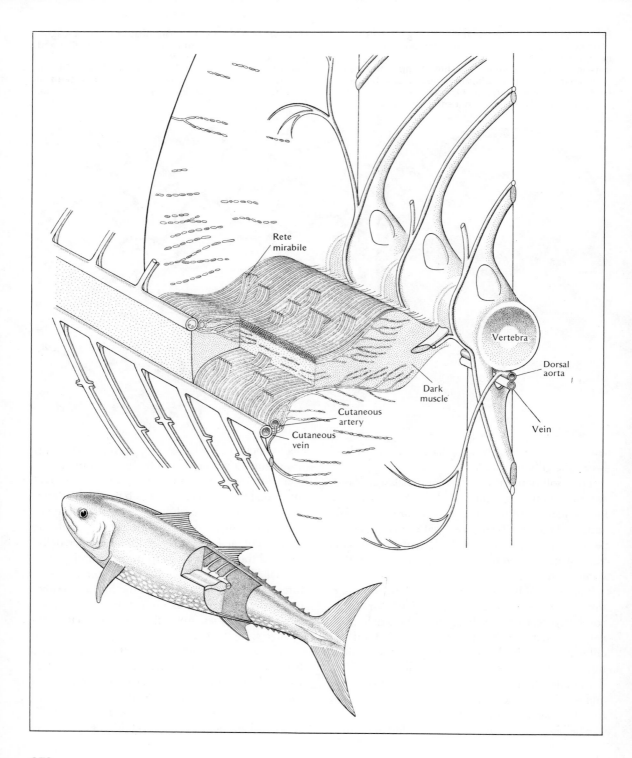

Rete
mirabile

Vertebra

Dorsal
aorta

Vein

Dark
muscle

Cutaneous
artery

Cutaneous
vein

FIGURE 8.29 The blood supply to the swimming muscle of the tuna comes through a heat exchanger, which is arranged so that these muscles retain a high temperature although the arterial blood is at water temperature. [Carey and Teal 1966]

The advantage of keeping the swimming muscles warm is that high temperature increases their power output.* A high power output gives the tuna a high swimming speed relatively independent of water temperature, and this in turn enables the tuna to pursue successfully prey that otherwise swims too fast to be caught, such as pelagic fish (e.g., mackerel) or squid.

Several of the large sharks, but by no means all of them, have similar heat exchangers that permit their muscles to be kept considerably warmer than the water in which they swim (Carey and Teal 1969; Cary et al. 1971).

The tuna enjoys an additional advantage from the substantial temperature differences between various parts of the body. Separate heat exchangers permit the digestive organs and the liver to be kept at a high temperature. The high power output of the muscles requires a high rate of fuel supply, and this puts a premium on a rapid rate of digestion, which is most readily achieved by a high temperature in the digestive tract.

Let us now compare the warm-bodied fish with marine mammals and consider the main differences and similarities between them. Seals and whales have heavily insulated surfaces (blubber); heat loss from the extremities is reduced by heat exchangers; and most importantly, they are air breathers whose blood escapes cooling to water temperature at the respiratory surface. The warm-bodied fish, in contrast, would gain no advantage from a better surface insulation, for all the arterial blood coming from the gills is already at water temperature. They do achieve independent temperature control in limited parts of the body by strategic location of efficient heat exchangers that help retain locally produced heat.

Terrestrial animals

The heat balance of terrestrial animals includes all the terms listed in the equation on page 270. To raise the body temperature (i.e., to increase heat storage), it is important to reduce evaporation and heat loss by conduction and to maximize heat gain by radiation and metabolic heat production. The only commonly available radiation source is the sun, and in the absence of solar radiation, metabolic heat production is the only way to increase the heat gain. Of course, external and internal sources of heat (i.e., sun and metabolism), can be used simultaneously, but it is more economical to use an external source rather than body fuel.

Solar radiation is used especially by insects and reptiles. To increase the amount of radiant energy absorbed, these animals depend both on their color and on their orientation relative to the sun. Many reptiles can change their color by dispersion or contraction of black pigment cells in their skin. Because about half the solar radiation energy is in the visible light, a dark skin substantially increases the amount of solar energy that is absorbed rather than reflected. The absorption in the near infrared is not much affected by color change, for the animal surface is already close to being "black" in this range of the spectrum (see earlier in this chapter, under "Radiation").

The other way of increasing heat gain from solar radiation is to increase the exposed area. This is done by orienting the body at right angles to the direction of the sun's rays and by spreading the legs and flattening the body. In this way a lizard can attain a temperature much higher than the surrounding air. When a suitable temperature has

---

* The force exerted by a contracting muscle is relatively independent of temperature, and the work performed in a single contraction (force × distance) is therefore also temperature-independent. However, at higher temperatures the muscle contracts faster, and because the number of contractions per unit time increases, the power output (work per unit time) increases accordingly.

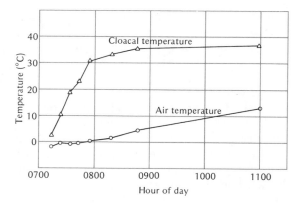

FIGURE 8.30 The Peruvian mountain lizard *Liolaemus* comes out in the morning while the air temperature is still below freezing and, by exposing itself to the sun, rapidly heats up to a body temperature that permits full activity. [Data from Pearson 1954]

it maintains its *preferred body temperature*, also known as the *eccritic temperature*.

It is, of course, important for the animal to heat up rapidly and become fully active as fast as possible. On the other hand, when the animal moves out of the sun retarding the inevitable cooling would be an advantage. Can the lizard do anything to increase the rate of heating and decrease the rate of cooling? The heating of the body core can be accelerated by increasing the circulation to the heated skin area, but what about retarding cooling?

This question has been studied in several species of lizards. It is of particular importance to the Galápagos marine iguana, which feeds in water where the temperature is 22 to 27 °C. When it is out of water, the animal sits on the rocks near the surf where it can warm itself in the sun. When it enters the water to feed on the seaweed, its temperature begins to fall toward that of the water, which is below the eccritic temperature. It would be advantageous for the marine iguana to reduce its rate of cooling while in water and thus extend the period during which it can feed actively and move fast enough to escape predators. Conversely, after leaving the sea, it would be advantageous to heat up as rapidly as possible.

A comparison of the rates of heating and cooling for the marine iguana shows that heating rates are about twice as high as cooling rates, both in air and in water (Figure 8.31). When a cold animal is immersed in water that is 20 °C above body temperature, the heating is twice as fast as the cooling when a warm animal is immersed in water 20 °C below its body temperature. In air both rates are lower, but again, heating is about twice as fast as cooling at the same temperature differential.

The different rates of heating and cooling can best be explained by changes in circulation, especially in the skin. During heating, the heart rate is high and increases with body temperature, but in

been reached, further heating is avoided by lightening the skin color and changing the orientation to a posture more parallel to the sun's rays, and eventually by moving into shade. The temperature of the substratum is also important, for a cool lizard can place itself in close contact with a warm rock and thereby increase its heat gain.

The temperature a lizard can attain by heating up in the sun can be spectacular. In the mountains of Peru, at an altitude of 4000 m where the air temperature is low even in summer, the lizard *Liolaemus* remains active and comes out at temperatures below freezing. One lizard that weighed 108 g was tethered in the sun in the morning while the air temperature was still at −2 °C. Within the next hour the air temperature increased to +1.5 °C, but the cloacal temperature of the lizard had increased from +2.5 °C to 33 °C (Figure 8.30). The lizard's body temperature was 30 °C higher than the air temperature, allowing it to be active in the near-freezing surroundings.

Once the lizard is warm, its body temperature remains more or less constant around 35 °C, while the air temperature slowly continues to rise. This is achieved because the lizard reduces the amount of absorbed solar energy by changing posture, so that

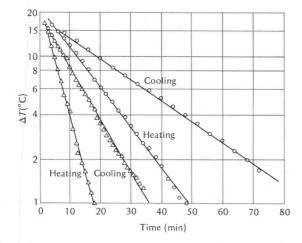

FIGURE 8.31 Heating and cooling rates of the Galápagos marine iguana (body weight 652 g) when immersed in water (open triangles) or in air (open circles). The *initial* temperature difference between animal and medium (ΔT) was, in all cases, 20 °C. [Bartholomew and Lasiewski 1965]

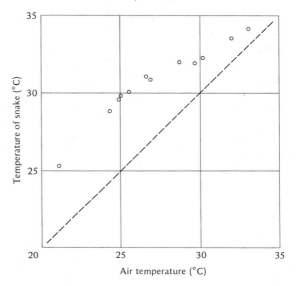

FIGURE 8.32 A brooding female python while incubating a clutch of eggs keeps her body temperature above that of the surroundings by strong muscular contractions reminiscent of shivering in mammals. The diagonal line indicates equality between the snake's temperature and air temperature. [Vinegar et al. 1970]

an animal being cooled the heart rate drops to half of what it otherwise is at the same body temperature. This ability to regulate the rates of heating and cooling through adjustment of the circulation is an obvious advantage for an animal that seeks its food by periodically entering cool water.

Not only lizards but other reptiles habitually bask in the sun. Both tortoises and snakes, and even the most aquatic reptiles, crocodiles, can often be seen sunning themselves at the edge of the water. How important this is for heating and temperature regulation is unknown for most of them.

Internal heat production (i.e., metabolic rate), can also be used to attain a high body temperature. It has been reported that some female snakes do this while incubating their eggs. This has been observed in the New York Zoological Park, where a 2.7-m-long female python started incubating a clutch of 23 eggs. She coiled herself around the eggs, and temperatures were measured with thermocouples placed between the tightly appressed coils of the snake. As the air temperature was lowered from 33 °C to 25 °C, the animal increased her rate of oxygen consumption and kept her body temperature at 4 to 5 °C above the air (Figure 8.32).

The increase in oxygen consumption of the brooding snake seemed to be caused by strong contractions of the body musculature, reminiscent of the shivering of mammals. At 25.5 °C the oxygen consumption was 9.3 times higher than during the nonbrooding period. When the temperature was decreased further to 21 °C, the animal seemed unable to increase her metabolic rate further, for her temperature fell and there was a drop in the oxygen consumption. After 30 days of incubation the animal had decreased in weight from 14.3 kg to 10.3 kg, a decrease of nearly 30 %, supposedly attributable to the fuel consumption needed to keep warm (Hutchison et al. 1966).

Flying insects

Most insects become increasingly sluggish and are unable to fly at low temperatures. Some insects, however, can warm up their flight muscles and be active in quite cold air. The flight muscles are located in the thorax and can produce large amounts of heat by a process similar to shivering in vertebrates. Such heating of the muscles before takeoff occurs mainly in large insects such as locusts, large moths, butterflies, and bumblebees, and also in wasps and bees that are strong and rapid fliers.

A bumblebee must have a thoracic temperature of at least 29 to 30 °C before it is able to fly (Heinrich 1972). If the flight muscles are at lower temperature, their speed of contraction is too slow to support flight, which requires a wing-beat frequency of 130 Hz.

The maintenance of a high thoracic temperature enables the bumblebee to forage for nectar at temperatures as low as 5 °C. However, remaining warm on a continuous basis is not feasible unless the bee can find food at a rate at least equal to the rate at which fuel is consumed. A bumblebee weighing 0.5 g may have an oxygen consumption of 50 ml O_2 per hour (100 liter O_2 kg^{-1} h^{-1}), which corresponds to the use of 60 mg sugar. This investment in fuel may be worthwhile, for at low temperature the competition for the available food supply is reduced because many other nectar-feeding insects are inactive.

The advantage of a small body size, relative to mammals, is not only that warmup can be achieved rapidly, but also that cooling is rapid. Therefore, the body temperature can easily be adjusted to the energy supplies, and the insect can choose to keep warm only when the energy supplies warrant the expense; in other words, it can be utterly opportunistic about being "warm-blooded."

The thermoregulation in bumblebees extends to the incubation of their brood. An incubating queen keeps her abdomen in close contact with the brood clump and regulates her abdominal temperature at 31 to 36 °C, which helps maintain the brood at a

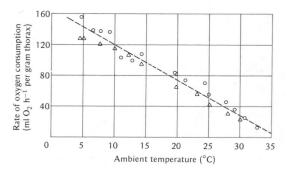

FIGURE 8.33 The rate of oxygen consumption of two bumblebee queens (*Bombus vosnesenskii*) during periods of uninterrupted incubation. The weight of the thorax, which contains the muscles that produce essentially all the heat, is about 0.2 g; the total body weight of a queen may be around 0.5 g, but varies with the contents of the honey stomach. [Heinrich 1974]

relatively high temperature. Even at an air temperature as low as 5 °C, the brood can be maintained nearly 20 °C higher by the incubating queen (Heinrich 1974). The oxygen consumption of the incubating queen increases linearly at lower temperature, so that her metabolic curve (Figure 8.33) resembles those we saw for mammals (e.g., Figure 8.7).

Another insect that has been carefully studied in regard to thermoregulation is the sphinx moth, *Manduca sexta* (whose larva is the tobacco hornworm). Sphinx moths are fairly large and weigh between 2 and 3 g. They feed on nectar while hovering in front of a flower, much like a hummingbird. Because they are nocturnal, they often fly at quite low air temperatures and flight would be impossible unless they could heat themselves. Hovering flight requires a muscle temperature of at least 35 °C; nevertheless, sphinx moths can fly and feed when the air temperature is as low as 10 °C. This is possible through preflight heating of the thorax. The lower the initial temperature, the lower is the rate of heating (Figure 8.34). During the warmup the animal seems to shiver as the wings vibrate lightly.

The heat is generated by contraction of the wing muscles, and the reason that the wings do not flap is that the muscles for upstroke and downstroke

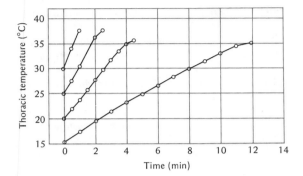

FIGURE 8.34 The increase in thoracic temperature of the sphinx moth (*Manduca sexta*) proceeds more rapidly when the initial temperature of the animal is higher. [Heinrich and Bartholomew 1971]

contract simultaneously, rather than alternately as in flight. We might expect that the warmup should progress faster and faster as the temperature of the moth increases, for more heat can be produced at higher temperatures. However, as the insect becomes warmer, the rate of heat loss to the air also increases, and for this reason the temperature curves in Figure 8.34 appear as more or less straight lines.

The thorax of the moth is covered by long, furry scales. This helps retain the generated heat in the thorax during the warmup. In free flight the sphinx moth maintains its thoracic temperature at about 40 to 41 °C over a wide range of air temperatures, but due to the high rate of heat production in the flight muscles it must be able to increase the rate of heat loss to keep the temperature from rising even higher. This is achieved through an increased circulation of blood between the thorax and the abdomen.

Heat loss from the abdomen is facilitated by its larger surface and a smaller amount of surface insulation. If we tie off the large dorsal blood vessel leading from the heart (located in the abdomen) so that the flight muscles cannot be cooled, the moth rapidly overheats and cannot continue to fly even at an air temperature of 23 °C. The cause could, of course, be that the blood supply to the muscles is necessary for supplying fuel. However, a simple experiment can answer this objection. If the scales are removed from the thorax of a moth with ligated circulation, the heat loss is sufficiently improved to enable the animal to fly again (Heinrich 1970).

Honeybees regulate not only their own individual temperature, but that of the entire colony as well. In summer the temperature of a beehive is maintained at about 35 °C and fluctuates very little. This provides the colony with an optimum temperature for development of the brood. Rapid reproduction permits a fast buildup of the population during the most favorable season, and an optimal temperature for the brood is therefore ecologically highly desirable. If the beehive temperature rises above the regulated level, the bees spread droplets of water over the comb, and evaporation is aided by air currents created by a large number of bees fanning their wings. Below 30 °C bees do not usually consume any water, but at higher temperatures their water intake increases enormously as water is used for cooling (Free and Spencer-Booth 1958).

At low outside temperatures the bees cluster together and maintain a temperature within the cluster of about 20 to 30 °C. The temperature at the center of the cluster is the highest, and it appears that the bees at the periphery force their way into the center where it is warmer, thus exposing others to the outside. The core temperature of the cluster is regulated between 18 and 32 °C at outside air temperatures ranging as low as −17 °C. In fact, there is a tendency for the cluster temperature to be higher at the very lowest ambient temperatures. For each 5 °C drop in the outside air temperature, the cluster temperature increases by about 1 °C (Southwick and Mugaas 1971). In this way a cluster of bees behaves much like a single large organism that maintains a well-regulated high temperature at external temperature conditions that would rapidly kill individual bees.

We have now discussed the temperature relations of animals, and in this last chapter we have seen how important water can be in maintaining proper temperature conditions. We shall now move on to a more detailed treatment of this important substance and its physiological role.

REFERENCES

Adolph, E. F. (1947) Tolerance to heat and dehydration in several species of mammals. *Am. J. Physiol.* *151*:564–575.

Andersen, H. T., Hammel, H. T., and Hardy, J. D. (1961) Modifications of the febrile response to pyrogen by hypothalamic heating and cooling in the unanesthetized dog. *Acta Physiol. Scand.* *53*:247–254.

Aschoff, J., Günther, B., and Kramer, K. (1971) *Energiehaushalt und Temperaturregulation.* Munich: Urban & Schwarzenberg. 196 pp.

Aschoff, J., and Wever, R. (1958) Kern und Schale im Wärmehaushalt des Menschen. *Naturwissenschaften* *45*:477–485.

Aschoff, J., and Wever, R. (1959) Wärmeaustausch mit Hilfe des Kreislaufes. *Dtsch. Med. Wochenschr.* *84*:1509–1517.

Augee, M. L., and Ealey, E. H. M. (1968) Torpor in the echidna, *Tachyglossus aculeatus.* *J. Mammal.* *49*: 446–454.

Bartholomew, G. A., and Hudson, J. W. (1962) Hibernation, estivation, temperature regulation, evaporative water loss, and heart rate of the pigmy possum, *Cercaertus nanus.* *Physiol. Zool.* *35*:94–107.

Bartholomew, G. A., and Lasiewski, R. C. (1965) Heating and cooling rates, heart rate and simulated diving in the Galápagos marine iguana. *Comp. Biochem. Physiol.* *16*:575–582.

Bartholomew, G. A., Lasiewski, R. C., and Crawford, E. C., Jr. (1968) Patterns of panting and gular flutter in cormorants, pelicans, owls, and doves. *Condor* *70*:31–34.

Bligh, J., Ingram, D. L., Keynes, R. D., and Robinson, S. G. (1965) The deep body temperature of an unrestrained Welsh mountain sheep recorded by a radiotelemetric technique during a 12-month period. *J. Physiol.* *176*:136–144.

Buettner-Janusch, J. (1966) *Origins of Man.* New York: Wiley. 674 pp.

Calder, W. A. (1964) Gaseous metabolism and water relations of the zebra finch, *Taeniopygia castanotis.* *Physiol. Zool.* *37*:400–413.

Calder, W. A., and Schmidt-Nielsen, K. (1967) Temperature regulation and evaporation in the pigeon and the roadrunner. *Am. J. Physiol.* *213*:883–889.

Carey, F. G., Teal, J. M. (1966) Heat conservation in tuna fish muscle. *Proc Natl. Acad. Sci. U.S.A.* *56*:1464–1469.

Carey, F. G., and Teal, J. M. (1969) Mako and porbeagle: Warm-bodied sharks. *Comp. Biochem. Physiol.* *28*:199–204.

Carey, F. G., and Teal, J. M., Kanwisher, J. W., and Lawson, K. D. (1971) Warm-bodied fish. *Am. Zool.* *11*:137–145.

Council of the Royal Society (1975) *Quantities, Units, and Symbols,* 2nd ed. London: The Royal Society. 55 pp.

Crawford, E. C., Jr. (1962) Mechanical aspects of panting in dogs. *J. Appl. Physiol.* *17*:249–251.

Crawford, E. C., Jr. (1972) Brain and body temperatures in a panting lizard. *Science* *177*:431–433.

Crawford, E. C., Jr., and Kampe, G. (1971) Resonant panting in pigeons. *Comp. Biochem. Physiol.* *40A*:549–552.

Dawson, W. R. (1954) Temperature regulation and water requirements of the brown and abert towhees, *Pipilo fuscus* and *Pipilo aberti. Univ. Calif. Publ. Zool.* *59*:81–124.

Dawson, T. J., and Hulbert, A. J. (1970) Standard metabolism, body temperature, and surface areas of Australian marsupials. *Am. J. Physiol.* *218*:1233–1238.

Dawson T., and Schmidt-Nielsen, K. (1966) Effect of thermal conductance on water economy in the antelope jack rabbit, *Lepus alleni. J. Cell. Physiol.* *67*:463–472.

Farner, D. S. (1956) The body temperatures of the North Island kiwis. *Emu* *56*:199–206.

Free, J. B., and Spencer-Booth, Y. (1958) Observations on the temperature regulation and food consumption of honeybees (*Apis mellifera*). *J. Exp. Biol.* 35:930–937.

Hainsworth, F. R., and Wolf, L. L. (1970) Regulation of oxygen consumption and body temperature during torpor in a hummingbird, *Eulampis jugularis*. *Science* 168:368–369.

Hales, J. R. S., and Bligh, J. (1969) Respiratory responses of the conscious dog to severe heat stress. *Experientia* 25:818–819.

Hales, J. R. S., and Findlay, J. D. (1968) Respiration of the ox: Normal values and the effects of exposure to hot environments. *Respir. Physiol.* 4:333–352.

Hammel, H. T. (1955) Thermal properties of fur. *Am. J. Physiol.* 182:369–376.

Hardy, J. D. (1949) Heat transfer. In *Physiology of Heat Regulation and the Science of Clothing* (L. H. Newburgh, ed.), pp. 78–108. Philadelphia: Saunders.

Hart, J. S. (1951) Calorimetric determination of average body temperature of small mammals and its variation with environmmental conditions. *Can. J. Zool.* 29:224–233.

Hart, J. S. (1956) Seasonal changes in insulation of the fur. *Can. J. Zool.* 34:53–57.

Hart, J. S., and Irving, L. (1959) The energies of harbor seals in air and in water, with special consideration of seasonal changes. *Can. J. Zool.* 37:447–457.

Heinrich, B. (1970) Thoracic temperature stabilization by blood circulation in a free-flying moth. *Science* 168:580–582.

Heinrich, B. (1972) Temperature regulation in the bumblebee *Bombus vagans*: A field study. *Science* 175:185–187.

Heinrich, B. (1974) Thermoregulation in bumblebees. 1. Brood incubation by *Bombus vosnesenskii* queens. *J. Comp. Physiol.* 88:129–140.

Heinrich, B., and Bartholomew, G. A. (1971) An analysis of pre-flight warm-up in the sphinx moth, *Manduca sexta*. *J. Exp. Biol.* 55:223–239.

Hensel, H., and Bock, K. D. (1955) Durchblutung und Wärmeleitfähigkeit des menschlichen Muskels. *Pflügers Arch.* 260:361–367.

Herreid, C. F., II, and Schmidt-Nielsen, K. (1966) Oxygen consumption, temperature, and water loss in bats from different environments. *Am. J. Physiol.* 211:1108–1112.

Hutchison, V. H., Dowling, H. G., and Vinegar, A. (1966) Thermoregulation in a brooding female Indian python, *Python molurus bivittatus*. *Science* 151:694–696.

Irving, L. (1969) Temperature regulation in marine mammals. In *The Biology of Marine Mammals* (H. T. Andersen, ed.), pp. 147–174. New York: Academic Press.

Irving, L., and Hart, J. S. (1957) The metabolism and insulation of seals as bare-skinned mammals in cold water. *Can. J. Zool.* 35:497–511.

Irving, L., and Krog, J. (1954) Body temperatures of arctic and subarctic birds and mammals. *J. Appl. Physiol.* 6:667–680.

Irving, L., and Krog, J. (1955). Temperature of skin in the arctic as a regulator of heat. *J. Appl. Physiol.* 7:355–364.

Johansen, K. (1961) Distribution of blood in the arousing hibernator. *Acta Physiol. Scand.* 52:379–386.

Johnston, D. W. (1971) The absence of brown adipose tissue in birds. *Comp. Biochem. Physiol.* 40A:1107–1108.

Kilgore, D. L., Jr., and Schmidt-Nielsen, K. (1975) Heat loss from ducks' feet immersed in cold water. *Condor* 77:475–478.

Kluger, M. J. (1977) Fever in the frog *Hyla cinerea*. *J. Thermal Biol.* 2:79–81.

Kluger, M. J., Ringler, D. H., and Anver, M. R. (1975) Fever and survival. *Science* 188:166–168.

Le Maho, Y., Delclitte, P., and Chatonnet, J. (1976) Thermoregulation in fasting emperor penguins under natural conditions. *Am. J. Physiol.* 231:913–922.

Lyman, C. P. (1948) The oxygen consumption and temperature regulation of hibernating hamsters. *J. Exp. Zool.* 109:55–78.

McNab, B. K. (1966) An analysis of the body temperatures of birds. *Condor* 68:47–55.

Morrison, P. (1966) Insulative flexibility in the guanaco. *J. Mammal.* 47:18–23.

Morrison, P. R., and Ryser, F. A. (1952) Weight and

body temperature in mammals. *Science* 116:231–232.

Øritsland, N. A. (1970) Temperature regulation of the polar bear (*Thalarctos maritimus*). *Comp. Biochem. Physiol.* 37:225–233.

Pearson, O. P. (1954) Habits of the lizard *Liolaemus multiformis multiformis* at high altitudes in Southern Peru. *Copeia* 1954(2):111–116.

Pinshow, B., Fedak, M. A., Battles, D. R., and Schmidt-Nielsen, K. (1976) Energy expenditure for thermoregulation and locomotion in emperor penguins. *Am. J. Physiol.* 231:903–912.

Reynolds, W. W., Casterlin, M. E., and Covert, J. B. (1976) Behavioural fever in teleost fishes. *Nature* 259:41–42.

Robinson, K. W., and Lee, D. H. K. (1946) Animal behaviour and heat regulation in hot atmospheres. *Univ. Queensland Papers, Dept. Physiol.* 1:1–8.

Robinson, K. W., and Morrison, P. R. (1957) The reaction to hot atmospheres of various species of Australian marsupial and placental animals. *J. Cell. Comp. Physiol.* 49:455–478.

Schmidt-Nielsen, K. (1963) Osmotic regulation in higher vertebrates. *Harvey Lect.* 58:53–93.

Schmidt-Nielsen, K. (1964) *Desert Animals: Physiological Problems of Heat and Water.* New York: Oxford University Press. 277 pp.

Schmidt-Nielsen, K., (1970) *Animal Physiology*, 3rd ed. Englewood Cliffs, N.J.: Prentice-Hall.

Schmidt-Nielsen, K., Dawson, T. J., and Crawford, E. C., Jr. (1966) Temperature regulation in the echidna (*Tachyglossus aculeatus*). *J. Cell. Physiol.* 67:63–72.

Schmidt-Nielsen, K., Schmidt-Nielsen, B., Jarnum, S. A., and Houpt, T. R. (1957) Body temperature of the camel and its relation to water economy. *Am. J. Physiol.* 188:103–112.

Scholander, P. F. (1955) Evolution of climatic adaptation in homeotherms. *Evolution* 9:15–26.

Scholander, P. F., Hock, R., Walters, V., Johnson, F., and Irving, L. (1950a) Heat regulation in some arctic and tropical mammals and birds. *Biol. Bull.* 99:237–258.

Scholander, P. F., and Krog, J. (1957) Countercurrent heat exchange and vascular bundles in sloths. *J. Appl. Physiol.* 10:405–411.

Scholander, P. F., and Schevill, W. E. (1955) Countercurrent vascular heat exchange in the fins of whales. *J. Appl. Physiol.* 8:279–282.

Scholander, P. F., Walters, V., Hock, R., and Irving, L. (1950b) Body insulation of some arctic and tropical mammals and birds. *Biol. Bull.* 99:225–236.

Shkolnik, A., and Schmidt-Nielsen, K. (1976) Temperature regulation in hedgehogs from temperate and desert environments. *Physiol. Zool.* 49:56–64.

Southwick, E. E., and Mugaas, J. N. (1971) A hypothetical homeotherm: The honeybee hive. *Comp. Biochem. Physiol.* 40A:935–944.

Suckling, J. A., Suckling, E. E., and Walker, A. (1969) Suggested function of the vascular bundles in the limbs of *Perodicticus potto. Nature, Lond.* 221:379–380.

Taylor, C. R. (1972) The desert gazelle: A paradox resolved. In *Comparative Physiology of Desert Animals* (G. M. O. Maloiy, ed.), Symposia, *Zoology Society of London*, no. 31, pp. 215–227. London: Academic Press.

Taylor, C. R., and Lyman, C. P. (1972) Heat storage in running antelopes: Independence of brain and body temperatures. *Am. J. Physiol.* 222:114–117.

Tucker, V. A. (1965a) Oxygen consumption, thermal conductance, and torpor in the California pocket mouse *Perognathus californicus J. Cell. Comp. Physiol.* 65:393–403.

Tucker, V. A. (1965b) The relation between the torpor cycle and heat exchange in the California pocket mouse *Perognathus californicus. J. Cell. Comp. Physiol.* 65:405–414.

Tucker, V. A. (1966) Diurnal torpor and its relation to food consumption and weight changes in the California pocket mouse *Perognathus californicus. Ecology* 47:245–252.

Vinegar, A., Hutchison, V. H., and Dowling, H. G. (1970) Metabolism, energetics, and thermoregulation during brooding of snakes of the genus *Python* (Reptilia, Boidae). *Zoologica* 55:19–50.

Weast, R. C. (1969) *Handbook of Chemistry and Physics*, 50th ed. Cleveland: Chemical Rubber.

ADDITIONAL READING

Fischer, K. C., Dawe, A. R., Lyman, C. P., Schönbaum, E., and South, F. E., Jr. (eds.) (1967) *Mammalian Hibernation*, vol. III. Edinburgh: Oliver & Boyd. 535 pp.

Gates, D. M. (1962) *Energy Exchange in the Biosphere*. New York: Harper & Row. 151 pp.

Gates, D. M., and Schmerl, R. B. (eds.) (1975) *Perspectives of Biophysical Ecology* New York: Springer-Verlag. 609 pp.

Hardy, J. D., Gagge, A. P., and Stolwijk, J. A. J. (eds.) (1970) *Physiological and Behavioral Temperature Regulation*. Springfield, Ill.: Thomas. 944 pp.

Heinrich, B. (1974) Thermoregulation in endothermic insects. *Science 185*:747–756.

Heller, H. C., and Glotzbach, S. F. (1977) Thermoregulation during sleep and hibernation. *Int. Rev. Physiol. 15*:147–188.

Maloiy, G. M. O. (ed.) (1972) *Comparative Physiology of Desert Animals*. Symposia, Zoological Society of London, no. 31. London: Academic Press. 413 pp.

Monteith, J. L. (1973) *Principles of Environmental Physics*. London: Edward Arnold. 241 pp.

Newburgh, L. H. (ed.) (1968) *Physiology of Heat Regulation and the Science of Clothing* (reprinted.). New York: Hafner. 457 pp.

Precht, H., Christophersen, J., Hensel, H., and Larcher, W. (1973) *Temperature and Life*. New York: Springer-Verlag. 779 pp.

Schmidt-Nielsen, K. (1964) *Desert Animals: Physiological Problems of Heat and Water*. New York: Oxford University Press. 277 pp.

Whittow, G. C. (ed.) (1970, 1971, 1973) *Comparative Physiology of Thermoregulation*: vol. 1, *Invertebrates and Nonmammalian Vertebrates* (1970), 333 pp.; vol. 2, *Mammals* (1971), 410 pp.; vol. 3, *Special Aspects of Thermoregulation* (1973), 278 pp. New York: Academic Press.

PART FOUR
WATER

CHAPTER NINE

Water and osmotic regulation

In a very crude way, the living organism can be described as an aqueous solution contained within a membrane, the body surface. Both the volume of the organism and the concentration of solutes should be maintained within rather narrow limits, for optimal function of an animal requires that its body fluids have a well-defined, relatively constant composition. Substantial deviations from the normal composition are usually incompatible with life.

The problem for animals is to maintain the proper concentrations of their body fluids when these almost invariably differ from those of the environment. Concentration differences tend to dissipate, upsetting the desired steady state of the internal conditions. Animals can minimize the difficulties by decreasing the gradients or by lowering the permeabilities, and both strategies are employed. Nevertheless, there will always be some diffusive leak, and steady-state internal conditions cannot be maintained unless the organism generates a counterflow exactly equal to the diffusive leak. Such a counterflow requires input of energy.

The problems of keeping water and solute concentrations constant vary with the environment and are entirely different in sea water, in fresh water, and on land. It is therefore convenient to treat these environments separately, to analyze the main physiological problems in each, and to see how different animals have solved their problems.

In this chapter we shall first deal with aquatic animals and then move on to terrestrial animals.

THE AQUATIC ENVIRONMENT

Before discussing the physiological problems peculiar to an environment, it is helpful to be familiar with that environment's most important physical and chemical characteristics.

More than two-thirds (71%) of the earth's surface is covered with water. Most of this is ocean; the total fresh water in lakes and rivers makes up less than 1% of the area and 0.01% of the volume of sea water (Sverdrup et al. 1942; Hutchinson 1957, 1967). On land, life exists in a thin film on, just below, and just above the surface; in water, organisms not only live along the solid bottom but extend throughout the water masses to the greatest depths of the oceans, in excess of 10 000 m.

All water contains dissolved substances – salts, gases, minor amounts of organic compounds, various pollutants – and the temperature of the water is of the greatest physiological importance. In this chapter we are concerned primarily with dissolved salts; the role of dissolved gases was discussed in Chapter 1 and the effects of temperature in Chapters 7 and 8. A brief summary of the properties of solutions, osmotic pressure, and related matters is given in Appendix E.

Sea water contains about 3.5% salts (i.e., 1 liter sea water contains 35 g salts).* The major ions are sodium and chloride, with magnesium, sulfate, and calcium present in substantial amounts (Table 9.1). The total salt concentration varies somewhat with the geographical location. The Mediterranean, for example, has a salt content of nearly 4% because the high evaporation is not balanced by an equal inflow of fresh water from rivers. In other areas, particularly in coastal regions, the salt content is somewhat lower than in the open ocean, but the relative amounts of the dissolved ions remain very nearly constant in the proportions given in Table 9.1.

Fresh water, in contrast to sea water, has a

*It is not feasible to determine the exact amount of dissolved salts by evaporating the water and weighing the residue. The reason is that the drying conditions affect the amount of water included as water of crystallization as well as the weight of the bicarbonate and carbonate salts.

TABLE 9.1 Composition of sea water. In addition to the ions listed, sea water contains small amounts of virtually all elements found on earth. [Potts and Parry 1964]

Ion	Amount per 1 liter sea water		Amount per 1 kg water[a]	
	mmol	g	mmol	g
Sodium	470.2	10.813	475.4	10.933
Magnesium	53.57	1.303	54.17	1.317
Calcium	10.23	0.410	10.34	0.414
Potassium	9.96	0.389	10.07	0.394
Chloride	548.3	19.440	554.4	19.658
Sulfate	28.25	2.713	28.56	2.744
Bicarbonate	2.34	0.143	2.37	0.145

[a] In thermodynamics concentrations are calculated per kilogram water and are referred to as the molality of the solution. A brief synopsis of the physics of solutions is found in Appendix E.

highly variable content of solutes. Minute amounts of salts are present already in rainwater, but its composition is greatly modified as it runs over and through the surface of the earth. The salts in rainwater are derived from the sea; droplets of ocean spray evaporate and salt particles are carried by air currents, often far inland, and deposited with the rain. If the water runs over hard, insoluble rock such as granite, it dissolves little additional material and is called *soft*. If, on the other hand, it percolates over and through porous limestone, it can dissolve relatively large amounts of calcium salts and is called *hard*. The total salt content in fresh water may vary from less than 0.1 mmol per liter to more than 10 mmol per liter, and the relative amounts of different ions can vary over a tremendous range. This is of physiological significance, especially if magnesium and sulfate are the major ions. The composition of various types of hard and soft waters is given in Table 9.2.

Natural rain is slightly acid. This is because carbon dioxide from the atmospheric air dissolves in the rainwater, which then can be expected to attain a pH of about 5.6. However, much of the rain that falls in the northeastern United States has a

TABLE 9.2 Typical composition of soft water, hard water, and inland saline water, given in millimoles per kilogram water and listed in the same order as in Table 9.1. [Recalculated from Livingstone 1963]

Ion	Soft lake water[a]	River water[b]	Hard river water[c]	Saline water[d]	Dead Sea[e]
Sodium	0.17	0.39	6.13	640	840
Magnesium	0.15	0.21	0.66	6	2302
Calcium	0.22	0.52	5.01	32	583
Potassium	—	0.04	0.11	16	152
Chloride	0.03	0.23	13.44	630	6662
Sulfate	0.09	0.21	1.40	54	8.4
Bicarbonate	0.43	1.11	1.39	3	Trace

[a] Lake Nipissing, Ontario.　[b] Mean composition of North American rivers.　[c] Tuscarawas River, Ohio.　[d] Bad Water, Death Valley, California.　[e] Dead Sea, Israel. This water also contains 118 mmol per kg H_2O of bromide.

pH as low as 4. It has been suggested that this is because sulfur dioxide and nitrous oxides from the combustion of fossil fuels form strong acids that lower the pH of the precipitation (Galloway et al. 1976).

The high acidity in rain is not unique to the United States. Large areas of Scandinavia downwind of the dense industrial areas of Central Europe receive acid rains that cause noticeable problems for animal life and in some areas have severe effects on the fish population. These effects have been particularly well documented in Norway, where the yield of the salmon fisheries in the southern parts of the country has declined precipitously, although there has been no corresponding decrease in more distant areas. Concurrently, many lakes in southern Norway have lost their population of brown trout, and the number of barren lakes is rapidly increasing.

Why is Norway particularly prone to damage from acid rain, when areas closer to the industries do not report similar effects? The explanation is that large parts of southern Norway are underlain by granitic bedrock that is highly resistant to weathering. As a result, rivers and lakes have extremely soft waters, with virtually no mineral content to buffer the acids.

The degree of acidity that can be tolerated by a given fish is not related to pH alone; the amount of dissolved salts is extremely important. The effect of acid is probably an inhibition of the active uptake of sodium, for sodium uptake has in experiments been shown to be inhibited when the water acidity is high (Maetz et al. 1976). This is in accord with the observation that the fish disappear first from lakes with extremely small amounts of dissolved salts and that fish kills have been particularly severe during the snow melt in spring when the run-off of melting snow carries a sudden surge of acid in virtually unbuffered water (Leivestad and Muniz 1976).

Some inland water has a very high salt content. The Great Salt Lake in Utah is saturated with sodium chloride, which crystallizes out on the shore. The Dead Sea in Israel is likewise saturated, but the predominant ions are magnesium and chloride, with calcium sulfate crystallizing out. In the Great Salt Lake no fish can live, but some animals thrive: the brine shrimp (*Artemia*), for example. The saturated salt solution of the Dead Sea has a different composition. It harbors no higher animal or plant life; only microorganisms survive. Some springs have high and unusual salt contents, but such habitats, although interesting, are of minor importance compared with the larger bodies of fresh and sea water.

Brackish water occurs in coastal regions where sea water is mixed with fresh water. At the mouth of a large river fresh water dilutes the ocean water for a considerable distance, and if the tides are large the estuary at the mouth may extend far up the river. In this area the salinity varies rapidly with the tides, often from nearly fresh to nearly undiluted sea water. In a larger enclosed area, such as the Baltic Sea, the situation is different. The Baltic's salinity at the west coast of Sweden is some 3% and gradually declines to less than 0.5% in the northernmost part. In this large area there is a relatively stable geographical gradient in the salinity with an almost imperceptible change from brackish to fresh water.

It is difficult to say exactly where sea water becomes brackish and, at the other extreme, where a very dilute brackish water for practical purposes is fresh. As a commonly accepted definition we can say that brackish water refers to salinities between 3.0 and 0.05%. Brackish water is physiologically extremely important. It forms a barrier to the distribution of many marine animals on one hand as well as fresh-water animals on the other, and it also forms an interesting transition between marine and fresh-water habitats. In geographical extent, however, brackish water covers less than 1% of the earth's surface.

Definitions

Some aquatic animals can tolerate wide variations in the salt concentration of the water in which they live; they are called *euryhaline* animals (Greek *eurys* = wide, broad; *halos* = salt). Other animals have a limited tolerance to variations in the concentration of the medium; they are called *stenohaline* (Greek *stenos* = narrow, close). A marine animal that can penetrate into brackish water and survive is euryhaline. An extremely euryhaline animal may even be able to tolerate

shorter or longer periods in fresh water. The term euryhaline is used also for fresh-water animals that can withstand considerable increases in the salt content of the water. A stenohaline organism, whether marine or fresh-water, can tolerate only a narrow range of changes in the salt concentration of the water in which it lives.

There is no sharp separation between euryhaline and stenohaline animals, and there is no commonly accepted definition that places a given animal in one or the other group.

Most marine invertebrates have body fluids with the same osmotic pressure as the sea water; they are *iso-osmotic* or *isosmotic* with the medium in which they live (Greek *isos* = equal). When there is a change in the concentration of the medium, an animal may respond in one of two ways. One way is to change the osmotic concentration of its body fluids to conform with that of the medium, thus remaining isosmotic with the medium; such an animal is an *osmoconformer*. The other way is to maintain or regulate its osmotic concentration in spite of external concentration changes; such an animal is called an *osmoregulator*. For example, a marine crab that maintains a high concentration of salts in its body fluids after it has been moved to dilute brackish water is a typical osmoregulator.

The concentrations of the various individual solutes in the body fluids of an animal usually differ substantially from those in the medium, even if the animal is isosmotic with the medium. The differences are usually carefully regulated, a subject dealt with under the term *ionic regulation*. Some degree of ionic regulation seems to occur in all living organisms, both in osmoregulators and in osmoconformers.

Fresh-water animals have body fluids that are osmotically more concentrated than the medium; these animals are *hyperosmotic*. If an animal has a lower osmotic concentration than the medium, as does a marine teleost fish, it is said to be *hypo-osmotic* or *hyposmotic*.

The concentration of a dissolved substance is usually expressed in units of molarity: moles per liter solution. In a biological context, it is often convenient to use the unit millimole (mmol). For example, a solution of 0.5 mol per liter is equal to 500 mmol per liter.

The osmotic concentration of a solution can be expressed as the osmolarity (osmoles per liter). The osmolarity of a solution depends on the number of dissolved particles and can be stated without knowledge of which specific solutes are present. The osmolarity of a solution of a nonelectrolyte (e.g., sucrose or urea) equals the molar concentration. A solution of an electrolyte (e.g., sodium chloride, which in solution dissociates into Na^+ and Cl^-) has a higher osmotic concentration than is expressed by its molarity. The degree of dissociation depends on the concentration of sodium chloride and on interactions with other ions. However, the total osmotic concentration can readily be determined. In biological work this is most commonly done by measuring the freezing point depression or the vapor pressure of the solution. It is the osmotic concentration, rather than a detailed list of all solutes, that is important in most considerations of osmotic regulation in animals.

Ordinary sea water, which contains about 470 mmol sodium and about 550 mmol chloride per liter plus substantial amounts of divalent ions (magnesium and sulfate), has an osmotic concentration of about 1000 mOsm per liter.

The term *isotonic* is used in a different sense. We say that a living cell is isotonic with a given solution if the cell neither swells nor shrinks in the solution. For example, if mammalian red blood cells are suspended in a sodium chloride solution of 150 mmol per liter (about 0.9%), the cells retain their size, shape, and volume. If, on the other hand, they are suspended in an isosmotic solution of urea (0.3 mol per liter), they rapidly swell and burst. The urea solution, although isosmotic, is not isotonic. The solute, urea, rapidly penetrates the red cell membrane so that the urea concentrations inside and outside the cell are equal, but the electrolytes do not move out of the cell, which behaves as if it were suspended in distilled water. Because of the osmotic concentration differences, water flows into the cell, which swells and bursts. *Isosmotic* is defined in terms of physical chemistry; *isotonic* is a descriptive word based on the behavior of cells in a given solution.

AQUATIC INVERTEBRATES

Marine animals

In most marine invertebrates the osmotic concentration of their body fluids equals that of the surrounding sea water; the animals are *osmoconformers*. From the osmotic view point, this eliminates one major physiological difficulty: They do not have to cope with the problem of osmotic movement of water.

Although as a rule marine invertebrates are osmoconformers, this does not mean that their body fluids have the same *solute* composition as sea water. On the contrary, there are characteristic differences that the animals must maintain, and this requires extensive regulation of ionic concentrations.

Ionic concentration in body fluids

The concentrations of the most important ions in the blood of some invertebrates are given in Table 9.3. In some animals the concentrations are similar to those in sea water, but in others they differ substantially. For example, several invertebrates have magnesium present in the same con-

TABLE 9.3 Concentrations of common ions (in millimoles per kilogram water) in sea water and in some marine animals. [Potts and Parry 1964]

	Na	Mg	Ca	K	Cl	SO$_4$	Protein (g liter^{-1})
Sea water	478.3	54.5	10.5	10.1	558.4	28.8	—
Jellyfish (*Aurelia*)	474	53.0	10.0	10.7	580	15.8	0.7
Polychaete (*Aphrodite*)	476	54.6	10.5	10.5	557	26.5	0.2
Sea urchin (*Echinus*)	474	53.5	10.6	10.1	557	28.7	0.3
Mussel (*Mytilus*)	474	52.6	11.9	12.0	553	28.9	1.6
Squid (*Loligo*)	456	55.4	10.6	22.2	578	8.1	150
Isopod (*Ligia*)	566	20.2	34.9	13.3	629	4.0	—
Crab (*Maia*)	488	44.1	13.6	12.4	554	14.5	—
Shore crab (*Carcinus*)	531	19.5	13.3	12.3	557	16.5	60
Norwegian lobster (*Nephrops*)	541	9.3	11.9	7.8	552	19.8	33
Hagfish (*Myxine*)	537	18.0	5.9	9.1	542	6.3	67

centration as in sea water, but others have much lower concentrations of this ion. The same holds true for sulfate.

Such differences can be maintained only if the body surface, including the thin surface membrane of the gills, is relatively impermeable to the ion in question. Some amounts of these ions will enter anyway, for no animal is completely impermeable and all ingested food contains some solutes. Therefore, the animals must have a mechanism for eliminating some ions while maintaining others at a higher level than in the water. The regulated elimination of solutes is a major function of excretory organs, such as the kidney.

Several of the animals listed in Table 9.3 regulate their concentrations of sulfate at less than half that in sea water. Such great differences obviously indicate both exclusion and elimination of this ion – in other words, active ion regulation. If the concentration in the animal differs only slightly from that in sea water, it is less clear whether the difference is attributable to regulation. In particular, proteins have a considerable influence on the distribution of ions across a semipermeable membrane (known as the Donnan effect, see Appendix E). A difference in ion concentrations, therefore,

does not necessarily indicate active regulation of the ion in question.

The difficulty of evaluating accurately the role of proteins and the Donnan effect on the various ions was circumvented by the British investigator J. D. Robertson (1957) in a simple way. He put the sample in a semipermeable cellophane bag, which he then placed in sea water. Because salts and water can pass through the cellophane but proteins cannot, the ion concentrations inside the bag will, at equilibrium, differ somewhat from those in the sea water, the differences being caused by the Donnan effect of the proteins. This procedure is called *dialysis*. The concentrations inside the bag serve as a baseline, and the concentration of each ion as found in the living animal is expressed as a percentage of this baseline. An appreciable difference between the observed value and the concentration reached passively by dialysis must be the result of active regulation of this particular ion. Some results of such experiments are listed in Table 9.4.

Echinoderms show no significant regulation of any ion. The coelenterate *Aurelia*, a jellyfish, regulates only sulfate, which is kept considerably below the concentration in sea water. In this ani-

TABLE 9.4 Ionic regulation in some marine invertebrates. Concentrations in plasma or coelomic fluid expressed as percentage of concentration in body fluid dialyzed against sea water. [Robertson 1957]

Animal	Na	Mg	Ca	K	Cl	SO$_4$
Coelenterates						
Aurelia aurita	99	97	96	106	104	47
Echinoderms						
Marthasterias glacialis	100	98	101	111	101	100
Tunicates						
Salpa maxima	100	95	96	113	102	65
Annelids						
Arenicola marina	100	100	100	104	100	92
Sipunculids						
Phascolosoma vulgare	104	69	104	110	99	91
Arthropods						
Maia squinado	100	81	122	125	102	66
Dromia vulgaris	97	99	84	120	103	53
Carcinus maenas	110	34	108	118	104	61
Pachygrapsus marmoratus[a]	94	24	92	95	87	46
Nephrops norvegicus	113	17	124	77	99	69
Molluscs						
Pecten maximus	100	97	103	130	100	97
Neptunea antiqua	101	101	102	114	101	98
Sepia officinalis	93	98	91	205	105	22

[a] This grapsoid crab is the only animal listed that is hyposmotic to sea water (ionic concentration 86% that of sea water).

mal the low sulfate is directly related to a problem of buoyancy: Exclusion of the heavy sulfate ion reduces the density of the jellyfish, which thus is kept from sinking (see Chapter 11).

The arthropods listed in Table 9.4 demonstrate the interesting feature that the magnesium level in the plasma is low in those that have the ability to move quickly. The crab *Pachygrapsus* and the lobster-like *Nephrops* are active and fast-moving animals; the spider crab *Maia*, on the other hand, is slow moving and has a high magnesium concentration. Magnesium is an anesthetic that depresses neuromuscular transmission, and one might hastily conclude that a high magnesium concentration is related to a low level of activity of these crustaceans. However, the further observation that the cuttlefish *Sepia*, which moves quickly and swims well, has a magnesium concentration as high as the clam *Pecten*, makes us suspicious of any cause-and-effect correlation between activity and magnesium concentration.

Intracellular concentrations and volume regulation

So far we have discussed the relationships between the surrounding water and the solute concentrations in blood and other body fluids. The concentrations of ions *inside* the cells are usually very different from those outside; for example, sodium and chloride concentrations are usually low inside the cells and high outside, and potassium is usually high inside and low outside. However, the cells are isosmotic with the surrounding tissue fluid and blood, although the concentrations of individual solutes differ.

To illustrate the problems of volume regulation,

we can consider the same cell membrane as permeable to water but impermeable to solutes. A change in the extracellular concentration will cause a corresponding change in cell volume. If the extracellular concentration is lowered, the cell takes up water and swells; if the extracellular concentration is increased, water is withdrawn and the cell shrinks. In reality, in most animals the cell volume is regulated so that the cell after an initial disturbance returns to its original volume, which then remains relatively constant.

It is now well established that intracellular concentration of free amino acids is an important factor in the control of cell volume in salinity-stressed environments. As the salinity in the water rises or falls, the amino acids in the cells increase or decrease so that the cells remain isotonic with the surroundings.

It has been suggested that the decrease in amino acid concentration could be achieved by the synthesis of protein and the concentration restored by degradation of protein. Another suggestion is that when an animal is stressed by low salinity the intracellular amino acids cross the initially distended cell membranes, accompanied by osmotically obligated water, and the cell volume thus is restored. According to this view, which is supported by studies of molluscan heart muscle, the volume regulation in the cells depends on the permeability characteristics of the cell membrane, which in turn are influenced by the extracellular osmotic concentration (Pierce and Greenberg 1973).

The changes in vertebrate cells during adjustment to changing salinities resemble those of the invertebrates. As the salinity of the medium increases or decreases, corresponding changes occur in the intracellular amino acid concentrations that are in accord with the maintenance of constant cell volume (Fugelli and Zachariassen 1976). In principle, therefore, cell volume regulation seems to be accomplished by similar mechanisms in vertebrates and invertebrates.

Animals in fresh and brackish water

If we transfer a variety of marine animals to somewhat dilute sea water, say 80% of the usual strength, most of them are likely to survive. When, after some time, we examine their body fluids, we find that they have adjusted to the dilution and have established new and lower concentrations of ions in their body fluids. The osmoconformers (e.g., starfish and oysters) have the same osmotic concentration as the dilute medium, although the concentrations of individual ions still differ from those in the diluted sea water. The osmoregulators, on the other hand, resist the dilution more or less successfully and remain hyperosmotic.

Thus, marine animals that penetrate into brackish water are of two types: passive osmoconformers or active osmoregulators. An osmoconformer, such as the oyster, can tolerate considerable dilution; it can also to some extent resist the effects of periodic dilution of the water in an estuary by keeping its shells closed. In the long run, however, the active regulators can better resist the fluctuations in the environment.

Let us see how the active regulators manage in dilute sea water. The relationship between the concentrations in the body fluids and the water is shown in Figure 9.1 for a number of osmoregulators. Even the good regulators meet certain limitations. An example is the European shore crab, *Carcinus*, which cannot survive in brackish water more dilute than about one-third of normal sea water. The limit of dilution tolerated by the shore crab, and by many other brackish-water animals, varies with the geographical locale. A shore crab from the North Sea is far less tolerant to dilute water than a shore crab from the Baltic Sea, where

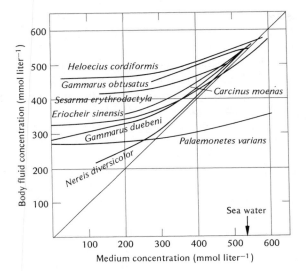

FIGURE 9.1 Relation between the concentrations of body fluids and of medium in various brackish-water animals. Full-strength sea water is indicated by an arrow. Diagonal line indicates equal concentrations in body fluid and medium. [Beadle 1943]

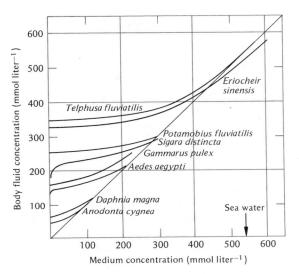

FIGURE 9.2 Relation between the concentrations of body fluids and of medium in various fresh-water animals. Full-strength sea water is indicated by an arrow. Diagonal line indicates equal concentrations in body fluid and medium. [Beadle 1943]

the salt concentration normally is much lower than in the open sea.

Another crab, the Chinese mitten crab *Eriocheir*, can tolerate far greater dilutions and, in fact, it is able to penetrate into fresh water. To reproduce, the mitten crab must return to the sea, and it therefore cannot establish itself permanently and carry out its complete life cycle in fresh water.

In principle, fresh-water animals behave osmotically in a way similar to the successful osmoregulators in brackish water, but there are great differences in the concentrations at which they maintain their body fluids (Figure 9.2). A crustacean such as the crayfish *Potamobius* maintains an osmotic concentration of about 500 mOsm per liter, but the fresh-water clam *Anodonta* maintains less than one-tenth of this, only about 50 mOsm per liter. *Anodonta* still is hyperosmotic, and no fresh-water animal is known that permits its concentration to decrease to the level of the fresh water in which it lives.

Most of the major animal phyla have representatives both in the sea and in fresh water, although the number of species in the sea is much greater. No echinoderms are found in fresh water, and among molluscs the entire class Cephalopoda (octopus and squid) is absent in fresh water.

Mechanism of osmoregulation

When an animal is hyperosmotic to the surrounding medium, it encounters two physiological problems: (1) water tends to flow into the animal because of the higher inside concentration, and (2) solutes tend to be lost because the inside concentration is higher and because the water that enters must be excreted and carries some solutes with it. The magnitude of these problems could be reduced by making all surfaces highly impermeable, but no completely impermeable animal is known. At the least, the respiratory surfaces must be thin and large enough to permit adequate diffusion of gases, and the respiratory surfaces are usually the site of the greatest loss of solutes and gain of water.

How can the animal compensate for the solute

loss? It might be possible to obtain the necessary ions in food, and this might suffice for a highly impermeable animal. However, many fresh-water animals are not particularly impermeable, and yet, even when starved they maintain their normal high concentrations. This is achieved by a direct uptake of ions from the medium.

The simplest way to demonstrate such uptake is to keep an animal, for example a crayfish, in running or frequently changed distilled water. This will gradually deplete the salts in the crayfish so that its concentration is reduced from, say, 500 mOsm per liter to 450, which it tolerates without difficulty. If the crayfish is now returned to ordinary fresh water, its blood concentration increases again, although the fresh water is 100 times more dilute than the blood (e.g., 5 mOsm per liter).

Because, in this experiment, the crayfish has removed ions from a dilute solution and moved them to a much higher concentration in the blood, the ions have been moved against a concentration gradient and the transport is an *active transport*. *

The organs responsible for active ion uptake are not always known. It has been assumed that the general body surface of some animals is the organ of uptake, but this is a conclusion based mostly on lack of any obvious recognizable organs that could be tested for this role. In other animals, notably crustaceans and insects, it is probable that the general body covering does not participate in the active transport. There is conclusive evidence that the gills of crustaceans are organs of active ion transport and that certain appendages of aquatic

---

* *Active transport* is defined as transport against an electrochemical gradient and as such is an energy-requiring process. It is often referred to as an "uphill" transport, as opposed to passive or "downhill" diffusion along the concentration gradient.

insect larvae, notably the "anal gills," are organs for ion uptake. There is no conflict between osmoregulation and respiration; the same organ can serve both functions, as do the gills of fresh-water crabs and crayfish. But the "anal gills" of insect larvae probably have no respiratory function and are exclusively organs of osmoregulation.

Active transport requires *energy*, and it would be interesting to know the increase in energy requirement as an animal moves to more dilute water. If the shore crab *Carcinus*, which is quite tolerant of brackish water, is moved from sea water to gradually more diluted solutions, its rate of oxygen consumption increases appreciably. When placed in sea water of one-quarter full strength, the oxygen consumption of *Carcinus* is increased by about 50%, and it could easily be concluded that this increase is attributable to the work of active ion transport necessary for the crab to maintain itself in the dilute water. In contrast, the mitten crab *Eriocheir* has the same metabolic rate in sea water, in brackish water, and in fresh water. The mitten crab therefore does not seem to spend measurable extra energy to maintain itself — not even in fresh water. This difference could be explained if the mitten crab were more impermeable than the shore crab, but there is no striking difference in the permeabilities of the two animals. Determining the change in oxygen consumption therefore does not give reliable information about the energy cost of osmoregulation (Potts and Parry 1964).

The energy required for osmoregulation can, however, be calculated from thermodynamic considerations. If we know how much solute is lost in a given time, we know how much must be taken up from the medium for the animal to remain in steady state. The osmotic work (W_{osm}) required depends on the solute concentration in the medium (C_{med}) and in the blood (C_{bl}), and the work

TABLE 9.5 Estimated metabolic cost of osmotic regulation for three animals, each weighing 60 g, in fresh water with a solute concentration of 6 mmol per liter. Observations on blood and urine concentrations (columns a and b) and urine volume (column c) are used to calculate the minimum theoretical cost of osmotic regulation (column d). In the last column the calculated osmotic work is compared with the observed metabolic rates of the animals. [Potts 1954]

Animal	(a) Blood concentration (mmol liter^{-1})	(b) Urine concentration (mmol liter^{-1})	(c) Urine volume (ml h^{-1})	(d) Osmotic work (cal h^{-1})	(e) Metabolic rate (cal h^{-1})	(f) Osmotic work (% of metabolic rate)
Mitten crab	320	320	0.1	0.076	14	0.5
Crayfish	420	124	0.1	0.037	10	0.3
Clam	42	24	0.5	0.015	1.2	1.3

to move 1 mol is given by the following equation, in which R is the universal gas constant and T is the absolute temperature:

$$W_{osm} = R\,T\,\ln \frac{C_{bl}}{C_{med}}$$

In our model we will assume that the surface of the animal is permeable to water but impermeable to solutes. In a hypotonic medium, water enters by osmosis, and the same volume of water must be removed as urine. The permeability of the animal determines the water influx, and the urine volume is therefore a measure of the permeability. Some solutes are lost with the urine, and the total amount of solutes lost is the product of the volume of the urine and its concentration, $V_u \times C_u$.

We now can see that the osmotic work depends on the permeability of the animal and the solute loss, according to:

$$W_{osm} = R\,T\,V_u\,C_u\,\ln \frac{C_{bl}}{C_{med}}$$

Some animals produce urine isotonic with the blood. Many others reabsorb some solutes from the urine and produce a hypotonic urine. In the latter case, less solute is lost, but work is required for solute reabsorption from the urine to make it hypotonic. This complicates the calculation of the minimum osmotic work required, but the result is

that the reabsorption of solutes from the urine constitutes a considerable saving.

Estimates of the minimum osmotic work required for osmoregulation in fresh water for three animal species are compiled in Table 9.5. The information about blood concentration, urine concentration, and urine volume is used to calculate the osmotic work.

The first animal, the mitten crab *Eriocheir sinensis* is a marine animal that readily penetrates into fresh water. Its urine concentration equals the blood concentration, which is common for marine invertebrates. The osmotic work required to recover lost solutes from the dilute medium is quite small: 0.5% of the metabolic rate of the animal.

The crayfish *Potamobius fluviatilis* is a fresh-water animal. It produces a dilute urine at about one-third of the blood concentration. The urine volume is the same as in the mitten crab, and because the blood concentration of the crayfish is higher, its permeability must be somewhat lower. The greatest saving, however, is derived from the recovery of solutes from the urine, so that the total osmotic work is less than one-half that for the mitten crab. It is only 0.3% of the metabolic rate of the crayfish.

The third animal, the fresh-water clam *Anodonta cygnea*, has a very low blood concentration, only one-tenth that of the crayfish. It also produces a dilute urine, at about one-half the blood concentration. However, the urine volume is high, which indicates a high permeability due to the large surface area of the soft tissues exposed to the medium. The high permeability makes it virtually mandatory for the clam to have a very low internal concentration; if it were to have a concentration similar to that of the crayfish, the osmotic inflow of water and thus the urine volume would be 10 times as high. This would give a corresponding increase in the amount of solutes to be recovered from the medium. The increase in the required work, however, would be much greater, because the uptake would be against the 10-fold higher blood concentration. For an animal with high permeability and low metabolic rate it is therefore essential that the blood concentration be as low as possible.

Although the animals described in Table 9.5 differ in their approach to osmoregulation, for all of them the minimum energy required for osmoregulation is a small fraction of the total metabolic rate. In reality the cost is higher, for the process does not have an efficiency of 100%. Nevertheless, we must conclude that the large increase in metabolic rate observed in some animals when they are moved to dilute media does not directly measure the increased energy requirements for osmoregulation.

Animals in saline habitats: hyporegulation

The shrimp *Palaemonetes* and a close relative, *Leander*, differ in one important way from all the other animals described in Figure 9.1. In normal, full-strength sea water they are hypotonic (i.e., their body fluids are osmotically more dilute than the medium), and this must require active osmoregulation. Hyporegulation is very uncommon for a marine invertebrate, and it is generally assumed that these two shrimps belong to a group

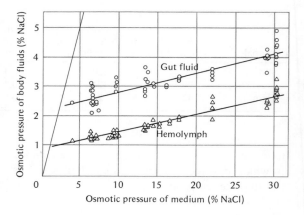

FIGURE 9.3 Osmotic pressure of body fluids of brine shrimp in highly concentrated NaCl solutions. The steep line running through the origin indicates equal concentrations in body fluid and medium. [Croghan 1958a]

that originally was at home in fresh water, and that they have secondarily invaded the sea while maintaining a lower concentration than is common in sea water.

Some saline waters are far more concentrated than sea water, and in such extreme environments hyporegulation is of greater importance. The best known example is probably the brine shrimp *Artemia*, which is found in tremendous numbers in salt lakes and in coastal evaporation ponds where salt is obtained for commercial use by evaporation of sea water.

Although the brine shrimp cannot survive in fresh water, it can adapt to media that vary from about one-tenth sea water to crystallizing brine, which contains about 300 g salt per liter. In dilute sea water, *Artemia* is hypertonic to its medium and behaves like a brackish-water organism. At higher concentrations, *Artemia* is an excellent hyporegulator; in concentrated brine it still maintains its body fluid at an osmotic pressure not much more than one-tenth that of the medium (Figure 9.3).

The brine shrimp maintains its low osmotic concentration, not by being exquisitely impermeable to water and ions, but by active regulation. It continuously swallows the medium, and the osmotic pressure of the gut fluid is appreciably greater than that of the hemolymph (blood). However, although the osmotic pressure of the gut fluid remains greater than the hemolymph's, the concentrations of sodium and chloride in the gut fluid are considerably below those in the hemolymph (Croghan 1958b). Sodium and chloride must therefore be removed from the gut by active uptake, and to eliminate these ions from the body, so that their blood concentration is kept low, excretion must take place elsewhere. It is probable that the epithelium of the gills has the major role in this process.

The larva of *Artemia* seems to use a different site for the active secretion of sodium, the "neck organ," a specialized structure located on the dorsal surface of the animal (Conte et al. 1972).

Several species of mosquito larvae can thrive equally well in fresh water and in saline water that is several times as concentrated as their hemolymph and can even tolerate salinites three times as high as sea water. Even more remarkable, the larvae of the mosquito *Aedes campestris* can thrive in alkaline salt lakes in which the dominant salt is sodium bicarbonate and the pH is greater than 10.

In low-salinity water the larvae of *Aedes* are hyperosmotic to the medium, but at higher concentrations they are hyposmotic. These larvae can adapt to concentrations that extend over a 500-fold range without changing the concentrations of the major ions in the hemolymph more than 2-fold.

The larvae of *Aedes* respond to rising external salinities by increasing their drinking rate several-fold. This seems to be the only way of obtaining water to compensate for the loss to the concentrated medium, but drinking also means ingestion of a heavy load of dissolved ions. The excess salt load is excreted with the aid of the Malpighian tubules and the rectum (to be discussed in Chapter 10). The so-called anal papillae, which in dilute media are the major sites of active ion uptake, seem to be insignificant for the elimination of excess salt.

The total ion balance in an *Aedes* larva in alkaline water is shown in Figure 9.4. In one day an 8-mg larva drinks 2.4 μl water, which is more than one-third of its total body water content (6.5 μl). The amount of sodium ingested in a day (1.2 μmol) is tremendous – more than the total sodium content of the animal (0.96 μmol). All this ingested sodium, together with a small amount that enters through the body surface (0.1 μmol), is excreted. The amount of water needed for this excretion (1.8 μl) is less than the amount ingested (2.4 μl), leaving enough to cover what is lost by diffusion to the concentrated medium through the body surface (0.2 μl) and from the anal papillae (0.4 μl). Thus, the hypotonic larva is able to osmoregulate and remain in balance in the highly concentrated solution.

Hyporegulation is, on the whole, the exception rather than the rule among invertebrates. As we now turn to the vertebrates we shall find that osmotic hyporegulation is much more widespread, although still not universally present, for there are other ways as well to solve osmotic problems.

AQUATIC VERTEBRATES

This section will deal primarily with fish and amphibians. Nobody would deny that whales and sea turtles are aquatic vertebrates, but we will treat them in a different context. They are descended from terrestrial ancestors and are air breathers, and it is more convenient to discuss them as terrestrial animals living in an environment where no fresh water is available.

The major strategies used by aquatic vertebrates are evident from an examination of Table 9.6. The

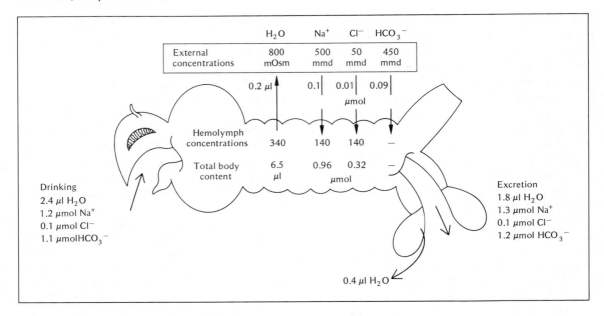

table lists examples of both marine and fresh-water vertebrates. The marine representatives fall into two distinct groups: those whose osmotic concentrations are the same as or slightly above sea water (hagfish, elasmobranchs, *Latimeria*, and crab-eating frog) and those whose osmotic concentrations are about one-third that of sea water (lamprey, teleosts). The former group has no major problem of water balance, for if the inside and outside concentrations are equal, there is no osmotic water flow. In contrast, those that remain distinctly hyposmotic live in constant danger of losing water to the osmotically more concentrated medium. The osmotic problems and the means to solve them thus differ drastically among marine vertebrates. Fresh-water vertebrates, on the other hand, uniformly have concentrations of about one-quarter to one-third that of sea water; thus

they are hyperosmotic to the medium and in principle similar to fresh-water invertebrates.

Cyclostomes

The cyclostomes are eel-shaped fish and considered the most primitive of all living vertebrates. They lack a bony skeleton, paired fins, and jaws (they are grouped in the class Agnatha, jawless vertebrates).

There are two groups of cyclostomes: lampreys and hagfish. Lampreys live both in the sea and in fresh water. Hagfish are strictly marine and stenohaline. Interestingly, lampreys and hagfish have employed two different solutions to the problem of life in the sea. Hagfish are the only true vertebrates whose body fluids have salt concentrations similar to that of sea water; in fact, the normal sodium concentration in hagfish blood slightly exceeds

TABLE 9.6 Concentrations of major solutes (in millimoles per liter) in sea water and in the blood plasma of some aquatic vertebrates.

| | Habitat | Solute | | | Osmotic concentration (mOsm liter^{-1}) |
		Na	K	Urea[a]	
Sea water		~450	10	0	~1000
Cyclostomes					
Hagfish (*Myxine*)[b]	Marine	549	11		1152
Lamprey (*Petromyzon*)[c]	Marine				317
Lamprey (*Lampetra*)[b]	Fresh water	120	3	<1	270
Elasmobranchs					
Ray (*Raja*)[b]	Marine	289	4	444	1050
Dogfish (*Squalus*)[b]	Marine	287	5	354	1000
Fresh-water ray (*Potamotrygon*)[d]	Fresh water	150	6	<1	308
Coelacanth (Latimeria)[b,e]	Marine	181		355	1181
Teleosts					
Goldfish (*Carassius*)[b]	Fresh water	115	4		259
Toadfish (*Opsanus*)[b]	Marine	160	5		392
Eel (*Anguilla*)[b]	Fresh water	155	3		323
	Marine	177	3		371
Salmon (*Salmo*)[b]	Fresh water	181	2		340
	Marine	212	3		400
Amphibians					
Frog (*Rana*)[f]	Fresh water	92	3	~1	200
Crab-eating frog (*R. cancrivora*)[g]	Marine	252	14	350	830[h]

[a] When no value is listed for urea, the concentration is of the order of 1 mmol per liter and osmotically insignificant. Values for ray, dogfish, and coelacanth include trimethylamine oxide.
[b] Bentley (1971).   [c] Robertson (1954).   [d] Thorson et al. (1967).   [e] Lutz and Robertson (1971).
[f] Mayer (1969).   [g] Gordon et al. (1961).
[h] Values for frogs kept in a medium of about 800 mOsm per liter, or four-fifths of normal sea water.

that in the medium. Nevertheless, hagfish have pronounced ionic regulation, but in being isosmotic and having high salt concentrations, they behave osmotically like invertebrates.

With the exception of hagfish, all marine vertebrates maintain salt concentrations in their body fluids at a fraction of the level in the medium. This situation has been cited in favor of the argument that the vertebrates originally evolved in fresh water and only later invaded the sea. Cyclostomes show many similarities to the ancestral forms of modern vertebrates, and knowledge of their anatomy has been of great importance to the interpretation of the fossil records of vertebrates and to the understanding of their early evolution.

The fact that hagfish deviate from the general vertebrate pattern of low salt concentrations means that the evolutionary theory of a fresh-water origin of all vertebrates is not supported by physiological evidence: A low salt concentration is not a universal vertebrate trait. However, present-day physiological characteristics cannot be used in evolutionary arguments, for, on the whole, physiological adaptation takes place more easily than morphological change. Anatomical structure and fossil records therefore remain more important to evolutionary hypotheses than physiological evidence.

The other group of cyclostomes, the lampreys, live both in fresh water and in the sea, but even the sea lamprey (*Petromyzon marinus*) is *anadromic** and ascends rivers to breed in fresh water.

The lampreys, whether fresh-water or marine, have osmotic concentrations about one-quarter to one-third the concentration of sea water. Their main osmotic problem is similar to that of teleost fish, whether marine or fresh-water species. These problems will be discussed in detail later in this chapter.

Marine elasmobranchs

The elasmobranchs – sharks and rays – are almost without exception marine. They have solved the osmotic problem of life in the sea in a very interesting way. Like most vertebrates they maintain salt concentrations in their body fluids at roughly one-third the level in sea water, but they still maintain osmotic equilibrium. This is achieved by adding to the body fluids large amounts of organic compounds, primarily urea, so that the total osmotic concentration of their blood equals or slightly exceeds that of sea water (Table 9.6).

In addition to urea, an osmotically important organic compound in elasmobranch blood is trimethylamine oxide (TMAO).

* A fish that ascends from the sea to spawn in fresh water is called *anadromic* (Greek *ana* = up; *dramein* = to run). Shad and salmon are well-known examples. *Catadromic* (Greek *kata* = down) refers to living in fresh water and descending to the sea. The common eel is catadromic; it grows to adult size in fresh water and descends to the sea to breed.

Urea is the end product of protein metabolism in mammals and some other vertebrates; it is excreted by the mammalian kidney, but the shark kidney actively reabsorbs this compound so that it is retained in the blood. TMAO is found in many marine organisms, but its origin and metabolism are still poorly understood. Whether TMAO is obtained by the sharks through the food chain or is produced in the body remains uncertain.

The blood urea concentration in marine elasmobranchs is more than 100 times as high as in mammals, and such concentrations could not be tolerated by other vertebrates. In the elasmobranchs urea is a normal component of all body fluids, and the tissues cannot function normally in the absence of a high urea concentration. The isolated heart of a shark can continue to contract normally for hours when perfused with a saline solution of ionic composition similar to the blood, provided urea is also present in a high concentration. If the urea is removed, however, the heart rapidly deteriorates and stops beating.

Although the elasmobranchs have solved the osmotic problem of life in the sea by being isosmotic, they still are capable of extensive ionic regulation. The sodium concentration, for example, is maintained at about half that in sea water. This means that sodium tends to diffuse from the medium into the shark, primarily through the thin gill epithelium; furthermore, some sodium is ingested with the food. Because the sodium concentration tends to increase but must be kept down, excess sodium must be eliminated.

Part of the sodium excretion is handled by the kidney, but a special gland, the *rectal gland*, is probably more important. This small gland opens via a duct into the posterior part of the intestine, the rectum. The gland secretes a fluid with high sodium and chloride concentrations – in fact, somewhat higher than in sea water. For example, in sharks maintained in sea water with a sodium concentration of 440 mmol per liter, the secretion from the rectal gland contained from 500 to 560 mmol Na per liter (Burger and Hess 1960).

The function of the rectal gland, however, does not fully explain the salt elimination in elasmobranchs. If the rectal gland is surgically removed in the spiny dogfish (*Squalus acanthias*), this shark is still able to maintain its plasma ion concentration at the usual level of about half that in sea water. As the gills are slightly permeable to salts, the blood concentrations should gradually increase unless some other means of excretion is available. It is likely that the kidney plays a major role in this excretion, but whether the gills are also sites for active transport of ions from elasmobranch blood is not known.

The fact that elasmobranchs are nearly in osmotic equilibrium with sea water eliminates the problem of a severe osmotic water loss (which for marine teleost fish is very important). Elasmobranchs do not need to drink, and thus they avoid the high sodium intake associated with drinking sea water.

It is an interesting fact, however, that elasmobranch blood is usually slightly more concentrated than sea water. This higher concentration inside causes a slight osmotic inflow of water via the gills. In this way, the elasmobranch slowly gains water osmotically, and this water is used for the formation of urine and for the secretion from the rectal gland. Because the excess osmotic concentration is attributable to urea, the retention of

urea can be considered a rather elegant solution to the otherwise difficult osmotic problem of maintaining a low salt concentration while living in the sea.

Fresh-water elasmobranchs

The overwhelming number of elasmobranchs belong in the sea, but a few enter rivers and lakes, and some may belong permanently in fresh water. Even those elasmobranchs that are thought of as typically marine include some species with a remarkable tolerance to low salinity in the external medium. In several parts of the world both sharks and rays enter rivers and apparently thrive in fresh water. A well-known case is the existence in Lake Nicaragua of the shark *Carcharhinus leucas*. It was previously assumed that this shark was landlocked, but recent evidence indicates that the shark in Lake Nicaragua is morphologically indistinguishable from the marine form and that it is able to move in free communication with the sea (Thorson et al. 1966).

Four elasmobranch species found in the Perak River in Malaysia probably do not live permanently in fresh water, but enter regularly from the sea. Their blood concentrations are lower than those of strictly marine forms; in particular, the urea is reduced to less than a third the value for marine sharks, although it remains far above the normal level for other vertebrates.

The low level of solutes in the blood reduces the problems of osmotic regulation, for the osmotic inflow of water is diminished and lower salt concentrations are easier to maintain. The reduced osmotic inflow of water gives less water to be eliminated by the kidney. Because the urine inevitably contains some solutes, a low urine flow in turn reduces the urinary salt losses. It is, of course difficult to say whether the lowered blood concentration is a primary adjustment or merely a passive

TABLE 9.7 Concentrations of solutes in the blood serum of the Amazon sting ray are similar to those of a teleost fish. Although the ray is an elasmobranch, urea is virtually absent from its body fluids. [Thorson et al. 1967]

Solute	Amount per liter	
Sodium	150	mmol
Potassium	5.9	mmol
Calcium	3.6	mmol
Magnesium	1.8	mmol
Chloride	149	mmol
Urea	0.5	mmol
TMAO	0	mmol
Osmolality	308	mOsm
Protein	18	g

result of the increased inflow of water and concomitant urinary losses (Smith 1931).

One elasmobranch, the Amazon sting ray *Potamotrygon*, is permanently established in fresh water. It is common in the Amazon and Orinoco drainage systems up to more than 4000 km from the sea. It does not survive in sea water, even when the transfer takes place through a gradual concentration increase (Pang et al. 1972). The average composition of its blood (Table 9.7) shows complete adaptation to fresh water, with a low blood urea concentration similar to that of fresh water teleosts.

The most striking feature is the low urea concentration – even lower than in most mammals. Clearly, the retention of urea is not a universal requirement for elasmobranchs – an interesting physiological observation, showing again that physiological function is far more changeable than most anatomical structures and that evolutionary arguments cannot be reliably based on physiological similarities or differences.

The coelacanth

Until 1938 it was believed that the group of fishes known as the Crossopterygii had been extinct for more than 75 million years, for they disappeared completely from the later fossil record.

Their evolutionary position is far away from modern fish, close to the lungfish, and in the ancestry of amphibians. In 1938 a living specimen of the coelacanth *Latimeria* was caught off the coast of South-East Africa, causing a worldwide scientific sensation. It was a large specimen, more than 1.5 m long, and weighed over 50 kg, but it was poorly preserved and no detailed information on its anatomy was obtained.

After intensive search, several additional specimens of *Latimeria* have been caught near Madagascar, and although no living specimen has been kept long enough to allow physiological experimentation, it is known that the coelacanth has solved its problems of osmoregulation in the same way as elasmobranchs. The data entered in Table 9.6 were obtained on a frozen specimen of *Latimeria*; the high urea concentration places it physiologically with the elasmobranchs.

Additional analyses have confirmed the high urea content and shown that the TMAO level is high in blood (>100 mmol per liter) and muscle (>200 mmol per liter) (Lutz and Robertson 1971). The figures given for sodium concentration in the plasma should perhaps be adjusted upward, for freezing and thawing cause exchange of sodium and potassium between blood plasma and the red cells, thus lowering the value for sodium in the plasma and increasing the potassium value (which indeed was found to be abnormally high, 51 mmol per liter) (Pickford and Grant 1967).

Teleost fish

Teleost fish maintain their osmotic concentrations at about one-quarter to one-third the level in sea water (Table 9.6). On the whole, marine and fresh-water fish are within the same range, although marine fish tend to have somewhat higher blood concentrations. Some fish can tolerate a

FIGURE 9.5 A marine teleost is osmotically more dilute than the water in which it lives. Because of the higher osmotic concentration in the medium, the fish constantly loses water (top diagram), primarily across the thin gill membranes. Additional water is lost in the urine. To compensate for the water loss, the marine teleost drinks substantial amounts of sea water. Of the ingested salts, sodium and chloride are absorbed in the intestine and eliminated via the gills by active transport (double arrow, bottom diagram); magnesium and sulfate are excreted by the kidney.

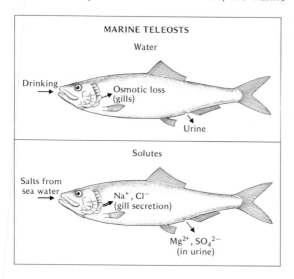

Although drinking restores the water content, large amounts of salts are also ingested and are absorbed from the intestinal tract together with the water. The salt concentration in the body increases, and the problem becomes one of eliminating the excess salt. To achieve a net gain in water from the ingestion of sea water, the salts must be excreted in a higher concentration than in the water taken in. The teleost kidney cannot serve this purpose, for it cannot produce a urine that is more concentrated than the blood.

Some other organ must therefore eliminate the excess salt. This is done by the gills, which thus have the dual function of participating in both osmotic regulation and gas exchange. The secretion of salt across the gill epithelium must be an active transport, for it takes place from a lower concentration in the blood to a higher concentration in the surrounding medium.

The main aspects of osmotic regulation in marine teleosts are summarized in Figure 9.5. The top diagram shows the movements of water: Water is lost osmotically across the gill membrane and in the urine. To compensate for the losses, the fish drinks sea water, absorbing both water and salts from the intestine. The lower diagram shows the movement of salts, which are taken in by mouth as the fish drinks from the surrounding sea water. The double arrow at the gills indicates the elimination, by active transport, of sodium and chloride. The excretion of sodium and chloride in the urine is of minor importance because teleost urine is usually more dilute than the body fluids. However, the kidney plays a major role in the excretion of divalent ions, magnesium and sulfate, which make up roughly one-tenth of the salts in sea water. These ions are not eliminated by the gills, which seem to transport only sodium and chloride.

Although marine fish drink sea water, measurements of the drinking rates have shown that only a wide range of salinities and move between sea, brackish, and fresh water.

Such moves are often associated with the life cycle; the salmon, for example, reproduces in fresh water, migrates to the sea, and after reaching maturity returns to fresh water to spawn. The life cycle of the common eel is the reverse: The eel breeds in the sea, the larvae drift with the currents and reach coastal areas where they ascend into fresh water, and when maturity approaches the eel returns to the sea to reproduce. The change from one environment to the other requires profound changes in the osmoregulatory processes.

Marine teleosts

Marine fish are hyposmotic and in constant danger of losing body water to the more concentrated sea water, for their body surfaces, in particular the large gill surfaces, are somewhat permeable to water. They must somehow compensate for the inevitable osmotic loss of water, and this they do by drinking sea water.

small part of the total sodium intake comes from drinking; the larger part of the sodium influx takes place elsewhere, presumably because the gills are somewhat permeable. Whether the general body surface or the gills are responsible, it is certain that fish adapted to sea water are relatively permeable to ions, and those adapted to fresh water are relatively impermeable (Motais and Maetz 1965).

The killifish (*Fundulus heteroclitus*), which readily adapts to both fresh and sea water, has been used to study the changes in permeability to sodium and chloride that take place during adaptation to various concentrations. The permeability decreases within a few minutes of transfer to fresh water, but the increase in permeability on return to sea water takes many hours (Potts and Evans 1967).

The advantage of a low permeability to ions in fresh water is obvious, but it is difficult to see the advantage of a higher permeability in sea water. Marine fish must expend work to maintain their osmotic steady state in sea water, and it seems that a low permeability would reduce this work. It takes the fish several hours to return to the higher permeability in sea water, and we can only wonder why they do not permanently retain the low permeability that seems to be within their physiological capacity.

It is unlikely that the general gill epithelium participates in the ion transport, which is probably carried out by some large cells known as *chloride cells*. Until recently it was uncertain whether the chloride ion is actively transported and sodium follows passively, or the sodium ion is actively transported and chloride follows passively. The cells were named chloride cells without definite knowledge of their function (Keys and Willmer 1932). However, it now appears that the name was quite appropriate, for eels kept in sea water transport the chloride ion actively (Maetz and Campanini 1966). The potential difference across the gill surface indicates active chloride transport, but sodium is not always in passive equilibrium and may be actively transported as well (House 1963).

FIGURE 9.6 A fresh-water teleost is osmotically more concentrated than the medium and therefore suffers a steady osmotic influx of water, mainly through the gills (top diagram). The excess water is eliminated as urine. Loss of solutes through the gills and in the urine is compensated for primarily through active uptake in the gills (double arrow, bottom diagram).

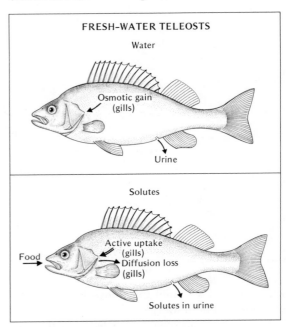

FRESH-WATER TELEOSTS

Fresh-water teleosts

The osmotic conditions for fish in fresh water resemble those for fresh-water invertebrates. The osmotic concentration in the blood, roughly in the range of 300 mOsm per liter, is much higher than in the surrounding fresh water.

The main events in the osmoregulation of fresh-water teleosts are outlined in Figure 9.6. The major problem is the osmotic water inflow. The gills are important in this gain because of their large surface and relatively high permeability; the skin is less important. Excess water is excreted as

FIGURE 9.7 By placing a fish in a chamber where the front and hind parts are separated by a rubber membrane it can be shown that the skin does not participate in the active uptake of ions, which is carried out by the gills. [Krogh 1937]

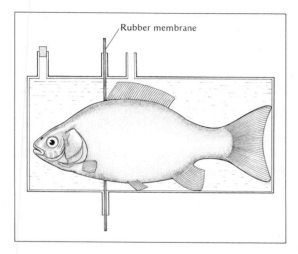

Rubber membrane

urine, which is very dilute and may be produced in quantities up to one-third of the body weight per day. Although the urine contains no more than perhaps 2 to 10 mmol per liter of solutes, the large urine volume nevertheless causes a substantial loss of solutes, which must be replaced. The gills are also slightly permeable to ions, and this loss must likewise be covered by ion uptake.

Some solutes are taken in with the food, but the main intake is by active transport in the gills. This has been shown by placing a fish in a divided chamber, in which the head and the remainder of the body can be studied separately (Figure 9.7). In such experiments active uptake of ions occurs only in the front chamber. It is therefore concluded that the skin does not participate in active absorption, and we assume that only the gills are responsible.

Catadromic and anadromic fish

Most teleost fish have a limited ability to move between fresh water and the sea; they are relatively stenohaline. We have already mentioned, however, that lampreys, salmon, and eels move between fresh and salt water as part of their normal life cycles (Koch 1968). Such moves between fresh and sea water expose the fish to cataclysmic changes in the demands on their osmoregulatory mechanisms.

If an eel is moved from fresh to sea water, the osmotic loss of water reaches 4% of the body weight in 10 hours (Keys 1933). If the eel is prevented from drinking sea water by placing an inflated balloon in the esophagus, it continues to lose water and dies from dehydration within a few days. If it can drink, however, it soon begins to swallow sea water, the weight loss subsides, and a steady state is reached within 1 or 2 days. If the eel is transferred in the opposite direction, from sea water to fresh water, there is an initial gain in weight, but as urine formation increases, a steady state is reached, again within 1 or 2 days.

When the eel is moved between fresh and sea water, not only does the osmotic flow of water change direction, but to achieve a steady state and compensate for the solute gain or loss, the active ion transport in the gills must change direction. How this change takes place is unknown, although we assume endocrine mechanisms are involved. Neither is it known whether different cell populations are responsible for the transport in the two directions, one or the other being activated as need arises. The other possibility is that the polarity of the transport mechanism in all participating cells can be reversed on demand. These questions remain unanswered.

Present knowledge indicates that it is unlikely that the direction of transport in the individual cell can be reversed. Of the many organs and cell types that participate in active transport of some sort, none is known that with certainty can carry out such reversals. The frog skin, which is analogous to the fish gill in performing active uptake from the dilute solutions of fresh water, does not seem able

FIGURE 9.8 Apparatus used for determining the sodium transport in an isolated piece of frog skin. The skin separates two chambers containing Ringer's solution, and sodium transport in the skin will produce a potential or voltage across the skin. If an external current is now applied in a direction opposite to the potential produced by sodium transport, this current will, when it is adjusted to give a zero voltage across the skin, be a direct measure of the sodium transport in the skin. A and A' indicate agar bridges to connect solutions to calomel electrodes; B and B' are agar bridges to provide electrical connections to the outside voltage source. [Ussing and Zerahn 1951]

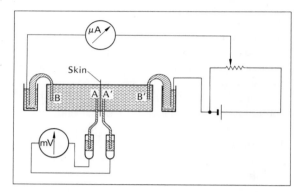

to reverse the direction of the transport in the one species known to live in sea water, the crab-eating frog (discussed in the next section).

Amphibians

Most amphibians are aquatic or semiaquatic. The eggs are laid in water, and the larvae are gill-breathing aquatic animals. At the time of metamorphosis, many amphibians (but not all) change to respiration by lungs. Some salamanders retain the gills and remain completely aquatic as adults; most frogs, on the other hand, become more terrestrial, although usually tied to the vicinity of water or moist habitats.

Some atypical frogs, at home in very dry habitats and highly resistant to water loss by evaporation, have recently been studied in Africa and South America. Their unusual physiological characteristics are described later in this chapter.

Fresh-water amphibians

With regard to osmotic regulation, amphibians are quite similar to teleost fish. Virtually all amphibians are fresh-water animals, and in the adult the skin serves as the main organ of osmoregulation. When the animal is in water, there is an osmotic inflow of water, which is excreted again as a highly dilute urine. There is, however, a certain loss of solutes, both in the urine and through the skin. This loss is balanced by active uptake of salt from the highly dilute medium. The transport mechanism is located in the skin of the adult, and the frog skin has become a well-known model for studies of active ion transport.

Pieces of frog skin can be readily removed and used as a membrane separating two chambers, which can be filled with fluids of various concentrations. By analyzing the changes in the two chambers, the transport function by the skin can be studied (Figure 9.8). Such isolated pieces of skin survive for many hours. This apparatus for the study of active transport processes was originally devised by Ussing and is known as an Ussing cell.

When frog skin in the Ussing cell separates two salt solutions * of the same composition, a potential difference of about 50 mV is rapidly established between the inside and the outside of the skin. The inside is positive, and it is therefore postulated that the potential is caused by an inward active transport of the positive sodium ion. When the potential difference has been established, the chloride ion passes through the skin by diffusion, accelerated by the electric field. An enormous amount of evidence has accumulated to indicate that this interpretation is correct, and the fact that the transport is active is clear from the developed potential and from the fact that metabolic inhibitors (e.g., cyanide) prevent the formation of a potential and inhibit the transport.

By applying an external potential across the

---

* The salt solution should have the same pH and osmotic concentration as the blood and have approximately the same concentrations of the major ions, Na^+, K^+, Ca^{2+}, and Cl^-. Such a balanced salt solution is called Ringer's solution, named after a British physiologist who discovered that survival of the isolated frog heart depended on balanced proportions of these ions.

skin, equal to the sodium potential but in the opposite direction, the potential across the skin can be reduced to zero. The current needed to maintain the potential at zero must equal the current generated by sodium transport in the skin. The current, called the short-circuit current, is therefore a direct measure of the inward transport of sodium. This method has, in similar form, become an extremely valuable tool in measuring active transport of ions in many other systems.

A salt-water frog

Frogs and salamanders are usually restricted to fresh water and die within a few hours if placed in sea water. There is one exception: the crab-eating frog (*Rana cancrivora*) of Southeast Asia. This small, ordinary-looking frog lives in coastal mangrove swamps, where it swims and seeks its food in full-strength sea water.

If a frog in sea water is to maintain the characteristic relatively low salt concentrations of vertebrates, it has two possible avenues for solving the problems. One strategy (the one used by marine teleosts) is to counteract the osmotic water loss through the skin and compensate for the inward diffusion of salt through the skin. The other strategy (that of marine elasmobranchs) is to retain urea and have body fluids in osmotic equilibrium with the medium, thus eliminating the problem of osmotic water loss. The salt-water frog has employed the elasmobranch strategy of adding large amounts of urea to the body fluids, which may contain as much as 480 mmol urea per liter (Gordon et al. 1961).

It seems reasonable that the salt-water frog uses this solution. Amphibian skin is relatively permeable to water, and it is therefore simpler to maintain the same osmotic concentration as that of the medium and eliminate osmotic water loss. To eliminate water loss solely by increasing the inter-

nal salt concentrations, the frog would need a salt tolerance unique among vertebrates (except the hagfish). If it were to use the teleost strategy and remain hyposmotic, the salt balance would be further impaired by the need to drink from the medium.

A crab-eating frog placed in sea water is not completely isosmotic with the medium; like sharks, it remains slightly hyperosmotic. The result is a slow osmotic influx of water, which is desirable because it provides the water required for formation of urine. This is certainly more advantageous than obtaining water by drinking sea water, which would inevitably increase the salt intake.

In the crab-eating frog, as in the elasmobranch, urea is an important osmotic constituent and not merely an excretory product. In addition to its osmotic importance, urea is necessary for normal muscle contraction, which rapidly deteriorates in the absence of urea (Thesleff and Schmidt-Nielsen 1962). Because urea is essential for the normal life of the animal, it should be retained and not be excreted in the urine. In sharks, urea is retained by active reabsorption in the kidney tubules (see Chapter 10). In the crab-eating frog, however, urea retention is achieved primarily by a reduction in urine volume when the frog is in sea water. It seems that urea is not actively reabsorbed, for the urea concentration in the urine consistently remains slightly above that in the plasma (Schmidt-Nielsen and Lee 1962).

The tadpoles of the crab-eating frog have an even greater tolerance for high salinities than the adults. Their pattern of osmotic regulation, however, is similar to that of teleost fish and thus differs from the elasmobranch pattern adopted by adult frogs (Gordon and Tucker 1965).

Although both the tadpole and the adult crab-eating frog are highly tolerant to sea water, this frog is not independent of fresh water, for both fertilization of the eggs and metamorphosis to the adult depend on a relatively low salt concentration in the water. Because of frequent torrential rains in the tropics, temporary fresh-water pools readily form near the shore, and spawning can therefore take place in dilute water. The tadpole is highly tolerant to salt, but metamorphosis is delayed for as long as the salinity remains high, and the frog goes through this critical stage only after the water has been diluted by heavy rain.

Although the crab-eating frog depends on fresh water for reproduction, its tolerance to sea water permits the exploitation of a rich tropical coastal environment that is closed to all other amphibians.

THE TERRESTRIAL ENVIRONMENT

The greatest physiological advantage of terrestrial life is the easy access to oxygen; the greatest physiological threat to life on land is the danger of dehydration. Successful large-scale evolution of terrestrial life has taken place in only two animal phyla, arthropods and vertebrates, which live and thrive in some of the driest and hottest habitats found anywhere. In addition, some snails thrive on land and are truly terrestrial; some even live in deserts.

The threat of dehydration is evidently a serious barrier to terrestrial life, for many "terrestrial" animals, except in the phyla just mentioned, depend on the selection of a suitably moist habitat and are terrestrial only in the technical sense of the word. An earthworm, for example, is highly dependent on a moist environment and soon succumbs to desiccation if exposed to the open atmosphere for any length of time. It has little resistance to water loss, depends on behavior, and seeks out microhabitats where air and surroundings are humid. Other examples are frogs and snails, which we can

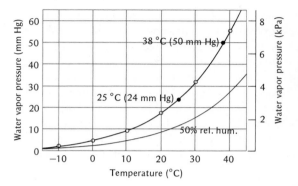

FIGURE 9.9 The water vapor pressure over a free water surface increases rapidly with temperature (open circles, fully drawn line). At the body temperature of mammals (38 °C) it is roughly twice as high as at room temperature (25 °C), and at higher temperature the rise is increasingly steep.

group together as moist-skinned animals that have a high rate of evaporation.

However, before we discuss animals, we must know something about the physical factors that influence evaporation. That subject covered, we will discuss first some moist-skinned terrestrial animals that depend heavily on water in their environment and then turn to arthropods and terrestrial vertebrates.

Evaporation

We know that evaporation from a free water surface increases with temperature and that evaporation is faster in a dry atmosphere than under humid conditions. In order to understand evaporation from animal surfaces, it is necessary to have a more precise concept of the physical laws that govern the rate of transfer of water from the liquid to the gas phase.

The water vapor pressure over a free water surface increases rapidly with temperature (Figure 9.9). As a rule of thumb we can remember that at mammalian body temperature (38 °C) the water vapor pressure (50 mm Hg) is about twice as high as at a room temperature of 25 °C (24 mm Hg) and more than 10 times as high as the vapor pressure at the freezing point (4.6 mm Hg).

If the air already contains some water vapor, the difference between the vapor pressure at the surface of liquid water and in the air will be less, and the driving force for water vapor to diffuse from the saturated boundary layer at the water surface into the air will be correspondingly reduced.

As a rough approximation to the measure of this driving force for evaporation, many investigators have used the *saturation deficit*. The saturation deficit is expressed as the difference between the vapor pressure over a free water surface at the temperature in question and the water vapor pressure in the air. If the relative humidity of the air is 50%, the saturation deficit increases with temperature, as shown by the difference between the two curves in Figure 9.9.

The saturation deficit has been useful in many ecological studies, both of plants and animals, but it is theoretically inadequate, for the saturation deficit is not the only physical factor that determines the rate of evaporation (Ramsay 1935; Edney 1957). For one thing, air movement caused by free or forced convection drastically alters the rate of evaporation. Another factor that is important under some circumstances is the loss of heat from and cooling of the surface from which evaporation takes place. The evaporating surface may therefore be cooler than other parts of the system, the temperature differences depending on the rate of evaporation as well as the transfer of heat from other parts, a parameter that is often difficult to measure. Furthermore, the rate of diffusion of water vapor into air increases with increasing temperature. And finally, the diffusion rate increases with decreasing barometric pressure. Although each variable is a complex function that in turn influences all the others, mathematical modeling can provide an extremely successful analysis of water exchanges in a terrestrial environment (Tracy 1976).

For animals under natural conditions, the extent of air movement is one of the most important variables. If there is wind (and in nature the air is virtually never completely still), the air layer close to the surface is rapidly renewed, and evaporation increases. This we feel as a cooling effect when moist skin is exposed to the moving air from wind or a fan. The cooling of a surface caused by evaporation also causes a change in the density of the adjoining air, producing convection currents. The extent of such convection differs between a horizontal and a vertical surface, and as a consequence, under otherwise identical physical conditions the evaporation from a moist surface varies with the orientation of the surface. Finally, evaporation also depends on the curvature of a surface.

Because of these complexities, it is difficult to describe accurately the variables that govern the rate of evaporation. The saturation deficit remains the most practical approximation we have, and expressing a certain rate of evaporation relative to the saturation deficit gives a reasonably good basis for comparisons. But we should realize that results on animals of various sizes and shapes obtained under seemingly similar conditions do not give an accurate basis for comparison and, in particular, cannot be directly compared with the evaporation from the horizontal surface of free liquid water.

MOIST-SKINNED ANIMALS

Consider the two extreme cases. From a moist animal surface the evaporation is high, and the rate of water loss is determined primarily by the transfer of water vapor into the surrounding air (diffusion aided by convection). In the opposite case (e.g., the dry insect cuticle), the greatest resistance to evaporation is in the surface itself, and any change in the permeability of the barrier greatly affects the rate of evaporation. In this case changes

TABLE 9.8 Evaporation of water from the body surface of various animals at room temperature. The data indicate orders of magnitude; exact figures vary with experimental conditions. All data refer to micrograms of water evaporated per hour from 1 cm² body surface at a saturation deficit of 1 mm Hg (0.13 kPa). [Schmidt-Nielsen 1969]

Earthworm	400
Frog	300
Salamander	600
Garden snail, active	870
Garden snail, inactive	39
Man (not sweating)	48
Rat	46
Iguana lizard	10
Mealworm	6

FIGURE 9.10 The earthworm, when placed in sodium chloride solutions of various concentrations, behaves like a typical fresh-water osmoregulator. It remains hyperosmotic to the medium at all concentrations, and more so at the lower concentrations. [Ramsay 1949]

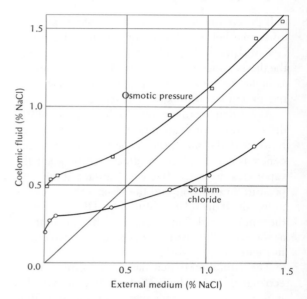

in the saturation deficit, convection, and so on are of minor importance and in some instances may even be disregarded. The two cases can be described as a *vapor-limited system* where the resistance to evaporation lies in water transport in the air, and a *membrane-limited system* in which the membrane is the major barrier to water evaporation (Beament 1961).

Earthworms

An earthworm kept in dry air rapidly loses weight because of evaporation. The rate of water loss (Table 9.8) is quite high, and in dry air the earthworm soon dies. If a partially dehydrated worm is placed in a U-shaped tube, covered with water but with the mouth and anus above the water surface, water is absorbed by the worm. Its skin is readily permeable to water in both directions. A more careful study will show that the earthworm behaves much like a typical fresh-water animal. If it is placed in salt solutions of various concentrations, its body fluids remain hypertonic to the medium (Figure 9.10). The urine, however, remains hypotonic to the body fluids, and also in this regard the earthworm resembles a typical fresh-water animal (Ramsay 1949).

These osmotic relations show that, with regard to water balance, the earthworm is really more of a fresh-water animal than a truly terrestrial animal.

It tunnels in the soil where the air is fully saturated, and it is in contact with soil particles covered with a thin film of free water. An earthworm is unable to live in completely dry soil, where it rapidly becomes dehydrated and dies.

Frogs and other amphibians

A frog, even when picked up on land, has a moist and cool skin. The rate of evaporation from the frog's skin is similar to that from the earthworm's (Table 9.8), and the twofold difference between frog and salamander is, for reasons explained above, not a meaningful difference. For both, the rate of evaporation is of the same magnitude as evaporation from a free water surface, and the amphibian skin does not seem to present any significant barrier to evaporation. Accordingly, those adult amphibians that are terrestrial and air-breathing usually live near water and in humid habitats where evaporation is low. When they enter water, they behave osmotically like the typical fresh-water animal.

Knowing that amphibians are moist-skinned and tied to the vicinity of free water makes it seem a paradox that some live successfully in deserts. Several species of frogs are found in the dry arid interior of Australia. These animals retreat to burrows several feet deep in the ground, where they estivate during the long periods of drought. Because reproduction takes place in water, they can breed only after rainfall, which gives an abundant but temporary supply of water.

When it rains these frogs reappear from their burrows, restore their water content, and deposit their eggs, which develop at exceptional speed into tadpoles that metamorphose before the pools dry up. When the frogs enter estivation again, their urinary bladders are filled with dilute urine, which is of considerable importance in their water balance. Some can store as much as 30% of their gross body weight as urine in the urinary bladder. The urine is extremely dilute and has an osmotic concentration corresponding to less than 0.1% NaCl. This urine is the main water reserve and is gradually depleted during estivation (Ruibal 1962).

It has been reported that the Australian aborigines use the desert frog (*Chiroleptes*) as a source of drinking water because the frog is so distended with dilute urine as to "resemble a knobbly tennis ball" (Buxton 1923). Once the animal has used up the water reserve in the bladder, further dehydration results in increased body concentrations, and the animal then undergoes progressive dehydration of blood and tissues.

The situation for frogs that remain deep in their burrows during dry periods is similar to that described for lungfish, which also estivate in the ground during drought. When a lake dries out, the lungfish wiggles into the mud, where it remains in a dry cocoon with a breathing channel to the rock-hard surface. It may remain alive in this condition for several years, and to keep its minimal metabolic rate going, it gradually uses up its body proteins. No water is available for the excretion of the urea formed in protein metabolism, and the concentration of urea in its body fluids may rise as high as 500 mmol per liter (Smith 1959).

Is the rate of evaporation from the skin of a desert amphibian less than in other amphibians? When determinations are made under similar conditions, including temperature and atmospheric humidity, the differences in the rates of evaporation seem to be minor and not sufficient to explain the great variations in habitat. It has been claimed that frogs lose water more rapidly than toads, which have thicker and more cornified skins, but this does not appear to be true (Bentley 1966). Therefore, resistance to cutaneous evaporation does not seem to be a major factor in the adaptation of amphibians to terrestrial life.

There are, however, some striking exceptions. The South African frog *Chiromantis* does not depend on a humid environment to prevent rapid dehydration. When placed in dry air it loses water by evaporation at a rate that is a small fraction of the rate in other frogs; in fact, its rate of water loss is similar to that of reptiles (Figure 9.11). Several frogs of the South American genus *Phyllomedusa* also exhibit rates of evaporation that are only about one-twentieth of those typical of other frogs (Shoemaker and McClanahan 1975). *Phyllomedusa* has specialized glands in the skin that secrete a waxy substance that apparently waterproofs the skin. This wax makes the skin surface hydrophobic so that droplets of water, instead of spreading on the skin, form beads that drop off (Blaylock et al. 1976). *Chiromantis* lacks glands that secrete wax, and its skin must be rendered impermeable by some other mechanism.

These unusual frogs have another physiological

FIGURE 9.11 When kept at 25 °C and 20 to 30% relative humidity, the frog *Chiromantis* loses water by evaporation at a very low rate, similar to that of a reptile (*Chamaeleo*). After 6 days in air its weight loss is only a few grams. Other amphibians lose weight rapidly. A toad (*Bufo*) loses over 40% of its body weight and dies within 2 days; an ordinary frog (*Rana*) dies in a day, and the aquatic *Xenopus* in less that 10 hours. [Loveridge 1970]

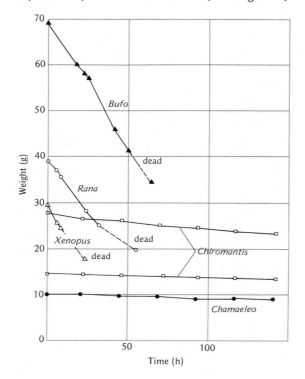

Snails

Snails and slugs are moist-skinned, and evaporation from their surfaces is high (Table 9.8). Naked slugs depend on the humidity of their habitat, and they are active mostly after rain and at night when the relative humidity is high. Otherwise they withdraw to and remain in microhabitats where the humidity is favorably high. Snails, in contrast, carry with them a water-impermeable shell into which they can withdraw.

It has been suggested that the *mucus* that covers a snail forms a barrier to evaporation, but this is difficult to understand. When snail mucus is removed from the animal, water evaporates from the isolated mucus at practically the same rate as from a free water surface, until the mucus is completely dried out. Nevertheless, this situation is not entirely clear, for if the surface of an otherwise undisturbed snail is lightly touched, fresh mucus is extruded from mucus glands in the skin and the rate of evaporation increases (Machin 1964).

A terrestrial snail that is inactive or estivates has an excellent barrier to water loss. It withdraws into its shell and covers the opening with a membrane, an epiphragm, which consists of dry mucus, but in some cases also contains a large amount of crystalline calcium carbonate.

The extremely low rate of water loss from an inactive snail in the shell and covered with an epiphragm permits some species to survive in hot, dry desert areas. One such snail, *Sphincterochila*, is found in the deserts of the Near East. Withdrawn into the shell and dormant, this snail can be found on the desert surface in midsummer, fully exposed to sun and heat. The snails become active after rains, which are concentrated in the winter months, from November to March, when they feed and reproduce.

During the long dry season the snails remain

characteristic completely different from other frogs. Instead of excreting urea, as all other amphibians do, they excrete uric acid, which is characteristic of reptiles. This will be discussed further in Chapter 10, in connection with nitrogen excretion.

The finding of reptilian physiological characteristics in amphibians illustrates the point we have made before. Physiological characteristics as a rule are much more plastic and adaptable to the environment than morphological characters and are therefore not suitable for establishing evolutionary relationships.

dormant. The water loss from a dormant *Sphincterochila* is less than 0.5 mg per day, and since one animal contains about 1.5 g water, it can survive for several years without becoming severely dehydrated. In fact, the snail itself (not including the calcareous shell), contains over 80% water. This high water content remains the same, even during the hottest and driest part of the year. This shows that these snails do not gradually deplete their water resources, but have solved the water problem by having a uniquely low water loss while remaining dormant (Schmidt-Nielsen et al. 1971).

ARTHROPODS

The most successful terrestrial animals are among the arthropods, which comprise insects, crustaceans, and several other animals characterized by a jointed, rigid exoskeleton. Insects are by far the most numerous animals in regard to number of species, for nearly 1 million different species have been described.

Insects and arachnids (spiders, ticks, mites, scorpions, etc.) are primarily terrestrial, and they are highly adapted to life on land and respiration in air. Only a small number have secondarily invaded fresh water, and almost no marine forms are found in these two classes. Two additional classes of arthropods are terrestrial: the millipedes (Diplopoda) and the centipedes (Chilopoda). In contrast, most crustaceans are aquatic, as are the Xiphosura, represented by the living fossil, the horseshoe crab (*Limulus*).

Crustaceans

Although most crustaceans are aquatic, appreciable numbers, including some decapods (crabs and crayfish), are terrestrial, even to the extent that they drown if kept submerged. However, they are mostly restricted to moist habitats and the vicinity of water. On the other hand, some isopods, notably sowbugs (or pill-bugs) and woodlice, are completely terrestrial, and a few even live in hot, dry deserts.

Crabs are common in the intertidal zone, and in the tropics many live permanently above the high-tide mark. They depend on a moist habitat, however, and often have burrows where a pool of water remains at the bottom. In the more terrestrial species emancipation from water is nearly complete, but the crabs do return to water for spawning.

The semiterrestrial and terrestrial crabs are generally excellent osmoregulators. Depending on conditions and need, they can usually carry out effectively both hyperregulation and hyporegulation. An interesting aspect of some land crabs (*Cardiosoma* and *Gecarcinus*) is their ability to take up water from damp sand or a moist substratum, even in the absence of visible amounts of free water. In this way they can survive for many months without direct access to water (Gross et al. 1966; Bliss 1966).

The ghost crab, *Ocypode quadrata*, which is often seen scurrying over sandy beaches, requires frequent immersion in sea water. Similarly, the well-known fiddler crabs, *Uca*, which are found along the coasts of all tropical seas and into temperate areas, dig deep burrows into the beach near or above high-tide mark, but depend on frequent returns to water.

The common terrestrial isopods known as pill-bugs, sowbugs, or woodlice are mostly found in humid habitats, well away from exposure. They remain hidden during the day and move around at night when the relative humidity is higher. These crustaceans are completely independent of free water for reproduction and are thus truly terrestrial.

Although woodlice mostly live in microhabitats with high humidity, they are sometimes exposed to drier air. Then they lose water by evaporation, and this water must be replaced. Normally, woodlice feed on moist decaying plant material, which is probably their normal source of water. It has been found, however, that several isopods can take up free water by drinking and also through the anus. When kept on a moist porous slab of plaster of Paris, they can even absorb water from this surface (Edney 1954).

Compared with that of an insect, the cuticle of an isopod is relatively permeable to water. The main reason for this difference seems to be that insect cuticle is covered by a thin layer of wax, which greatly reduces water loss; all attempts at demonstrating a wax layer in isopods have been unsuccessful (Edney 1954).

It may seem strange to find a land crustacean in a desert, but a woodlouse, *Hemilepistus*, is quite common in desert areas. These animals still breathe by gills, and they survive by digging narrow vertical holes about 30 cm deep, where they spend the hot part of the day. Temperature and humidity measurements show that their hideouts remain much cooler than the desert surface, and most importantly, the relative humidity is high, around 95%. The cuticle of *Hemilepistus* is somewhat less permeable to water than that of other woodlice, but the animal's major adaptation to life in the desert seems to be a question of behavior (Edney 1956).

Insects and arachnids

If numbers indicate success, insects are undoubtedly the most successful animals on earth. There are more species of insects than of all other animals combined, and furthermore, many insect species are extremely numerous. They live in almost every conceivable habitat on land and in fresh water; only the polar regions and the ocean seem to be relatively free from them. Dryness, desert, and absence of free water seem to be no barrier to these animals. A clothesmoth that thrives on a woolen garment or a flour beetle that completes its life cycle from egg to adult in dry flour, lives, grows, and thrives virtually without any free water in its environment. Nevertheless, as the adult insect consists of more than two-thirds water, these animals must have extraordinary capacities to retain water and reduce losses.

Because there are considerable similarities between insects and arachnids, these two classes will be discussed together.

Water balance: gains and losses

For an organism to remain in water balance, all loss of water must, over a period of time, be balanced by an equal gain of water. The components in the water balance are:

Water loss
  Evaporation
    from body surface
    from respiratory organs
  Feces
  Urine
  Other (specialized secretions)

Water gain
  Drinking
  Uptake via body surface
    from water
    from air
  Water in food
  Oxidation water (metabolic water)

The problem of maintaining balance may work either way. There may be an excess of water or a shortage of water, and the physiological mechanisms must therefore be able to cope with both situations. For example, a fresh-water insect may swallow large amounts of water with its food and also be subject to osmotic inflow of water. The excess water must be eliminated, a task normally carried out by the renal organ or kidney, which then must produce a large volume of dilute urine. A terrestrial insect living in a dry habitat, on the other hand, has a very limited water intake, and all losses must be reduced to such a level that their sum in the long run does not exceed the total gain.

Any organism can tolerate a certain variation in its water content, and some are more tolerant than others. Mammals, for example, can usually withstand losing 10% of their body water, although they are then in rather poor condition; a loss of 15 or 20% is probably fatal to most. Many lower organisms can withstand greater losses. Some frogs, for example, can withstand the loss of 40% of their body water, but very few animals can tolerate losing half their body water.*

When the water content of an animal is expressed as a percent of its body weight, the results are not always easy to interpret. For example, the fat content of animals varies a great deal, and fatty tissues have a low water content, about 10% or less. An animal with a large fat storage therefore will have a lower overall percent water content in

* Two methods are available for the determination of total body water. The water content can be obtained by killing the animal, drying the body completely (e.g., in a drying oven), and determining the weight loss. If such a destructive method is employed, some individuals are used to establish a normal or baseline value, and others are used for experimentation. A nondestructive method is the isotope dilution technique. A small volume of water labeled with an isotope (deuterium or tritium) is injected into the animal. After the labeled water is evenly distributed in all body water, a sample of blood or any other body fluid is withdrawn and the concentration of the label determined. It is then simple to calculate the water volume into which the isotope label was distributed. The method permits repeated determinations on the same individual animal.

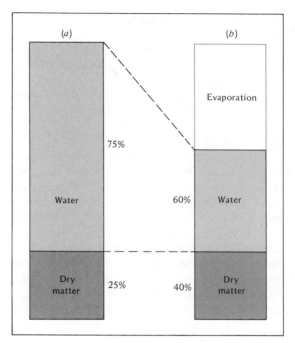

FIGURE 9.12 Diagram to illustrate that it can be deceptively misleading to express body water content in percent of body weight. If an animal originally has a water content of 75% of its body weight (a) and loses exactly half its body water, the water content of the dehydrated animal will be 60% of the body weight (b). Thus a water loss that in reality is very serious appears not very great.

the body than a lean animal in which muscles and internal organs have exactly the same degree of hydration. If the water content is expressed in relation to the fat-free tissue mass, the figures are usually much less variable. The disadvantage is that this method requires an estimate of fat content and therefore becomes much more complicated.

The custom of expressing water content as a percent of the body weight tends to be misleading in another way that is best explained by an example.

Consider an insect which initially contains 75% water. If it loses exactly half of all water in its body, it will contain 60% water. The result may seem surprising, but an examination of Figure 9.12 shows the arithmetic to be correct. At first glance, a reduction in water content from 75% to 60% does not seem excessive, although in reality half of all body water was lost. The sometimes deceptive effect of percentage values can be avoided by expressing results in absolute units. If the insect in the above example weighed 100 mg, it would initially contain 75 mg water and after dehydration 37.5 mg water, and it is immediately apparent that half its water was lost.

Water loss: evaporation

Evaporation takes place both from the respiratory organs and from the general body surface, but it is often experimentally difficult to determine these two variables separately. The respiratory organs of insects and vertebrates differ profoundly in structure and function. The lining of the vertebrate lung is always moist, the respiratory air is saturated with water vapor, and the water loss from the respiratory tract is substantial. The respiratory organs of insects, on the other hand, consist of tubes lined with chitin, and only the finest branches of these are relatively permeable to water. Nevertheless, for insects the water loss from the respiratory system is important, and the possible means to reduce this loss were discussed under insect respiration in Chapter 2.

The general body surface of an insect is covered by a hard, dry cuticle, and it is deceptively easy to assume that this cuticle is completely impermeable to water. It is not. There are enormous differences in evaporation between species that normally live in moist environments and others that live in completely dry surroundings (Table 9.9). The evaporation from the aquatic larvae of the marsh fly, when in air, is of the same order of magnitude as evaporation from a moist-skinned animal (see Table 9.8). This means that the cuticle essentially is no barrier to evaporation. On the other hand, in insects that live in very dry habitats the permeability of the cuticle can be lower than in any other animals about which we have information.

TABLE 9.9 Evaporation of water from the body surface of insects and arachnids. Water loss from respiratory organs is not included. All data refer to micrograms of water evaporated per hour from 1 cm^2 body surface at a saturation deficit of 1 mm Hg (0.13 kPa). [Schmidt-Nielsen 1969]

Marsh fly larvae (*Bibio*)	900
Cockroach (*Periplaneta*)	49
Desert locust (*Schistocerca*)	22
Tsetse fly (*Glossina*)	13
Mealworm (*Tenebrio*)	6
Flour mite (*Acarus*)	2
Tick (*Dermacentor*)	0.8

The insect cuticle is a highly complex organ, which consists of several layers. The hard chitin itself is not particularly impermeable to water; the resistance to evaporation resides in a thin covering layer of wax. If this layer is scratched by abrasives (e.g., alumina particles) the evaporation increases greatly and insects that can otherwise live in very dry material, such as stored grain, die of dehydration (Wigglesworth 1945). This is of some practical interest, for abrasives have been used as nontoxic insecticides that can later be separated from the grain with relative ease because of differences in density. Not only sharp particles, but also materials such as dry clay dust have a similar effect, presumably because they too disturb the structural arrangement of the wax layer.

The evaporation from insect cuticle increases with rising temperatures (Figure 9.13). The increase is far greater than the increase in water vapor pressure and seems related to physical changes in the wax layer. If the permeability is plotted against the insect's surface temperature, rather than air temperature, a sharp transition point is observed in evaporation. This transition point differs from species to species. The surface waxes have been isolated and their physical characteristics show a transition point at the same temperature as the observed transition in water loss (Beament 1959).

For arthropods a wax layer is not a universal prerequisite for successful terrestrial life. It was

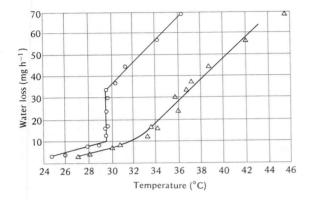

mentioned above that many isopods are adapted to terrestrial life and that some even live in deserts. No wax has been found on their cuticle, and their water loss from evaporation is directly related to the saturation deficit in the atmosphere and is otherwise temperature-independent. In insects, in contrast, evaporation changes with temperature even if the vapor pressure deficit is kept constant.

Water loss in feces and urine

Insects eliminate their feces and urine through the same opening, the anus. The urine is formed by the Malpighian tubules, which open into the posterior part of the gut. Liquid urine, as well as fecal material from the intestine, enter the rectum, where water reabsorption takes place. This matter will be discussed in greater detail in Chapter 10.

Storage excretion

If excretory products, instead of being eliminated in the urine, are withheld in the body, no water is expended for their excretion. Because uric acid is a highly insoluble compound, its retention in the body is in fact a feasible approach to the problem of "excretion," and the deposition of uric acid in various parts of the insect body seems to be a regular feature. For example, in several species of cockroaches, as much as 10% of the total dry weight of the body may be uric acid.

Many insects store uric acid in the fat body; others store it in the cuticle. Uric acid deposited in the cuticle is probably never mobilized again. In the fat body, however, the situation may be different. It is possible that this uric acid represents a storage or depot from which nitrogen can be mobilized during periods of nitrogen deprivation and used for metabolic purposes. So far evidence for such a function is inadequate.

"Active" transport of water

An insect that lives in very dry surroundings (e.g., a mealworm) withdraws water from the rectal contents until the fecal pellets are extremely dry. This withdrawal of water has the appearance of an active transport of water, for water is moved from a high osmotic concentration in the rectum to a lower concentration in the blood. However, because of the complexity of the rectum and its surrounding structures, alternate hypotheses are possible. In general, most uphill transport of water in animal systems can be explained by a primary transport of a solute, with water following passively due to osmotic forces, acting as explained by the *three-compartment theory* proposed by Curran (1960).

The Curran hypothesis can be understood by reference to the artificial model in Figure 9.14, which shows a cylinder divided into three compartments. Compartments A and B are separated by a cellophane membrane, which is readily permeable to water but not to sucrose. Compartments B and C are separated by a porous glass disc (a sheet of filter paper would do if given suitable mechanical support). Let us initially fill compartment A with 0.1 molar sucrose solution, compartment B

FIGURE 9.14 The Curran model to demonstrate net transfer of water against an osmotic gradient between chambers A and C. During experiments the exit tube from chamber B is closed. For details, see text. [Curran and MacIntosh 1962]

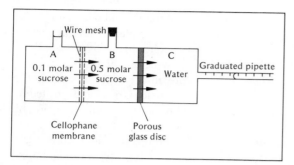

FIGURE 9.15 Three-compartment model for water transport across an epithelial cell. Heavy arrows indicate active Na^+ transport. Dotted area corresponds to compartment B in Figure 9.14.

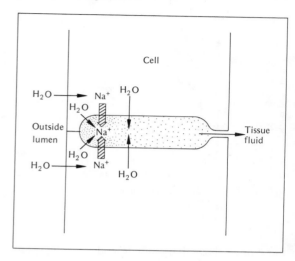

with 0.5 molar sucrose, and compartment C with distilled water. The concentrated solution in B will immediately draw water osmotically from A and C, but if the liquid in B is kept from expanding, it will flow through the porous disc into compartment C. The overall result is therefore a net movement of water from A to C, although the osmotic concentration in A is higher than in C. The movement of water against the overall osmotic gradient obviously cannot continue forever, for the concentration differences will gradually disappear and the system run down. However, if somehow we can maintain a high solute concentration in compartment B, the water movement will continue.

Movement of water through a variety of biological membranes can be explained according to the Curran model. The diagram in Figure 9.15 represents a typical epithelial cell, which has deep infoldings of the membrane on the side facing the tissue fluid. Active transport of sodium establishes a high osmotic concentration within the lumen of the infolding. Then, because of the osmotic forces, water diffuses into the lumen, and the increased hydrostatic pressure causes a bulk flow of liquid through the opening into the tissue fluid. The result is an overall movement of water from the outside lumen to the tissue fluid. This water transport, which can be against an overall osmotic gradient from the outside of the cell to the tissue

fluid, depends on the active transport of sodium as the primary driving force.

Transport as explained by this mechanism may be a common way of moving water. Long, narrow intercellular or intracellular spaces are characteristic of many epithelia, and available experimental evidence supports the hypothesis (Diamond 1962; Diamond and Bossert 1967).

Water reabsorption from the insect rectum has been studied in cockroaches, and the results conform to the three-compartment hypothesis. Samples of fluid, as small as 0.1 nl ($0.0001 mm^3$) have been withdrawn from the rectal pads. This fluid is more concentrated than the fluid in the rectal lumen, while the blood is less concentrated. The fluid thus corresponds to compartment B in the Curran model (Wall and Oschman 1970).

Water gain: drinking

The most obvious form of water intake is drinking free water. Such water is, of course, available to all fresh-water insects and to insects under many other circumstances, such as when there is dew or

TABLE 9.10 Amounts of water formed in the oxidation of various foodstuffs. [King 1957; Schmidt-Nielsen 1964]

Foodstuff	Water formed (g H_2O g^{-1} food)	Metabolic energy value (kcal g^{-1})	Water formed (g H_2O kcal^{-1})
Starch	0.56	4.2	0.13
Fat	1.07	9.4	0.11
Protein (urea excretion)	0.39	4.3	0.09
Protein (uric acid excretion)	0.50	4.4	0.11

rain. For most insects, however, free water is available only intermittently and at irregular intervals, and many live in dry habitats where no free water is available at any time. These must obtain water from elsewhere.

Water in the food

Insects that eat plants may obtain large amounts of water in the food, for fresh vegetable material has a high water content. Succulent fruits, leaves, and so on may contain over 90% water, but even the driest plant material contains some free water. Dry grains and seeds, flour, wool, and other seemingly completely dry substances on which insects are able to subsist, grow, and reproduce may contain about 5 to 10% free water.

When water is plentiful in the food, the problem is to eliminate the excess. This is the normal function of the kidney, which then produces dilute urine in large quantities. Under dry conditions, in contrast, the goal is to eliminate the excretory products with a minimum of water and produce a urine as concentrated as the renal mechanism is able to achieve.

Oxidation water

For animals living under dry conditions, the most significant water gain is from water formed in the oxidation of organic materials. We can properly call it *oxidation water*, but the term *metabolic water* is also common.

Everybody knows that water is formed when organic materials burn; we can see water condense on the outside of a cold pot placed over a gas flame, or water dripping from the exhaust pipe of a car on a cold morning. For glucose the oxidation reaction is:

$$C_6H_{12}O_6 + 6\ O_2 \rightarrow 6\ CO_2 + 6\ H_2O$$
$$(180g) + (192g) \rightarrow (264g) + (108g)$$

In the overall reaction water can be regarded as formed by oxidation of the hydrogen. In the oxidation of food, the amount of water depends on the amount of hydrogen present in the foodstuff in question. From the above equation it can be readily calculated that 1 g glucose yields 0.60 g water, and no amount of physiological maneuvering can obtain more water than what is indicated by the equation.

In the metabolism of other carbohydrates, such as polysaccharides and starch, slightly less water is formed because of the lower hydrogen content of these materials. The oxidation of starch, for example, is as follows:

$$(C_6H_{10}O_5)_n + n\ O_2 \rightarrow 6n\ CO_2 + 5n\ H_2O$$

In this case, the amount of water formed in oxidation is 0.56 g water per gram starch metabolized (Table 9.10).

Oxidation of fat gives more oxidation water than is obtained from carbohydrates, about 1.07 g water per gram fat, a figure that varies slightly with the

TABLE 9.11 Arthropods that can absorb water from atmospheric air. The last column gives the limiting relative humidity below which a net gain in water is no longer possible.

Animal	Limiting relative humidity (%)
Tick (*Ornithodorus*)	94
Tick (*Ixodes*)	92
Mealworm (*Tenebrio*)	90
Mite (*Echinolaelaps*)	90
Desert roach (*Arenivaga*)	83
Grasshopper (*Chortophaga*)	82
Flour mite (*Acarus*)	70
Flea (*Xenopsylla*)	50
Firebrat (*Thermobia*)	45

composition of the fat and the degree of saturation of the fatty acids. Thus, about twice as much water is formed in fat oxidation as in the oxidation of starch. In some respects, however, this figure is misleading, for fat also gives more energy per gram (9.4 kcal against 4.2 for starch). For a given metabolic rate, an animal therefore uses less than half the amount of fat, with a corresponding reduction in the yield of oxidation water. The amount of oxidation water formed, relative to a given metabolic rate, is therefore slightly more favorable for starch than for fat. This is shown in the last column of Table 9.10, which lists the amount of water formed in relation to the energy value of the food.

Protein metabolism is somewhat more complex, because the nitrogen contained in the protein yields excretory products that contain some hydrogen; this hydrogen is excreted and not oxidized to water. The amount of oxidation water depends on the nature of the end product of protein metabolism. If it is urea, the amount of oxidation water formed is 0.39 g water per gram protein; if the end product is uric acid, which is common in insects, the amount of oxidation water formed is higher.

Urea (CH_4ON_2) contains two hydrogen atoms per nitrogen atom; uric acid ($C_5H_4O_3N_4$) contains only one hydrogen per nitrogen atom, that is, only half as much. This increases the yield of oxidation water to 0.50 g water per gram protein when uric acid is the end product. To give an accurate account of the amount of oxidation water, we must therefore know not only the exact amounts and composition of the foodstuffs oxidized, but also the nature of the metabolic end product of protein metabolism. This is important only if a very accurate account is needed, for the differences in oxidation water formed for the various foodstuffs are relatively small (Table 9.10 last column).

Water uptake via body surface

In aquatic insects the uptake of water through the body surface is the same as in other fresh-water animals: The higher osmotic concentration of solutes in the body fluids causes an osmotic inflow of water. For aquatic insects the problem is to eliminate excess water, and, as mentioned above, this is usually done by the kidney or equivalent excretory organ.

In terrestrial insects, however, the situation is different; they are often in water shortage. A very interesting phenomenon has been observed in some terrestrial insects and arachnids: They are able to absorb water vapor directly from atmospheric air. When first reported, this observation was met with considerable doubt because it is extremely difficult to suggest any plausible mechanism for water uptake directly from the air. The early reports have been confirmed by several competent investigators, and some well-documented examples are listed in Table 9.11. The table also contains information about the lowest relative humidity at which each animal can carry out such absorption of water vapor.

The characteristics of the uptake mechanism can best be described with the aid of an example, and we will use the desert roach *Arenivaga*. *Arenivaga* takes up water from the atmosphere only after it has been partly dehydrated. If it has been placed in a very dry atmosphere until, say, 10% of the body weight has been lost, uptake of water begins when the animal is moved to any relative humidity

FIGURE 9.16 Relative proportions of water and dry material in the desert roach *Arenivaga* before and after dehydration in dry air, and after subsequent rehydration by absorption of water vapor from air at 95% relative humidity. The animals, which initially weighed about 250 mg, were not fed during the experiment and therefore lost dry substance. As a result the percentage of water was higher at the end of the experiment (73.2%) than initially (67.2%). The ordinate refers all measurements to the initial weight, normalized to 100% (Edney 1966).

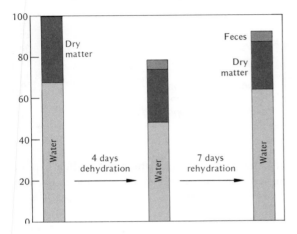

above 83%. The uptake continues until the animal is fully rehydrated and then ceases (i.e., the absorption process is accurately regulated according to the need for water). Figure 9.16 depicts the decrease and subsequent increase in the water content of the desert roach after dehydration and subsequent rehydration.

We saw that the atmospheric humidity below which uptake cannot take place differs from species to species (Table 9.11); this species-specific limit is not influenced by temperature. Therefore, the relative humidity rather than the vapor pressure deficit limits the uptake (i.e., the process appears to be similar to a hygroscopic effect). This point is difficult to understand, and any hypothesis ought to explain it adequately.

One hypothesis for the uptake of water vapor is that temperature microgradients within the animal lead to condensation of water vapor. To achieve condensation from an atmosphere of 90% relative humidity requires a gradient in excess of 2 °C, which is rather unthinkable when experiments are done under constant temperature conditions. At 50% relative humidity, roughly the lowest humidity from which uptake has been reported in any species, the air temperature would have to be lowered from 25 °C to 14 °C to achieve condensation. Obviously, in experiments under controlled constant temperature conditions, such temperature differences are out of the question (Edney 1966).

The anatomical location for the uptake of water vapor is difficult to establish. It might seem simple to determine whether the respiratory system is involved, for it is relatively easy to seal the openings of the tracheae with wax. In some insects such sealing stops water absorption, but in the tick *Ixodes* uptake still continues after the spiracles have been sealed. In those animals in which uptake ceases, this could be explained as a secondary effect of sealing the respiratory system, for this prevents oxygen uptake, and anoxia as such also stops water vapor uptake.

The many uncertainties are in part attributable to the very small size of some of the experimental animals (the house dust mite weighs 0.01 mg), but present information indicates that there is not one uniform mechanism in all those animals that are able to absorb water vapor. An oral site for absorption is likely for ticks and mites; the rectum almost certainly is the site of uptake in the firebrat and the mealworm (Noble-Nesbitt 1977). In the desert roach *Arenivaga* the mechanism is associated with a well-defined bubblelike structure in the hypopharynx that is extruded and visible during periods of active water absorption (O'Donnell 1977).

TERRESTRIAL VERTEBRATES

Reptiles

There are four major orders of living reptiles. Of these, the crocodilians are always associated with water. The other three – snakes, lizards, and tortoises – are considered well adapted to dry habitats, but they also have some representatives that are aquatic or semiaquatic. All the aquatic reptiles

HOUSE MITE The house mite (*Dermatophagoides farinae*) can absorb water vapor from air. In very dry air it becomes dehydrated, but if the relative humidity exceeds 70% it can take up water directly from the atmosphere. The top photo shows a dehydrated mite that weighs 8 μg (0.008 mg); the bottom photo shows the mite after it has increased its weight to 13 μg by absorption of water vapor from the air. The length of the animal is about 0.4 to 0.5 mm. [Courtesy of L. G. Arlian, Wright State University, Dayton, Ohio]

TABLE 9.12 Evaporation of water from the body surface of reptiles at 23 to 25 °C. All data refer to micrograms of water evaporated per hour from 1 cm² body surface at a saturation deficit of 1 mm Hg (0.13 kPa). [Schmidt-Nielsen 1969]

Caiman (*Caiman*)	65
Water snake (*Natrix*)	41
Pond turtle (*Pseudemys*)	24
Box turtle (*Terrapene*)	11
Iguana (*Iguana*)	10
Gopher snake (*Pituophis*)	9
Chuckawalla (*Sauromalus*)	3
Desert tortoise (*Gopherus*)	3

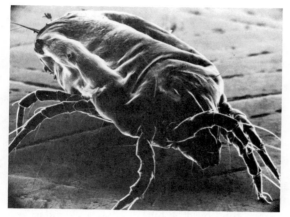

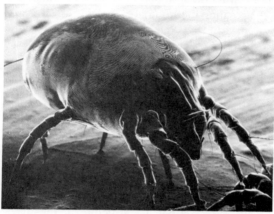

have lungs, are air breathers, and are obviously descended from terrestrial stock.

The skin of a reptile is dry and scaly and has been assumed to be impermeable to water. Let us therefore examine evaporation from the skin in a number of reptiles (Table 9.12). To make a comparison easier, we will use the same units as we used before (Tables 9.8 and 9.9). We can now see that cutaneous evaporation in a dry-habitat reptile is only a small fraction of that in an aquatic reptile

and that even in the aquatic reptile (when kept in air) evaporation is one magnitude lower than in moist-skinned animals such as frogs (see Table 9.8).

We might expect that in a reptile evaporation from the moist respiratory tract would be much greater than evaporation from the dry skin, but this is not so. The contribution made by the skin to the total evaporation always exceeds the respiratory evaporation by a factor of two or more (Figure 9.17). When the water snake *Natrix* is kept in air, the skin contributes nearly 90% of the total evaporation. The relationships are similar for turtles and lizards: The skin remains more important than the respiratory tract in the water loss. Even in the chuckawalla, a desert lizard, two-thirds of the total evaporation is from the skin and only one-third from the respiratory tract.

There is a close correlation between evaporation and habitat. The drier the normal habitat, the lower the rate of evaporation. The total evaporation, combining body surface and respiratory evaporation, from a desert rattlesnake is less than 0.5% of its body weight per day, and the snake could probably survive for 2 or 3 months at this rate. If it remained in an underground burrow or tunnel where the humidity is higher, it could undoubtedly last even longer.

In addition to water lost through evaporation, water is also needed for urine formation. Reptiles

FIGURE 9.17 In reptiles, evaporation from the dry skin exceeds evaporation from the respiratory tract, ranging in these animals from 66 to 88% of the total evaporation. The figure in parentheses next to each animal's name is the total evaporation per day (skin and respiratory tract combined) from that animal in grams water per 100 grams body mass. The total evaporation is habitat-related and is more than 10 times as high in a water snake as in a desert tortoise. All observations in dry air at 23 °C. [Schmidt-Nielsen 1969]

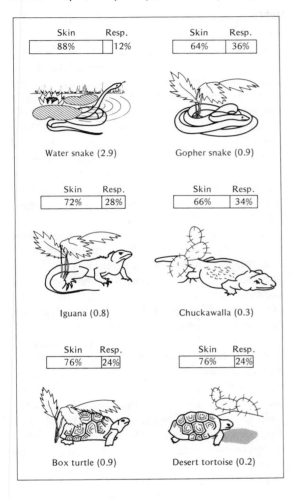

Skin	Resp.
88%	12%

Water snake (2.9)

Skin	Resp.
64%	36%

Gopher snake (0.9)

Skin	Resp.
72%	28%

Iguana (0.8)

Skin	Resp.
66%	34%

Chuckawalla (0.3)

Skin	Resp.
76%	24%

Box turtle (0.9)

Skin	Resp.
76%	24%

Desert tortoise (0.2)

excrete mainly uric acid as an end product of protein metabolism, and as this compound is highly insoluble it requires only small amounts of water for excretion. The relationship between nitrogen excretion and water metabolism will be discussed in greater detail in Chapter 10, which deals with excretion and excretory organs.

Marine reptiles have a special problem because the water they live in and much of their food contain large amounts of salt. How this affects excretion will be discussed later in this chapter.

The gain in water that is necessary to balance the losses is, of course, the same for reptiles as for other animals. There is no indication that reptiles have the ability to obtain water by absorption from the atmosphere, as some insects can. Therefore, when drinking water is unavailable, their total water intake must be derived from food and from oxidation water.

Birds and mammals

Until now we have discussed water balance without considering problems of temperature regulation, but some animals, in particular birds and mammals, use water to keep cool in hot surroundings. Man and some other mammals sweat; dogs and many other mammals and birds pant; and the increased evaporation cools the animal. This aspect of water expenditure was discussed in Chapter 8, and we shall not discuss it further here.

Mammals the size of rodents are convenient for a discussion of the basic aspects of water balance, for they do not pant and they lack skin glands in sufficient numbers to be of importance in heat regulation. This permits us to examine the basic, unavoidable components of their water expenditure – evaporation and the losses in urine and feces – without regard for a large and highly variable amount used for heat regulation. We can then see whether the available sources of water cover the needs.

In most deserts free water is available only on the rare occasions of rain, yet birds and small rodents live in many deserts where rain falls once a year or even less often. These animals must obtain all their water from the food, for dew seems to play a minimal role, at least for higher vertebrates. Many animals obtain more than sufficient water

TABLE 9.13 Overall water metabolism, balancing gains and losses, during a period in which a kangaroo rat consumes and metabolizes 100 g barley (usually about 4 weeks). Air temperature, 25 °C; relative humidity, 20%. [Schmidt-Nielsen 1964]

Water gains	ml	Water losses	ml
Oxidation water	54.0	Urine	13.5
Absorbed water	6.0	Feces	2.6
		Evaporation	43.9
Total water gain	60.0	Total water loss	60.0

from green leaves, stems, fruits, roots, and tubers; carnivorous animals obtain much water from the body fluids of their prey, which may contain from 50 to 80% water.

Some desert rodents depend on such moist food, but others live primarily on dry seeds and other dry plant material, and their intake of free water is minimal. The kangaroo rats and pocket mice which are abundant in the North American deserts are well-known examples of this type. They can live indefinitely on dry food, and yet never drink. An examination of their water balance will therefore explain the mechanisms through which a mammal can live and remain in water balance seemingly without any water intake.

A kangaroo rat is not an exceptionally "dry" animal; it contains as much water as other mammals (about 66%). Even when a kangaroo rat has lived on a diet of only dry barley or oats for weeks or months, its water content remains the same. It maintains its body weight or may even gain in weight. This shows that the animal remains in water balance on the dry food, in other words, that water loss does not exceed water gain.

An account of the overall water metabolism of a kangaroo rat is given in Table 9.13. The account refers to a period during which a kangaroo rat would eat and metabolize 100 g barley. For a kangaroo rat that weighs about 35 g, this amount of food might be consumed in about 1 month. The exact time is of no consequence, however, for the balance would look exactly the same if the animal

had a higher metabolic rate and the same amount of food were consumed in, say, 2 weeks.

On the gain side we find that 54 g water is formed in the oxidation of the food. This figure can be calculated from the composition of the grain. By using the figures for oxidation water formed in the metabolism of starch, protein, and fat (as given in Table 9.10) we arrive at the listed figure. The grain also contains a small amount of free water, the exact amount depending on the humidity of the air. At 20% relative humidity there are 6 g water in 100 g of barley, giving a total amount on the gain side of 60 g. This has to suffice for all the animal's needs.

On the loss side we find that nearly one-quarter of the available water goes for urine formation. In amount, the most important excretory product is urea, formed from the protein in the grain. This excretory product must be eliminated (there is no evidence for the kind of storage exretion described for insects), and the more concentrated the urine, the less water is used. The kangaroo rat has a remarkable renal concentrating ability, far better than most nondesert mammals (see Table 10.2). If we know the amount of protein in 100 g barley, we can calculate how much urea must be excreted and, in turn, how much water is needed. The amount of water lost in the feces can be determined by collecting the feces and measuring their water content.

The most important avenue for water loss is evaporation, occurring mostly from the respiratory tract. In Table 9.13, water gains and losses just balance. If the atmospheric humidity is lower, the water gain is slightly reduced because less free water is absorbed in the grain, and at the same time, in the drier air more water is evaporated from the respiratory tract. In a very dry atmosphere, therefore, the kangaroo rat cannot maintain water balance. In more humid air, above 20% relative

KANGAROO RAT This rodent (*Dipodomys spectabilis*) is common in the deserts of North America. It does not drink and subsists mostly on seeds and other dry plant material. Through economical use of water for urine formation and evaporation from the respiratory tract, it manages, using the water formed in metabolic oxidation processes as its main water source. [K. Schmidt-Nielsen, Duke University]

humidity, less water evaporates from the respiratory tract because the inhaled air contains more water and the kangaroo rat now readily maintains water balance. In nature, kangaroo rats spend much time in their underground burrows where the air humidity is somewhat higher than in the outside desert atmosphere, and this aids in their water balance.

The amount of evaporation from the respiratory tract depends on how much air is brought into the lungs (the ventilation volume) and on the fact that the exhaled air is always saturated with water vapor. The amount of water already present in inhaled air determines how much additional water is needed to saturate the air, for there is no evidence that any mammal exhales air at less than full saturation.

Ventilation volume is determined by the rate of oxygen consumption (which in turn determines how much oxidation water is available). Mammals in general remove about 5 ml O_2 from 100 ml alveolar air before it is exhaled. If the oxygen extraction could be increased, the volume of respired air could be reduced (for a given oxygen consumption), and the amount of water evaporated would be correspondingly reduced. This avenue for reducing the respiratory evaporation apparently has not been used by any mammal. To extract more oxygen in the lungs the hemoglobin of the blood would have to have a higher affinity for oxygen, but kangaroo rat blood does not differ from the blood of other rodents in this regard. Increased oxygen extraction would also lead to an increase in carbon dioxide in the blood, which would be reflected in the acid–base base balance, but this system shows no such deviation from the usual mammalian pattern.

However, an important reduction in respiratory evaporation is achieved by exhalation of air at a lower temperature than the body core. Although the lung air is at core temperature and saturated, it is cooled as it passes out through the nose. The mechanism is simple (Figure 9.18): During inhalation the walls of the passages lose heat to the air flowing over them; the wall temperature decreases and because of evaporation may fall below the

FIGURE 9.18 Model of heat exchange in the nasal passages. Ambient air is 28 °C and saturated; body temperature is 38 °C. As inhaled air flows through the passages (left), it gains heat and water vapor and is saturated and at 38 °C before it reaches the lungs. On exhalation (right), the air flows over the cool walls and gives up heat, and water recondenses. As heat and water exchange approaches completion, exhaled air approaches 28 °C, saturated. [Schmidt-Nielsen 1972]

FIGURE 9.19 Temperature of exhaled air in seven species of birds, measured in dry air. The body temperatures of the birds were between 40 and 41 °C. Two of the birds, cactus wren and budgerigar (parakeet), are desert species, but there is no difference in the extent of cooling (and thus water recovery) between these and birds from other habitats. The less effective cooling of exhaled air in the duck comes with the larger size of its respiratory passages. [Schmidt-Nielsen et al. 1970]

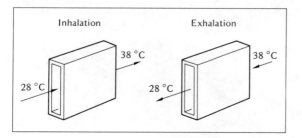

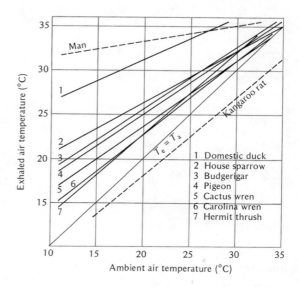

temperature of the inhaled air. On exhalation, as warm air from the lungs passes over the cool surfaces, the air is cooled and water condenses on the walls. The extent of cooling (i.e, the final temperature of the exhaled air) varies with the temperature and humidity of the inhaled air.

How important is the cooling for the water balance? That varies, of course, with the conditions of the air the kangaroo rat breathes, but let us use a reasonable example with the air at 30 °C and 25% relative humidity. This air when inhaled into the lungs becomes heated to body temperature (38 °C), and is saturated with water vapor so that it contains 46 mg H_2O per liter. On exhalation, this air is cooled to 27 °C, and although still saturated, it contains only half as much water (i.e., half of the water is retained in the nasal passages and is used to humidify the next inhaled breath). But note that the exhaled air is saturated and always contains more water vapor than inhaled air; in other words, the nose is not a system to wring water out of the air, and there is always a net loss of water from the respiratory tract.

Heat exchange in the nasal passages is not unique for kangaroo rats; it occurs in all animals. It is an inevitable result of heat exchange between air and the nasal walls, and this heat exchange is more complete when the passages are narrow and the surface is large, as in small rodents. In birds, which have shorter and wider nasal passages, the

cooling is less effective (Figure 9.19). Kangaroo rats, in dry air, exhale air at temperatures below the inhaled air; birds, although they do cool the air appreciably, have higher exhaled-air temperatures. Man, with his much wider nasal passages, has a very incomplete heat exchange, and the exhaled air is commonly only a few degrees below body temperature. In man and other large animals, therefore, heat exchange in the nose and the associated water conservation are of lesser importance, and larger amounts of water are lost in the exhaled air.

MARINE AIR-BREATHING VERTEBRATES

The higher vertebrates – reptiles, birds, and mammals – are typically terrestrial, and, as we have seen, some are at home in the most arid deserts of the world. However, several lines of terrestrial vertebrates have secondarily invaded the sea and remained air breathers. In regard to problems of water and salt, they are essentially terrestrial animals and, compared with a fish, are physiologically isolated from the surrounding sea water. In contrast to fish, which have gills that are relatively permeable to water, the higher marine vertebrates have lungs and thus escape the osmotic problem of an intimate contact with the sea water over a large gill surface. The higher marine vertebrates differ physiologically from their terrestrial relatives primarily in that they have only sea water to drink and that much of their food has a high salt content.

The sea contains enough water; the problem is that all this water contains about 35 g salt per liter and has an osmotic concentration of some 1000 mOsm. Much of the food also has a high salt content, although in ionic composition it often differs substantially from sea water. Plants and invertebrate animals, when eaten, present roughly the same osmotic problem as the drinking of sea water. Marine teleost fish, as we have seen, contain much less salt, and animals that subsist primarily on fish have a less severe salt problem than those that feed on plants or invertebrates. In any event, all of them must excrete the end products of protein metabolism, which in mammals is mainly urea and in birds and reptiles is uric acid.

If a vertebrate drinks sea water, the salts are absorbed and the concentration of salt in the body fluids increases. Unless the salts are eliminated with a smaller volume of water than that which was taken in, there can be no net gain of water. In other words, the salts must be excreted in a solution at least as concentrated as sea water, otherwise the body will become more and more dehydrated.

The reptilian kidney cannot produce urine that is more concentrated than the body fluids, and the bird kidney can usually produce urine no more than twice as concentrated as the blood. Because the concentration of their body fluids is about 300 to 400 mOsm per liter, the urine cannot reach the concentration of sea water (1000 mOsm per liter). The kidney therefore does not have a sufficient concentrating ability to permit these animals to drink sea water or eat food with a high salt content, and if they do, they must have other mechanisms for salt excretion.

Marine reptiles

Three orders of reptiles – turtles, lizards, and snakes – have marine representatives. Some of the sea snakes are completely independent of land, even in their reproduction, for they bear live young, never leave the sea, and would in fact be quite helpless on land. Sea turtles spend most of their lives in the open ocean, but they return to sandy tropical beaches for reproduction. Only the female turtles go on land to lay their eggs; the males never set foot on land after they as hatchlings enter the sea.

The marine lizards are more tied to land. An example is the Galápagos marine iguana, *Amblyrhynchus cristatus*. It lives in the surf of the Galápagos Islands, where it climbs on the rocks and feeds exclusively on seaweed. The fourth living order of reptiles, the crocodiles, probably have no truly marine representative. The salt-water crocodile *Crocodylus porosus*, is primarily estuarine in its habits; it subsists mainly on fish and probably cannot survive indefinitely in a truly marine environment.

The excretion of excess salt, which the reptilian kidney is unable to handle, is carried out by glands in the head called *salt-excreting glands* or simply *salt glands*. The salt glands produce a highly concentrated fluid that contains primarily sodium and chloride in concentrations substantially higher than in sea water. The glands do not function continually, as the kidney does; they secrete only intermittently in response to a salt load that increases

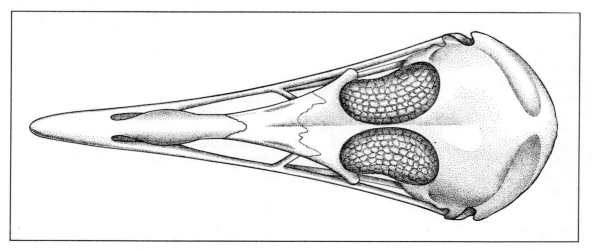

the plasma salt concentrations. Similar salt-secreting glands are found in marine birds, in which they have been studied in greater detail.

In the marine lizard the salt glands empty their secretion into the anterior portion of the nasal cavity, and a ridge keeps the fluid from draining back and being swallowed. A sudden exhalation occasionally forces the liquid as a fine spray of droplets out through the nostrils. The Galápagos iguana eats only seaweed, which has a salt concentration similar to that of sea water. It is therefore a necessity for this animal to have a mechanism for excretion of salts in high concentration (Schmidt-Nielsen and Fänge 1958).

A marine turtle, whether plant eating or carnivorous, has a large salt-excreting gland in the orbit of each eye. The duct from the gland opens into the posterior corner of the orbit, and a turtle that has been salt-loaded cries salty tears. (The tears of man, which everybody knows have a salty taste, are isosmotic with the blood plasma. Human tear glands therefore play no particular role in salt excretion.)

Sea snakes, which also excrete a salty fluid in response to a salt load, have salt glands that open into the oral cavity, from which the secreted liquid is expelled (Dunson 1968). The sea snakes are close relatives of the cobras and are highly venomous, and physiological investigation of their salt metabolism has certain exciting aspects which have tended to slow the study.

Although the marine reptiles have the physiological mechanism necessary to eliminate salt as a highly concentrated fluid, the question of whether many of them normally drink sea water in appreciable quantities remains unresolved.

Marine birds

Many birds are marine, but most of them live on and above, rather than in, the ocean. Many are coastal, but some are truly pelagic. The young albatross that is hatched on a Pacific island spends 3 or 4 years over the open ocean before it returns to the breeding grounds. The penguin, the most highly adapted marine bird, has lost its power of flight. It is an excellent swimmer and is well advanced in the evolution toward a fully aquatic life. Nevertheless, in the physiological sense it has remained essentially a terrestrial air-breathing animal that reproduces on land. The saying that the emperor penguin does not even breed on land is a play on words, for it hatches its egg while standing on the ice during the long, dark Antarctic winter.

All marine birds have paired nasal salt glands that, through a duct, connect with the nasal cavity (Fänge et al. 1958). Nasal glands are found also in most terrestrial birds, although in these they are very small; in marine birds the glands are without exception large. Most frequently they are located on top of the skull, above the orbit of each eye, in shallow depressions in the bone (Figure 9.20). In birds that regularly eat a diet high in salt or receive salt solutions to drink, the glands increase in size and become even larger than normal (Schmidt-Nielsen and Kim 1964).

The salt glands usually remain inactive and start secreting only in response to an osmotic stress (e.g., when sea water or salty food is ingested). Otherwise, the glands remain at rest, and in this respect they differ from the kidney, which produces urine continuously. The secreted fluid has a simple composition; it contains mostly sodium and chloride in rather constant concentrations. Also in this regard the salt gland differs drastically from the kidney, which changes the concentrations and the relative proportions of the secreted components over a very wide range. Urine also contains a variety of organic compounds, whereas the salt gland secretion has no more than a trace amount of non-electrolytes (Schmidt-Nielsen 1960).

Although the salt concentration in the secreted fluid is consistently high, there are characteristic species differences related to the normal ecology of the birds and their feeding habits. For example, the cormorant, which is a coastal bird and a fish eater, secretes a fluid with a relatively low salt con-

TABLE 9.14 Concentration of sodium in the nasal secretion of different species of birds. The chloride concentration in a given sample is nearly identical to the sodium concentration, and other ions are found only in small amounts. [Schmidt-Nielsen 1960]

Species	Sodium concentration (mmol liter^{-1})
Duck, mallard	400–600
Cormorant	500–600
Skimmer, black	550–700
Pelican, brown	600–750
Gull, herring	600–800
Gull, black-backed	700–900
Penguin, Humboldt's	725–850
Guillemot	750–850
Albatross, blackfooted	800–900
Petrel, Leach's	900–1100

tent, about 500 to 600 mmol Na per liter (Table 9.14). The herring gull, which eats more invertebrate food and consequently ingests more salt, has a concentration of 600 to 800 mmol Na per liter. The petrel, a highly oceanic bird that feeds on planktonic crustaceans, has sodium concentrations in the nasal fluid up to 1100 mmol per liter.

The salt gland has an extraordinary capacity to excrete salt. In an experiment in which a gull received nearly one-tenth of its body weight of sea water (which would correspond to 7 liters for a man), the entire salt load was eliminated in about 3 hours. The details of this experiment are recorded in Table 9.15. The bird weighed 1420 g and was given 134 ml sea water by stomach tube. During the following 3 hours the combined volume of excreted liquid, 131.5 ml, equaled that ingested, with less than half (56.3 ml) coming from the salt gland and the remainder (75.2 ml) from the cloaca. The amount of sodium excreted by the nasal gland was, however, about 10 times as high as the amount appearing in the cloacal discharge (urine mixed with some fecal material).

Examination of Table 9.15 shows that the volume of nasal secretion was highest during the second hour, after which it gradually tapered off. The concentration of the nasal secretion was quite con-

TABLE 9.15 Nasal and cloacal excretion by a black-backed gull during 175 minutes following the ingestion of sea water in an amount nearly one-tenth of its body weight. [Schmidt-Nielsen 1960]

Time (min)	Nasal excretion			Cloacal excretion		
	Volume (ml)	Sodium concentration (mmol liter^{-1})	Sodium amount (mmol)	Volume (ml)	Sodium concentration (mmol liter^{-1})	Sodium amount (mmol)
15	2.2	798	1.7	5.8	38	0.28
40	10.9	756	8.2	14.6	71	1.04
70	14.2	780	11.1	25.0	80	2.00
100	16.1	776	12.5	12.5	61	0.76
130	6.8	799	5.4	6.2	33	0.21
160	4.1	800	3.3	7.3	10	0.07
175	2.0	780	1.5	3.8	12	0.05
Total	56.3		43.7	75.2		4.41

stant, in spite of changes in flow rate. The volume of cloacal discharge (mostly urine) exceeded the volume of nasal secretion, but its salt concentration was quite low. The bird kidney can produce urine with a maximum sodium concentration of about 300 mmol per liter, but in this experiment the urine concentration was only a small fraction of this. Also, the urine sodium concentration varied about eight-fold during the experiment, by no means unusual for the kidney, but in sharp contrast to the very constant concentration in the nasal secretion.

The volume of fluid secreted by the salt gland is remarkably high, particularly in view of the exceptional osmotic work performed in producing the concentrated salt solution. The flow from the salt glands, estimated per kilogram body weight, is about twice as high as the urine production of a man in maximum water diuresis (Table 9.16). Calculated per gram gland, the difference is even more striking. One gram of the salt gland of a gull can produce 0.6 ml fluid per minute, whereas the kidney of man in maximum water diuresis produces only one-twentieth as much. For the human kidney, when it produces a concentrated urine (which is still only half as concentrated as the se-

cretion from the gull's salt gland), the flow rate may decline to only 1% of that in water diuresis. Clearly, the salt gland is one of the most effective ion transport systems known.

The fluid secreted from the salt gland of a marine bird always contains mostly sodium and very little potassium. The ratio between these two ions remains at about 30:1, and if the amount of potassium in the diet of a gull is increased, the Na/K ratio does not change much (Schmidt-Nielsen 1965). Sea turtles and sea snakes have similarly high Na/K ratios, but in the marine iguana we find a lower ratio and the relative amount of potassium excreted by the salt gland is higher (Table 9.17). This is easy to understand; the diet of the marine iguana is primarily marine algae, and in general plants contain large quantities of potassium. The marine iguana therefore lives on a diet with a relatively high potassium content, and its need for potassium excretion is correspondingly high.

A typical land lizard from a dry habitat, the false iguana (*Ctenosaura*), has a nasal secretion with a very high potassium concentration and a low sodium concentration, giving a Na/K ratio of 0.15.

Another way in which in the salt gland secretion of a land reptile differs from that of its marine

TABLE 9.16 Secretion rates of the salt gland of the herring gull and of the human kidney during maximum water diuresis.

Secretion rate	Salt gland	Kidney
Per kilogram body weight (ml min⁻¹)	0.5	0.24
Per gram gland (ml min⁻¹)	0.6	0.03

TABLE 9.17 The concentrations of sodium and potassium (in millimoles per liter) in sea water and in nasal salt gland secretion. The Na^+/K^+ ratio is high in the marine species and low in terrestrial plant eaters.

	Na^+	K^+	Na^+/K^+ ratio
Sea water	470	10	47
Herring gull (*Larus*)[a]	718	24	30
Sea turtle (*Lepidochelys*)[b]	713	29	25
Sea snake (*Pelamis*)[b]	607	28	24
Marine iguana (*Amblyrhynchus*)[b]	1434	235	6.7
False iguana (*Ctenosaura*)[c]	78	527	0.15

[a] Schmidt-Nielsen (1960).
[b] Dunson (1969).
[c] Templeton (1967).

counterpart is that the composition of the secreted fluid changes with the nature of the load. If a land reptile is given a sodium load, the secretion changes in favor of this ion; if it is given a potassium load, the secretion contains mainly potassium. In terrestrial reptiles the composition of the secreted fluid is regulated according to need, while the salt glands of marine birds and reptiles appear to be organs highly specialized specifically for the excretion of sodium and chloride.

Marine mammals

Three orders of mammals – seals, whales, and sea cows – are exclusively marine in the sense that they spend practically their entire lives in the sea. Seals return briefly to land to bear and nurse their young, but whales and sea cows (manatee and dugong) even bear their young in water.

The food these mammals eat varies greatly in salt content. Seals and whales are carnivorous and feed on fish, certain large invertebrates, and marine plankton organisms. Those that feed on fish obtain food with a rather low salt content (less than 1%), but with a relatively high protein content. The crab-eating seal of the Antarctic and the walrus (which feeds on clams and other bottom organisms) live on food organisms that are isosmotic with sea water. The baleen whales feed on crustacean plankton organisms with the high salt content characteristic of marine invertebrates. If sea water is incidentally ingested with the food, this adds further to the salt load. Dugongs and manatees are herbivorous and feed on plants that are in

osmotic equilibrium with sea water; their salt intake is therefore high.

Do marine mammals have some physiological mechanism that corresponds to the salt-secreting glands of birds and reptiles? This should not be necessary, for the kidneys of whales and seals can produce urine more concentrated than sea water. The highest chloride concentration reported in whale urine is 820 mmol per liter (Krogh 1939). This is well above the concentration of sea water (about 535 mmol Cl per liter), and a whale that takes food with a high salt content, or even sea water, should be able to eliminate the salts without difficulty.

It is well known that sea water is toxic to man and that a castaway at sea only hastens the dehydration processes if he drinks sea water. Table 9.18 shows the effect of drinking sea water on the water balance of a man and of a whale. A whale can drink 1 liter of sea water and have a net gain of about one-third liter of pure water after the salts are excreted. The kidney of man is less powerful; the maximum urine concentration is below that of sea water, and if he drinks 1 liter of sea water, he inevitably ends up with a net water loss of about one-third liter and is worse off than if he had not

TABLE 9.18 Effect on the water balance of ingesting 1 liter of sea water in a man and in a whale.

	Sea water consumed		Urine produced		
	Volume (ml)	Cl concentration (mmol liter^{-1})	Volume (ml)	Cl concentration (mmol liter^{-1})	Water balance: gain or loss (ml)
Man	1000	535	1350	400	−350
Whale	1000	535	650	820	+350

drunk at all. His dehydration is further aggravated by the large amount of magnesium and sulfate in sea water, which acts as a laxative and causes diarrhea, thus increasing the water loss.

There is still inadequate information about whether seals and whales ingest any appreciable amounts of sea water, either by drinking or incidentally with the feed. There is good evidence, however, that seals do not have to resort to drinking sea water. California sea lions (*Zalophus californianus*) have been kept in captivity and given nothing but fish to eat; even after 45 days without access to water, these animals were perfectly normal and in positive water balance (Pilson 1970). It is interesting to note that if sea lions eat invertebrate animals, such as squid, they probably manage equally well.

Although the blood of the squid is in osmotic equilibrium with sea water and has a high salt content, the osmotically active substances within the cells of the squid are only in part salts, for organic compounds are important tissue constituents and make up a substantial fraction of the total osmotic concentration. Thus, squid and many other invertebrates actually impose a lower salt load than we would expect merely from the osmotic concentrations of their body fluids.

Mammals have in their water balance an item that does not apply to birds and reptiles: The female nurses her young, and large quantities of water are required for production of milk. One way of reducing this loss of water would be to produce a more concentrated milk. It has long been known that seal and whale milk has a very high fat content and a higher protein content than cow's milk (Table 9.19). This has usually been interpreted as necessary for the rapid growth of the young, and particularly as a means of transferring a large amount of fat to be deposited as blubber and serve as insulation.

The high fat content of seal milk can also be viewed in light of the limited water resources of the mother. In the Weddell seal, which has been better studied than other seals, the fat content of the milk gradually increases during the lactation period, while the water content decreases correspondingly. The highest fat content found in Weddell seal milk is 57.9% (nearly twice as much fat as is contained in whipping cream), with a water content in the same sample of only 27.2% (which contrasts to a water content in ordinary lean meat of about 65%). Indeed, seals provide nutrients for their young with a minimal expenditure of water. For each gram of water used, seals transfer nourishment more than 10 times as effectively as land mammals (Table 9.19, last column). In accord with these findings is the observation that female Weddell seals in captivity do not suffer dehydration during the lactation period (Kooyman and Drabek 1968).

In our discussions of osmotic regulation we have over and over again seen how important the organs of excretion are for the maintenance of the relative

TABLE 9.19 Composition and energy value of mammalian milk.

Animal	Composition (g per 100 g milk)				Energy value (kcal per g H₂O)
	H_2O	Fat	Protein	Carbohydrate	
Cow[a]	87.3	3.7	3.3	4.8	0.8
Human[a]	87.6	3.8	1.2	7.0	0.8
Harp seal[b]	45.3	42.7	10.5	0	9.9
Weddell seal					
Mean value[c]	43.6	42.2	(14.1)[d]	—	10.5
Extreme value[c]	27.2	57.9	(19.5)[d]	—	23

[a] Kon and Cowie (1961). [b] Sivertsen (1935). [c] Kooyman and Drabek (1968).
[d] Figures refer to total nonfat solids, but as seal milk is virtually carbohydrate-free, most of it is protein.

constancy, or steady state, of internal concentrations and water content of living organisms. We shall now move on to a more detailed discussion of excretory organs and their function.

REFERENCES

Beadle, L. C. (1943) Osmotic regulation and the faunas of inland waters. *Biol. Rev.* 18:172–183.

Beament, J. W. L. (1958) The effect of temperature on the waterproofing mechanism of an insect. *J. Exp. Biol.* 35:494–519.

Beament, J. W. L. (1959) The waterproofing mechanism of arthropods. 1. The effect of temperature on cuticle permeability in terrestrial insects and ticks. *J. Exp. Biol.* 36:391–422.

Beament, J. W. L. (1961) The water relations of insect cuticle. *Biol. Rev.* 36:281–320.

Bentley, P. J. (1966) Adaptations of amphibia to arid environments. *Science* 152:619–623.

Bentley, P. J. (1971) *Endocrines and Osmoregulation: A Comparative Account of the Regulation of Water and Salt in Vertebrates.* New York: Springer-Verlag. 300 pp.

Blaylock, L. A., Ruibal, R., and Platt-Aloia, K. (1976) Skin structure and wiping behavior of Phyllomedusine frogs. *Copeia* 1976:283–295.

Bliss, D. E. (1966) Water balance in the land crab, *Gecarcinus lateralis,* during the intermolt cycle, *Am. Zool.* 6:197–212.

Burger, J. W., and Hess, W. N. (1960) Function of the rectal gland in the spiny dogfish. *Science* 131:670–671.

Buxton, P. A. (1923) *Animal Life in Deserts: A Study of the Fauna in Relation to the Environment.* London: Arnold. 176 pp.

Conte, F. P., Hootman, S. R., and Harris, P. J. (1972) Neck organ of *Artemia salina* nauplii: A larval salt gland. *J. Comp. Physiol.* 80:239–246.

Croghan, P. C. (1958a) The osmotic and ionic regulation of *Artemia salina* (L.). *J. Exp. Biol.* 35:219–33.

Croghan, P. C. (1958b) The mechanism of osmotic regulation in *Artemia salina* (L.): The physiology of the gut. *J. Exp. Biol.* 35:243–249.

Curran, P. F. (1960) Na, Cl, and water transport by rat ileum *in vitro. J. Gen. Physiol.* 43:1137–1148.

Curran, P. F., and MacIntosh, J. R. (1962) A model system for biological water transport. *Nature, Lond.* 193:347–348.

Diamond, J. M. (1962) The mechanism of water transport by the gall-bladder. *J. Physiol.* 161:503–527.

Diamond, J. M., and Bossert, W. H. (1967) Standing-gradient osmotic flow. *J. Gen. Physiol.* 50:2061–2083.

Dunson, W. A. (1968) Salt gland secretion in the pelagic sea snake *Pelamis. Am. J. Physiol.* 215:1512–1517.

Dunson, W. A. (1969) Electrolyte excretion by the salt gland of the Galápagos marine iguana. *Am. J. Physiol.* 216:995–1002.

Edney, E. B. (1954) Woodlice and the land habitat. *Biol. Rev.* 29:185–219.

Edney, E. B. (1956) The micro-climate in which wood-lice live. *Proc. 10th Int. Congr. Entomol.* 2:709–712.

Edney, E. B. (1957) *The Water Relations of Terrestrial Arthropods.* Cambridge: Cambridge University Press. 109 pp.

Edney, E. B. (1966) Absorption of water vapour from unsaturated air by *Arenivaga* sp. (Polyphagidae, Dictyoptera). *Comp. Biochem. Physiol.* 19:387–408.

Fänge, R., Schmidt-Nielsen, K., and Osaki, H. (1958) The salt gland of the herring gull. *Biol. Bull.* 115:162–171.

Fugelli, K., and Zachariassen, K. E. (1976) The distribution of taurine, gamma-aminobutyric acid, and inorganic ions between plasma and erythrocytes in flounder (*Platichthys flesus*) at different plasma osmolalities. *Comp. Biochem. Physiol.* 55A:173–177.

Galloway, J. N., Likens, G. E., and Edgerton, E. S. (1976) Acid precipitation in the northeastern United States: pH and acidity. *Science 194*:722–724.

Gordon, M. S., Schmidt-Nielsen, K., and Kelly, H. M. (1961) Osmotic regulation in the crab-eating frog (*Rana cancrivora*). *J. Exp. Biol.* 38:659–678.

Gordon, M. S., and Tucker, V. A. (1965) Osmotic regulation in the tadpoles of the crab-eating frog (*Rana cancrivora*). *J. Exp. Biol.* 42:437–445.

Gross, W. J., Lasiewski, R. C., Dennis, M., and Rudy, P., Jr. (1966) Salt and water balance in selected crabs of Madagascar. *Comp. Biochem. Physiol* 17:641–660.

House, C. R. (1963) Osmotic regulation in the brackish water teleost *Blennius pholis*. *J. Exp. Biol.* 40:87–104.

Hutchinson, G. E. (1957) *A Treatise on Limnology*, vol. I, *Geography, Physics, and Chemistry.* New York: J. Wiley. 1015 pp.

Hutchinson, G. E. (1967) *A Treatise on Limnology*, vol. II, *Introduction to Lake Biology and the Limnoplankton.* New York: Wiley. 1115 pp.

Keys, A. B. (1933) The mechanism of adaptation to varying salinity in the common eel and the general problem of osmotic regulation in fishes. *Proc. R. Soc. Lond. B. 112*:184–199.

Keys, A. B., and Willmer, E. N. (1932) "Chloride secreting cells" in the gills of fishes, with special reference to the common eel. *J. Physiol.* 76:368–378.

King, J. R. (1957) Comments on the theory of indirect calorimetry as applied to birds. *Northwest Sci.* 31:155–169.

Koch, H. J. A. (1968) Migration. In *Perspectives in Endocrinology: Hormones in the Lives of Lower Vertebrates* (E. J. W. Barrington and C. B. Jørgensen, eds.), pp. 305–349. London: Academic Press.

Kon, S. K., and Cowie, A. T. (1961) *Milk: The Mammary Gland and Its Secretion*, vol. II. New York: Academic Press. 423 pp.

Kooyman, G. L., and Drabek, C. M. (1968) Observations on milk, blood, and urine constituents of the Weddell seal. *Physiol. Zool.* 41:187–194.

Krogh, A. (1937) Osmotic regulation in fresh water fishes by active absorption of chloride ions. *Z. Vergl. Physiol.* 24:656–666.

Krogh, A. (1939) *Osmotic Regulation in Aquatic Animals.* Cambridge: Cambridge University Press. 242 pp. Reprinted by Dover Publications, New York, 1965.

Leivestad, H., and Muniz, I. P. (1976) Fish kill at low pH in a Norwegian River. *Nature, Lond.* 259:391–392.

Livingstone, D. A. (1963) Chemical composition of rivers and lakes. In *Data of Geochemistry*, 6th ed. (M. Fleischer, ed.). Geological Survey Professional Paper 440, Chapter G, 64 pp. Washington, D.C.: GPO.

Loveridge, J. P. (1970) Observations on nitrogenous excretion and water relations of *Chiromantis xerampelina* (Amphibia, Anura). *Arnoldia* 5:1–6.

Lutz, P. L., and Robertson, J. D. (1971) Osmotic constituents of the coelacanth *Latimeria chalumnae* Smith. *Biol. Bull 141*:553–560.

Machin, J. (1964) The evaporation of water from *Helix aspersa*. L. The nature of the evaporating surface. *J. Exp. Biol.* 41:759–769.

Maetz, J., and Campanini, G. (1966) Potentiels trans-épithéliaux de la branchie d'anguille *in vivo* en eau douce et en eau de mer. *J. Physiol. (Paris)* 58:248.

Maetz, J., Payan, P., and de Renzis, G. (1976) Controversial aspects of ionic uptake in freshwater animals. In *Perspectives in Experimental Biology*, vol. 1, *Zoology* (P. S. Davies, ed.), pp. 77–92. Oxford: Pergamon Press.

Mayer, N. (1969) Adaptation de *Rana esculenta* à des milieux variés: Etude speciale de l'excrétion rénale de l'eau et des électrolytes au cours des changements de milieux. *Comp. Biochem. Physiol.* 29:27–50.

Motais, R., and Maetz, J. (1965) Comparaison des échanges de sodium chez un téléostéen euryhalin (le flet) et un téléostéen sténohalin (le serran) en eau de mer: Importance rélative du tube digestif et de la branchie dans ces échanges. *C. R. Acad. Sci. Paris* 261:532–535.

Noble-Nesbitt, J. (1977) Absorption of water vapour by *Thermobia domestica* and other insects. In *Comparative Physiology: Water, Ions and Fluid Mechanics* (K. Schmidt-Nielsen, L. Bolis, and S. H. P. Maddrell, eds.), pp. 53–66. London: Cambridge University Press.

O'Donnell, M. J. (1977) Site of water vapor absorption in the desert cockroach, *Arenivaga investigata. Proc. Natl. Acad. Sci. U.S.A.* 74:1757–1760.

Pang, P. K. T., Griffith, R. W., and Kahn, N. (1972) Electrolyte regulation in the fresh water stingrays (Potamotrygonidae). *Fed. Proc.* 31:344.

Phillips, J. E., Bradley, T. J., and Maddrell, S. H. P. (1977) Mechanisms of ionic and osmotic regulation in saline-water mosquito larvae. In *Comparative Physiology: Water, Ions and Fluid Mechanics* (K. Schmidt-Nielsen, L. Bolis, and S. H. P. Maddrell, eds.), pp. 151–171. London: Cambridge University Press.

Pickford, G. E., and Grant, F. B. (1967) Serum osmolality in the coelacanth, *Latimeria chalumnae:* Urea retention and ion regulation. *Science* 155:568–570.

Pierce, S. K., Jr., and Greenberg, M. J. (1973) The initiation and control of free amino acid regulation of cell volume in salinity-stressed marine bivalves. *J. Exp. Biol.* 59:435–440.

Pilson, M. E. Q. (1970) Water balance in sea lions. *Physiol. Zool.* 43:257–269.

Potts, W. T. W. (1954) The energetics of osmotic regulation in brackish- and fresh-water animals. *J. Exp. Biol.* 31:618–630.

Potts, W. T. W., and Evans, D. H. (1967) Sodium and chloride balance in the killifish *Fundulus heteroclitus. Biol. Bull.* 133:411–425.

Potts, W. T. W., and Parry, G. (1964) *Osmotic and*

Ionic Regulation in Animals. Oxford: Pergamon Press. 423 pp.

Ramsay, J. A. (1935) Methods of measuring the evaporation of water from animals. *J. Exp. Biol.* 12:355–372.

Ramsay, J. A. (1949) The osmotic relations of the earthworm. *J. Exp. Biol.* 26:46–56.

Robertson, J. D. (1954) The chemical composition of the blood of some aquatic chordates, including members of the Tunicata, Cyclostomata and Osteichthyes. *J. Exp. Biol.* 31:424–442.

Robertson, J. D. (1957) Osmotic and ionic regulation in aquatic invertebrates. In *Recent Advances in Invertebrate Physiology* (B. T. Scheer, ed.), pp. 229–246. Eugene: University of Oregon Press.

Ruibal, R. (1962) The adaptive value of bladder water in the toad, *Bufo cognatus. Physiol. Zool.* 35:218–223.

Schmidt-Nielsen, K. (1960) The salt-secreting gland of marine birds. *Circulation* 21:955–967.

Schmidt-Nielsen, K. (1963) Osmotic regulation in higher vertebrates. *Harvey Lect.* 58:53–95.

Schmidt-Nielsen, K. (1964) *Desert Animals: Physiological Problems of Heat and Water.* Oxford: Clarendon Press. 277 pp. To be reprinted by Dover Publications, New York.

Schmidt-Nielsen, K. (1965) Physiology of salt glands. In *Funktionelle und morphologische Organisation der Zelle: Sekretion und Exkretion,* pp. 269–288. Berlin: Springer-Verlag.

Schmidt-Nielsen, K. (1969) The neglected interface: The biology of water as a liquid gas system. *Q. Rev. Biophys.* 2:283–304.

Schmidt-Nielsen, K. (1972) *How Animals Work.* Cambridge: Cambridge University Press. 114 pp.

Schmidt-Nielsen K., and Fänge, R. (1958) Salt glands in marine reptiles. *Nature, Lond.* 182:783–785.

Schmidt-Nielsen, K., Hainsworth, F. R., and Murrish, D. E. (1970) Countercurrent heat exchange in the respiratory passages: Effect on water and heat balance. *Respir. Physiol.* 9:263–276.

Schmidt-Nielsen, K., and Kim, Y. T. (1964) The effect of salt intake on the size and function of the salt gland of ducks. *Auk.* 81:160–172.

Schmidt-Nielsen, K., and Lee, P. (1962) Kidney function in the crab-eating frog (*Rana cancrivora*). *J. Exp. Biol.* 39:167–177.

Schmidt-Nielsen, K., Taylor, C. R., and Shkolnik, A. (1971) Desert snails: Problems of heat, water and food. *J. Exp. Biol.* 55:385–398.

Shoemaker, V. H., and McClanahan, L. L., Jr. (1975) Evaporative water loss, nitrogen excretion and osmoregulation in Phyllomedusine frogs. *J. Comp. Physiol.* 100:331–345.

Sivertsen, E. (1935) Ueber die chemische Zusammensetzung von Robbenmilch. *Magazin for Naturvidenskaberne* 75:183–185.

Smith, H. (1931) The absorption and excretion of water and salts by the elasmobranch fishes. 1. Fresh water elasmobranchs. *Am. J. Physiol.* 98:279–295.

Smith, H. W. (1959) *From Fish to Philosopher*. Summit, N.J.: CIBA. 304 pp. Reprinted (1961), Garden City, N.Y.; Doubleday.

Sverdrup, H. U., Johnson, M. W., and Fleming, R. H. (1942) *The Oceans: Their Physics, Chemistry, and General Biology.* Englewood Cliffs, N.J.: Prentice-Hall. 1087 pp.

Templeton, J. R. (1967) Nasal salt gland excretion and adjustment to sodium loading in the lizard, *Ctenosaura pectinata. Copeia* 1967:136–140.

Thesleff, S., and Schmidt-Nielsen, K. (1962) Osmotic tolerance of the muscles of the crab-eating frog *Rana cancrivora. J. Cell. Comp. Physiol.* 59:31–34.

Thorson, T. B., Cowan, C. M., and Watson, D. E. (1967) *Potamotrygon* spp.: Elasmobranchs with low urea content. *Science* 158:375–377.

Thorson, T. B., Watson, D. E., and Cowan, C. M. (1966) The status of the freshwater shark of Lake Nicaragua. *Copeia* 1966:385–402.

Tracy, C. R. (1976) A model of the dynamic exchanges of water and energy between a terrestrial amphibian and its environment. *Ecol. Monogr.* 46:293–326.

Ussing, H. H., and Zerahn, K. (1951) Active transport of sodium as the source of electric current in the short-circuited isolated frog skin. *Acta Physiol. Scand.* 23:110–127.

Wall, B. J., and Oschman, J. L. (1970). Water and sol-

ute uptake by rectal pads of *Periplaneta americana. Am. J. Physiol.* 218:1208–1215.

Wigglesworth, V. B. (1945) Transpiration through the cuticle of insects. *J. Exp. Biol.* 21:97–114.

ADDITIONAL READING

Bayly, I. A. E. (1972) Salinity tolerance and osmotic behavior of animals in athalassic saline and marine hypersaline waters. *Annu. Rev. Ecol. Syst.* 3:233–268.

Bentley, P. J. (1971) *Endocrines and Osmoregulation: A Comparative Account of the Regulation of Water and Salt in Vertebrates.* New York: Springer-Verlag. 300 pp.

Bentley, P. J. (1976) Osmoregulation. In *Biology of the Reptilia,* vol. 5, *Physiology A* (C. Gans and W. R. Dawson, eds.), pp. 365–412. London: Academic Press.

Burton, R. F. (1973) The significance of ionic concentrations in the internal media of animals. *Biol. Rev.* 48:195–231.

Dunson, W. A. (1975) Salt and water balance in sea snakes. In *The Biology of Sea Snakes* (W. A. Dunson, ed.), pp. 329–353. Baltimore: University Park Press.

Edney, E. B. (1977) *Water Balance in Land Arthropods.* Berlin: Springer-Verlag. 282 pp.

Gupta, B. L., Moreton, R. B., Oschman, J. L., and Wall, B. J. (eds.) (1977) *Transport of Ions and Water in Animals.* London: Academic Press.

Keynes, R. C. (1971) A discussion on active transport of salts and water in living tissues. *Philos. Trans. R. Soc. Lond.* (*Biol.*) 262:83–342.

Kirschner, L. B., et al. (1970) Refresher course on ionic regulation in organisms. *Am. Zool.* 10:329–436.

Krogh, A. (1965) *Osmotic Regulation in Aquatic Animals.* New York: Dover. 242 pp. Republication of 1st ed. (1939).

Maetz, J., and Bornancin, M. (1975) Biochemical and biophysical aspects of salt excretion by chloride cells in teleosts. *Fortschr. Zool.* 23:322–362.

Peaker, M., and Linzell, J. L. (1975) *Salt Glands in Birds and Reptiles.* Cambridge: Cambridge University Press. 307 pp.

Potts, W. T. W., and Parry, G. (1964) *Osmotic and*

Ionic Regulation in Animals. Oxford: Pergamon Press. 423 pp.

Schmidt-Nielsen, K., Bolis, L., and Maddrell, S. H. P. (eds.) (1978) *Comparative Physiology: Water, Ions and Fluid Mechanics.* Cambridge: Cambridge University Press. 360 pp.

Shoemaker, V. H., and **Nagy, K. A.** (1977) Osmoregulation in amphibians and reptiles. *Annu. Rev. Physiol.* 39:449–471.

Smith, H. W. (1961) *From Fish to Philosopher.* Garden City, N.Y.: Doubleday.

10

Excretion

In the previous chapter we saw that excretory organs have an important role in osmoregulation and in the maintenance of steady-state water and solute concentrations. This chapter will discuss how excretory organs work.

Excretory organs have a number of functions, all related to the maintenance of a constant internal environment in the organism. The maintenance of a constant composition entails one basic requirement: Any material an organism takes in must be balanced by an equal amount removed. This, in turn, requires that the excretory paths must have a variable capacity that can be adjusted to remove judiciously controlled amounts of each of a tremendous variety of different substances. The major functions of excretory systems can be listed as follows:

1. Maintenance of proper concentrations of individual ions (Na^+, K^+, Cl^-, Ca^{2+}, H^+, etc.)
2. Maintenance of proper body volume (water content)
3. Maintenance of osmotic concentrations (follows from 1 and 2)
4. Removal of metabolic end products (e.g. urea, uric acid, etc.)
5. Removal of foreign substances and/or their metabolic products.

The first three functions were repeatedly brought up in Chapter 9. The fourth concerns metabolic excretory products. One major end product of metabolic activity is carbon dioxide, and the bulk of it is removed by the respiratory organs; most other metabolic end products are removed by the excretory organs. These also eliminate a wide variety of foreign substances, either unchanged or after some modification that makes them harmless (*detoxification*) or more easily excreted.

The main role of excretory organs, therefore, is

removing from the body accurately regulated amounts of materials that are in excess, thus helping to maintain a steady state in response to all those influences that tend to impose a change.

There is a large variety of excretory organs, but there are, in principle, only two basic processes responsible for the formation of the excreted fluid: ultrafiltration and active transport.

In *ultrafiltration*, pressure forces a fluid through a semipermeable membrane that withholds protein and similar large molecules but allows water and small molecular solutes to pass (see Appendix E).

Active transport is the movement of solute against its electrochemical gradient by processes requiring the expenditure of metabolic energy. If active transport is directed from the animal into the lumen of the excretory organ or organelle, we call it an *active secretion*. If the active transport is in the opposite direction, from the lumen back into the animal, we speak about an *active reabsorption*.

In those excretory organs where the initial fluid is formed by ultrafiltration, this fluid is modified as it passes through the canals of the excretory organ, certain substances being removed from the filtrate by active reabsorption and others added to the ultrafiltrate by active secretion. Thus active transport is often, if not always, superimposed on the filtration system.

Excretory organs and organelles show a great variety of morphological structure and anatomical location, yet they can be classified into a relatively small number of functional types. Some are generalized, or nonspecialized, excretory organs and, in a general sense, can be regarded as kidneys and their excretory product as urine. Other excretory organs have more specialized roles in that they carry out one particular function. Some examples will help:

Generalized excretory organs
Contractile vacuoles of protozoans
Invertebrate nephridial organs
Malpighian tubules of insects
Vertebrate kidneys

Specialized excretory organs
Gills (crustaceans, fish)
Rectal glands (elasmobranchs)
Salt glands (reptiles, birds)
Liver (vertebrates)

The vertebrate kidney is the best known of the generalized excretory organs. It has been studied extensively, and its function is well understood. The contractile vacuole is only a part of a cell (i.e., an organelle) and is so small that study of the fluid it produces has been extremely difficult; nevertheless, a few facts about its function are known. The Malpighian tubules of insects have been carefully studied in a few species and are quite well understood. The excretory organs of other invertebrates, however, are much less well understood, and we have only approximate ideas about how some of them work. For most of them we do not even know whether a secretory process or ultrafiltration is the initial step in the formation of the excreted fluid.

Several specialized excretory organs were discussed in the preceding chapter on osmotic regulation. However, no mention was made of the vertebrate liver. One of its excretory functions is as follows. Red blood cells that have reached their normal life span (in man ca. 100 days) are sequestered from circulation and broken down; the porphyrin of the hemoglobin molecule is transformed by the liver to compounds known as bile pigments. These are excreted in the bile, discharged into the intestine, and finally eliminated with the feces. Thus, in addition to its many other functions, the liver is an excretory organ specialized for porphyrin excretion. It also plays a major role in the metabolism and detoxification of a wide variety of foreign substances and may have other minor roles in excretion.

ORGANS OF EXCRETION

The function of various excretory organs is easier to discuss if we first classify the organs into groups or types.* The major types and their distribution in the animal kingdom are approximately as follows:

None demonstrated
    Coelenterates
    Echinoderms

Contractile vacuoles
    Protozoans
    Sponges

Nephridial organs
    Protonephridium, closed
        Platyhelminthes
        Aschelminthes
    Metanephridium, open-ended
        Annelids
    Nephridium
        Molluscs

Antennal gland (green gland)
    Crustaceans

Malpighian tubules
    Insects

Kidneys
    Vertebrates

---

* This functional grouping of the renal organs is widely used, but is not in accord with the morphological grouping of excretory organs of Goodrich (1945), whose classification is based on whether the excretory organ is derived from a tube that develops from the outside and inward (a *nephridium*) or from the inside

No specific excretory organs have been identified in coelenterates and echinoderms. This is curious, for fresh-water coelenterates are distinctly hypertonic to the medium, and they undoubtedly gain water by osmotic influx. How the excess water is eliminated is not known. Echinoderms, on the other hand, have no problem of osmoregulation, for echinoderms do not occur in fresh water and the marine forms are always isosmotic with sea water.

Contractile vacuoles

Two animal groups have contractile vacuoles: protozoans and sponges. It seems that all fresh-water protozoans have contractile vacuoles. Whether contractile vacuoles are present in all marine forms is more uncertain, but they have been demonstrated in at least some marine ciliates. The occurrence of contractile vacuoles in fresh-water sponges, although previously doubted, has been confirmed beyond doubt (Jepps 1947).

Because a fresh-water form is always hyperosmotic to the medium in which it lives, and its surface is permeable to water, it must continually bail out the water. It must not only eliminate excess water, but also replace lost solutes, presumably by active uptake of salts from the external medium. Estimates of the water permeability of the large ameba *Chaos chaos* indicate that the calculated osmotic influx of water is in good agreement with the observed volume of fluid eliminated by the contractile vacuole. This confirms the widely

---

(the coelom) and outward (a *coelomoduct*). The renal organs of molluscs, arthropods, and vertebrates are derived from coelomoducts; those of other invertebrates (except protozoans, sponges, coelenterates, and echinoderms) are nephridia. Goodrich also recognizes a complex variety of mixed renal organs. The renal organs of insects, the *Malpighian tubules*, arise from the posterior part of the gut and do not fit inoto the Goodrich classification.

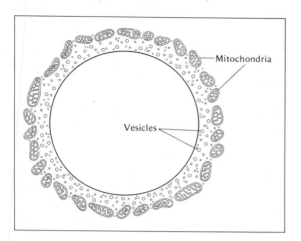

FIGURE 10.1 The contractile vacuole of *Amoeba proteus* is enclosed in a membrane and surrounded by a layer of tiny vesicles that are filled with fluid and appear to empty into the vacuole. Around this structure is a layer of mitochondria, which presumably provide the energy for the secretory process. [Mercer 1959]

Mitochondria

Vesicles

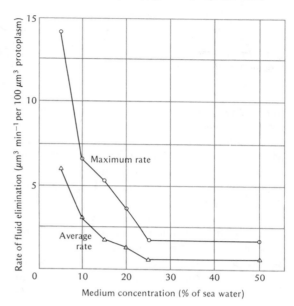

FIGURE 10.2 Rate of fluid elimination by the contractile vacuole of *Amoeba lacerata* in relation to the concentration of the medium. The amebas were tested in the solution in which they were grown. [Hopkins 1946]

Maximum rate

Average rate

Rate of fluid elimination (μm^3 min^{-1} per 100 μm^3 protoplasm)

Medium concentration (% of sea water)

held opinion that the primary function of the contractile vacuole is in osmotic and volume regulation (Løvtrup and Pigon 1951).

Microscopic observation of the contractile vacuole in fresh-water protozoans reveals continuous cyclic changes. It collects fluid and gradually increases in volume until it reaches a critical size. It then suddenly expels its contents to the outside and decreases in size, whereupon it begins to enlarge again, and the cycle is repeated.

The lumen of the contractile vacuole of an ameba is surrounded by a single thin membrane. Surrounding this membrane is a thick layer (0.5 to 2 μm thick) densely packed with small vesicles, each some 0.02 to 0.2 μm in diameter. Around this layer of small vesicles is a layer of mitochondria, which presumably provide the energy required for the osmotic work of producing the hypotonic contents of the vacuole. In electron micrographs it appears that the small vesicles empty into the contractile vacuole by fusion of their membranes (Figure 10.1).

The role of the contractile vacuole in osmotic regulation has been well demonstrated in the eury-

haline ameba, *Amoeba lacerata*. This ameba is originally a fresh-water organism, but it has a high salt tolerance and can eventually be adapted to 50% sea water. The rate of emptying of the contractile vacuole of this ameba, when it is adapted to various salt concentrations, varies inversely with the osmotic concentration of the medium (Figure 10.2)

Apparently, the contractile vacuole works to eliminate water as fast as it enters by osmotic inflow, for as the concentration of the medium increases, the amount of entering water decreases. In a marine habitat, where the inside and outside osmotic concentrations must be assumed to be nearly equal, contractile vacuoles (in those forms where they have been observed) empty at a very low rate. In these cases we must assume that they are not primarily organelles of osmoregulation, but carry out other excretory functions as well.

If the primary function of the contractile vac-

TABLE 10.1 Solute concentrations in plasma and in the contractile vacuole of a fresh-water ameba. Mean volume of the vacuole was about 0.2 nl. [Riddick 1968]

Concentration	Medium	Cytoplasm	Vacuole	Ratio: vacuole cytoplasm
Osmotic (mOsm liter^{-1})	<2	117	51	<0.49
Na$^+$ (mmol liter^{-1})	0.2	0.60	19.9	33
K$^+$ (mmol liter^{-1})	0.1	31	4.6	0.15

uole of fresh-water protozoans is to remove water, the contents of the vacuole should be hypotonic to the remainder of the cells. This is indeed the case. Minute samples of fluid withdrawn from the contractile vacuole have an osmotic concentration about one-third that of the cytoplasm, yet several times as high as the surrounding medium's (B. Schmidt-Nielsen and Schrauger 1963).

The contractile vacuole can eliminate a hypotonic fluid and serve to remove water. Nevertheless, because the eliminated fluid has a higher osmotic concentration than the medium, solutes are continuously lost and it is necessary to postulate that the ameba can take up the needed solutes, presumably by active transport directly from the medium.

How can the vacuole increase in volume and yet contain a fluid less concentrated than the cytoplasm? This could be explained in several ways. One is active water transport into the vacuole. However, for a number of reasons this is an unlikely hypothesis. Another alternative is that the vacuole originally contains an isotonic fluid from which osmotically active substances are withdrawn before the fluid is discharged to the surface. This proposal is contrary to the observation that the fluid concentration is hypotonic and relatively constant in composition throughout the growth of the vacuole.

Information about the composition of the vacuolar fluid permits us to propose a third mechanism. Table 10.1 shows that the osmotic concentration of the vacuolar fluid is about half that of the cytoplasm, but more than 25 times as high as the surrounding medium's. The sodium concentration in the vacuolar fluid is relatively high – in fact, 33 times as high as the cytoplasmic concentration of sodium. The potassium concentration in the vacuolar fluid, on the other hand, is relatively low, and substantially below the potassium concentration in the cytoplasm. The sum of sodium and potassium in the vacuolar fluid is nearly 25 mmol per liter, and if the anion is chloride, almost the total osmotic concentration of the fluid (51 mOsm per liter) would be accounted for.

The most likely mechanism for formation of the contractile vacuole is as follows. The small vesicles that surround the contractile vacuole are initially filled with a fluid isotonic with the cytoplasm. The vesicles then pump sodium into this fluid by active transport and remove potassium, also by active transport, in such a manner that removal of potassium exceeds sodium accumulation. The membrane of the vesicles must be relatively impermeable to water to permit the formation of a vesicle fluid hypotonic to the cytoplasm. If the hypotonic vesicles now fuse and open into the contractile vacuole, as the electron micrographs indicate, the contractile vacuole becomes a receptacle for fluid produced by the vesicles, the necessary energy for the osmotic work being provided by the layer of mitochondria adjacent to the vesicles. Because sodium is continuously lost through the activity of the contractile vacuole, we must assume it is replaced by active uptake at the cell surface (Riddick 1968).

Invertebrate excretory organs

Protonephridia and metanephridia

True organs of excretion are found only in those animal phyla that show bilateral symmetry. The most common type, widely distributed among invertebrates, is a simple or branching tube that opens to the outside through a pore (nephridial pore). There are two major types: the *protonephridium*, whose internal end is closed and terminates blindly, and the *metanephridium*, which connects to the body cavity through a funnel-like structure called a nephridiostome or nephrostome.

Protonephridia occur mainly in animals that lack a true body cavity (coelom). An animal may have two or more protonephridia, which are often extensively branched. The closed ends terminate in enlarged bulblike structures, each with a hollow lumen in which is found one or several long cilia (Figure 10.3). If there is a single cilium, the terminal cell is called a *solenocyte*; if there are numerous (often several dozen) cilia projecting into the lumen, the structure is called a *flame cell* because the tuft of cilia, as it undulates, has some resemblance to the flickering flame of a candle. No information is available about functional differences between flame cells and solenocytes.

Metanephridia are characteristically unbranched, and the inner ends open through funnels into the coelomic cavity. Metanephridia are found only in animals with a coelom, but the reverse is not true; some animals with a coelom have protonephridia, which otherwise are characteristic of acoelomate and pseudocoelomate animals (Figure 10.4).

The function of the metanephridium (often called simply nephridium) is reasonably clear, primarily thanks to studies by J. A. Ramsay, who suc-

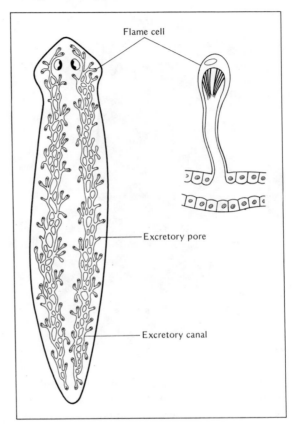

FIGURE 10.3 The excretory system of a planarian is a widely branched system. The excreted fluid is initially formed at the solenocytes or flame cells, and then passed down the nephridial ducts and discharged through excretory pores.

Flame cell

Excretory pore

Excretory canal

ceeded in removing minute samples of fluid from various parts of the earthworm nephridium. The results corroborate the following view. Fluid from the coelom drains into the nephridium through the funnel-shaped nephrostome, and as it passes down through the extensively looped duct, its composition is modified. When originally entering the nephridium the fluid is isotonic, but salt is withdrawn in the terminal parts of the organ, and a dilute urine is discharged (Ramsay 1949). (The im-

FLAME CELL The flame cell of the excretory organ of a larval liver fluke (*Fasciola hepatica*). The photograph (a) shows, in cross section, that the "flame" (F) is a densely packed bundle of cilia. As also indicated in the drawings (b and c), the flame is surrounded by a thin wall (W) of pillarlike rods. The initial urine is presumably formed by ultrafiltration through this wall. [Courtesy of G. Kümmel, Freie Universität Berlin]

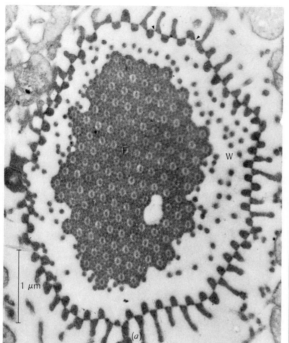

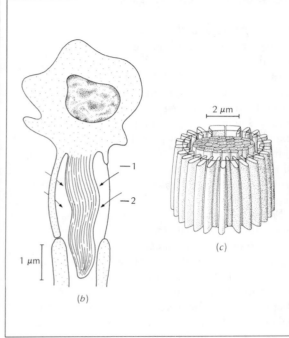

portance of a dilute urine for the osmoregulation of the earthworm was discussed in Chapter 9.) The metanephridium therefore functions as a filtration – reabsorption kidney, in which an initial fluid is formed by ultrafiltration and is modified as it passes through a uriniferous tubule.

How the protonephridium functions is more uncertain, for it has been assumed that the closed end is not a suitable structure for ultrafiltration. Furthermore, the closed end is located in tissue between the body cells, rather than in contact with coelomic fluid, as in the case of metanephridia. The function of solenocytes and flame cells is obscure, although it has been suggested that their constant beating could produce sufficient negative pressure to cause ultrafiltration. Direct evidence

for this hypothesis may be difficult to obtain because of the small size of these structures.

There is evidence, however, that the protonephridium of a rotifer, *Asplanchna*, functions on the basis of filtration and reabsorption. *Asplanchna* has body fluids hypertonic to the medium and produces a dilute urine; if the animal is moved to a more dilute medium, more dilute urine is formed. This shows at least that the protonephridium is involved in osmoregulation and the excretion of water (Braun et al. 1966).

The most important question concerning the protonephridium is whether ultrafiltration takes place. The generally accepted method for demonstrating ultrafiltration is to inject the substance inulin into the body and observe whether it appears

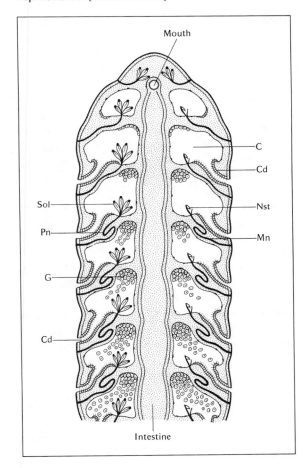

FIGURE 10.4 Diagram of the front end of a primitive annelid in longitudinal section, showing relations of nephridia and coelomoducts to segmental coelomic cavities (C). On left side, protonephridia (Pn) with solenocytes (Sol); on right side, metanephridia (Mn) with funnel-shaped nephrostomes (Nst). The gonadal products (G) empty through the coelomoducts (Cd) at the nephrostome. [Goodrich 1945]

in the urine. Inulin (which will be further discussed in connection with the vertebrate kidney) is a soluble polysaccharide with a molecular weight of about 5000. Inulin is not metabolized in the body, and it can appear in the urine only if filtration takes place. It is never excreted by cellular transport or secretion, for it is inert to all known active transport processes.

When injected into *Asplanchna*, inulin does ap-

pear in the urine, and this suggests a filtration process. Furthermore, the concentration of inulin in the urine of *Asplanchna* is higher than in the body, indicating that water was reabsorbed before the urine was discharged to the exterior. The wall of the flame cell of *Asplanchna* is extremely thin and seems suitable for ultrafiltration (Pontin 1964).

Obviously, other protonephridia might function differently, but since it has been shown in *Asplanchna* that there is an initial filtration, the need to distinguish functionally among the several types of nephridia seems less important.

The molluscan kidney

The major groups of molluscs are cephalopods (octopus and squid), bivalves (clams, etc.), and gastropods (snails). Octopus and squid are strictly marine, but clams and snails occur in both sea and fresh water, and snails include terrestrial species as well. Renal function has been studied quite well in representatives of all these groups. In all, an initial fluid is formed by ultrafiltration of the blood. This fluid contains the same solutes as are present in the blood, except the proteins, and in virtually identical concentrations.* The ultrafiltrate therefore contains not only substances to be excreted, but also valuable substances such as glucose and amino acids. It would be very wasteful to discharge the ultrafiltrate to the outside were not these valuable substances reabsorbed before the urinary fluid is eliminated.

In addition to ultrafiltration followed by selective reabsorption, there is also active secretion of certain substances, which are added to the urinary fluid in specific portions of the kidney. Two com-

---

* The small concentration differences between blood and ultrafiltrate are attributable to the Donnan effect of the blood proteins (see Appendix E).

pounds that are actively secreted in a wide variety of kidneys (including vertebrate kidneys), are *para-aminohippuric acid* (often abbreviated *PAH*) and the dye *phenol red* (phenolsulfonphthalein). Phenol red has the advantage that, because of its color, its presence is readily observed and its concentration is easily measured. Phenol red and PAH will be mentioned again in the discussion of the verbetrate kidney because of their importance in demonstrating secretory processes.

One aspect of molluscan excretion seems rather peculiar: The function of the two kidneys is not always identical. For example, the following information concerns the abalone, *Haliotis*. Both kidneys have an initial ultrafiltration, for inulin appears in the urine from both kidneys, and its concentration is the same in blood, in pericardial fluid, and in the final urine from both kidneys. This indicates that the volume of urine is the same as the filtered volume and that water is not reabsorbed to any appreciable extent.

It is reasonable that a marine animal, which is isotonic with sea water and thus has no major problem of water regulation, does not reabsorb water in its kidney. It must, however, regulate its ionic composition and excrete metabolic products, and in this regard the abalone's two kidneys differ in function, for PAH and phenol red are actively secreted primarily by the right kidney, while glucose reabsorption seems to take place primarily in the left kidney (Harrison 1962).

The antennal gland of crustaceans

The renal organ of crustaceans is the *antennal gland* or *green gland*. The paired glands are located in the head. Each consists of an initial sac, a long coiled excretory tubule, and a bladder, and opens into an excretory pore near the base of the antennae. Hence the name, antennal gland.

Urine is formed in the antennal gland by filtration and reabsorption, with tubular secretion added. Ultrafiltration can be demonstrated by injection of inulin, which then appears in the urine. The lobster, a typical marine animal, produces urine with the same inulin concentration as the blood (i.e., the urine/blood concentration ratio is 1.0), which shows that water is not reabsorbed. PAH and phenol red, however, are found in higher concentrations in the urine than in the blood. This shows that they are secreted into the urine, and that, in addition to the filtration—reabsorption process, secretion is involved in urine formation (Burger 1957).

In contrast to the lobster, the common shore crab *Carcinus* reabsorbs water from the ultrafiltrate. *Carcinus* is a good osmoregulator and can penetrate into brackish water. The inulin concentration in its urine may exceed that in the blood by several-fold and this must be the result of water reabsorption from the initial filtrate. The most likely explanation for the water movement is active reabsorption of sodium followed by passive reabsorption of water. The result is a urine with a lower sodium concentration than the blood, and this is of obvious value to an animal that penetrates into a dilute environment where it will have difficulty maintaining high blood concentrations (Riegel and Lockwood 1961).

The fresh-water amphipod *Gammarus pulex* produces urine that is not only very dilute, but may even be more dilute than the medium (Figure 10.5). A close relative, *G. duebeni*, a brackish-water species that can tolerate fresh water, has a urine concentration closer to the blood concentration. This species is unable to form a very dilute urine, and in fresh water it cannot compete successfully with *G. pulex*.

The Malpighian tubules of insects

For many invertebrate excretory organs, the question of whether the initial urine is formed by ultrafiltration is unresolved. For the excretory

FIGURE 10.5 The relation between blood, urine, and medium concentrations in two species of amphipod. Top: the fresh-water species, *Gammarus pulex*, can form a highly dilute urine. Bottom: the closely related brackish water species, *G. duebeni,* is unable to produce urine more dilute than the medium. [Lockwood 1961]

FIGURE 10.6 The excretory system of the bug *Rhodnius*. Only one of the four Malpighian tubules is shown in full. The Malpighian tubules have two distinct parts: upper and lower. They empty into the rectum at its junction with the midgut. [Wigglesworth 1931]

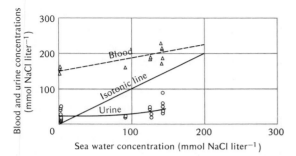

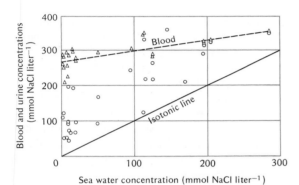

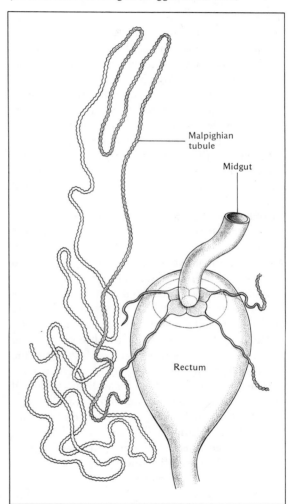

organs of insects, however, the answer seems to be unequivocally negative. This conclusion is supported by the fact that inulin, considered the most reliable indicator of ultrafiltration, does not appear in insect urine, thus indicating the absence of ultrafiltration (Ramsay and Riegel 1961). How, then, is insect urine formed?

The excretory system of insects consists of tubules, known as *Malpighian tubules*, which may number anywhere from two to several hundred. Each tubule opens into the intestine between the midgut and the hindgut; the other end is blind and, in most insects, lies in the hemocele (Figure 10.6). Some insects, however, notably beetles that feed on dry substances (e.g., the mealworm, *Tenebrio*), have a particular arrangement associated with a remarkable ability to withdraw water from the

excrement. In these, the blind end of the tubule lies closely associated with the rectum, the entire structure being surrounded by a membrane (the *perirectal membrane*). The space formed by this membrane is filled with fluid (*perirectal fluid*), which surrounds both the Malpighian tubule and the rectal epithelium, but is separated from the general hemolymph.

The Malpighian tubule functions as follows. Potassium is actively secreted into the lumen of the tubule and water follows passively, due to osmotic forces. As a result a copious amount of a potassium-rich fluid is formed in the tubule, from which it enters the hindgut. In the hindgut solutes and much of the water are reabsorbed, and uric acid (which entered the fluid as water-soluble potassium urate) is precipitated. This facilitates the further withdrawal of water, for the precipitated uric acid does not contribute to the osmotic activity of the rectal contents. What remains in the rectum is eventually deposited as mixed urine and feces.

An insect that lives on fresh vegetation eats food with a high water content and excretes copious amounts of liquid urine. An insect that lives on dry food can produce very dry excreta and therefore may lose virtually no water in feces and urine. The mealworm, the larva of the flour beetle (*Tenebrio molitor*), is a good example. It spends its entire life cycle in dry flour and yet has all the water it needs.

The mealworm has the characteristic perirectal membrane described above, and it can produce excreta so dry that they absorb water from air with 90% relative humidity (Ramsay 1964). Let us see how the mechanism of water withdrawal can be explained. The osmotic concentration of the fluid in the perirectal space can be very high, its freezing point depression (ΔT) reaching as much as 8 °C. This is very much greater than the concentration in the hemolymph ($\Delta T = 0.7$ to 1.4 °C). The difference is more pronounced if the mealworm has been kept at particularly low humidity.

The rectal complex of the mealworm could function in either of two ways: (1) active transport of water from the lumen of the rectal complex into the hemolymph, or (2) active transport of a solute

(presumably potassium chloride) from the hemolymph into the perirectal space, a high concentration in this space in turn accounting for the osmotic withdrawal of water from the rectal lumen (Grimstone et. al 1968).

The second of these alternatives for water reabsorption is, in principle, the same as Curran's three-compartment theory for fluid transport (Chapter 9). The study of cockroaches has given further support for this hypothesis. When cockroaches are provided with water, they excrete a dilute urine; if deprived of water, they produce dry fecal pellets and their rectal contents are highly hyperosmotic to the hemolymph. A detailed analysis of osmotic concentrations in the rectal complex suggested the mechanism for water reabsorption shown in diagrammatic form in Figure 10.7.

In conclusion, it appears that the excretory system of insects operates without any initial ultrafiltration, that it is based on a primary secretion of potassium into the Malpighian tubules, followed by passive movement of water, and that water as well as solutes are withdrawn in the hindgut and the rectal complex. There is ample evidence that the movement of water is based on a primary solute transport.

Vertebrate kidneys

The kidneys of all vertebrates – fish, amphibians, reptiles, birds, and mammals – are similar in that they function on the *filtration–reabsorption principle* with tubular secretion added. A few teleost fish differ from this general pattern; they lack the ultrafiltration mechanism and depend entirely on a secretory-type kidney.

What are the advantages and disadvantages of a filtration mechanism? The initial ultrafiltrate contains all the compounds present in the blood, except substances of large molecular size, such as proteins. Many of the filtered compounds are valu-

FIGURE 10.7 Model of fluid movements in the rectal wall of the roach *Periplaneta americana*. Movement of solute is shown by black arrows; of water by open arrows. They are shown separately for reasons of clarity, although both occur in the same intracellular sinus. Solute is pumped actively into the intracellular sinus, both from the rectal lumen and from the fluid flowing into the subepithelial sinus. The high osmotic concentration in the intracellular sinus causes water to enter osmotically, and as a result there is a bulk flow of water and solutes toward the subepithelial sinus. The system is similar to Curran's three-compartment model (Figure 9.14). [Oschman and Wall 1969]

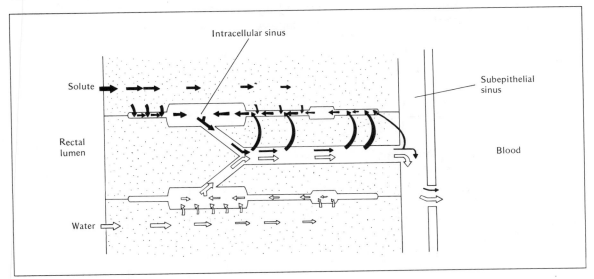

able and should not be lost. Reabsorption mechanisms must therefore be present so that compounds such as glucose, amino acids, and vitamins are conserved.

A filtration–reabsorption kidney can process large volumes of fluid, and often more than 99% of the filtered volume is reabsorbed and less than 1% is excreted as urine. It might seem more advantageous to design a kidney that functions by tubular secretion only, for this would save a great deal of the work of reabsorption. We know that this solution in the evolutionary sense is feasible, for some teleost fish manage this way, and, as we have seen, the insect kidney also lacks ultrafiltration.

There is one unique consequence of ultrafiltration: Any substance that has been filtered remains in the urine unless it is reabsorbed. An organism that encounters "new" substances to be excreted can, with the aid of a filtration kidney, eliminate these without having to develop a specialized secretory mechanism for each particular new substance it may meet. This gives an organism much greater freedom in exploring new environments, changing food habits, and so on. A secretory kidney is in this sense much more restrictive. It is in accord with this viewpoint that, among vertebrates, secretory-type kidneys are found primarily in a few marine fish that live in a stable and "conservative" environment.

All vertebrates can produce urine that is isotonic with, or hypotonic to, the blood, but only birds and mammals can produce urine more concentrated than the body fluids. In fresh water a dilute urine serves to bail out excess water while withholding solutes. Animals in sea water cannot use a dilute or isotonic urine to eliminate excess salt, and they have accessory organs for salt excretion (e.g., gills, rectal glands, salt glands). Marine mammals have kidneys with exceptional concentrating ability, and they manage their salt problems by renal excretion. For land mammals the ability to produce a concentrated urine is of great impor-

FIGURE 10.8 Schematic diagram of a mammalian kidney. The kidney contains a large number, up to several million, of single nephrons. Only one nephron is indicated in this diagram and is shown enlarged to the right. The outer layer of the kidney, the *cortex,* contains the *Malpighian bodies* and the proximal and the distal convoluted tubules. The capillary network within the Malpighian body is known as the *glomerulus.* The inner portion, the *medulla,* contains *Henle's loops* and *collecting ducts.*

The urine is initially formed by ultrafiltration in the Malpighian bodies, and the filtered fluid is modified and greatly reduced in volume as it passes down the renal tubule and into the collecting ducts, which empty the urine into the renal *pelvis,* from where it is conveyed via the *ureter* to the bladder.

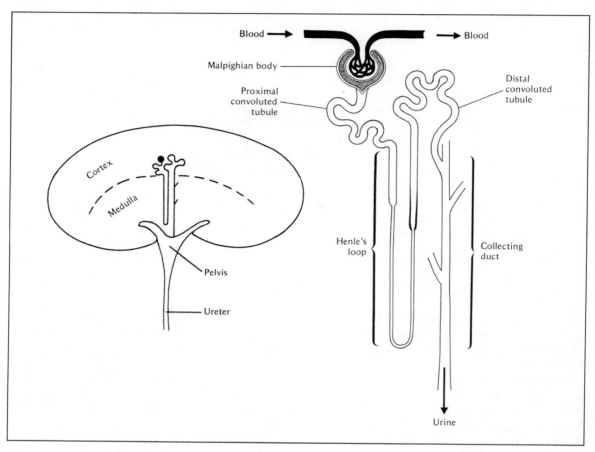

tance in their water balance. The fact that birds and reptiles excrete uric acid (rather than urea) enables them to produce a semisolid urine that puts only moderate demands on the use of water for excretion (see under "Nitrogen excretion," later in this chapter).˙

Structure

To understand how the kidney functions, we must know its structure (Figure 10.8). All ver-
tebrate kidneys consist of a large number of units, *nephrons.* A small fish may have only a few dozen nephrons in its kidneys; a large mammal may have several million in each kidney. Each nephron begins with a *Malpighian body,* in which ultrafiltration of the blood plasma takes place. A small artery leads into each Malpighian body, where it splits up into a bundle or tuft of capillaries, the *glomerulus.* It is through the walls of the glomerular capillaries that fluid is forced out by the blood

pressure; this fluid, the glomerular filtrate, enters the tubule that leads from the Malpighian body. Within the tubule the fluid is modified, both by tubular reabsorption and by tubular secretion, to form the final urine.

The tubule can be divided into two parts: the first or *proximal tubule* in which many solutes such as salt and glucose as well as water are reabsorbed, and a *distal tubule* that continues the process of changing the tubular fluid into urine. The distal tubules join to form collecting ducts, and these empty the urine into the renal pelvis. From here, the urine passes through the ureters to the urinary bladder, from which, at intervals, it is discharged to the exterior. In the nephron of mammals the proximal and distal segments of the tubule are separated by a characteristic thin segment that forms a loop, like a hairpin, known as *Henle's loop*. A similar structure, although not as well developed, is found in the bird kidney. The loop is the particular structure in the nephron that is responsible for the formation of a urine which is more concentrated than blood plasma. The loop is missing in fish, amphibians, and reptiles, and these can produce urine no more concentrated than the blood plasma.

How renal function is studied

The methods used to study renal function can give a surprising amount of information without any direct experimentation on the kidney itself. From the intact animal, we can obtain information about the amount of fluid formed by ultrafiltration, the amount of fluid reabsorbed in the renal tubule, the process of tubular secretion, and even the rate of blood flow to the kidney. The methods are routinely used, without undue hazard or discomfort, to study kidney function in man. We should therefore understand the principles involved in such renal studies.

Much information about renal function has also been obtained by direct experimentation on the kidney. This is especially true of studies of the mechanism responsible for formation of a concentrated urine. Through so-called micropuncture methods, in which fine pipettes are inserted into various segments of the nephron to withdraw minute samples of fluid, information has been gathered about the function of various parts of the nephron. Such methods require extensive surgery and are not suitable for use on man.

Filtration. Ultrafiltration, which takes place in the Malpighian body, is caused by the blood pressure, which forces fluid out through the thin walls of the capillaries. Because the wall does not permit proteins to pass, the plasma proteins are withheld, and only substances of low molecular weight are filtered with the water.

For ultrafiltration to take place, the blood pressure must exceed the osmotic pressure of the blood proteins (the colloidal osmotic pressure). If the artery to a dog kidney is gradually constricted with a clamp so that the blood pressure in the kidney is decreased, urine production ceases when the blood pressure in the kidney falls to the colloidal osmotic pressure. This was shown many years ago by Starling. When the clamp is gradually loosened again so the blood pressure in the capillaries is increased, filtration begins again when the blood pressure just exceeds the colloidal osmotic pressure, and the higher the excess pressure, the higher the filtration rate (Starling 1899).

The Malpighian bodies of the mammalian kidney are located deep within the organ, but in the amphibian kidney they are visible at the surface. The distinguished American renal physiologist A. N. Richards succeeded in inserting micropipettes into the Malpighian bodies of amphibian kidneys, and he showed that the concentrations of chloride, glucose, urea, and phosphate, as well as the total osmotic pressure in the filtrate, were the same as in the blood, except for the minor differences caused by the Donnan effect of the proteins. These experiments gave the essential proof that the initial urine is indeed an ultrafiltrate of the blood (Richards 1935).

The total amount of fluid formed by ultrafiltration can be determined following injection of the polysaccharide inulin, without exposing the kidney or performing any other drastic manipulation. Inulin has a molecular weight of about 5000 and therefore passes through the capillary wall, which only withholds molecules larger than about 70 000 molecular weight. Inulin is not metabolized and is neither reabsorbed nor secreted by the renal tubule; all the inulin found in the final urine is therefore derived only from the filtration process.

Let us write this fact as a simple equation:

inulin in filtrate = inulin in urine

The amount of inulin in the filtrate equals the volume of filtrate ($\dot{V}_{filtr}$) times the inulin concentration in the filtrate (C_{filtr}); likewise, the amount of inulin in the urine equals the urine volume ($\dot{V}_{urine}$) times the inulin concentration in the urine (C_{urine}). The two amounts are equal, and we therefore have:

$$\dot{V}_{filtr} \times C_{filtr} = \dot{V}_{urine} \times C_{urine}$$

or

$$\dot{V}_{filtr} = \frac{\dot{V}_{urine} \times C_{urine}}{C_{filtr}}$$

The three terms on the right side can readily be determined. $\dot{V}_{urine}$ and C_{urine} are obtained by measuring the urine volume in a given period of time and analyzing the urine for its inulin content. C_{filtr} is equal to the inulin concentration in the blood plasma and can be determined from a blood sample. We thus have all the information necessary to calculate the glomerular filtration rate ($\dot{V}_{filtr}$).

Let us try an example. Say that we measure a urine flow of 1.3 ml per minute in a man and that the inulin concentration in the urine is 2% and the plasma inulin concentration is 0.02%. This gives a glomerular filtration rate of 130 ml per minute, which is a normal filtration rate for man.

We can now see that in order to determine the filtration rate in an intact animal we need only (1) inject a suitable amount of inulin,* (2) collect the urine during a known period of time and measure its volume, and (3) determine the inulin concentration in the urine and in the plasma. It turns out that in man the filtration rate is rather constant, about 130 ml per minute, and variations in urine volume are attributable primarily to variations in the amount of water reabsorbed. In many other animals, particularly in the lower vertebrates, the filtration rate can vary considerably, and in the frog the filtration rate can, without harm, cease completely for long periods.

A filtration rate in man of 130 ml per minute, or 0.13 liter per minute means that in 1 hour 7.8 liters of filtrate is formed. This is roughly twice the total volume of blood plasma in man, and this volume has been subject to removal of excretory products. If glucose, amino acids, and other important substances in the filtrate were lost, the filtration process would soon deplete the organism of these valuable compounds. The saving process is tubular reabsorption.

Tubular reabsorption. One important substance that is reabsorbed in the proximal tubule is glucose, and it can serve as an example to illustrate the general principle of tubular reabsorption.

* Usually, inulin is infused continuously throughout an experiment. In this way the plasma concentration of inulin is maintained constant, instead of declining continuously during the observation period.

FIGURE 10.9 Excretion of glucose by the dog. Below a certain limit (2.3 mg glucose per milliliter plasma) no glucose appears in the urine; this is because all the filtered glucose has been removed by tubular reabsorption. Above this limit the reabsorption mechanism is fully saturated, and although it continues to work at full capacity, glucose appears in the urine in amounts that increase with further increases in the plasma glucose concentration. The amount reabsorbed remains constant above the saturation point and thus represents the tubular maximum for glucose reabsorption. [Shannon and Fisher 1938]

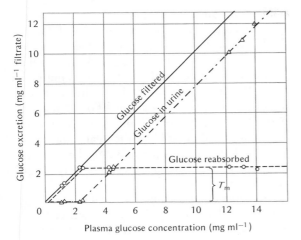

found in the urine. Should the plasma glucose exceed the threshold value, however, reabsorption is incomplete and some glucose remains in the urine. In other words, the maximum capacity of the reabsorption mechanism has been reached. This amount is often designated the *tubular maximum* (T_m) for glucose. Above the threshold value, larger and larger amounts of glucose are found in the urine (Figure 10.9). The line for filtered glucose and the line for glucose in the urine run parallel; this means that the tubular maximum remains constant, as indicated by the constant distance between these two lines.

The tubular maximum for any other reabsorbed substance can be determined in a similar way by increasing the plasma concentration to above the threshold value for that substance, so that the reabsorption mechanism is saturated and the substance begins to appear in the urine.

Tubular secretion. In addition to filtration and reabsorption, a third process, tubular secretion, is important in urine formation. It was mentioned above that phenol red and PAH are secreted by the renal tubules. Because these substances are also filtered, the amount appearing in the urine is the sum of the filtered and the secreted amounts.

If we inject one of these substances into the bloodstream, the total amount secreted will depend on the plasma concentration, as shown in Figure 10.10. The amount filtered increases linearly with the plasma concentration. At low plasma concentrations the amount added by tubular secretion exceeds the filtered amount. However, at a given plasma concentration, the transport mechanism becomes saturated. The total excretion now increases parallel to the filtered amount, and the difference between the total excretion and the filtered amount (the distance between the two parallel lines) gives the amount added by tubular secretion. In other words, this

Reabsorption of glucose is an active transport, and normally all filtered glucose is reabsorbed.

If the amount of glucose in the plasma is increased above the normal level, the reabsorption mechanism is presented with larger than usual amounts of glucose, and if the amount exceeds the capacity of the transport mechanism, not all the glucose is reabsorbed and some remains in the urine. This is what happens to persons suffering from diabetes mellitus; their blood glucose readily increases to higher than normal levels, and when the filtered amount exceeds the capacity for reabsorption, glucose appears in the urine.

Let us examine the situation by reference to Figure 10.9. Glucose is continually filtered, and because the filtration rate remains constant, the filtered amount increases linearly with the plasma concentration of glucose. At low plasma glucose levels, all filtered glucose is reabsorbed, and none appears in the urine. This is the normal situation, for the plasma glucose value usually remains about 1 mg per milliliter. If plasma glucose increases, all filtered glucose is still reabsorbed up to the threshold value, about 2.3 mg per milliliter, and none is

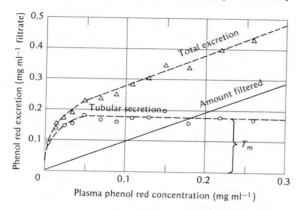

FIGURE 10.10 Excretion of phenol red by the bullfrog. The amount filtered increases in proportion to the plasma concentration. In addition, phenol red is added to the urine by active tubular transport, increasing the amount in the urine. The amount added by tubular secretion remains constant at concentrations above 0.05 mg per milliliter plasma; this indicates the tubular maximum for phenol red has been reached. [Forster 1940]

amount is the tubular maximum for phenol red excretion.

Other substances that are foreign to the body are also eliminated by tubular secretion. Often this is the case with phenolic compounds and their detoxification products. One well-known substance rapidly eliminated by tubular secretion is penicillin, a disadvantage that is counteracted by repeated injections of large amounts.

Renal blood flow. The process of tubular secretion can, in a very interesting way, be used to determine how much blood flows to the kidney (i.e., to determine the renal blood flow). For this we can use a substance that is excreted by tubular secretion so readily that it is completely removed from the blood that passes through the kidney. One such substance is para-aminohippuric acid (PAH), which is removed so completely that none is present in the blood that leaves the kidney (unless its tubular maximum is exceeded). The method depends on the fact that the renal artery is the only blood supply to the mammalian kidney; all the blood first flows through the glomerular capillaries, and then on to the capillaries that surround the renal tubules. We inject some PAH and deter-

mine the amount of PAH that appears in the urine in a given time. From the concentration of PAH in the blood plasma, we can calculate the volume of plasma that has flowed through the kidney in that period. In man the renal plasma flow is normally about 0.7 liter per minute. If the hematocrit is 45%, this corresponds to a blood flow to the kidneys of 1.25 liters per minute. This is between one-quarter and one-fifth of the total cardiac output at rest (5 to 6 liters per minute), showing that the kidneys of mammals receive a surprisingly large blood supply.

The same method cannot be used to determine arterial blood flow to the kidneys of lower vertebrates, because the kidneys in these animals have a different type of circulation and receive a double blood supply. In addition to receiving blood from the renal artery, these kidneys also receive venous blood from the posterior part of the body (called the renal portal system). The importance of the renal portal circulation is not clear. It may be related to the fact that when a renal portal circulation is present urine production can continue even if glomerular filtration ceases. In the frog, for example, filtration may stop, but phenol red that arrives in the kidney via the renal portal system is still secreted by tubular secretory activity.

Excretion in fish

The kidneys of most teleosts consist of nephrons with typical vertebrate characteristics. Fresh-water fish are hypertonic to the medium, and excess water that enters the body due to the osmotic gradient is eliminated as a hypotonic urine.

Marine teleosts, which suffer a water shortage because they are hypotonic to the medium, have a low rate of urine production. They cannot produce a concentrated urine, and excess salts are excreted by the gills. To make up for the osmotic water loss, they drink sea water, and one major function of the kidneys is to excrete the divalent ions, magnesium and sulfate, found in sea water.

Some marine teleosts have kidneys that lack glomeruli. This holds true for the goosefish (*Lophius piscatorius*), the toadfish (*Opsanus tau*), and some pipefish (*Syngnathus*). Aglomerular marine teleosts have been very important in the study of renal function, for they permit the analysis of secretory processes in a kidney where no filtration can take place. If these fish are injected with inulin, no inulin appears in the urine. The absence of inulin from the urine is an excellent functional test for the filtration mechanism.

Among the fish that lack glomeruli, and thus the initial filtration process in urine formation, are several Antarctic genera. For these fish discarding the filtration process makes a great deal of sense. We saw in Chapter 7 that many Antarctic fish avoid freezing because they possess glycoproteins with antifreeze properties. The molecular weight of these glycoproteins ranges from a few thousand up to about 30 000. Because the glomerulus withholds proteins of about 70 000 molecular size, but permits smaller molecules to pass, the glycoproteins would rapidly be lost in a filtration kidney unless a suitable recovery mechanism were present. In any event, a mechanism for reabsorbing the glycoproteins would involve a considerable energy expenditure, which is avoided by never permitting these substances to enter the tubular fluid. It therefore appears that an aglomerular kidney is a great bonus to those fish that use antifreeze substances as a protection against damage from ice formation (Dobbs et al. 1974).

It is peculiar that a few fresh-water pipefish also seem to lack glomeruli, but these fish and their renal function have not been studied in detail. A fresh-water fish should have a considerable os-motic inflow of water, and it is difficult to explain how a nonfiltering kidney can eliminate excess water. Obviously, these fish deserve further study.

The elasmobranchs have glomerular kidneys, and as we saw earlier, urea is essential for elasmobranchs to maintain their osmotic balance. Urea, being a small molecule (molecular weight = 60), is filtered in the glomerulus, but it is reabsorbed by the tubules. This means that there is, in the elasmobranch kidney, a specific mechanism for active transport of urea. Details of this mechanism are discussed later in this chapter.

Excretion in amphibians

The first unequivocal evidence for an initial ultrafiltration process in the kidney was obtained on frogs. This is because many of the glomeruli in the amphibian kidney are rather large and located so close to the surface of the kidney that it is possible to insert a fine micropipette and withdraw minute samples of fluid for analysis. Another advantage of experimentation on cold-blooded animals is that the kidney functions quite well at room temperature, as opposed to the mammalian kidney, which must be elaborately maintained at body temperature.

Most amphibians live at or near fresh water. When they are in water there is an osmotic influx of water, and to eliminate the water they produce a large volume of highly dilute urine. The inevitable loss of sodium is compensated for by cutaneous active uptake of sodium from the dilute medium.

In the frog kidney urea is eliminated, not only by glomerular filtration, but by tubular secretion as well (Marshall 1933). This has certain advantages, for the following reason. When a frog is exposed to dry air, it reduces its urine output (water loss) by reducing the glomerular filtration rate and increasing the tubular reabsorption of water. However, urea excretion may still remain high. In fact, of the total urea excreted by the kidney, as little as one-seventh may be filtered and the remainder added to the urine by tubular secretion (B. Schmidt-Nielsen and Forster 1954). This means that the excretion of urea can continue at a high rate even when glomerular filtration has almost ceased. Thus, in contrast to the situation in the elasmobranch, where urea is actively reabsorbed, in the amphibian kidney urea is actively secreted.

In Chapter 9, on osmoregulation, we saw that one amphibian, the crab-eating frog of Southeast Asia (*Rana cancrivora*), lives and seeks its food in full-strength sea water. Physiologically this frog resembles elasmobranchs in that it carries a high level of urea in the blood, thus achieving osmotic equilibrium with the surrounding sea water and yet retaining the relatively low salt concentrations characteristic of vertebrates in general. However, the crab-eating frog differs from sharks in the way its kidney handles urea.

The kidney of the crab-eating frog does not reabsorb urea, but when the animal is in sea water, very little water enters its body and a very low rate of urine flow combined with a cessation of urea secretion reduces the loss of urea. If the crab-eating frog is transferred to more dilute water, the osmotic influx of water increases, urine volume goes up, and more urea is lost. This reduces the osmotic difference between the frog and the medium, which in turn reduces the osmotic influx of water (Schmidt-Nielsen and Lee 1962).

Excretion in reptiles

The reptilian kidney conforms to the typical vertebrate pattern. It can produce dilute or isotonic urine, but it cannot produce urine more concentrated than the blood plasma. For aquatic freshwater reptiles (e.g., crocodiles and aquatic turtles)

this is adequate, for a dilute urine serves to eliminate excess water.

Marine reptiles have the opposite problem: water shortage and an overabundance of salt. Their kidneys cannot handle excess salts, which are eliminated instead by nasal or orbital salt glands.

Terrestrial reptiles that live in dry habitats and have limited water supplies excrete their nitrogenous waste in the form of uric acid, which is highly insoluble and therefore precipitates in the urine. As a result, urine can be eliminated as a paste or as a semisolid pellet, requiring very little water.

The excretion of uric acid provides an extra dividend. Not only uric acid, but also its sodium and potassium salts, have very low solubilities. By incorporating Na^+ and K^+ in the precipitated fraction of the urine, these important ions can be eliminated without additional expense of water, as is required when they are excreted by the mammalian kidney or by the avian nasal salt gland (McNabb and McNabb 1975).

Many terrestrial reptiles do have nasal salt glands that eliminate excess sodium and potassium, but in desert reptiles that lack salt glands (e.g., *Amphibolurus*) and in terrestrial snakes, the uric acid excretion provides a mechanism for eliminating salts with a minimal loss of water (Minnich 1972).

Urine concentration
in birds and mammals

Uniquely among vertebrates, birds and mammals can produce urine hyperosmotic to the blood plasma. In birds this ability is not very pronounced; the maximum urine osmotic concentration can reach about two times the plasma concentration. The mammalian kidney, in contrast, can produce urine up to perhaps 25 times the plasma

DOG KIDNEY The arterial system of a dog kidney. The photograph is of a preparation made by injecting a silicone rubber compound into the renal artery. The glomeruli can be discerned as numerous tiny, beadlike dots in the outer layer of the kidney, the renal cortex. [Courtesy of A. Clifford Barger, Harvard University]

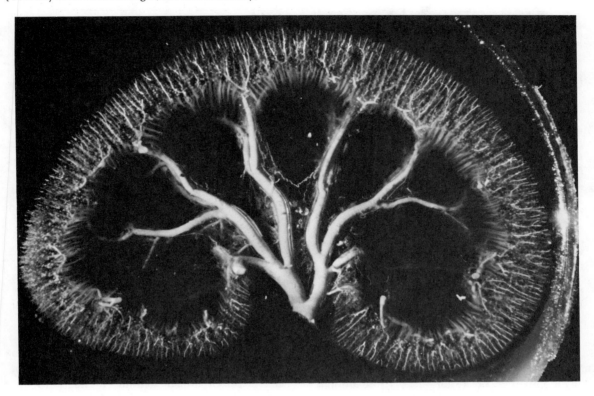

concentration (MacMillen and Lee 1969). The highest urine concentrations are found in mammals from desert habitats; animals such as beavers, which always have a superabundance of water, have only a moderate concentrating ability (Table 10.2).

The mechanism responsible for formation of a concentrated urine in birds and mammals is based on an ingenious geometric arrangement of the renal tubule. It would seem that the production of a concentrated urine would require either reabsorption of water from the tubular fluid or active secretion of solute into the tubule. Neither is the case. The mechanism responsible for forming a concentrated fluid in the renal tubule is based on the reabsorption of sodium and chloride from the tubular fluid, which, paradoxically, in the end serves to produce a highly concentrated urine. The mechanism is found in birds and is developed to the highest perfection in those mammals that form the most concentrated urine.

The concentrating ability of the mammalian kidney is closely related to the length of Henle's loop, the hairpin loop formed by the thin segment interposed between the proximal and the distal tubule. Most mammalian kidneys have two types of

DOG KIDNEY Close-up view of glomeruli from a dog kidney after arterial injection of silicone rubber. The glomerulus in the center of the photo shows the slightly thicker vessel leading into the glomerulus and the somewhat thinner vessel leaving it. The diameter of these vessels is about 15 to 20 μm; the diameter of the glomerulus is about 150 μm. [Courtesy of A. Clifford Barger, Harvard University]

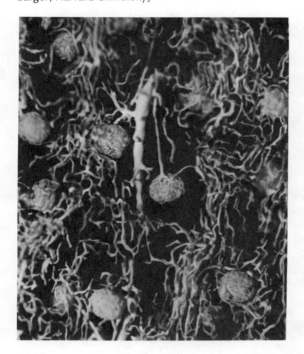

TABLE 10.2 The maximum concentrating ability of the kidney of various mammals is correlated with the normal habitat of the animal, desert animals having the highest urine concentrations and fresh-water animals the lowest.

Animal	Urine maximum osmotic concentration (Osm liter^{-1})	Urine/plasma concentration ratio
Beaver[a]	0.52	2
Pig[a]	1.1	3
Man[b]	1.4	4
White rat[b]	2.9	9
Cat[b]	3.1	10
Kangaroo rat[b]	5.5	14
Sand rat[b]	6.3	17
Hopping mouse[c]	9.4	25

[a] B. Schmidt-Nielsen and O'Dell (1961).
[b] K. Schmidt-Nielsen (1964).
[c] MacMillen and Lee (1967).

nephrons (Figure 10.11). Some nephrons have long loops; others have short ones. Those animals that produce the most highly concentrated urine have only long-looped nephrons with loops so long that they extend into the renal papilla. Others, such as the beaver and the pig, have only short loops. These are the animals that cannot produce urine more than about twice as concentrated as the plasma.

The role of the loop structure in the formation of a concentrated urine has been clarified by the Swiss investigator Wirz and his collaborators, and investigators in Europe and the United States have added further evidence. The hypothesis proposed by Wirz was based on an initial active transport of sodium out of the renal tubule, but recent evidence indicates that the primary pump may be a chloride pump. For the understanding of the concentrating mechanism, this makes little difference: whichever ion is transported actively is followed by the appropriate counter-ion, so we can speak about an initial active transport of NaCl.

The mechanism for the formation of a concentrated urine can best be understood by reference to Figure 10.12 (Kokko and Tisher 1976). The site of active reabsorption of sodium chloride is the ascending limb of the loop of Henle, specifically in the thick segment of the ascending loop. In this segment chloride is transported into the interstitium, followed by sodium as the counter-ion. The next site to examine is the descending loop of Henle, which is permeable to water but impermeable to solutes. As the chloride pump has made the interstitium more concentrated, water is withdrawn from the descending limb of the loop, making its contents more concentrated. The tubular fluid that makes the hairpin turn has thus become more concentrated and is now presented to the pump; more chloride is pumped out; additional water is withdrawn; and so on. In this way the single effect of the ion pump is multiplied as more

FIGURE 10.11 Diagram showing two typical nephrons, one "long-looped" (located close to the border of the medulla and therefore called juxtamedullary) and one "short-looped" (or cortical) nephron. The long-looped nephron is paralleled by a loop formed by the blood capillary. The short-looped nephron is surrounded by a capillary network. Most mammalian kidneys contain a mixture of the two types of nephrons, but some species have only one or the other kind. [Smith 1951]

FIGURE 10.12 Diagram of the concentrating mechanism in the loop of Henle in the mammalian kidney during the formation of concentrated urine. Active transport of chloride ion is indicated by heavy arrows; passive flux of water and urea by light arrows. [Kokko and Tisher 1976]

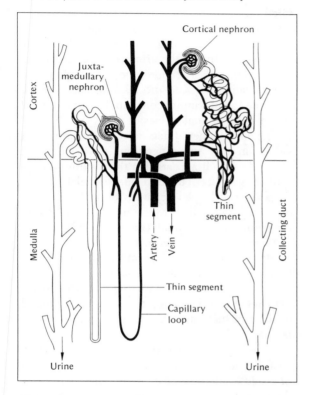

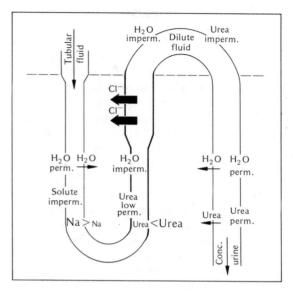

water is withdrawn and more and more concentrated fluid is presented to the ascending limb.

The next question is: What happens to the water? The important structures in this regard are the capillary loops that run parallel to the loop of Henle. These capillaries, known as the *vasa recta*, are freely permeable to water and solutes. Their contents therefore display the same increased solute concentrations toward the hairpin turn as do the tubular fluid and the interstitium. However, the capillary wall is not permeable to protein, and the colloidal osmotic pressure of the plasma proteins therefore causes the withdrawal of water into the bloodstream, as it does in other capillaries.

From there excess water is carried into the venous system. Consider the overall picture: If the kidney is to remove a concentrated fluid (i.e. urine) from the blood, the outgoing venous blood must, of necessity, end up more dilute than the incoming arterial blood.

The next area to consider is the distal portion of the tubule. The fluid that flows from the thick ascending segment into the distal tubule has had sodium chloride removed from it, and its concentration is less than that in the plasma. The sodium chloride that was removed in the thick segment remains around the hairpin loop, where it accumulates in high concentration.

As the dilute tubular fluid enters the collecting duct, it is low in salt but carries urea, which is an important factor in the system. The descending collecting duct is permeable to water and to urea. Because of the high osmotic concentration in the interstitium, water leaves the collecting duct, and as urea becomes more concentrated, it also diffuses out. This urea contributes significantly to the interstitial osmotic concentration. As the tubular fluid that flows in the descending collecting duct achieves osmotic equilibrium with the high concentration of the medullary interstitium, it becomes urine with a high solute concentration. The high concentration in the medulla results from urea derived from the collecting ducts and salt derived from the loops of Henle, and passive diffusion of water out of the collecting duct provides the mechanism for formation of a concentrated urine.

This model for the formation of a concentrated urine depends on highly selective permeabilities of the various segments of the nephron. In spite of the minute size of the tubules, it has been possible to carry out microperfusion of isolated fragments of loops of Henle, and the observed permeabilities are in accord with the requirements of the model (Kokko 1974).

The final question to be answered is: How can a dilute urine be formed? In water diuresis the tubular fluid that enters the collecting ducts from the distal convoluted tubule remains dilute. This results from a decreased permeability to water in the collecting ducts so that water is not withdrawn. The change in permeability to water is controlled by the antidiuretic hormone (ADH). In the absence of ADH the collecting ducts are impermeable to water, and a dilute urine is formed; in its presence the collecting ducts are water-permeable, water is withdrawn, and a concentrated urine is formed.

The fluids in the descending and the ascending limbs flow in opposite directions; the flow is a countercurrent flow. In any one place along the loop the concentration difference between the ascending limb and the descending limb is moderate, but along the length of the loop this moderate concentration difference is additive (i.e., there is a multiplication effect). The entire system is therefore called a *countercurrent multiplier system*. The final concentration that can be reached in this system depends on the length of the loop; the longer the loop, the higher the urine concentrations that can be achieved. Therefore, there is a high degree of correlation between the relative length of the loop and the maximum concentration of the urine in any given animal species.

The bird kidney differs from the mammalian kidney in that some nephrons have loops but others do not. It appears that the maximum concentrating ability of the bird kidney is related to the proportion of looped nephrons to nephrons without loops, rather than to the maximum length of the loops, as in mammals (Poulson 1965).

Many birds live under relatively dry conditions, eat dry food such as seeds, and have limited access to water. It is reasonable to ask why their kidneys cannot match the concentrating ability of mammalian kidneys. Perhaps the answer is to be found in the fact that birds excrete uric acid as an end product of their nitrogen metabolism.

If the bird kidney withdrew water to a much greater extent than it does, the highly insoluble uric acid would precipitate in the tubules and clog them, unless sufficient liquid remained to carry the precipitate along. Although bird urine contains precipitated uric acid, it must remain relatively liquid in the nephrons and in the ureter until it enters the cloaca, which is the very last portion of the intestine, where urine and fecal material accumulate until, at intervals, they are emptied through the single exterior opening. After the urine has entered the cloaca, further water can be withdrawn before the cloacal contents are voided to the exterior.

Role of the cloaca

As the urine of a bird or a reptile enters the cloaca, it is liquid or semiliquid. If an animal needs to conserve water, however, the urine is dropped as a semisolid paste or pellet (i.e., water has been withdrawn). This reabsorption of water may, at least in part, take place outside the cloaca, for it has been shown that urine from the cloaca may migrate into the lower part of the intestine, where both salt and water are reabsorbed (Nechay et al. 1968; Skadhauge 1967, 1976).

What is the nature of the process of water withdrawal from the cloacal contents? It can be a solute-linked water transport, and, at least in lizards, it can be a passive process, resulting from the colloidal osmotic pressure of the plasma proteins. In the lizard cloaca water is absorbed until the force necessary to withdraw more water corresponds to the colloidal osmotic pressure of the plasma proteins; then further water absorption ceases. If the osmotic pressure of the plasma proteins is counterbalanced by placing within the cloaca a protein solution with the same colloidal osmotic concentration as the plasma, water reabsorption is prevented. Thus, it seems that in lizards a passive osmotic removal of water suffices for the production of a semidry urine within the cloaca. Nevertheless, it remains possible that an active transport of sodium or other solutes also takes place in the lizard cloaca, causing solute-linked water withdrawal (Murrish and Schmidt-Nielsen 1970).

When salt is withdrawn from the cloaca together with water, the animal will have difficulty eliminating excess salt. However, lizards (and many birds) can excrete excess salt via the nasal glands. Thus, lizards that live in a dry habitat can use the kidney primarily for the excretion of uric acid and lose a minimal amount of water, and they can use the nasal glands for the excretion of salts in a highly concentrated solution, again with only a small loss of water (Schmidt-Nielsen, et al. 1963).

Regulation of urine concentration and volume

If an animal is in caloric balance, its body weight, and thus volume, remain remarkably constant. A major factor in this constancy is the accurate regulation of the rate of urine production. The total volume of urine depends on the need for water excretion. If there is an excess intake of water, a large volume of dilute urine is excreted; if there is a shortage of water in the organism, the urine volume decreases and its concentration increases toward the maximum possible.

In man, the urine flow may vary more than 100-fold. When a man is deprived of water, urine production may be as low as 10 ml per hour, but it will not decrease further until the situation becomes critical and renal failure sets in. In the opposite case, when the water intake is high, urine flow may exceed 1 liter per hour.

One might think that by drinking more, urine production could be increased further, but this is not possible. The maximum urine volume is determined by the amount of fluid filtered, which, in man, remains quite constant and does not increase when we drink more. In the proximal tubule about 85% of the filtered volume of water is invariably withdrawn, leaving 15% of the filtered water as the maximum volume of urine that can be excreted. The reason for the reabsorption of water in the proximal tubule is that major solutes such as glucose and sodium chloride are reabsorbed here. As solutes are removed, the tubular content becomes hyposmotic, and due to the osmotic effect, water follows passively through the tubular wall. Thus, in the proximal tubule the reabsorption of water is a direct and inevitable result of the reabsorption of glucose and sodium. This is called the *obligatory water reabsorption*.

The distal mechanism (Henle's loop and the collecting duct) is, as described above, the structure responsible for formation of a concentrated urine. The concentrating process is controlled by the *antidiuretic hormone* (ADH), now usually called *vasopressin*. In the absence of this hormone, which is secreted from the posterior hypophysis, a large volume of highly dilute urine is produced. Under the influence of vasopressin, more water is reabsorbed, urine volume decreases, and urine concentration increases.

Vasopressin acts by influencing the permeability of the collecting ducts. In its absence, the permeability of the collecting duct to water is low. Thus, the dilute fluid that enters the collecting duct from the distal tubule undergoes no further loss of water, even if the osmotic concentration in the surrounding tissues is high. With increasing amounts of vasopressin, the permeability of the collecting ducts increases. Thus, because of the high osmotic concentration in the tissue of the papilla (produced by the countercurrent multiplier mechanism), water is osmotically withdrawn from the fluid within the collecting ducts, leaving a small volume of concentrated urine.

This view of the role of the collecting duct has been verified by introducing filament-thin polyethylene tubes through the tip of the papilla in rats and withdrawing samples from various levels of the collecting duct (Hilger et al. 1958). When vasopressin is present in the blood, the collecting duct urine is in osmotic equilibrium with the fluid in the loop, but in the absence of vasopressin such equilibrium does not exist because of the lowered permeability of the collecting duct (Gottschalk 1961).

It may seem paradoxical that a hormone which causes the formation of a concentrated urine acts by increasing the permeability of the collecting ducts. Another paradoxical aspect is that the transport mechanism that leads to the formation of a concentrated urine is based on the withdrawal of sodium and chloride from the tubular fluid, rather than its addition. In the end, however, it is precisely this reabsorption which, in combination with the high permeability of the collecting ducts under the influence of vasopressin, causes the final urine to reach a maximum concentration.

It is interesting that the effect of vasopressin in increasing duct permeability is similar to its effect on frog skin, a much used model in studies of salt transport. In the frog skin the sodium transport takes place in the same direction as the sodium pump in the kidney tubule, by returning sodium to the organism. In both systems vasopressin increases the permeability to water. In the mammalian kidney it has a water-saving effect; in the frog skin the increased permeability leads to an increased osmotic influx of water.

NITROGEN EXCRETION

Most of the food animals eat contains three major nutrient components: carbohydrates, fats, and proteins (plus smaller amounts of nucleic acids). When carbohydrates and fats are metabolized, they give rise to carbon dioxide and water as the only end products of oxidation. Proteins and nucleic acids also yield carbon dioxide and water, but in addition, the chemically bound nitrogen in these foodstuffs gives rise to the formation of some relatively simple nitrogen-containing excretory products. The three compounds of greatest interest are ammonia, urea, and uric acid.

Ammonia	Urea	Uric acid (keto-form)
NH_3	CH_4ON_2	$C_5H_4O_3N_4$

When amino acids are metabolized, the amino group ($-NH_2$) is removed by the process of deamination and forms ammonia (NH_3). The ammonia is excreted unchanged by many animals, in particular by aquatic invertebrates, but by some

TABLE 10.3 Metabolic end products of the major groups of foodstuffs. Ammonia from protein metabolism may be excreted as such or may be synthesized into other N-containing excretory products; purines from nucleic acids may be excreted as such or as any of a number of degradation products, including ammonia.

Foodstuff	End product
Carbohydrate $\rightarrow$	$CO_2 + H_2O$
Fat $\rightarrow$	$CO_2 + H_2O$
Protein $\rightarrow$	$NH_3 \rightleftarrows$ Urea / Uric acid
Nucleic acids $\rightarrow$	Purines + Pyrimidines
	$\downarrow$ $\downarrow$
	Uric acid β-Amino acids
	$\downarrow$ $\downarrow$
	Allantoin NH_3
	$\downarrow$
	Allantoic acid
	$\downarrow$
	Urea
	$\downarrow$
	NH_3

animals it is synthesized to urea and by others to uric acid before being excreted. The nitrogen compounds contained in nucleic acids are either purines or pyrimidines, and they may be excreted as any of the degradation products listed in Table 10.3. The bulk of the excreted nitrogen compounds is derived from protein (amino acids) by deamination, and we shall discuss these first.

Other nitrogen-containing products are also excreted, some derived from the diet, others arising in metabolic pathways. For example, the urine of vertebrates normally contains a small amount of creatinine. This compound is normally formed by the muscles and is of particular interest because it can be used in studies of renal function in a manner similar to inulin. It is formed in the body at a relatively constant rate, and monitoring its excretion allows study of renal function without injection of a foreign substance. Creatinine is apparently derived from the metabolism of the high-energy phosphate compound, phosphocreatine.

Another group of nitrogen compounds frequently found in urine comes from detoxification processes. For example, benzoic acid (present in certain plant foods), when combined with the amino acid glycine, forms the conjugation product hippuric acid (named after the Greek *hippos* = horse, because it was originally discovered in horse urine). Other phenolic compounds are detoxified in a similar manner and are then excreted in the urine. Hippuric acid and the closely related compound, para-aminohippuric acid (PAH), are important in studies of renal physiology because they are readily secreted by tubular activity and can be used to determine renal blood flow, as described earlier in this chapter.

One important aspect of nitrogen metabolism is the assumption that nitrogen gas never enters into the metabolism of animals. It has been universally accepted that only nitrogen-fixing bacteria can assimilate molecular nitrogen (N_2) into chemical compounds, and that only denitrifying bacteria have metabolic pathways that lead to the formation of molecular nitrogen gas.*

Whether the nitrogen from protein metabolism is excreted as ammonia, urea, or uric acid is closely related to the normal habitat of the animal and the availability of water (Table 10.4). Ammonia is a highly toxic compound, even in minute concentrations, and must therefore be removed rapidly, either to the exterior or by synthesis into less toxic compounds (urea or uric acid).

Ammonia

Most aquatic invertebrates excrete ammonia as the end product of protein metabolism. Because of its high solubility and small molecular size, ammonia diffuses extremely rapidly. It can be lost, to a great extent, through any surface in contact with

---

* There is convincing evidence that this is not true under all circumstances. Formation of gaseous nitrogen in mammals on a high-protein diet has been reported by several investigators (e.g., Costa et al. 1968, 1974). These findings are bound to upset some of our present ideas about protein metabolism and nitrogenous end products.

TABLE 10.4 Major nitrogen excretory products in various animal groups.

Animal	Major end product of protein metabolism	Adult habitat	Embryonic environment
Aquatic invertebrates	Ammonia	Aquatic	Aquatic
Teleost fish	Ammonia, some urea	Aquatic	Aquatic
Elasmobranchs	Urea	Aquatic	Aquatic
Crocodiles	Ammonia, some uric acid	Semiaquatic	Cleidoic egg
Amphibians, larval	Ammonia	Aquatic	Aquatic
Amphibians, adult	Urea	Semiaquatic	Aquatic
Mammals	Urea	Terrestrial	Aquatic
Turtles	Urea and uric acid	Terrestrial	Cleidoic egg[a]
Insects	Uric acid	Terrestrial	Cleidoic egg
Land gastropods	Uric acid	Terrestrial	Cleidoic egg
Lizards	Uric acid	Terrestrial	Cleidoic egg
Snakes	Uric acid	Terrestrial	Cleidoic egg
Birds	Uric acid	Terrestrial	Cleidoic egg

[a] The role of cleidoic eggs is discussed later in this chapter.

water and need not be excreted by the kidney. In teleost fish most of the nitrogen is lost as ammonia from the gills. In the carp and goldfish the gills excrete 6 to 10 times as much nitrogen as the kidneys do, and only 10% of this is urea; the remaining 90% is ammonia (Smith 1929).

Urea

Urea is easily soluble in water and has a moderately low toxicity. The synthesis of urea in higher animals has been clarified by the British biochemist Hans Krebs, the same man whose name is attached to the Krebs cycle of oxidative energy metabolism (the tricarboxylic acid cycle).

In the synthesis of urea, ammonia and carbon dioxide are condensed with phosphate to form carbamyl phosphate, which enters a synthetic pathway to form citrulline, as shown in Figure 10.13. A second ammonia is added from the amino acid aspartic acid, leading to the formation of the amino acid arginine. In the presence of the enzyme arginase, arginine is decomposed into urea

and ornithine. This frees the ornithine for renewed synthesis of citrulline, repeating the entire cycle; the total pathway is therefore known as the *ornithine cycle* for urea synthesis. The presence of arginase in an animal shows that the animal has the ability to produce urea and often indicates that urea is the major nitrogenous excretory product. This is not invariably so, however, for arginase can be present with the remaining pathway absent.

Urea in vertebrates

The vertebrates that excrete mainly urea and possess the ornithine cycle enzymes for urea synthesis are indicated in Figure 10.14. Some urea is excreted by teleost fish, and in elasmobranchs, amphibians, and mammals it is the main nitrogenous excretory product. In elasmobranchs (sharks and rays), as well as in the crab-eating frog and in the coelacanth *Latimeria*, urea is retained and serves a major role in osmotic regulation, and is therefore a valuable metabolic product. Urea is filtered in the glomerulus of the elasmobranch kid-

FIGURE 10.13 Urea is synthesized from ammonia and carbon dioxide by condensation with the amino acid ornithine. Through several more steps arginine is formed, which, with the aid of the enzyme arginase, splits off urea, forming ornithine, which can reenter the cycle.

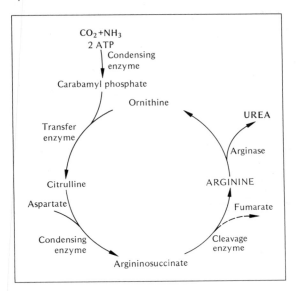

TABLE 10.5 Urea is actively transported by the kidney tubule of both shark (active reabsorption) and frog (active secretion). Three closely related compounds, however, are not transported alike by the two kinds of animals. This provides evidence that the cellular transport mechanism is not identical in the two kidneys. [B. Schmidt-Nielsen and Rabinowitz 1964]

Compound	Shark	Frog
Urea	Active	Active
Methylurea	Active	No
Thiourea	No	Active
Acetamide	Active	No

ney, but, because of its importance in osmoregulation, must not be lost in the urine. Accordingly, in elasmobranchs urea is recovered by active tubular reabsorption. In amphibians the situation is different. Urea is filtered, but in addition a substantial amount is added to the urine by active tubular secretion. Thus, both elasmobranchs and amphibians have active tubular transport of urea, but the transport is in the opposite direction in the two groups. Apparently, the mechanism of the pump is not metabolically identical, for some closely related urea derivatives are treated differently by the two animal groups (Table 10.5). This serves as an excellent example that a physiological function can arise independently in two groups and that it does not necessarily have to use an identical mechanism to achieve the same end (in this case the active transport of urea).

The crab-eating frog, which also retains urea for osmoregulation, does not show evidence of active tubular reabsorption of urea (Schmidt-Nielsen and Lee 1962). The crab-eating frog has a low rate of urine production, and the renal tubules are highly permeable to urea. Urea therefore diffuses from the tubular fluid back into the blood and appears in the urine in approximately the same concentration as in the blood. Thus, only small amounts of urea are lost in the urine.

If ordinary frogs have active tubular secretion of urea, why has the crab-eating frog not utilized this pump by simply reversing its direction? The question is not readily answered, but it appears that in general the direction of an active transport mechanism is a conservative physiological function not readily reversed. As we have seen, the frog skin and the mammalian kidney both retain the direction of active transport of sodium chloride from the exterior into the organism. In the mammalian kidney, however, the inward transport of sodium chloride from the tubular fluid back into the organism has been utilized in the countercurrent multiplier system in such a way that the end result nevertheless is the formation of a concentrated urine.

The common concept of urea excretion in the mammalian kidney is that urea is filtered in the glomerulus and then treated passively by the tubules, although some urea, because of its high diffusibility, reenters the blood by passive diffusion. There is, however, convincing evidence that urea is an important element in the function of the countercurrent multiplier system and that the excretory pattern for urea is an essential element in the function of the mammalian kidney.

FIGURE 10.14 Nitrogen excretion in relation to the phylogeny of vertebrates. The lines enclose groups of animals that produce ammonia, urea, and uric acid, respectively, as the major nitrogenous excretory product. [B. Schmidt-Nielsen 1972]

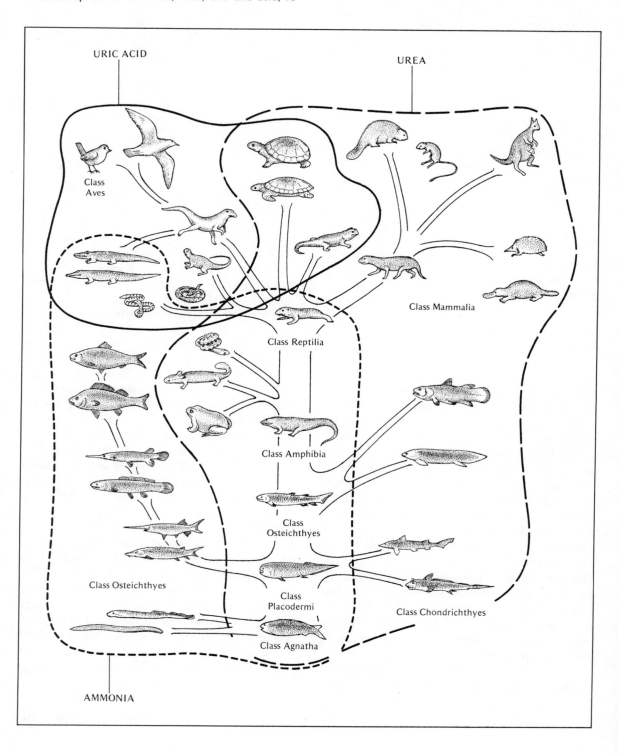

TABLE 10.6 Ammonia excretion by the terrestrial toad *Bufo bufo* and by the fully aquatic frog *Xenopus laevis*. The figures give the excretion of free·ammonia, expressed in percent of the total ammonia and urea excretion at various stages of development. [Munro 1953]

Stage	*Bufo*	*Xenopus*
No hindlimbs		85
Hindlimbs three-fourths developed	80	
Hindlimbs functional	85	83
Forelimbs free, tail remaining	50	81
Tail atrophying	36	
Tailless	20	77
Adult	15	81

Urea and amphibian metamorphosis

The tadpoles of frogs and toads excrete mostly ammonia; the adults excrete urea. At metamorphosis there is a well-defined changeover from ammonia to urea excretion in frog (*Rana temporaria*), toad (*Bufo bufo*), newt (*Triturus vulgaris*), and other amphibians. The South African frog *Xenopus*, however, which remains aquatic during adult life, continues to excrete ammonia as an adult (Table 10.6).

The change to urea excretion at the onset of metamorphosis in the semiterrestrial amphibians is associated with a marked increase in the activities of all the liver enzymes of the ornithine cycle (Brown et al. 1959).

It is interesting that adult specimens of the aquatic *Xenopus*, if kept out of water for several weeks, accumulate urea in the blood and tissues. Accumulation of urea can be induced by keeping the animals in 0.9% NaCl solution. When *Xenopus* adults were kept out of water but in moist moss to prevent dehydration, the blood urea concentration increased between 10- and 20-fold, rising to nearly 100 mmol per liter. When the animals were returned to water, the excess urea was excreted (Balinsky et al. 1961).

A group of *Xenopus* which was found naturally estivating in the mud near a dried pool had urea concentrations raised similarly by a factor of 15 to 20. Of the enzymes responsible for urea synthesis, the enzyme carbamyl phosphate synthetase, responsible for the first synthetic step in Figure 10.13, was increased about sixfold, but the activities of the other enzymes of the cycle were unchanged. The synthesis of carbamyl phosphate may be the rate-limiting step in urea synthesis, and an increase in this enzyme is probably responsible for keeping the plasma ammonia low in animals out of water (Balinsky et al. 1967).

Urea in the lungfish

The changes in the African lungfish, *Protopterus*, are exactly analogous to those in amphibians. Normally, when the lungfish lives in water, it excretes large amounts of ammonia (and some urea), but when it estivates in a cocoon within the dried mud, it channels the entire waste nitrogen into urea, which accumulates in the blood and may reach concentrations as high as 3% (500 mmol per liter) at the end of 3 years of estivation (Smith 1959).

The presence of all five enzymes of the ornithine cycle has been demonstrated in the liver of the African lungfish (Janssens and Cohen 1966). The levels of the two enzymes that are rate-limiting in urea synthesis are similar in the lungfish and in the tadpole of the frog *Rana catesbeiana*, and considerably lower than the levels reported for adult frogs. This is consistent with the predominant excretion of ammonia by the lungfish when it is in water. It has been calculated, however, that the amounts of the ornithine cycle enzymes present in the liver of a nonestivating lungfish are sufficient to account for the accumulation of urea actually observed in estivating lungfish (Forster and Goldstein 1966).

In the Australian lungfish, *Neoceratodus*, the concentrations of the ornithine cycle enzymes are low. This is in accord with the life habits of the Australian lungfish, which uses its lung only as an accessory respiratory organ and cannot survive in air for any length of time (see discussion of lungfish respiration in Chapter 2). The rate of synthesis of urea by liver slices of the Australian lungfish is only one-hundredth of the rate observed in the African lungfish. This again is consistent with the completely aquatic nature of the Australian lungfish (Goldstein et al. 1967).

Uric acid

Excretion of uric acid prevails in insects, land snails, most reptiles, and birds. All these are typical terrestrial animals, and the formation of uric acid can be considered a successful adaptation to water conservation in a terrestrial habitat. Because uric acid and its salts are only slightly soluble in water (the solubility of uric acid is about 6 mg per liter water), the withdrawal of water from the urine causes uric acid and its salts to precipitate.

Uric acid in birds and insects

The semisolid white portion of bird droppings is urine and consists mostly of uric acid; very little water is used for excretion of the nitrogenous excretory products in these animals. Some insects have carried the reduction in urine water loss so far that they do not excrete the uric acid at all, but deposit it in various parts of the body, mainly in the fat body. In these forms, therefore, no water whatsoever is required for the elimination of nitrogenous end products (Kilby 1963).

It has been suggested that birds gain a further advantage by using uric acid as their main excretory product. Because little water is needed for urine formation, uric acid excretion has been said to save weight for flying birds. This argument, however, is not convincing, for birds that have access to water,

both fresh-water and marine species, often eliminate large quantities of liquid urine.

The cleidoic egg

It has been suggested by Joseph Needham that the difference between those higher vertebrates that form urea (mammals and amphibians) and those that form uric acid (reptiles and birds) is primarily correlated with their mode of reproduction. The amphibian egg develops in water, and the mammalian embryo develops in the liquid environment of the uterus, where waste products are transferred to the blood of the mother. The embryonic development in reptiles and birds, on the other hand, takes place in a closed egg, a so-called *cleidoic** egg, where only gases are exchanged with the environment and all excretory products remain within the eggshell. In the cleidoic egg the embryo has a very limited water supply, and ammonia is, of course, too toxic to be tolerated in large quantities. If urea were produced, it would remain inside the egg and accumulate in solution. Uric acid, however, can be precipitated and thus essentially eliminated; this happens when it is deposited as crystals in the allantois, which thus serves as an embryonic urinary bladder.

Uric acid in reptiles

Lizards and snakes excrete mostly uric acid; many turtles excrete a mixture of uric acid and urea; and crocodiles excrete mainly ammonia (Cragg et al. 1961). This fits the generalization that nitrogen excretion is closely related to the availability of water in the environment.

Crocodiles and alligators excrete ammonia in the urine, where the principal cation is NH_4^+ and the principal anion is HCO_3^- (Coulson et al. 1950; Coulson and Hernandez 1955). It is possible

* Greek *kleistos* = closed, from *kleis* = key.

TABLE 10.7 Partition of nitrogen in the urine of turtles (in percent of total nitrogen excretion). The most aquatic species excrete almost no uric acid, whereas this compound dominates in the most terrestrial species. [Moyle 1949]

Species	Habitat	Urine component[a]				
		Uric acid	Ammonia	Urea	Amino acids	Unaccounted for
Kinosternon subrubrum	Almost wholly aquatic	0.7	24.0	22.9	10.0	40.3
Pelusios derbianus	Almost wholly aquatic	4.5	18.5	24.4	20.6	27.2
Emys orbicularis	Semiaquatic; feeds on land in marshes	2.5	14.4	47.1	19.7	14.8
Kinixys erosa	Damp places; frequently enters water	4.2	6.1	61.0	13.7	15.2
K. youngii	Drier than above	5.5	6.0	44.0	15.2	26.4
Testudo denticulata	Damp, swampy ground	6.7	6.0	29.1	15.6	32.1
T. graeca	Very dry, almost desert conditions	51.9	4.1	22.3	6.6	4.0
T. elegans	Very dry, almost desert conditions	56.1	6.2	8.5	13.1	12.0

[a] There were small amounts of allantoin, guanine, xanthine, and creatinine, and a variable amount not accounted for.

that the presence of these ions in the urine permits an improved retention of sodium and chloride in these fresh-water animals, which, incidentally, also lose very little sodium and chloride in the feces.

There is little doubt that a close correlation exists between the habitats of turtles and their nitrogen excretion. Table 10.7 shows the composition of urine samples from eight species of turtles obtained in the London Zoo. The most aquatic species excrete considerable amounts of ammonia and urea and only traces of uric acid; the most terrestrial species excrete over half their nitrogen as uric acid.

There have been conflicting reports about whether turtles excrete mostly urea or uric acid. The reason is that not only do differences exist among species, but also different individuals of one species may excrete mainly uric acid, mainly urea, or a mixture of both (Khalil and Haggag 1955). One individual may even in the course of time change from one compound to the other. The fact that some precipitated uric acid may remain in the cloaca while the liquid portion of the urine is voided to the exterior makes it unreliable

to determine the amount of uric acid formed by analyzing a single or a few urine samples. An incomplete emptying of the cloaca may give an entirely too low amount of uric acid, and an evacuation that includes precipitate accumulated over a period of time will give too large an amount.

The cause of the shift between urea and uric acid in the excretion of the tortoise, *Testudo mauritanica*, seems to be a direct function of temperature and hydration of the animal. Uric acid excretion increases when the water balance is unfavorable, but the mechanism that controls the shift in biochemical activity from urea to uric acid synthesis is not understood.

Two unusual frogs

It was mentioned in Chapter 9 that the African frog, *Chiromantis xerampelina*, loses water from the skin very slowly, at rates similar to those in reptiles. *Chiromantis* resembles reptiles also in that it excretes mainly uric acid, rather than urea as adult amphibians normally do. This is a sensational finding, for it challenges our commonly accepted view of nitrogen excretion in amphibians. There is no doubt about the accuracy of the report, for uric acid was determined in *Chiromantis* urine by means of an enzymatic method specific for uric acid, and uric acid was found to make up 60 to 75% of the dry weight of the urine (Loveridge 1970).

A South American frog, *Phyllomedusa sauvagii*, shows similar reptile-like traits. Its cutaneous water loss is of the same magnitude as from the dry skin of reptiles, and its urine contains large amounts of a semisolid urate precipitate (Shoemaker et al. 1972). The fraction of total nitrogen excreted as urate was found to be 80% in *Phyllomedusa*, and increased water intake did not change the rate of urate production. Thus, this species continues to produce mainly uric acid, even when water is freely available. When the frog is in need of water conservation, the excretion of uric acid, rather than urea, becomes very important. It has been estimated that this frog, if urea were its excretory product, would require about 60 ml water per day per kilogram body weight for urine formation. Instead, *P. sauvagii*, because it excretes uric acid, uses as little as 3.8 ml water per day per kilogram body weight for urine formation (Shoemaker and McClanahan 1975).

Ammonia and renal function

The preceding discussion makes it appear that ammonia is excreted mainly by aquatic animals; but this is not completely true. Ammonia is normally found in the urine of terrestrial animals, where it serves to regulate the pH of the urine. If urine becomes acid due to the excretion of acid metabolic products, ammonia is added to neutralize the excess acid.

Excess acid is normally formed in protein metabolism, for sulfuric acid is the end product of oxidation of the sulfur-containing amino acid cysteine. The more acid the urine is, the more ammonia is added. The ammonia used to neutral-

TABLE 10.8 Nitrogenous end products of purine metabolism in various animals. [Keilin 1959]

Animal	End product
Birds	Uric acid
Reptiles, terrestrial	Uric acid
Insects	Uric acid
Man, apes, dalmatian dog	Uric acid
Mammals (except those above)	Allantoin
Gastropod molluscs	Allantoin
Diptera	Allantoin
Some teleost fish	Allantoic acid
Amphibians	Urea
Teleosts	Urea
Elasmobranchs	Urea
Lamellibranch molluscs (fresh-water)	Urea
Lamellibranch molluscs (marine)	Ammonia
Most other aquatic invertebrates	Ammonia

ize an acid urine is produced in the kidney and is derived from the amino acid glutamine. Glutaminase is found in the kidney, and its presence there serves the particular purpose of producing ammonia. The ammonia in mammalian urine, therefore, has no direct connection with the ammonia produced in the liver by the deamination of amino acids and in this sense should not be considered a normal end product of protein metabolism.

Nucleic acids and nitrogen excretion

Nucleic acids contain two groups of nitrogen compounds: purines (adenine and guanine) and pyrimidines (cytosine and thymine). In some animals purines are excreted as uric acid (which is, itself, a purine); in other animals the purine structure is degraded to a number of intermediaries or to ammonia, any one of which may be excreted.

The metabolic degradation and excretion of purines have not been as carefully studied as the metabolism of protein nitrogen. The main features, however, are listed in Table 10.8. Birds, terrestrial reptiles, and insects degrade purines to uric acid and excrete this compound. These are the same animals that synthesize uric acid from amino nitrogen; it would obviously be meaningless for an animal to synthesize uric acid and also to have mechanisms for its degradation. We therefore cannot expect to find further breakdown of purines in animals in which uric acid is the end product of protein metabolism.

Among mammals, man, apes, and the dalmatian dog form a special group: they excrete uric acid, whereas mammals in general excrete allantoin. Allantoin is formed from uric acid in a single step in the presence of the enzyme uricase. This enzyme is absent in man and apes. Because of its low solubility, uric acid is at times deposited in the human organism, causing swelling of the joints and the very painful disease gout. If man had retained the enzyme uricase, gout would be unknown.

Although the dalmatian dog excretes uric acid in amounts much higher than other dogs, this is not the result of a metabolic defect. The dog's liver contains uricase and some allantoin is formed. However, the dalmatian has a kidney defect that prevents tubular reabsorption of uric acid (as occurs in other mammals, including man); uric acid is therefore lost in the dog's urine faster than it can be converted to allantoin by the liver (Yü et al. 1960). There is considerable evidence that uric acid in the dalmatian dog, in addition to being filtered in the glomerulus, is also excreted by active tubular transport (Keilin 1959).

The structure of the purines adenine and guanine is similar to that of uric acid: each contains one six-membered and one five-membered ring. The pyrimidines (cytosine and thymine), however, are single six-membered rings that contain two nitrogen atoms. In higher vertebrates the pyrimidines are degraded by opening this ring, forming one molecule of ammonia and one molecule of a β-amino acid. These are then metabolized in the normal metabolic scheme by further deamination.

The most striking feature of nucleic acid metabolism is that the "higher" animals listed at the top of Table 10.8 completely lack the enzymes necessary to degrade the purines. Among the "lower" animals we find an increasing complexity in the biochemical and enzymatic systems for the further degradation of the purines. Thus the "lowest" animals in this case possess the most complete enzyme systems.

Other nitrogen compounds

In spiders *guanine* is the major excretory product. It appears to be synthesized from amino nitrogen, although the complete pathway is unknown. Some spiders, notably the bird-eating tarantulas, after a meal excrete more than 90% of the total nitrogen as guanine (Peschen 1939). In the common garden spider, *Epeira diadema*, the identification of guanine has been confirmed by a highly specific enzymatic method (Vajropala 1935).

Guanine is also found rather widely in a variety of other animals. For example, the silvery sheen in fish scales is caused by a deposit of guanine crystals. The garden snail *Helix* excretes guanine, but only to the extent of about 20% of the total purine excretion, the remainder being uric acid. It is possible that this fraction is derived from nucleic acid metabolism, while the remainder of the nitrogen is derived from protein metabolism.

Amino acids are not major end products of nitrogen metabolism, but they are found in small amounts in the urine of many animals. It would seem more economical for an animal to deaminate the amino acid, excrete the ammonia in the usual way, and use the resulting organic acid in energy metabolism. Because of the minor role excretion

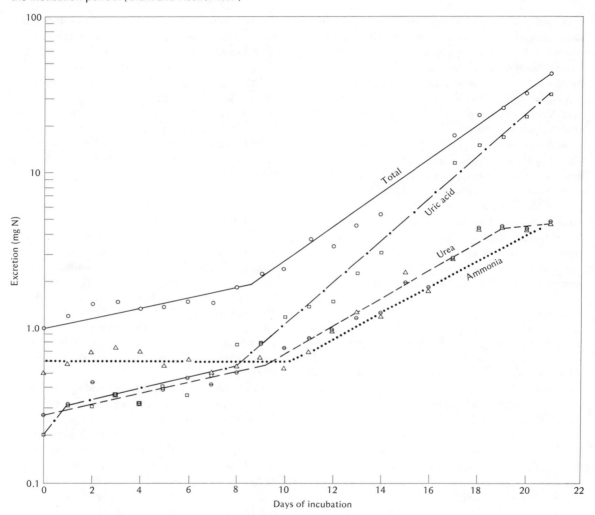

FIGURE 10.15 The accumulation of nitrogenous products in the chick embryo during incubation of the egg. Urea and uric acid are formed from the beginning of incubation; uric acid, however, dominates later. Ammonia, also present from the beginning, remains constant for about 10 days and then increases further throughout the incubation period. [Clark and Fischer 1957]

of amino acids plays, it will not be discussed further here.

The recapitulation theory

The nitrogen excretion of the developing chick embryo is said to change with time and go through a series of peaks, with first ammonia as the main end product, then urea, and finally uric acid. This development would *recapitulate* the evolutionary events that terminate in uric acid excretion in birds. Ammonia production in the chick embryo is reported to culminate at 4 days, urea at 9 days, and uric acid at 11 days of incubation (Baldwin 1949).

A more recent study claims that the nitrogen excretion in the chick embryo differs sharply from these reported peaks (Clark and Fischer 1957). All the three major excretory products – ammonia, urea, and uric acid – are present and are formed from the beginning of embryological development. Toward the end of the incubation period, uric acid is by far the dominant excretory product. However, the amounts of both urea and ammonia continue to increase throughout the incubation, and at the time of hatching both are present in approximately equal quantities. At the end of incubation the chick has excreted 40 mg nitrogen, of which 23% is divided equally between urea and ammonia, and the remainder is present as uric acid (Figure 10.15).

What is the reason for the conflicting reports? The older results may have been less accurate because of more primitive analytical techniques, but this could hardly explain the observation of distinct peaks. The main reason is simply that the earlier results were expressed as the amount of each excretory product relative to the weight of the embryo. Because the embryo continuously increases in size, and at a rising rate, an artificial peak is created by dividing the amount of each excretory product by the weight of the embryo.

In fact, all three excretory products are present from the beginning and increase in amount throughout embryological development, but ammonia does not increase appreciably until after the tenth day of incubation. The urea produced by the chick embryo is not synthesized from amino acid nitrogen by the ornithine cycle, but from the action of arginase upon arginine (Eakin and Fisher 1958). Thus, neither ammonia production nor urea production in the chick embryo supports the claim that embryonic biochemical development recapitulates the evolutionary history of nitrogen excretion.

We have now discussed a wide range of excretory organs and have seen the characteristics they have in common. Excretory organs eliminate metabolic waste products, aid in the maintenance of proper concentrations of salts and other solutes, and regulate the water content of the organism by a judiciously regulated balance between conserving a scarce commodity and disposing of an excess amount.

REFERENCES

Baldwin, E. (1949) An Introduction to Comparative Biochemistry. Cambridge: Cambridge University Press. 164 pp.

Balinsky, J. B., Choritz, E. L., Coe, C. G. L., and Van Der Schans, G. S. (1967) Amino acid metabolism and urea synthesis in naturally aestivating Xenopus laevis. Comp. Biochem. Physiol. 22:59–68.

Balinsky, J. B., Cragg, M. M., and Baldwin, E. (1961) The adaptation of amphibian waste nitrogen excretion to dehydration. Comp. Biochem. Physiol. 3:236–244.

Braun, G., Kümmel, G., and Mangos, J. A. (1966) Studies on the ultrastructure and function of a primitive excretory organ, the protonephridium of the Rotifer Asplanchna priodonta. Pflügers Arch. 289:141–154.

Brown, G. W., Jr., Brown, W. R., and Cohen, P. P. (1959) Comparative biochemistry of urea synthesis. 2. Levels of urea cycle enzymes in metamorphosing Rana catesbeiana tadpoles. J. Biol. Chem. 234: 1775–1780.

Burger, J. W. (1957) The general form of excretion in the lobster, Homarus. Biol. Bull. 113:207–223.

Clark, H., and Fischer, D. (1957) A reconsideration of nitrogen excretion by the chick embryo. J. Exp. Zool. 136:1–15.

Costa, G., Kerins, M. E., Kantor, F., Griffith, K., and Cummings, W. B. (1974) Conversion of protein nitrogen into gaseous catabolites by the chick embryo. Proc. Natl. Acad. Sci. U.S.A. 71:451–454.

Costa, G., Ullrich, L., Kantor, F., and Holland, J. F. (1968) Production of elemental nitrogen by certain mammals including man. Nature, Lond. 218:546–551.

Coulson, R. A., and Hernandez, T. (1955) Renal excretion of carbon dioxide and ammonia by the alligator. *Proc. Soc. Exp. Biol. Med.* 88:682–687.

Coulson, R. A., Hernandez, T., and Brazda, F. G. (1950) Biochemical studies on the alligator. *Proc. Soc. Exp. Biol. Med.* 73:203–206.

Cragg, M. M., Balinsky, J. B., and Baldwin, E. (1961) A comparative study of nitrogen excretion in some amphibia and reptiles. *Comp. Biochem. Physiol.* 3:227–235.

Dobbs, G. H., III, Lin, Y., and DeVries, A. L. (1974) Aglomerularism in antarctic fish. *Science* 185:793–794.

Eakin, R. E., and Fisher, J. R. (1958) Patterns of nitrogen excretion in developing chick embryos. In *The Chemical Basis of Development* (W. D. McElroy and B. Glass, eds.), pp. 514–522. Baltimore: Johns Hopkins Press.

Forster, R. P. (1940) A renal clearance analysis of phenol red elimination in the frog. *J. Cell Comp. Physiol.* 16:113–122.

Forster, R. P., and Goldstein, L. (1966) Urea synthesis in the lungfish: Relative importance of purine and ornithine cycle pathways. *Science* 153:1650–1652.

Goldstein, L., Janssens, P. A., and Forster, R. P. (1967) Lungfish *Neoceratodus forsteri*: Activities of ornithine-urea cycle and enzymes. *Science* 157:316–317.

Goodrich, E. S. (1945) The study of nephridia and genital ducts since 1895. *Q. J. Microsc. Sci.* 86:113–392.

Gottschalk, C. W. (1961) Micropuncture studies of tubular function in the mammalian kidney. *Physiologist* 4:35–55.

Grimstone, A. V., Mullinger, A. M., and Ramsay, J. A. (1968) Further studies on the rectal complex of the mealworm, *Tenebrio molitor*, L. (Coleoptera, Tenebrionidae). *Philos. Trans. R. Soc. Lond. B.* 253:343–382.

Harrison, F. M. (1962) Some excretory processes in the abalone, *Haliotis rufescens*. *J. Exp. Biol.* 39:179–192.

Hilger, H. H., Klümper, J. D., and Ullrich, K. J. (1958) Wasserrückresorption und Ionentransport durch die Sammelrohrzellen der Säugetierniere (Mikroanalytische Untersuchungen). *Pflügers Arch.* 267:217–237.

Hopkins, D. L., (1946) The contractile vacuole and the adjustment to changing concentration in fresh-water amoebae. *Biol. Bull.* 90:158–176.

Janssens, P. A., and Cohen, P. P. (1966) Ornithine-urea cycle enzymes in the African lungfish, *Protopterus aethiopicus*. *Science* 152:358–359.

Jepps, M. W. (1947) Contribution to the study of the sponges. *Proc. R. Soc. Lond. B.* 134:408–417.

Keilin, J. (1959) The biological significance of uric acid and guanine excretion. *Biol. Rev.* 34:265–296.

Khalil, F., and Haggag, G. (1955) Ureotelism and uricotelism in tortoises. *J. Exp. Zool.* 130:423–432.

Kilby, B. A. (1963) The biochemistry of the insect fat body. *Adv. Insect Physiol.* 1:111–174.

Kokko, J. P. (1974) Membrane characteristics governing salt and water transport in the loop of Henle. *Fed. Proc.* 33:25–30.

Kokko, J. P., and Tisher, C. C. (1976) Water movement across nephron segments involved with the countercurrent multiplication system. *Kidney Int.* 10:64–81.

Lockwood, A. P. M. (1961) The urine of *Gammarus duebeni* and *G. pulex*. *J. Exp. Biol.* 38:647–658.

Loveridge, J. P. (1970) Observations on nitrogenous excretion and water relations of *Chiromantis xerampelina* (Amphibia, Anura). *Arnoldia* 5:1–6.

Løvtrup, S., and Pigon, A. (1951) Diffusion and active transport of water in the amoeba *Chaos chaos* L. C. R. *Trav. Lab. Carlsberg* (Ser. Chim.) 28: 1–36.

MacMillen, R. E., and Lee, A. K. (1967) Australian desert mice: Independence of exogenous water. *Science* 158:383–385.

MacMillen, R. E., and Lee, A. K. (1969) Water metabolism of Australian hopping mice. *Comp. Biochem. Physiol.* 28:493–514.

Marshall, E. K., Jr. (1933) The secretion of urea in the frog. *J. Cell. Comp. Physiol.* 2:349–353.

McNabb, R. A., and McNabb, F. M. A. (1975) Urate excretion by the avian kidney. *Comp. Biochem. Physiol.* 51A:253–258.

Mercer, E. H. (1959) An electron microscopic study of *Amoeba proteus*. *Proc. R. Soc. Lond. B.* 150:216–232.

Minnich, J. E. (1972) Excretion of urate salts by reptiles. *Comp. Biochem. Physiol.* 41A:535–549.

Moyle, V. (1949) Nitrogenous excretion in chelonian reptiles. *Biochem. J. 44*:581–584.

Munro, A. F. (1953) The ammonia and urea excretion of different species of Amphibia during their development and metamorphosis. *Biochem. J. 54*:29–36.

Murrish, D. E., and Schmidt-Nielsen, K. (1970) Water transport in the cloaca of lizards: Active or passive? *Science 170*:324–326.

Nechay, B. R., Boyarsky, S., and Catacutan-Labay, P. (1968) Rapid migration of urine into intestine of chickens. *Comp. Biochem. Physiol. 26*:369–370.

Needham, J. (1931) *Chemical Embryology*, vol. I, pp. 1–614; vol. II, pp. 615–1254; vol. III, pp. 1255–1724. Cambridge: Cambridge University Press.

Oschman, J. L., and Wall, B. J. (1969) The structure of the rectal pads of *Periplaneta americana* L. with regard to fluid transport. *J. Morphol. 127*:475–510.

Peschen, K. E. (1939) Untersuchungen über das Vorkommen und den Stoffwechsel des Guanins im Tierreich. *Zool. Jahrb. 59*:429–462.

Pontin, R. M. (1964) A comparative account of the protonephridia of *Asplanchna* (Rotifera) with special reference to the flame bulbs. *Proc. Zool. Soc. Lond. 142*:511–525.

Poulson, T. L. (1965) Countercurrent multipliers in avian kidneys. *Science 148*:389–391.

Ramsay, J. A. (1949) The site of formation of hypotonic urine in the nephridium. *J. Exp. Biol. 26*:65–75.

Ramsay, J. A. (1964) The rectal complex of the mealworm *Tenebrio molitor*, L. (Coleoptera, Tenebrionidae). *Philos. Trans. R. Soc. Lond. B. 248*:279–314.

Ramsay, J. A., and Riegel, J. A. (1961) Excretion of inulin by Malpighian tubules. *Nature, Lond. 191*:1115.

Richards, A. N. (1935) Urine formation in the amphibian kidney. *Harvey Lect. 30*:93–118.

Riddick, D. H. (1968) Contractile vacuole in the amoeba, *Pelomyxa carolinensis. Am. J. Physiol. 215*:736–740.

Riegel, J. A., and Lockwood, A. P. M. (1961) The role of the antennal gland in the osmotic and ionic regulation of *Carcinus maenas. J. Exp. Biol. 38*:491–499.

Schmidt-Nielsen, B. (1972) Mechanisms of urea excretion by the vertebrate kidney. In *Nitrogen Metabolism and the Environment* (J. W. Campbell and L. Goldstein, eds.), pp. 79–103. London: Academic Press.

Schmidt-Nielsen, B., and Forster, R. P. (1954) The effect of dehydration and low temperature on renal function in the bullfrog. *J. Cell. Comp. Physiol. 44*:233–246.

Schmidt-Nielsen, B., and O'Dell, R. (1961) Structure and concentrating mechanism in the mammalian kidney. *Am. J. Physiol. 200*:1119–1124.

Schmidt-Nielsen, B., and Rabinowitz, L. (1964) Methylurea and acetamide: Active reabsorption by elasmobranch renal tubules. *Science 146*:1587–1588.

Schmidt-Nielsen, B., and Schrauger, C. R. (1963) *Amoeba proteus*: Studying the contractile vacuole by micropuncture. *Science 139*:606–607.

Schmidt-Nielsen, K. (1964) *Desert Animals: Physiological Problems of Heat and Water*. Oxford: Clarendon Press. 277 pp. To be reprinted by Dover Publications, New York.

Schmidt-Nielsen, K., Borut, A., Lee, P., and Crawford, E. C., Jr. (1963) Nasal salt excretion and the possible function of the cloaca in water conservation. *Science 142*:1300–1301.

Schmidt-Nielsen, K., and Lee, P. (1962) Kidney function in the crab-eating frog (*Rana cancrivora*). *J. Exp. Biol. 39*:167–177.

Shannon, J. A., and Fisher, S. (1938) The renal tubular reabsorption of glucose in the normal dog. *Am. J. Physiol. 122*:766–774.

Shoemaker, V. H., Balding, D., and Ruibal, R. (1972) Uricotelism and low evaporative water loss in a South American frog. *Science 175*:1018–1020.

Shoemaker, V. H., and McClanahan, L. L., Jr. (1975) Evaporative water loss, nitrogen excretion and osmoregulation in Phyllomedusine frogs. *J. Comp Physiol. 100*:331–345.

Skadhauge, E. (1967) *In vivo* perfusion studies of the cloacal water and electrolyte resorption in the fowl (*Gallus domesticus*). *Comp. Biochem. Physiol. 23*:483–501.

Skadhauge, E. (1976) Cloacal absorption of urine in birds. *Comp. Biochem. Physiol. 55A*:93–98.

Smith, H. W. (1929) The excretion of ammonia and urea by the gills of fish. *J. Biol. Chem.* 81:727–742.

Smith, H. W. (1951) *The Kidney: Structure and Function in Health and Disease.* New York: Oxford University Press. 1049 pp.

Smith, H. W. (1959) *From Fish to Philosopher.* Summit, N.J.: CIBA. 304 pp. Reprinted (1961), Garden City, N. Y.: Doubleday.

Starling, E. H. (1899) The glomerular functions of the kidney. *J. Physiol.* 24:317–330.

Vajropala, K. (1935) Guanine in the excreta of arachnids. *Nature, Lond.* 136:145.

Wigglesworth, V. B. (1931) The physiology of excretion in a blood-sucking insect, *Rhodnius prolixus* (Hemiptera, Reduviidae). 2. Anatomy and histology of the excretory system. *J. Exp. Biol.* 8:428–442.

Yü, T. F., Berger, L., Kupfer, S., and Gutman, A. B. (1960) Tubular secretion of urate in the dog. *Am J. Physiol.* 199:1199–1204.

ADDITIONAL READING

Cochran, D. G. (1975) Excretion in insects. In *Insect Biochemistry and Function* (D. J. Candy and B. A. Kilby, eds.), pp. 179–281. New York: Wiley.

Dantzler, W. H. (1976) Renal function (with special emphasis on nitrogen excretion). In *Biology of the Reptilia,* vol. 5, *Physiology* A (C. Gans and W. R. Dawson, eds.), pp. 447–503. London: Academic Press.

Forster, R. P. (1973) Comparative vertebrate physiology and renal concepts. In *Handbook of Physiology,* Sect. 8, *Renal Physiology,* (J. Orloff, R. W. Berliner, and S. R. Geiger, eds.), pp. 161–184. Washington, D.C.: American Physiological Society.

Kirschner, L. B. (1967) Comparative physiology: Invertebrate excretory organs. *Annu. Rev. Physiol.* 29:169–196.

Maddrell, S. H. P. (1971) The mechanism of insect excretory systems. *Adv. Insect Physiol.* 8:199–331.

Orloff, J., Berliner, R. W., and Geiger, S. R. (eds.) (1973) *Handbook of Physiology,* Sect. 8, *Renal Physiology.* Washington, D.C.: American Physiological Society. 1082 pp.

Peaker, M., and Linzell, J. L. (1975) *Salt Glands in Birds and Reptiles.* Cambridge: Cambridge University Press. 307 pp.

Riegel, J. A. (1972) *Comparative Physiology of Renal Excretion.* Edinburgh: Oliver & Boyd. 204 pp.

Smith, H. W. (1951) *The Kidney: Structure and Function in Health and Disease.* New York: Oxford University Press. 1049 pp.

Wall, B. J., and Oschman, J. L. (1975) Structure and function of the rectum in insects. *Fortschr. Zool.* 23 (2/3):193–222.

Wessing, A. (ed.) (1975) Excretion. *Fortschr. Zool.* 23:1–362.

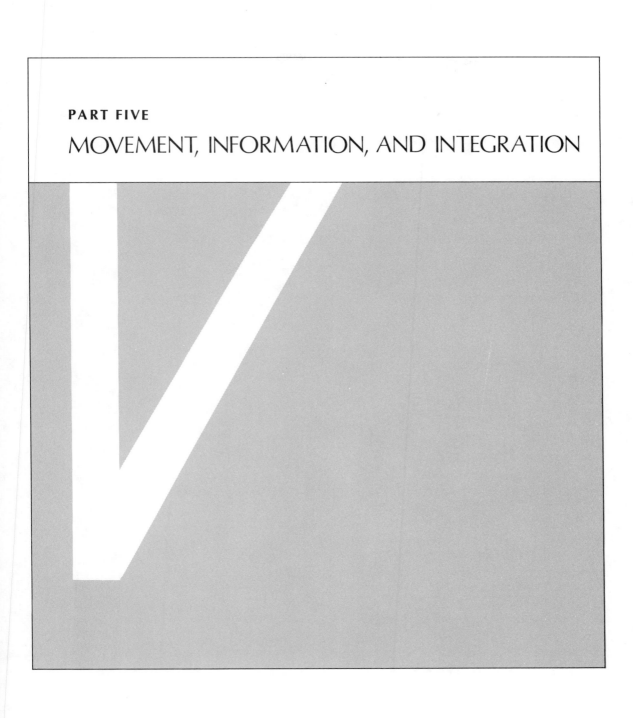

PART IA

ALIGNMENT, INFORMATION AND INTEGRATION

CHAPTER ELEVEN

Muscle, movement, locomotion

Most animals, even those that remain attached and never move about (e.g., corals and sponges), show a great deal of movement or motility. Movements serve not only for locomotion (i.e., moving from place to place) but also for many other purposes. The latter include such varied activities as the pumping of blood in the vascular system, the transport of food through the intestinal tract, and the moving of the external medium, such as pumping air in and out of the lungs or water over the gills.

Only a limited number of basic mechanisms is used in achieving motility, but the variation in how these are used is very great indeed. We will therefore concentrate on basic mechanisms and avoid detailed descriptions of the wide variety of designs employed by different animals. The three basic mechanisms are *ameboid, ciliary,* and *muscular* movements.

Ameboid movement derives its name from the motion of the ameba, the unicellular organism described in every biology textbook. This form of locomotion involves extensive changes in cell shape, flow of cytoplasm, and pseudopodal activity. It is the most poorly understood type of locomotion, although it has received a great deal of attention.

Ciliary locomotion is the characteristic way in which ciliated protozoans such as *Paramecium* move, but cilia are found in all animal phyla and serve a variety of purposes. For example, the respiratory passages of air-breathing vertebrates are lined with ciliated cells, which move foreign particles that lodge on their surfaces, thus slowly removing these particles. Cilia serve to set up currents or move fluid in internal structures such as the water-vascular system of echinoderms, and in an external medium such as water, flowing over the gills of lamellibranchs. The sperm of virtually all animals are motile, and most move with the aid of a tail, which in principle acts in a fashion similar to a cilium.

The overwhelming majority of animal movements, however, depend on the use of muscle, which throughout the animal kingdom has one characteristic in common: the ability to exert a force by shortening (contraction).

We have already discussed the use of muscle for pumping fluids, whether the pumps are peristaltic (intestinal transport of foodstuffs) or valved (the heart). In this chapter we shall be concerned with mechanisms of animal locomotion, the use of muscle being by far the most widespread and most important mechanism. In Chapter 6, on energy metabolism, we discussed the energy requirements for locomotion; at the end of this chapter we shall discuss ways in which some aquatic animals can reduce their use of muscular energy.

AMEBOID, CILIARY, AND FLAGELLAR LOCOMOTION

Ameboid movement

What we designate as ameboid movement is characteristic of some protozoans, slime molds, and vertebrate white blood cells. The movement of these cells is connected with cytoplasmic streaming, change in cell shape, and extension of pseudopodia. These changes are easily observed in a microscope, but the mechanisms involved in achieving the movement are not well understood.

When an ameba moves, its cytoplasm flows into newly formed armlike extensions of the cell (*pseudopodia*), which gradually extend and enlarge so that the entire cell occupies the space where previously only a small pseudopodium began to form. As the cell moves, new pseudopodia are formed in the direction of movement, while the posterior parts are withdrawn.

In an ameba the outermost layer, the *ectoplasm*, is a somewhat stiffer gel-like layer. As a pseudopodium is formed, the more liquid *endoplasm* streams into it, and new ectoplasm is formed on the surface. In the rear part of the advancing cell the ectoplasmic gel should then be converted to a more liquid endoplasmic sol by a gel–sol transformation.

What is the driving force in the streaming of the endoplasm? It has been suggested that contraction of material at the posterior end of an ameba drives endoplasm forward and forces it into the extending pseudopodium. A lower viscosity in certain parts of the endoplasm would facilitate streaming, and the driving force would be minute pressure differences.

The hypothesis that hydrostatic pressure plays the major role in pseudopodium extension is contradicted by experimental evidence. If one pseudopodium of an ameba is sucked into a capillary connected to a reduced pressure, the extension of other pseudopodia is not prevented, as would be expected if the pressure hypothesis were correct (Allen et al. 1971). If the endoplasmic streaming could be attributed to pressure gradients along the length of the stream, streaming should be reversed under an applied pressure gradient of opposite sign.

Ameboid movement has obvious similarities to cytoplasmic streaming (*cyclosis*), a commonly observed phenomenon in all sorts of cells, both plant and animal, which plays an important role in intracellular transport. Ideas about how intracellular forces arise have been revolutionized during the last few years because two important force-generating proteins in muscle, actin and myosin, have been found in all eukaryotic cells (i.e., all cells – animal, plant, and protozoan – except bacteria and blue-green algae). The actins from the most varied sources are remarkably similar in structure

FIGURE 11.1 Diagram showing the timing of the ciliary beat of the ctenophore *Pleurobrachia*, drawn from a film taken at 17 °C. The position of one cilium is shown at 5-ms intervals. The power stroke takes 10 ms, and the complete return stroke takes 50 ms. The cilium then remains in the initial position for 20 ms, until the next stroke begins at 65 ms. [Sleigh 1968]

FIGURE 11.2 The typical beat of a flagellum (left) propels water parallel to the main axis of the flagellum (arrow). The beating of a cilium (right) propels water parallel to the surface to which the cilium is attached. [Sleigh 1974]

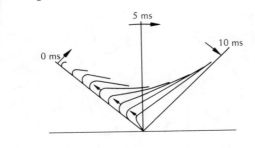

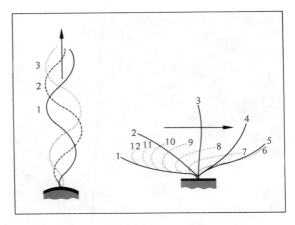

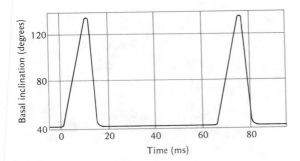

and occur as thin filaments within the cells. It thus appears that both cytoplasmic streaming and the formation of pseudopodia may involve the same fundamental mechanism as muscle contraction. The myosins from various sources are more diverse than the actins, but all bind reversibly to actin filaments and all catalyze the hydrolysis of adenosine triphosphate (ATP), the universal energy source for movement (Pollard 1977).

The existence of a force-generating system in the ameba helps us understand how the ameboid movement can be generated. However, further experimental work will be difficult because we are unable to measure directly the mechanical forces involved, partly as a result of the small size of amebas and partly because of the absence of mechanical structures that can be subjected to direct manipulation and measurement.

Movement by cilia and flagella

Cilia and flagella have a similar internal structure; the difference between them relates to their different beating patterns. A flagellum, like the tail of a sperm, beats with a symmetrical undulation that is propagated as a wave along the flagellum. A cilium, in contrast, beats asymmetrically with a fast or dashlike stroke in one direction, followed by a slower recovery motion in which the bending cilium returns to its original position (Figure 11.1). In flagellar motion water is propelled parallel to the long axis of the flagellum; in ciliary motion water is propelled parallel to the surface that bears the cilia (Figure 11.2).

A flagellated cell usually carries only one or a few flagella; a ciliated cell such as a paramecium may have several thousand cilia distributed on the surface. It is, however, difficult to maintain a sharp distinction between flagella and cilia: Their internal structures are identical, the motions are generated within the organelles themselves, and intermediate patterns of movement are common.

The "flagella" of bacteria are quite different. They are thinner (about 0.02 μm in diameter, against 0.25 μm for true flagella and cilia), short

CILIATED PROTOZOAN The coordinated beating of the cilia of *Paramecium caudatum* gives the appearance of waves passing over its surface. The length of the entire animal is about 150 μm. [Courtesy of Robert L. Hammersmith, Indiana University]

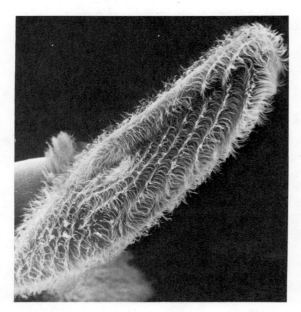

FLAGELLA This electron micrograph of the flagella of the protozoan *Trichonympha* shows clearly the characteristic two-plus-nine arrangement of the filaments in each flagellum. [Courtesy of Dr. A. V. Grimstone, Cambridge University]

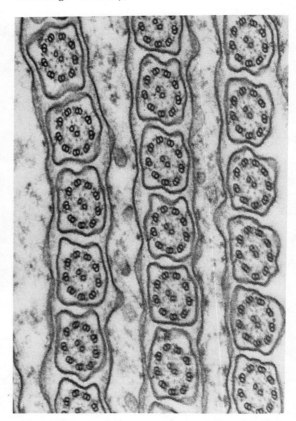

and relatively rigid, and are rotated by forces at the base where they are attached to the cell (Berg and Anderson 1973).

Cilia and flagella are found in many protozoans and are of primary importance in locomotion. The sperm of a vast number of animals swim by means of flagella. Ciliated gills and tentacles, common in many invertebrates, serve two main functions: respiratory exchange and filtering of water to obtain food particles. Cilia are common also in more highly organized animals, where they serve, for example, in moving fluid through tubes, such as reproductive and excretory systems (e.g., the nephridia of annelids). In mammals ciliated epithelia aid in the transport of material on internal surfaces, such as moving mucus in the respiratory tract and the egg within the oviduct.

Cilia are found in all animal phyla. Insects were formerly considered an exception, for they do not

have functional cilia. However, modified ciliary structures, recognized by the characteristic nine-plus-two arrangement of the internal filaments, occur in insect eyes, as well as in a majority of other sense organs in most animal phyla. Main exceptions are the eyes of some invertebrates and the taste receptors of vertebrates. In many cases where a ciliary structure is recognized in a sense organ, it is difficult to determine the importance of the ciliary structure in sensory reception, whether of light or of chemical molecules. These problems are difficult to solve, and only studies on a broad compar-

FIGURE 11.3 The flagellate *Ochromonas* swims with the
single flagellum extending in the direction of movement. The undular motion of the flagellum consists of a
wave that travels from its base to its tip (i.e., in the same
direction as the direction of swimming). This paradox-

ical situation is attributable to the fixed protrusions extending at a 90° angle from the flagellum, causing a
movement of water as shown by the curved arrows. The
organism therefore is propelled in the opposite direction. [Jahn et al. 1964]

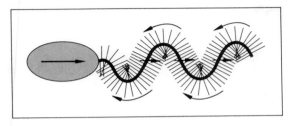

ative basis are likely to yield acceptable generalizations (Barber 1974).

Examination in the electron microscope has revealed that cilia and flagella have a common structure: a central pair of filaments surrounded by an
additional nine thin filaments. This arrangement
of two plus nine filaments is striking; it occurs from
protozoans to vertebrates and in the overwhelming
majority of animal sperm. The particular combination of nine plus two is almost universal, although not essential, for a few sperm are known
that have three central filaments, one central filament, or none at all (Blum and Lubliner 1973).

Cilia used for movement can function only in
an aqueous medium and are therefore found only
on cell surfaces that are covered by fluid or a film
of an aqueous medium such as mucus.

We intuitively expect an organism that swims
with the aid of a single flagellum, such as a sperm,
to move in the opposite direction from the wave
that travels down the tail. This is correct for a
smooth, undulating filament, which propels the
organism away from the filament or "tail." Therefore, it seems peculiar that some flagellates move
with the long flagellum beating anteriorly, in the
direction of movement, and that the waves move
out from the body to the anterior tip of the flagellum (Jahn et al 1964).

This seemingly paradoxical situation, that wave
propagation and locomotion are in the same direction, may occur if the filament is rough or covered

with projections. The flagella of species that move
in this way have tiny appendages in the form of
thin lateral projections that are responsible for the
unexpected direction of movement (Figure 11.3).

The mechanism of movement of cilia and
flagella has long been a matter of speculation.
Three types of mechanisms have been suggested:
(1) that the flagellum is moved passively, much
like a whip, by forces exerted at its base; (2) that elements along the inner curvature of a propagating
wave contract while the opposite side does not; and
(3) that the thin filaments inside the cilium slide
relative to each other due to forces between them
of a nature similar to the sliding filaments of muscle contraction.

The idea that the flagellum is a passive element
driven from its base is incompatible with the form
of the waves (Machin 1958). In the sperm tail the
bending waves pass along without decrease in amplitude, and in some cases with increasing amplitude (Rickmenspoel 1965). It is therefore necessary
to assume that there are active elements within the
flagellum and that the energy is generated locally.

Assuming that the filaments are responsible for
the movement of the cilium, two mechanisms
present themselves as possible explanations for the
motion: (1) bending of the cilium could be
achieved by contraction of the filaments on one
side or (2) movement could be achieved by a longitudinal sliding between the filaments. The first
hypothesis is contradicted by the fact that the ciliary tubules maintain constant length during the
bending cycle (Satir 1968, 1974), and present evidence is entirely in accord with the second alternative.

The characteristic fine structure of a flagellum
or cilium is shown in Figure 11.4. The doublet
tubules are made up of a protein, *tubulin*, with a
molecular weight of about 55,000. Each A-tubule
has attached to it a doublet of "arms," consisting of

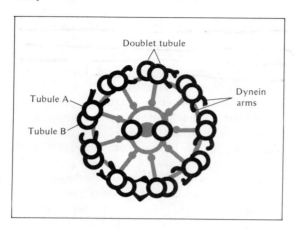

FIGURE 11.4 Diagram showing the internal structure of a cilium or flagellum. The doublet tubules consist of the protein tubulin. Attached to the A-tubules are "arms" containing the protein dynein. [Brokaw and Gibbons 1975]

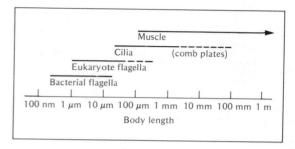

FIGURE 11.5 The range of body lengths of organisms that use flagella, cilia, or muscles for swimming. There is considerable overlap; some organisms propelled by muscles are smaller than the largest that use flagella, and ciliated organisms overlap both these groups. Comb plates are the comblike groups of cilia that ctenophores (comb jellies) use for swimming. [Sleigh 1977]

the protein *dynein*. This protein shows little resemblance to any of the contractile proteins of muscle, except for its ability to split ATP enzymatically, a property it shares with the important muscle protein myosin (Gibbons and Rowe 1965; Gibbons 1977).

Bending of the flagellum occurs when the extending dynein arms attach to the neighboring B-tubule, inducing active sliding movements at the expense of ATP. The process is similar to the sliding of the filaments in muscle, but the exact molecular movements in the flagellar movement have not been as well clarified.

It is possible to prepare an experimental model of the ciliary apparatus of a paramecium that can be used to study the control of the ciliary movement. The cell is extracted with a detergent that leaves the ciliary apparatus functional, but the cell membrane disrupted. Such an extracted paramecium can be reactivated to swim in a solution of ATP containing magnesium ions. It can then be shown that the concentration of calcium ions determines the direction in which the cilia beat (Naitoh and Kaneko 1972). When a living paramecium meets an obstacle, it retreats and swims

backward because the ciliary beat is reversed; the initial process in this reversal is an increase in the permeability of the cell to calcium ions (Eckert 1972). As we shall see later, calcium ions are important in the control of a variety of processes, including muscle contraction.

The overall efficiency of conversion of metabolic energy to mechanical energy in the swimming sperm has been estimated to be at least 19%, and probably closer to 25% (Rikmenspoel et al. 1969). This value is strikingly similar to the efficiency of muscles in performing external work.

Body size and method of propulsion

Small organisms use cilia or flagella for swimming; larger organisms use muscle (Figure 11.5).

There are some very good reasons for this distribution of propelling mechanisms. One is a matter of simple geometry. The small organism has a much larger surface area relative to its volume and obtains forward thrust by moving the cilia against the surrounding surface layer of water, using the viscosity of the water to obtain the necessary thrust. Imagine a whale with a ciliated surface depending on the viscosity in a layer of water, only a fraction of a millimeter thick, along its surface to push against. In relation to its bulk, the thrust it could obtain would be infinitesimal.

Another reason is found in the laws of fluid dynamics. For a very small organism that swims at low speed, the movements are completely dominated by viscous forces – a situation analogous to a man swimming in syrup. For a large animal, by contrast, swimming depends on inertial forces – the thrust obtained by accelerating a mass of water in the opposite direction. For this the animal uses suitably large surfaces, such as fins and flukes, and muscle to provide the power.

We shall now turn to the function of muscle as it is used not only in locomotion, but also in achieving a variety of movements which all are based on the fact that muscle exerts a force by contraction.

MOVEMENT AND MUSCLE

Muscle, what it is: structure

As far as we know, the biochemical mechanism of muscle contraction is the same in all muscles. Two proteins, *actin* and *myosin*, are part of the machinery, and ATP is the immediate energy source for the contraction. The detailed arrangement, however, varies a great deal, and it is convenient to classify various kinds of muscles accordingly. The classification is based primarily on vertebrate muscles for the simple reason that we know and understand more about these than about invertebrate muscles.

The broadest classification is based on the presence or absence in the muscle of regular cross-striations that can be seen in an ordinary light microscope. Vertebrate skeletal muscles and heart muscle are *striated*; muscles of the internal organs – in the walls of the bladder, intestine, blood vessels, uterus, and so on – are *unstriated* (also called *smooth*). The heart muscle, although striated, is often considered as a separate type because it differs from skeletal muscle in characteristic ways, the functionally most important being that a beginning contraction of the heart muscle spreads to the entire organ.

Skeletal muscles are usually called *voluntary*, for the muscles of limb and trunk are under control of the will. This does not mean that we are always aware of or decide about our movements; on the contrary, locomotion, breathing, and so on take place without conscious knowledge of the muscles involved. Voluntary muscle is always striated, but the term is very poor and, at best, useful only for man.

Vertebrate smooth muscle almost always occurs in the walls of hollow internal organs, is not under control of the conscious mind, and is called *involuntary*. The contractions are usually much slower than those of striated muscle, and normally we are completely unaware of the state of contraction of the smooth muscle in our blood vessels, stomach, intestine, and so on.

Striated muscle

The organization of striated muscle is shown in diagram form in Figure 11.6. The muscle is made up of a large number of parallel *fibers*, which are between 0.1 and 0.01 mm in diameter, but may be several centimeters long. These fibers in turn are made up of thinner *fibrils*. These fibrils have characteristic cross-striations, the so-called Z-*lines*, which are repeated at completely regular intervals of about 2.5 μm in a relaxed muscle. From the narrow Z-line very thin *filaments* extend in both directions, and in the center these thin filaments are interspersed with somewhat thicker filaments. The result is a number of less conspicuous bands located between the Z-lines; the appearance of these bands changes with the state of contraction of the muscle. The thin filaments are about 0.005 μm in diameter; the thick filaments are about

FIGURE 11.6 Schematic diagram of a vertebrate striated muscle. The whole muscle is composed of fibers, which in the light microscope appear cross-striated. The fibers consist of fibrils, which have lighter and darker bands. The electron microscope reveals that these bands result from a repeating pattern in the regular arrangement of thick and thin filaments.

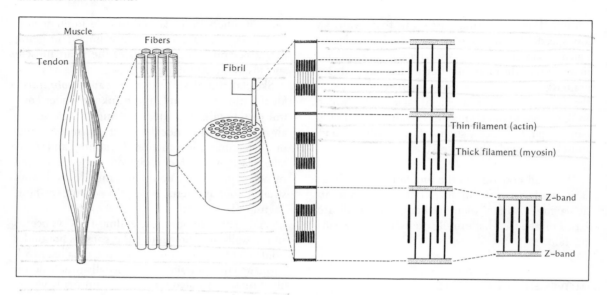

Muscle
Fibers
Tendon
Fibril
Thin filament (actin)
Thick filament (myosin)
Z–band
Z–band

twice that size, about 0.01 μm in diameter. The length of the thick filaments is about 1.5 μm. The thin filaments vary somewhat more in length and are often between 2 and 2.6 μm from tip to tip (or about half this length measured from the Z-line to the tip). The arrangement of the filaments is extremely regular and well ordered.

In striated muscle the thick filaments consist of *myosin* and the thin filaments of *actin*. The thick and thin filaments are linked together by a system of molecular cross linkages, and when the muscle contracts and shortens, these cross linkages are rearranged so that the thick filaments slide in between the thin filaments, reducing the distance between the Z-lines (Huxley 1969).

This model of muscle structure is well supported by electron micrographs of the filaments in various states of contraction. Neither the thick filaments nor the thin filaments change in length during contraction, but it can clearly be seen that they move relative to each other. What is even more

important in support of the sliding filament model is that the force developed by a muscle is related to the amount of overlap of the filaments (page 389).

Cardiac and smooth muscle

Cardiac muscle is striated, like skeletal muscle, but its properties differ for reasons related to its structure. The muscle fibers, instead of being arranged like a bundle of parallel cylindrical fibers, are branched and connected somewhat like a meshwork. An important result of this structure is that, when a contraction starts in one area of the heart muscle, it rapidly spreads throughout the muscle mass. Another important property of cardiac muscle is that a contraction is immediately followed by a relaxation. As a result, a long-lasting contraction, like the sustained contraction of a skeletal muscle, does not occur. These two properties are essential for the normal rhythmic contraction of the heart.

MUSCLE FIBER This electron micrograph of a fiber from the flight muscle of a giant water bug (*Lethocerus cordofanus*) shows the very regular arrangement of thick and thin muscle filaments. The two dark bands crossing the fiber are Z-lines; the distance between them is about 3 μm and shows the length of a single sarcomere. The darker bodies to the upper left and lower right of the fiber are mitochondria. [Courtesy of Barbara M. Luke, Oxford University]

Smooth muscle lacks the cross-striations characteristic of skeletal muscle, but the contraction depends on the same proteins as in striated muscle, actin and myosin, and on a supply of energy from ATP. The lack of cross-striations seems attributable to an absence of a regular pattern in the arrangement of thick and thin filaments.

Smooth muscle has not been as extensively studied as striated muscle. There are several reasons for this. One is that smooth muscle is often interspersed with connective tissue fibers. Another is that smooth muscle fibers do not form neat parallel bundles that can readily be isolated and studied. Also, smooth muscle consists of much smaller cells; the fibers are often only a fraction of a millimeter long. Finally, the rate of contraction is usually much, much slower than in striated muscle; in fact, virtually all smooth muscle is incapable of producing rapid contractions.

Substances for energy storage

The immediate source of energy for a muscle contraction is *adenosine triphosphate* (ATP), a compound that is the primary energy source for nearly every energy-requiring process in the body. ATP is the only substance the muscle proteins can use directly. When the terminal phosphate group of this compound is split off, the high energy of the bond is available for the energy of muscle contraction.

In spite of its great importance, ATP is present in muscle in very small amounts. The total quantity may be sufficient for no more than ten rapid contractions; it follows, therefore, that ATP must be restored again very rapidly, for otherwise the muscles would soon be exhausted. The source is another organic phosphate compound, creatine phosphate, which is present in larger amounts. Its phosphate group is transferred to adenosine diphosphate (ADP) and thus the supply of ATP is restored.

The creatine phosphate must, however, eventually be replenished, and the ultimate energy source is the oxidation of carbohydrates or fatty acids. Carbohydrate is stored in the muscle in the form of glycogen, which is often present in an amount of between 0.5 and 2% of the wet weight of the muscle and provides an amount of energy perhaps 100 times as great as the total quantity of creatine phosphate. In the absence of sufficient oxygen, the glycogen can still yield energy by being split into lactic acid, but the amount of energy is then only a small fraction, about 7%, of that available from complete oxidation (see Chapter 6).

Creatine phosphate has been found in the muscles of all vertebrates examined, but it is conspicuously absent from the muscles of some invertebrates. In these animals we find other organic phosphates. One such compound is arginine phosphate, which is never found in vertebrate muscle.

These various organic phosphate compounds all provide energy through the splitting off of a phosphate group, the high bond energy being available for energy-requiring processes. The fact that creatine phosphate is found in only some invertebrate groups but is universally present in vertebrates has led to the theory that the biochemical similarity between these invertebrate groups and the vertebrates indicates which invertebrate phylum gave rise to the vertebrates. Present knowledge suggests, however, that the distribution of arginine and creatine phosphate among invertebrates is very irregular and not related to the classification of these animals. In particular, there seems to be no pattern in the occurrence of creatine phosphate among invertebrates that can give a clue to the evolution of vertebrates.

Muscle, how it works: contraction

In muscle physiology *contraction* refers to a state of mechanical activity. It may involve a shortening of the muscle, but if the muscle is kept from shortening by having its ends solidly attached, we still use the word contraction to describe the active state. In the latter case, contraction of the muscle causes a force to be exerted on the points of attachment; because no change in length takes place, this is called an *isometric* contraction. If, on the other hand, we attach to one end of the muscle a weight it can lift, the muscle shortens during contraction; because the load remains the same throughout the contraction, this is called an *isotonic* contraction.

Virtually no movements of muscles in the body are purely isometric or purely isotonic, for usually both the length of a muscle and the load change during contraction. For example, when an arm lifts a certain weight, both the length of the contracting muscles and the load on them change

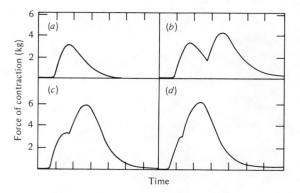

FIGURE 11.7 The force developed in isometric contraction by the gastrocnemius muscle from a cat's leg: *a,* after a single stimulus; *b–d,* after double stimuli applied at shorter and shorter intervals. Time marks are 20 ms apart. Temperature 34.5 °C. [Cooper and Eccles 1930]

continuously, for the leverage and the angle of attack change throughout the movement.

Although pure isometric or pure isotonic contractions are unimportant for muscle function in the body, it is convenient in the study of isolated muscles and their function to distinguish between isometric and isotonic contractions. From isometric contractions, we obtain a great deal of information about the forces a muscle can exert; in the study of isotonic contractions, the amount of work performed by the muscle can easily be calculated. In the laboratory we therefore usually choose one method or the other, according to the purpose of our study.

Force

A muscle can be stimulated to contract by applying a single brief electric pulse, and if the muscle is kept from shortening (isometric contraction), the force it produces gives a record as in Figure 11.7a. Immediately after the stimulus is applied, there is a very short period (the *latent period*) of a few milliseconds before contraction starts. Then the force exerted by the muscle rises rapidly to a maximum and declines again somewhat more slowly. Such a single contraction in response to a single stimulus is called a *twitch*.

FIGURE 11.8 Force developed in isometric contraction by an eye muscle of a cat in response to repeated stimulation: *a*, at a rate of 125 per second (8 ms interval); *b*, at a rate of 210 per second. The latter gives almost complete fusion to a tetanus. Temperature 36 °C. [Cooper and Eccles 1930]

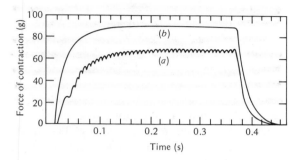

The time required to reach maximum tension (the *contraction time*) varies from muscle to muscle. The contraction time for the major muscles of locomotion of a cat may be as long as 100 ms; the fastest muscles, such as the eye muscles, may have contraction times of less than 10 ms.

If a second stimulus is applied before a contraction is ended, a new contraction is superimposed on the first twitch (Figure 11.7*b*). The second contraction reaches a higher peak of force than the single twitch. If we continue to apply two stimuli spaced more closely, the two contractions become more and more fused (Figure 11.7*c, d*).

Repeated stimuli applied at suitably long intervals give a series of separate twitches, but if we space the stimuli more and more closely, the twitches become fused and eventually yield a smooth, sustained contraction (Figure 11.8*b*). Such a smooth, sustained contraction is called a *tetanus*.

The maximum force developed in a tetanus depends on the initial length of the muscle. If a muscle is made to contract isometrically (i.e., with its ends fixed), the force is reduced if the distance between the ends is less than the resting length of the muscle, and the force is also reduced if the length is greater than the resting length. The maximum force is obtained when the muscle is kept at its resting length.

FIGURE 11.9 The force developed by a contracting muscle (ordinate) in relation to its initial length (abscissa). The force is expressed in percent of the maximal isometric force the muscle can develop. The amount of overlap between thick and thin filaments is shown at the top by diagrams located above the corresponding place on the force curve. [Gordon et al. 1966]

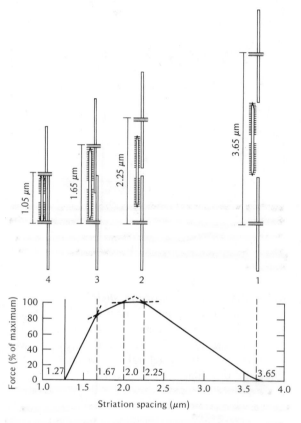

This is closely related to the position of the filaments relative to each other (Figure 11.9). If the muscle is initially stretched so that there is no overlap between thick and thin filaments (position 1), no force is developed. This is consistent with the complete inability of the filaments to form cross bridges in this position. If the muscle is permitted to be shorter, however, the force increases as the filament overlap is greater. The maximum force is developed when the spacing is such that all the cross bridges along the full length of the thick filaments are in contact with the thin filaments (posi-

FIGURE 11.10 The thin filament (above) consists of a double helix of the globular protein, actin. In the groove between the actin strands are located two thin strands of the protein tropomyosin and, for every seven actin molecules, one molecule of troponin. The thick filament (below) consists of the rod-shaped protein, myosin, with the larger "heads" protruding toward the actin strands.

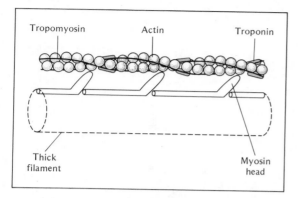

tion 2). If the muscle is even shorter before contraction begins, the force is again diminished and rapidly decreases to zero as the ends of the thick filaments make contact with the Z-lines (positions 3 and 4).

What makes the filaments slide?

The molecular events during muscle contraction are better known than most other processes in the living organism. To describe these events, we must refer to the structural composition of the thick and thin filaments.

The thick filaments consist of the protein *myosin*. Each myosin molecule resembles a thin rod with a globular "head." One thick filament contains several hundred myosin molecules, lined up as shown in Figure 11.10 with the heads facing in both directions away from the center of the filament. This leaves a bare zone in the middle and the heads protruding all along the remainder of the thick filament.

The thin filaments are more complex, with three important proteins. The backbone of the filament is the protein *actin*, a relatively small globular protein, arranged in the filament as a twisted double strand of beads. Each actin molecule is slightly asymmetrical, and in each filament all the actin is lined up facing in the same direction. This directionality is important for muscle contraction.

The next important protein in the thin filament is *tropomyosin*, long and thin molecules attached to each other end to end, forming a very thin threadlike structure that lies in the grooves between the double helix of the actin molecules. The length of one tropomyosin molecule is such that it extends over seven actin molecules. The single thin filament is about 1 μm long and contains about 400 actin molecules and 60 tropomyosin molecules.

A third protein molecule, *troponin*, is attached to each tropomyosin molecule. Troponin, a calcium-binding protein, is a key to the contraction process. When troponin binds calcium ions, it undergoes a conformational change that is essential for the interaction between the myosin heads of the thick filaments and the actin of the thin filaments (Squire 1975).

The interaction between thick and thin filaments consists of a cyclic attachment and detachment of cross bridges between the two filaments. The myosin heads attach to the thin filaments at a certain angle; they undergo a conformational change that makes the bridge swivel to a different angle, pulling the thin filament past the thick (Huxley 1973). The cross bridges on opposite sides of the bare zone in the middle of the thick filament swivel in opposite directions, pulling at the opposing ends of the thin filaments, thereby shortening the distance between the Z-lines. This decrease in the distance between the Z-lines causes the shortening of the muscle.

For any significant shortening to occur, each cross bridge must be attached, swivel, detach, and then attach again at a point further along the thin filament. Each cycle of attachment requires the expenditure of energy in the form of ATP. The ATP molecule binds temporarily to the myosin head, forming an active complex that will attach to the actin strand, provided the appropriate site is available and not blocked by the thin tropomyosin strand (Weber and Murray 1973).

The role of Ca^{2+}. The tropomyosin is located in the grooves of the actin filaments, where it blocks the interaction of the myosin head with the actin filament. The controlling protein, troponin, has a high affinity for calcium ion. Immediately after stimulation of the muscle, the calcium ion concentration within the muscle fiber rises abruptly; the calcium ion binds to troponin, which in turn undergoes a conformational change that allows tropomyosin to move out of its blocking position on the actin strands.

A direct demonstration of the increase of the calcium ion concentration immediately after stimulation has been made with the aid of the protein *aequorin*. This protein, which can be isolated from a luminescent jellyfish, emits light in the presence of calcium ions. Aequorin injected into the muscle fibers of the giant barnacle *Balanus nubilis* yields a faint glow of light when the fibers are stimulated electrically, indicating an increase in calcium ion concentration within the muscle fibers. Measurement of this light emission is a sensitive indicator of very rapid and transient changes in the intracellular calcium concentration (Ridgway and Ashley 1967).

How is contraction triggered?

Normally, muscle contraction is initiated when a nerve impulse arrives at the neuromuscular junction (the *motor end-plate*). The impulse spreads rapidly as an electric depolarization over the surface of the muscle fiber (the *sarcolemma*), momentarily abolishing the normal surface potential of about 60 mV. This depolarization is almost instantaneously communicated to the interior of the muscle cell, causing contraction to take place.

As we shall see later, the message from the motor end-plate to the muscle fiber consists in the diffusion of a messenger chemical, acetylcholine, across the minute distance between the motor end-plate and the sarcolemma. However, the distance from the surface to the interior of the muscle fiber is so great that, if the communication were based on the diffusion of a chemical, the impulse could not be communicated to all the filaments simultaneously. One of the classic problems in muscle physiology has been to explain how communication throughout the muscle fiber could be rapid enough to achieve simultaneous contraction of all the filaments.

The answer can be found in the ultrastructure of the muscle fiber. The sarcolemma (the muscle cell membrane) connects with a complex system of transverse tubules that run across the muscle cells, typically near the Z-lines, known as the *T-system* (Figure 11.11). When an impulse arrives at the motor end-plate, it causes depolarization of the sarcolemma; the depolarization continues within the T-system, and this triggers the further events of contraction.

The next important structure to consider is a system of flattened vesicles, the *sarcoplasmic reticulum*, which surrounds the muscle fiber like a sleeve (Figure 11.11).

We now have the structural basis for the rapid communication throughout the muscle fibers. As an impulse spreads over the sarcolemma and the T-system, the change is transmitted to the membrane of the sarcoplasmic reticulum, causing an increase in its permeability. As a result, calcium ions that were sequestered inside the sarcoplasmic reticulum of the resting muscle can escape (Figure 11.12). The sudden appearance of calcium triggers the change in the configuration of the troponin, which permits the interaction between myosin and actin that is the basis of contraction. The calcium

FIGURE 11.11 The system of T-tubules and the sarcoplasmic reticulum that surrounds the fibers of striated muscle. [Peachey 1965. Courtesy of Lee D. Peachey and The Rockefeller University Press.]

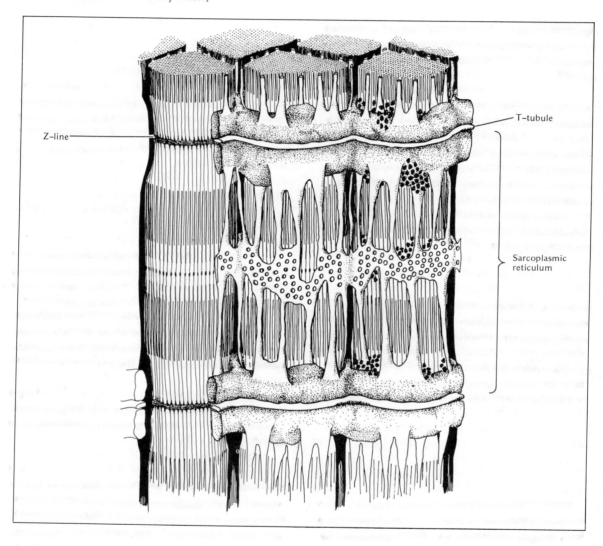

Z–line

T–tubule

Sarcoplasmic
reticulum

concentration necessary to activate the contractile events is no more than about 0.01 mmol per liter. The calcium that was released is immediately taken up again by the sarcoplasmic reticulum, where it is held until the next impulse arrives; in the relaxed skeletal muscle the calcium concentration is believed to be as low as 0.0002 mmol per liter.

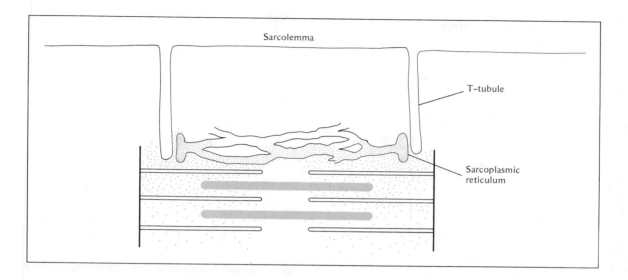

FIGURE 11.12 The sliding between the thick and thin filaments of muscle is triggered by the release of Ca^{2+} ions (indicated by dots) from the sarcoplasmic reticulum.

We are now in a position to list the major events in muscle contraction (Table 11.1).

Force and work

If a muscle is free to contract when it is stimulated, it may shorten by, say, one-third. However, if a load is attached to the muscle, the shortening is smaller. If the load is gradually increased, the muscle will be able to lift the load over a shorter and shorter distance until the load is so heavy that the muscle is unable to lift it at all. We thus obtain a curve for shortening versus force, as shown in Figure 11.13. In this particular case, a load of just below 100 g completely prevented shortening.

Do the muscles of various animals differ much in strength? When we see an ant carry in its jaws a seed that weighs more than the animal itself, we gain the impression that insect muscles must be inordinately strong. An objective comparison of muscle strength shows that this is not so. In order to compare different muscles we must, of course, take into consideration their different sizes.

The force a muscle of a given kind can exert is directly related to its cross-sectional area, but not to its length. We can therefore compare widely different muscles if we relate the force they can exert to their cross section, as Table 11.2 does for muscles from a variety of animals. Although many muscles are less powerful than those listed, it appears that the maximum limit is roughly the same for all muscles, irrespective of their origin: some 4 to 6 kg per cm² cross-sectional area.

The fact that force of contraction is similar in such a wide variety of muscle may at first seem surprising, but it is understandable in view of the basic similarity of the contractile mechanism itself. All muscle contraction is based on sliding filaments of actin and myosin, with the energy supplied by ATP. It would be unreasonable to expect that this mechanism could be improved to provide a greater force per cross-sectional area, for the maximal force should be related to the number of filaments that can be packed within that area, and

TABLE 11.1 Sequence of events in stimulation and contraction of muscle.

Stimulation
1. Sarcolemma depolarized
2. T-system depolarized
3. Ca^{2+} released from sarcoplasmic reticulum
4. Ca^{2+} diffuses to thin filament

Contraction
5. Ca^{2+} binds to troponin
6. Troponin–Ca^{2+} complex removes tropomyosin blockage of actin sites
7. Heads of thick filament (containing preexisting myosin–ATP complex) form cross bridges to actin strand
8. Cross bridges swivel as ATP is hydrolyzed and ADP is released

Relaxation
9. Ca^{2+} sequestered from thin filament by sarcoplasmic reticulum
10. Ca^{2+} diffuses from thin filament toward sarcoplasmic reticulum
11. Ca^{2+} released from troponin–Ca^{2+} complex
12. Troponin permits tropomyosin return to blocking position
13. Myosin–actin cross bridges break
14. ATP–myosin complex re-formed in heads of thick filament

this again depends on the size of the protein molecules that make up the filaments.*

When a muscle lifts a weight, the external work performed is the product of the load and the distance over which it is moved. This is shown by the arched curve in Figure 11.13. When the load is zero, the external work performed is also zero, although the muscle shortens maximally. At the other extreme, when the load is too heavy to be lifted at all, the external work is also zero. Between these extremes, the external work is at a maximum

* The diameter of the thick and thin filaments limits the number of filaments per cross-sectional area. However, the force exerted could conceivably be increased by lengthening the overlap between thin and thick filaments (i.e., increasing the distance between the Z-lines and having longer filaments). This would give a larger number of cross bridges per filament and thus increase the force. This avenue for increasing the force may be unavailable because of limitations on the structural strength of the filaments that must support the increased force.

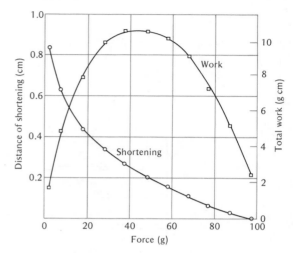

FIGURE 11.13 Isotonic contraction of the sartorius muscle of the toad *Bufo marinus*. The shortening of the muscle decreases to zero as the load is increased to 97 g. The work performed in the contraction increases to a maximum at a load of about 40% of the maximum, and decreases again as the load is increased. [Courtesy of Frans Jöbsis]

when the weight is about 40% of the maximal load the muscle can lift and the muscle shortens by about one-third of the maximal shortening.

The work that can be performed by a muscle is obviously related to its size. Consider two muscles that have the same dimensions and can develop the same maximum force per unit cross-sectional area. If they shorten to the same extent, the work they perform must be the same. Now consider two muscles of the same cross section but different lengths, one twice as long as the other, that can both contract by the same fraction of their resting length, say, by one-third. The maximum work each can perform is directly related to its initial length (i.e., twice as high in one as in the other), although the force of contraction is the same (same cross-sectional area).

We can extend this argument by saying that because work is the product of force and distance, and the volume of the muscle is proportional to the product of cross section and length, the work a muscle can perform must be directly related to its volume. This, of course, is a generalization from which there are exceptions, but as a general rule it is a useful fact to keep in mind.

TABLE 11.2 The maximum force of contraction measured in muscles from a variety of animals.

Animal	Muscle	Contraction force (kg cm^{-2})
Annelid (*Arenicola*)[a]	Body wall	3
Bivalve (*Anodonta*)[b]	Adductor	5
Bivalve (*Mytilus*)[c]	Anterior byssus retractor	4.5
Octopus (*Octopus*)[d]	Funnel retractor	5.1
Insect (*Locusta*)[e]	Hindlegs	4.7
Insect (*Decticus*)[e]	Flexor tibiae	5.9
Insect (*Drosophila*)[f]	Wing muscles	5
Frog (*Rana*)[g]	Anterior tibial (20 °C)	4.4
Rabbit[h]	"Skeletal muscle" (37 °C)	5.0
Man[i]	Ankle flexors	4.2

[a] Trueman (1966).
[b] Weber and Portzehl (1954).
[c] Abbott and Lowy (1953).
[d] Lowy and Millman (1962).
[e] Wigglesworth (1972).
[f] Roeder (1953).
[g] Casella (1950).
[h] Wilson (1972).
[i] Haxton (1944); voluntary contraction.

For power output (work per unit time), the situation is quite different. Because a fast muscle contracts in a short period of time, the power produced is greater than in a slow muscle. Consequently, if we compare the muscles of elephants and shrews, we can expect (1) that the muscles have approximately the same contractile force per square centimeter of cross-sectional area, (2) that they can shorten to approximately the same fraction of their resting length, and (3) that the work performed during the contraction of 1 g of muscle will be similar in the two animals. However, because the contraction of a shrew muscle takes place much faster (i.e., in a much shorter time), the power output per gram shrew muscle is correspondingly much higher than for elephant muscle. This is precisely what is reflected in the metabolic rates of the animals (see Chapter 6), which are usually expressed in power units (work or energy per unit time) and are much higher in the small animal.

Muscle, how it is used

The way muscle is used by various animals differs a great deal (depending primarily on the function of the particular muscle). The demands on the flight muscles of an insect, which contract several hundred times per second, and on the muscle that closes the shells of a clam and remains contracted perhaps for several hours, are very different indeed. The best way to describe how muscle can serve different purposes is to examine some characteristic types of muscles and how they are modified to meet specific demands.

Vertebrate fast and slow muscles

Vertebrate striated muscle is composed of muscle fibers that fall into two (or more) distinct classes and are frequently referred to as fast and slow fibers. This terminology can easily lead to confusion, and it is now common to designate the fast fibers as *twitch* fibers and the slower as *tonic* fibers. Usually there is a difference in the amount of myoglobin present in the two kinds of fibers. The twitch fibers, which have a lower myoglobin content, have been known as pale or white; the tonic fibers, with a higher myoglobin content, as red. Any one muscle may consist of only twitch fibers, only tonic fibers, or a mixture of both.

The distinction between twitch (fast) and tonic (slow) muscle fibers was first made for frog muscles. The main functional difference is that twitch fibers are used for rapid movements, and tonic fibers are used to maintain low-force prolonged contractions. The twitch system is associated with large nerve fibers (10 to 20 μm diameter) with conduction velocities about 8 to 40 m s^{-1}, which lead to quick contractile responses. This twitch system is used, for example, in jumping. The tonic (slow) system has small nerve fibers (about 5 μm diameter), which conduct at 2 to 8 m s^{-1} and lead to slower, graded muscular contractions accompanied by nonpropagated muscle potentials of small amplitude and long duration. This is the tonic system used, for example, to maintain the posture of the body (Kuffler and Williams 1953a).

The sartorius muscle of the frog, which runs along the thigh and is used primarily in jumping, consists entirely of twitch fibers. The nerve to the muscle contains large, fast-conducting nerve fibers, and each makes contact with the muscle fibers through a single terminal (the end-plate). The twitch fibers respond according to the all-or-none rule, which means that when a stimulus exceeds a certain minimum (the *threshold value*) the fibers respond with a complete contraction and a further increase in stimulus intensity gives no further increase in response. Thus, the muscle fiber either does not respond or responds maximally. This is known as an *all-or-none response*.

Tonic muscle is different. The muscle fibers, whose behavior in many respects bears little resemblance to that of the twitch fibers, are innervated by nerve fibers, all of small diameter, whose terminal contacts are distributed along the length of the muscle fiber. In contrast to the fast twitch-producing muscle fibers, the response of the tonic fibers does not necessarily spread throughout the fiber to give a complete all-or-none response, and there is a summation on repetitive stimulation. The response to a single stimulus is small, and the tension rises with repeated stimulus frequency. Thus, repeated stimulation of small nerves is necessary to

cause a significant tension rise in tonic fibers. Increased frequencies speed the rate of tension rise. Relaxation after slow muscle fiber contraction is at least 50 to 100 times slower than after twitch fiber action. This type of response plays an important role in the general postural activity of the frog (Kuffler and Williams 1953b).

When we wish to study separately the function of the different types of muscle fibers, we do not search for muscles that contain only one type. The observed differences between a muscle consisting of only twitch fibers and one with only tonic fibers could just as well result from the different locations and uses of the two muscles. It is therefore better to examine different fiber types as they are found in a single muscle.

The gastrocnemius muscle of the cat (which runs from the knee to the heel tendon and is used for stretching the foot) can be used for this purpose. It has not only two, but three, distinct functional types of muscle fibers. This is shown in the following way. A single motor neuron in the appropriate region of the spinal cord is stimulated with the aid of an inserted microelectrode. This activates only one motor unit (which consists of many muscle fibers jointly innervated by the single nerve axon). This may give one of three different types of contraction, depending on which particular neuron is stimulated.

Figure 11.14 shows the response of the three types of fibers: (1) a fast-contracting and fast-fatiguing type, (2) a fast-contracting and fatigue-resistant type, and (3) a slow-contracting type which does not fatigue even during prolonged stimulation. The slow-contracting (tonic) fibers have contraction times more than twice as long as the contraction time for the fast-contracting, fast-fatiguing (twitch) fibers (Table 11.3). The difference in the force developed by the muscle is even more striking: The fast-contracting fibers develop a force

FIGURE 11.14 The gastrocnemius muscle of a cat contains three types of fibers that respond to stimulation (40 per second) in three different ways. They may be fast contracting-fast fatiguing (top); fast contracting-fatigue resistant (center); or slow contracting-nonfatiguing (bottom). [Burke et al. 1971]

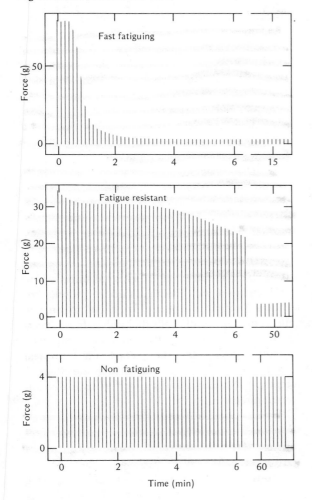

TABLE 11.3 Contraction time and maximum tetanic force developed by three different functional types of muscle fibers in the gastrocnemius muscle of a cat. [Burke et al., 1971]

	Fiber type		
	Fast contracting, fast fatiguing	Fast contracting, fatigue-resistant	Slow-contracting
Twitch contraction time (ms)	34	40	73
Tetanic force (gram force)	60	20	5

of muscles in the hindlimb of a rabbit, it has been possible to make tonic muscle acquire characteristics more nearly those of twitch muscle.

In the newborn cat all muscles seem to be equally slow, and the differentiation takes place as the muscles are used. By severing the nerves to two muscles and then reconnecting each nerve to the other muscle, it has been shown that when a nerve from a fast (twitch) motor neuron is connected to a slow (tonic) muscle, the muscle is transformed into a fast (twitch) muscle. Likewise, slow (tonic) motor neurons make twitch muscles change toward tonic (Buller et al. 1960). It has become apparent, however, that cross innervation does not completely convert one type of muscle to the other, but to some mixed form (Buller and Lewis 1964).

The muscles of fish also have twitch and tonic fibers. Pelagic fish such as mackerel and tuna, which swim continuously at relatively low speeds, have the two types of fibers separated in different muscle masses, which have strikingly different appearances. The tonic muscle is deep red because of a high concentration of myoglobin. It is located along the sideline and stretches in toward the vertebral column. The basal swimming during cruising is entirely executed by this red muscle; the large mass of white muscle (twitch type) represents a reserve of power for short bursts of high-speed activity (Figure 11.15).

more than 10 times as high as the slow-contracting fibers. This is evidently paid for in the much faster fatigue of these fibers.

Whether a muscle is twitch or tonic is not necessarily a constant inherent characteristic; somehow it seems to depend on how the muscle is used. By changing the position of insertion of the tendon

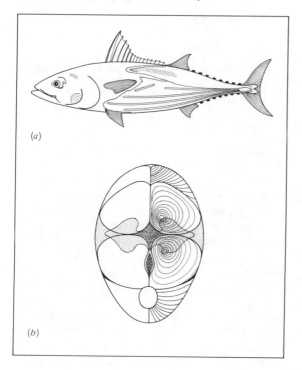

FIGURE 11.15 In the skipjack tuna (*a*) the red (tonic) swimming muscle, which is used during steady cruising, is located along the sideline and in toward the vertebral column (stippled in *b*). The remaining large mass of white (twitch) muscle is used in bursts of high-speed swimming. [Rayner and Kennan 1967]

(*a*)

(*b*)

The swimming muscles of sharks are also divided into similar types, with less than 20% of the muscle mass made up of tonic fibers. Again, the tonic fibers are for cruising and the twitch fibers are used during vigorous swimming such as in the pursuit of prey. The twitch fibers are capable of a high level of anaerobic power output, as would be expected during maximal activity, and accordingly have a high glycogen content. After vigorous activity the glycogen level in the fast fibers falls markedly. During prolonged periods of slow swimming, however, there is no change in their glycogen level; this indicates that they are not used at all during cruising (Bone 1966).

Smooth muscle

In contrast to striated muscle (in vertebrates the skeletal and heart muscle), the muscles of internal organs are smooth (also called unstriated, nonstriated, plain, or involuntary muscle). In vertebrates this type of muscle occurs in the stomach, intestine, bladder, ureters, uterus, bronchi, blood vessels, and so on. Smooth muscle also occurs in a vast number of invertebrates, but it does not form coherent functional groups in these.

Vertebrate smooth muscle is innervated from the autonomic nervous system by two sets of nerves. One set is stimulatory and the other is inhibitory (i.e., the two sets act in an antagonistic manner). Smooth muscle differs from voluntary muscle in that the muscle layers, mostly located in the walls of hollow organs, contain numerous nerve fibers and nerve cells. This has impeded experimentation with isolated smooth muscle, for when a stimulus is applied it is difficult to separate effects on the muscle cells themselves from effects on the nerve cells. A further difficulty in studying smooth muscle is that the fibers do not form neat bundles; the cells are smaller and shorter than in striated muscle and frequently run in many different directions.

Smooth muscle need not be stimulated through the nerves in order to contract. It shows spontaneous rhythmic contractions that can vary greatly in both frequency and intensity. The internal neurons and nerve fibers in the muscle may have considerable influence on this activity, but the details of such interaction remain to be clarified.

Sudden stretching of smooth muscle causes an immediate contraction. Distention of a hollow organ is therefore frequently followed by contraction. An isolated piece of smooth muscle behaves somewhat differently, however. If a piece is loaded slightly, it stretches slowly and may, without change in the load, adopt different lengths at different times. Thus, smooth muscle has *no particular resting length*.

A distinctive feature of vertebrate smooth muscle is the slowness of response. Its most characteristic property is that it can maintain contraction for prolonged periods with very little energy expenditure. The ability to maintain a slight contraction or *tonus* at various different lengths is most important to the economy of maintaining such tension, for in general the cost of maintaining tension is inversely related to the speed of contraction (Rüegg 1971).

The reason no striations are visible in smooth muscle is that the filaments, which presumably are present, are not regularly aligned to form visible bands. Vertebrate smooth muscle contains thin (actin) filaments, but, although the muscles do contain myosin, it has so far been impossible to demonstrate thick filaments of the kind found in striated muscle.

Many invertebrate muscles are smooth. One kind that will be discussed later is the slow closing muscle of bivalves, which not only pulls the shells together, but can maintain the contraction for hours or days. Not all mollusc smooth muscle is slow, however; the muscle of the octopus and squid mantle, which is responsible for the jet propulsion these animals use in swimming, is a fast-contracting muscle. The duration of a full contraction in these muscles in the living animal may be 0.1 or 0.2 second, which for a smooth muscle is very fast indeed (Trueman and Packard 1968).

In structure and composition invertebrate smooth muscle differs somewhat from vertebrate smooth muscle. For example, mollusc smooth muscle contains a protein, tropomyosin A, which is not found elsewhere. Nevertheless, it appears that the basic mechanism of contraction always depends on a sliding filament displacement, which is similar in both smooth and striated muscle throughout the animal kingdom.

Molluscan catch muscle

Clams and mussels protect themselves by closing the shells. One or more adductor muscles pull the shells together and keep them closed against the springy action of the elastic hinge. Species that live in the intertidal zone must keep their shells closed when the water recedes, and the muscle is kept contracted for hours with the shells tightly closed under conditions that are probably completely anaerobic. If a starfish attacks a mussel, it attaches its tube feet to the two shells and tries to pull them apart; the catch mechanism helps the mussel withstand the strong pull in this endurance test.

In some species, but not all, the closing muscles are divided into two portions, one of smooth and one of striated fibers. Functionally, the striated portion contracts quickly and is referred to as the fast or phasic (twitch) portion; the nonstriated is much slower and is known as the slow or tonic portion. The long-maintained contractions that keep the shells closed for extended periods are produced primarily by the slow portion of the closing muscle (Hoyle 1964).

It has been suggested that the closing muscles, once they have shortened, enter into a state of *catch*. This means that the contraction is supposedly maintained without further expenditure of energy, in other words, that the relaxation phase is immensely prolonged and lasts for periods of up to many hours. The problem of how prolonged contractions are maintained in catch muscle has been only partly resolved, and there is still much uncertainty about the mechanism.

The hypotheses to explain catch are basically of two kinds. One postulates a change in the muscle at the molecular level that prevents the breakage of established cross bridges between the filaments. The other postulates the need for repeated reactivation of the contractile system through the nervous system or through internal action potentials in the muscle.

The most studied muscle of the catch type is the anterior byssus retractor muscle (ABRM) of the blue mussel (*Mytilus edulis*), which contains only smooth fibers. It can be made to contract by electric stimuli, but, depending on the nature of the stimulus, relaxation may be complete within a matter of seconds or contraction may be maintained for minutes or hours (Johnson and Twarog 1960).

Maintaining the state of tension without the expenditure of energy would be particularly advantageous to the animal under anaerobic conditions. The oxygen consumption of the ABRM is too small to be discovered as an increase in the oxygen consumption of the whole animal, in part because it is difficult to exclude as energy sources the energy-rich phosphate compounds and the contribution of anaerobic processes.

The oxygen consumption of the isolated ABRM slowly diminishes with a time course roughly similar to that of the decrease in tension in the muscle. During contraction of this muscle, the oxygen consumption is reduced if the external tension on the muscle is released, and increases again when the tension is restored by stretching the muscle back to its original length. This confirms that the tension in the muscle determines the intensity of oxygen consumption during the catch. Study of the excess oxygen consumption (that above the resting level) shows that its relation to the tension in the muscle is linear (Figure 11.16).

These results are compatible with the hypothesis that the tonic tension in the catch muscle is maintained as an active state that has a metabolic energy

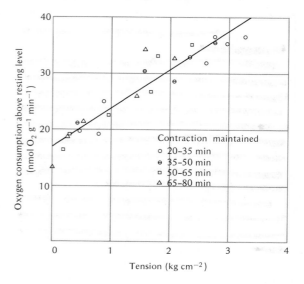

FIGURE 11.16 The excess oxygen consumption (above the resting level) during contraction of the closing muscle of the blue mussel, *Mytilus*. The excess respiration is linearly related to the tension in the muscle and seems to be independent of the length of time the contraction has been maintained. [Baguet and Gillis 1968]

requirement. Owing to the inherent slowness of the muscle, however, the energy requirement for maintained contraction is more than 1000 times lower than for the maintenance of tetanic contractions of vertebrate fast muscle.

Crustacean muscle

Crustacean muscles show in pure form some characteristics that to varying degrees are present in the muscles of a number of other animals. We shall discuss two points: (1) the gross arrangement of the muscle fibers into parallel-fibered or pinnate muscles, and (2) the multiple innervation of a single muscle fiber and its responses to the different nerves.

Nearly everybody who has handled live crabs or lobsters has made contact with the impressive force these animals can exert with their claws. This is not because the strength of the muscle is extraordinarily high, but rather because of the anatomical arrangement of the fibers. In the claw of the crab

FIGURE 11.17 The muscles in the claw of a crab or lobster are arranged so that a mechanical advantage is gained. This pinnate arrangement also permits the muscle to thicken during contraction, which otherwise would not be possible within the confined space of the claw. [Alexander 1968]

FIGURE 11.18 Top: a crustacean muscle is innervated from a few neurons, and the nerve fibers branch extensively and supply the entire muscle. Vertebrate muscle is innervated from a large number of neurons in the spinal cord, each one supplying a relatively small number of muscle fibers. Bottom: a crustacean muscle fiber supplied by three different nerve fibers, one for fast, one for slow contraction, and one for inhibition.

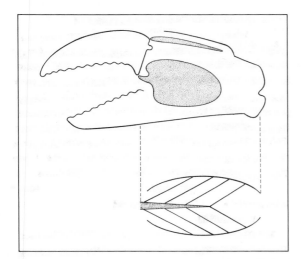

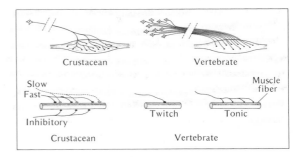

the muscle is *pinnate*; this means that the fibers, instead of being parallel to the direction of the pull, are arranged at an angle, which greatly increases their mechanical advantage (Figure 11.17).

In the enclosed space of the claw, a pinnate muscle of a given volume can have more and shorter fibers than a parallel-fibered muscle. The increased force, attributable to this mechanical arrangement, is gained at the expense of the distance over which the attachment can be moved. In the crab claw, the closing muscle can exert, because of the pinnate arrangement, about twice as much force as it could if it were parallel-fibered (Alexander 1968).

Another advantage of the pinnate muscle in the crab claw is related to the confined space in which it works. As a parallel-fibered muscle contracts, its cross section increases, and in the rigid enclosure of the claw this would cause difficulties. Because of the arrangement of the pinnate muscle, however, the cavity of the crab claw can be almost completely filled with muscle, for as the fibers

shorten, their angle is changed so that there is space for the thickening. This is impossible with a parallel-fibered muscle in a limited space.

A pinnate muscle can perform only the same amount of work as a parallel-fibered muscle of equal volume because the increased force of contraction is offset by a corresponding decrease in shortening distance. It should be noted that pinnate muscles are not unique to crustaceans; they also occur quite commonly in vertebrates.

The most interesting aspect of crustacean muscle is the multiple innervation: Individual muscle fibers may be innervated by two or more nerve fibers. In addition to multiple nerves that stimulate the muscle to contract, many arthropod muscles also have inhibitory nerves whose stimulation causes the muscle to relax if it is already in a state of contraction.

Another crustacean characteristic is that a whole muscle is often innervated by only a few or a single axon. This makes the whole muscle act much as a single unit, and gradation of response results from variations in the nerve impulses in combination with the balance between excitatory and inhibitory impulses. One muscle may, therefore, depending on the impulses it receives, act as a fast or as a slow muscle.

The difference in the innervation of vertebrate and crustacean muscle is shown in Figure 11.18.

As stated above, the most striking characteristic of crustacean muscle is that it is supplied by two or more different nerve fibers. Often, there is a fast excitatory fiber, a slow excitatory fiber, and finally an inhibitory nerve fiber. In many muscles the majority of fibers is innervated by the fast nerve, while a smaller number is innervated by the slow nerve. In combination with the inhibitory fibers this permits a wide range in the gradation of the contractions.

The other characteristic of crustacean muscle that differs from vertebrate muscle is that a whole muscle is usually innervated from a very few neurons in the central nervous system and is reached by a correspondingly small number of axons that branch and connect to the fibers of the entire muscle.

Vertebrate muscle, in contrast, is innervated from a large number of neurons in the central nervous system, and each nerve fiber reaches only a small number of the fibers in the entire muscle. In the fast (twitch) vertebrate muscle a gradation in contraction of a muscle depends on the number of axons that are carrying excitatory impulses to their respective small groups of fibers. Inhibition does not occur in vertebrate muscle; in vertebrates, inhibition is achieved at the level of the neurons of the central nervous system and not at the level of the muscle itself (Hoyle 1957; Wiersma 1961).

Insect flight muscle

Many insects, such as dragonflies, moths, butterflies, and grasshoppers, have relatively low wingbeat frequencies, and each muscle contraction occurs in response to a nerve impulse. Because the muscle contraction is synchronized with the nerve impulse, this type is known as *synchronous muscle*.

Many small insects, such as bees, flies, and mosquitoes, beat their wings at frequencies from 100 to more than 1000 beats per second (Sotavalta 1953). This is much too fast for one nerve impulse to arrive at the muscle for each contraction, exert its effect, and decay before the next impulse arrives. It follows that for the muscles to contract at the high rates, they must be stimulated by some other mechanism. These fast muscles do have nerves, but the nerve impulses arrive at a lower frequency than the contractions, and the muscles are therefore known as *asynchronous muscles*. This type of muscle is found in four insect orders: flies and mosquitoes (Diptera), wasps and bees (Hymenoptera), beetles (Coleoptera), and some true bugs (Hemiptera).

The first surprise is that the muscles inside the thorax of such a fast-beating insect are not attached to the wings at all, but rather to the wall of the thorax. There are two sets of muscles, which for convenience we can call vertical and horizontal. The primary effect of muscle contraction is to distort the thorax, and because the amount of shortening is very small, only a few percent of the muscle length, it can take place very fast. The most important property of asynchronous muscle is, however, that it contracts in response to being stretched.

When the vertical muscles contract, the thorax is distorted and springs into a new position with a click, so that the tension is suddenly removed from the contracting vertical muscles (Figure 11.19). However, the sudden change in shape of the thorax causes the horizontal set of flight muscles to be stretched, and this in turn acts as a stimulus for these to contract. This contraction now distorts the thorax in the opposite direction, and it suddenly clicks back to the initial position, releasing the tension on the horizontal muscles. This stretches the vertical muscles, which respond with a new contraction, and so on.

The elastic thorax with its muscles constitutes a

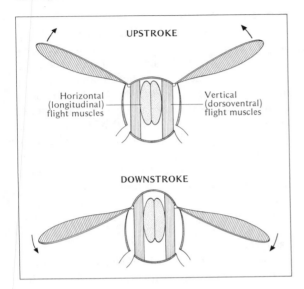

FIGURE 11.19 In some insects the flight muscles together with the elastic thorax form an oscillating system that permits the wingbeat to occur at a much higher frequency than the frequency of nerve impulses. For details, see text.

UPSTROKE

Horizontal (longitudinal) flight muscles

Vertical (dorsoventral) flight muscles

DOWNSTROKE

mechanical oscillator, which by the energy input from the muscles is kept oscillating without being damped out by frictional forces. The most important damping is, of course, the drag on the wings as they propel the insect, and most of the muscle work is used to overcome this drag. The flight continues although the nerve impulses arrive at intervals of several muscle contractions. In some cases as many as 40 wingbeats have been recorded for each nerve impulse. Nerve impulses are needed, however, both for the initiation of flight and for maintenance of continued flight activity (Pringle 1949, 1957).

SKELETONS

To exert their force muscles must be connected to some mechanical structure; they cannot do much for an animal unless they have something to pull on. The appropriate mechanical structure for transmission of force is the skeleton. It usually consists of a rigid, jointed structure, although a fluid can also be used to transmit force, the same way the hydraulic system is used in an automobile. We have the following major categories to consider:

Rigid skeleton
  Endoskeleton: vertebrates
  Exoskeleton: insects, crustaceans

Hydraulic skeleton
  Fluid + soft walls: worms, octopus, starfish
  Fluid + rigid elements: spider legs

A vertebrate has a rigid *internal skeleton*, the bones; an insect or a crustacean uses its stiff cuticle as an *external skeleton*. Some animals (e.g., earthworms) use their internal fluids for transmission of force; they have *hydraulic skeletons*.

Rigid skeletons

A vertebrate has most of the soft tissues of the body draped around the skeleton. Is this the best arrangement? An arthropod has its skeleton on the outside, and all the soft parts are well protected. Is this a better arrangement?

Let us compare the two kinds of structures. For reasons of simple mechanics, a hollow cylindrical tube can support a much greater weight without buckling than can a solid cylindrical rod made from the same amount of material. Therefore, if an animal can afford only a certain weight of skeleton, it is advantageous to use the material for a tube. This is precisely what arthropods have done, and they seem to have an advantage over vertebrates in this regard.

Why, then, do not vertebrates utilize a similar arrangement? First, for a constant weight of material, the rigidity of a hollow cylinder decreases very rapidly as its radius is increased, and eventually it will be so thin that the entire structure collapses.

To prevent buckling of the hypothetical exoskeleton of a very large animal, it would be necessary to increase the thickness and thus the weight. An exoskeleton has another mechanical disadvantage: Even if it is strong enough to provide the necessary support, it will be very sensitive to impact, with the risk of buckling as well as being punctured. This danger, which is of little significance for small animals such as insects, increases rapidly with increasing body size.

Another difficulty with an exoskeleton is growth. Most adult insects never grow; they retain the same rigid shape and size throughout their lifetime, and the larval growing phase is equipped with a softer cuticle that can be stretched. Even so, the larval skin is shed at intervals to permit adequate size increase.

Aquatic arthropods that molt periodically go through a period of increase in size when the old exoskeleton has been shed and the new exoskeleton is still soft and can be stretched. Aside from the lack of protection and the inability to move during this "soft" period, the mechanical disadvantage of a periodic loss of rigidity is not too inconvenient for an aquatic animal, for its body weight is supported by the water. For a land animal, the problem would be enormous: Any sizable animal would collapse under its own weight if it were to go through a molting process.

Nevertheless, some of the protective advantages of an exoskeleton are, in fact, utilized by vertebrates: The brain, certainly a vital organ, is enclosed in a rigid bone structure. The braincase is not part of the locomotory skeleton, however.

One advantage of an exoskeleton stems from its elastic properties. Elastic deformation of skeletal elements constitutes a storage of energy that is utilized for purposes of locomotion; some examples are insect flight (discussed above) and the jump of fleas (to be examined later).

Hydraulic skeletons

There is one important difference between rigid and hydraulic skeletons. A rigid skeleton consists of elements that usually act around a pivot point (a joint). The muscles are attached at both ends, and the force they exert is transmitted through the rigid element and acts somewhere else. For the hydraulic skeleton to work, a fluid must be enclosed in a limited space, and the muscle force must be used to produce pressure in this fluid (i.e., the muscle must usually enclose the fluid space in circular layers). In this case, the muscle does not have a point of attachment; a ring of muscle pulls only on itself.

Earthworms

The role of the hydraulic skeleton in the locomotion of worms has been studied in detail. The common earthworm is a familiar example. Its body wall has layers of muscles that run in two distinct directions, either in circular sheets around the body or in the longitudinal direction.

The movement of the earthworm when it crawls forward is approximately as follows. A wave of contraction of the circular muscles begins at the anterior end. This causes the body to become thinner and elongated, pushing the anterior end forward. The circular contraction passes down the body as a wave, followed by a wave of contraction of the longitudinal muscles. Tiny bristles on the side of the body prevent backsliding, and the anterior end pulls the adjacent part of the worm in its direction. In this way a portion of the worm is pulled forward, the motion passing as a wave backward on the heels of the wave of circular contraction. The cycle is repeated with a new elongation at the anterior end, the worm moving forward as the waves of contraction move backward (Figure 11.20).

When the earthworm moves, the muscles exert their force on the fluid contents of the animal; the

FIGURE 11.20 The crawling earthworm uses its body fluids as an internal hydraulic skeleton. Contraction of circular muscles pushes the front end forward and is followed by a contraction in longitudinal muscles that thickens the body. The thickened segments remain in place relative to the ground as the other parts move forward. The track of individual points on the worm's body and their movements relative to each other are shown by the lines running obliquely forward from left to right of the diagram. Diagram was prepared from a movie film. [Gray and Lissmann 1938]

JUMPING SPIDER The jumping spider (*Sitticus pubescens*), like other spiders, lacks muscles for extension of its legs. When this spider leaps on its prey, it uses the hind pair of legs, which are extended hydraulically by blood pressure. [Courtesy of G. A. Parry, Cambridge University]

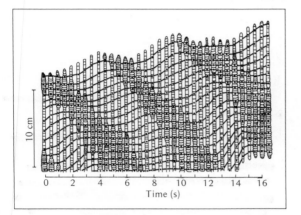

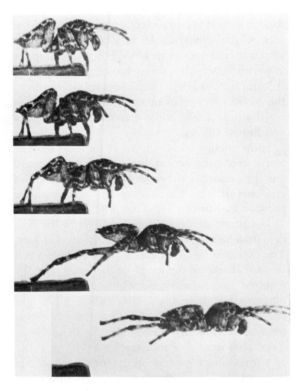

circular muscles build up a pressure that causes the longitudinal muscles to be stretched, and when the longitudinal muscles in turn contract, the pressure stretches the circular muscles and the worm thickens.

The body of the earthworm is separated into compartments by septa. This helps a great deal in locomotion, for it makes the various regions of the worm relatively independent of the rest of the worm. Many worms, the lugworm (*Arenicola*), for example, lack such septa (Seymour 1971). One disadvantage of the lack of septa is that if the worm is wounded and body fluid is lost, it is virtually unable to move. In contrast, as nearly everybody knows, if an earthworm is cut in two, each part can move about more or less like an intact worm.

Spider legs

Spiders cannot extend their legs with the aid of muscles because they have no extensor muscles. To flex the legs they use muscles, but how are the legs extended? There are two possibilities: the use of elastic forces or hydraulic forces.

Elastic hinge joints in the legs could be used if they were arranged to extend the legs whenever the flexing muscles relax. This would be similar to the hinge of bivalves, in which the shells gape when the closing muscle relaxes. We can exclude this possibility by a simple observation: The joints of a detached spider leg remain neutral over a wide angle.

The other possibility, the use of hydraulic pressure, is what spiders do. This can be shown in various ways, most simply by detaching a leg and pinching the open end with a pair of forceps. This closes the opening and raises the fluid pressure inside, and as a result the leg extends. Furthermore, there is a direct relationship between the internal pressure applied to the leg and the joint angles of the leg. Also, if a spider is wounded so that blood fluid is lost, it is completely unable to extend the legs, which remain flexed in a disorderly fashion up against the body. The blood pressure in a spider is surprisingly high – up to 400 mm Hg, or about 0.5 atm (ca. 50 kPa). This differs very much from the blood pressure of other arthropods, which is usually only a few millimeters of mercury (Parry and Brown 1959a).

Some small spiders, known as jumping spiders, can leap on their prey at distances of more than 10 cm. For jumping they use a sudden extension of one pair of legs. Can the spiders use hydraulic pressure for this very rapid motion?

The jump itself is produced entirely by the sudden straightening of the last (fourth) pair of legs. The extension torque necessary for the jump can be estimated from the length of the jump, and the required hydraulic forces, which are difficult to measure directly in the jumping spider (it weighs only about 10 mg), fall within the range of blood pressures commonly measured in spiders. High-speed photographs of jumping spiders support the view that blood pressure is involved, for at the moment of the jump the minute spines on the legs become erect precisely in the way they do when there is an increase in fluid pressure in the leg. The exact release mechanism is not fully understood, however (Parry and Brown 1959b).

LOCOMOTION: BIOMECHANICS

The skeletal muscles are used primarily to provide mechanical energy for animals to move about. Animals can walk and run and jump, fly in the air, and swim in the water. With the exception of ameboid, ciliary, and flagellar locomotion, and a few more unusual means of moving about, muscles are universally used in animal locomotion. In Chapter 6 we discussed the energy requirements for moving about; in the following pages we shall consider some mechanical aspects of these problems.

Running

An animal that runs on a horizontal surface uses power in three major ways:

1. Moving through the air requires power to overcome the frictional drag of the air.
2. As the animal runs, its center of mass is constantly lifted and falls again. Work is done each time it is raised, increasing its potential energy, and most of this is lost again as the center of mass is lowered.
3. As the animal moves, the limbs are constantly accelerated and decelerated again. Work is required to accelerate a limb and provide kinetic energy; to decelerate the limb requires additional work.

The mechanical power to overcome friction in joints is minimal, but some goes into increased breathing movements and pumping activity of the heart. The amount, however, is fairly small, compared with the power required for the mechanical movements of the body.

The air resistance can be calculated from fluid dynamics considerations and from direct measurements of wind pressure on shapes of similar area to the running animal. It can also be measured from the increase in oxygen consumption during running in winds of various speeds. The results show that for humans the energy cost of overcoming air resistance in track running at middle distance speed may be 7.5% of the total energy cost. The air resistance can be expected to increase approximately with the square of the speed, and at sprint speed the cost may be 13% of the total energy cost of running. Running at low speed or walking involves negligible air resistance. It has been found that running 1 m behind another runner virtually eliminates air resistance (Pugh 1971). For other mammals, except those that run at extremely high speeds, air resistance is of minor importance in the total cost.

The friction against a firm substratum is likewise neglible (most of us know from experience that walking in loose sand is an entirely different matter). Thus, the work done on the environment (air resistance plus frictional resistance against the ground) is a small fraction of the total output, and as a result, we must conclude that most of the energy used in running is dissipated internally as heat. (It was mentioned in Chapter 6 that the situation is different for a flying bird, where perhaps 25% of the power output goes into work done on the environment.)

Running uphill

Additional power is required to run uphill. A body that moves uphill acquires potential energy, the product of its weight and the vertical distance through which it has moved. We can therefore predict that the energy required to move one unit of body weight for one unit of vertical distance should be the same for animals of widely different body sizes.

To lift 1 kg of body weight 1 m vertically increases its potential energy by 1 kg m (which equals 9.8 J or 2.34 cal and corresponds to an oxygen consumption of 0.49 ml O_2). If the efficiency of the muscles of large and small animals is similar, say 20%, they would all use five times this amount. This expected similarity in the relative metabolic cost of moving vertically has been amply documented by many investigators on animals ranging in size from mice to horses.

This similarity in the cost of moving vertically has an interesting consequence: It must be much easier for a small animal to run uphill than for a large animal. Why?

The answer is simple. The normal specific metabolic rate of a 30-g mouse is about 15 times higher than for a 1000-kg horse. Because the vertical component of moving one unit body weight is the same for the two animals, the relative increase in metabolic rate attributable to the vertical component should be only one-fifteenth as great in the mouse as in the horse.

This surprising conclusion has been experimentally tested by Taylor et al. (1972). Mice were trained to run on a treadmill at a 15° incline, which indeed is a rather steep hill. The results showed no statistically significant difference in the oxygen consumption of mice running uphill, on the level, or downhill. For chimpanzees (17 kg body weight) the oxygen consumption nearly doubled on the uphill slope, relative to running horizontally. For a horse, the increase, relative to moving horizontally, must be even greater.

Taylor concluded that moving vertically at 2 km per hour requires an increase in oxygen consumption of 23% for a mouse (hardly noticeable), of 189% for a chimpanzee (nearly doubling), and of 630% for a horse (heavy work). This explains the ease with which a squirrel moves up and down a tree trunk, apparently without effort. For an animal of such small size it makes little difference whether it runs up or down.

Energy saved by elasticity

A tennis ball bouncing on a hard surface bounces repeatedly, but the height of the bounce gradually decays. When the ball hits the surface, its kinetic energy is taken up in elastic forces and recovered again as the ball bounces off. As the ball rises, its kinetic energy is transformed into potential energy, only to be changed into kinetic energy on the way down again. Energy is gradually lost because of friction in the air as well as friction during elastic deformation of the tennis ball. However, a small input of energy, just enough to overcome the frictional losses, can keep the ball bouncing.

Does this have any similarity to animal locomotion? We saw that a great deal of energy goes into accelerating and decelerating the limbs and in moving the center of mass in the gravitational field. If elastic elements could act as springs and take up some of the kinetic energy, a great deal of energy could be saved.

The elastic elements we must consider are the tendons and the muscles. At first thought it may seem surprising that muscles can be considered as elastic, for their main function is to contract and exert a force. However, consider a running man and the moment his foot hits the ground. If the leg muscles were flaccid and limp, the man would collapse; obviously, the muscles must resist stretching and thus already be contracted in advance of the precise moment we consider. The kinetic energy of the body provides the force to stretch the muscles, and these resist, which requires work. Can some of the kinetic energy be taken up in elastic elements and be recovered?

The most significant studies of these problems have been made by a group of Italian investigators in Milan, among them Rodolfo Margaria and Giovanni Cavagna, and their results are most helpful in understanding the energetics of locomotion.

The fluctuations in potential and kinetic energy during running and measurements of the forces exerted on the ground can be analyzed with the use of a *force platform*, an instrument that in principle is quite simple. The subject under study, whether animal or man, runs over a solid platform of known mass, suspended on very stiff springs, while force transducers measure the minute movements of the platform in three dimensions. The recorded movements can then be used to calculate the forces and their direction during the entire time a foot is in contact with the force platform. From the information, it is possible to compute the total mechanical work needed for a single step. It turns out that the work performed per kilometer is independent of speed, the total mechanical work for a running man being about 0.40 to 0.50 kcal kg^{-1} km^{-1} (Cavagna et al. 1964).

Simultaneous measurement of the oxygen consumption shows how much metabolic energy the man uses. A calculation of the efficiency (the ratio of work output to energy input) gives an apparent efficiency of approximately 50%. This value seems quite unreasonable, for all available evidence shows that the mechanical efficiency of muscles in transforming chemical energy to external work cannot exceed 25%. This discrepancy, however, could vanish if one-half of the mechanical work recorded by the force platform were obtained from stored kinetic energy recovered by the elastic recoil from the stretched leg muscles. The conclusion that elastic storage is important in human and animal locomotion has been amply confirmed by later investigations.

Anyone can easily convince himself that elastic recoil is important in saving energy. When doing deep knee bends from a standing position, it is much easier to raise the body if the legs are stretched immediately than if one waits a second or so in the low position before standing straight again. During the bending movement of the knees, the extensor muscles are kept tensed, controlling the downward movement. If they shorten again immediately, elastic recoil helps, but if the

TABLE 11.4 Oxygen consumption and mechanical efficiency of doing knee bends. In one case, the subjects made a short pause in the squatting position; in the other case, they took advantage of elastic rebound by straightening immediately. [Thys et al. 1972]

	Oxygen consumption (liter O_2 min^{-1})	Efficiency
No rebound	1.89	0.19
Rebound	1.49	0.26
Difference (%)	−22%	+37%

tension in the muscles is allowed to decay before raising the body, there is no elastic rebound. Is the apparent difference in effort real?

An objective way of finding out is to measure the oxygen consumption. When the knee-bending exercise is carried out on a force platform, the mechanical work performed in the stretching can be measured accurately, assuring that there is no difference in the two exercises. A difference could hardly be expected, because a subject lifts the same mass (his body) to the same height each time he stretches, and the measurements confirm this conclusion.

In a series of experiments the subjects performed exactly 20 knee bends per minute, in one case stretching immediately after flexing the knees, and in the other stretching after an interval of 1.5 seconds in the squatting position; the total number of knee bends per minute being the same (Thys et al. 1972). The results confirm the subjective impression (Table 11.4). When the exercise was performed with the advantage of elastic recoil, the oxygen consumption decreased by 22%. The mechanical work performed was measured with the force platform, and it was therefore possible to calculate the efficiency of the work, which increased by an amazing 37%.

A similar exercise is to stand on tiptoes with stiff knees and perform repetitive small jumps on both feet. In this exercise elastic energy is stored in the contracted muscles of the calf as these are forcibly stretched. Thus, only part of the positive work done to lift the body is derived from energy-yielding chemical processes in the muscles; the other part is derived from elastic energy stored in the contracted muscles. The latter accounts for as much as 60% of the total, with only 40% coming from metabolic processes (Thys et al. 1975).

The fact that in exercise and locomotion muscles often shorten actively immediately after being stretched by an external force is extremely important. Experiments on isolated muscles show that the force developed when the muscle shortens after being stretched is greater than that developed, at the same speed and length, when the muscle shortens starting from a state of isometric contraction. There is evidence that in addition to the elastic characteristics of the muscle, there may also be some change in the contractile process, for the force-velocity characteristics change during contraction after stretching (Cavagna et al. 1968; Cavagna and Citterio 1974).

A different approach has been used in studying animal performance. A trained German shepherd performed a single running jump, or long jump, clearing about 3 m over a series of hurdles. In these experiments a force platform was placed at the spot of takeoff, giving information about direction and forces throughout the jump. The muscles we are concerned with are the combined gastrocnemius and plantaris, the only muscles exerting a force on the ankle (Figure 11.21) (Alexander 1974). These muscles must exert a force throughout the period of contact of the hindpaws with the ground, and from the force on the platform and the moment arm of the bone at the ankle joint, the total force exerted by the two muscles can be computed.

The change in force during landing and takeoff is shown in Figure 11.22. In this graph the force exerted by the combined gastrocnemius and plan-

FIGURE 11.21 Outline tracings of a dog making a running long jump, taking off from a force platform. The numbers correspond to film frames, taken at 56 frames per second. The arrow superimposed on each outline represents the direction and the magnitude of the re- sultant force in that frame. Dots mark joints used for calculation of forces. To the right is an outline of the skeleton of the hindlimb at the moment of takeoff, showing the location of the gastrocnemius and plantaris muscles. [Alexander 1974]

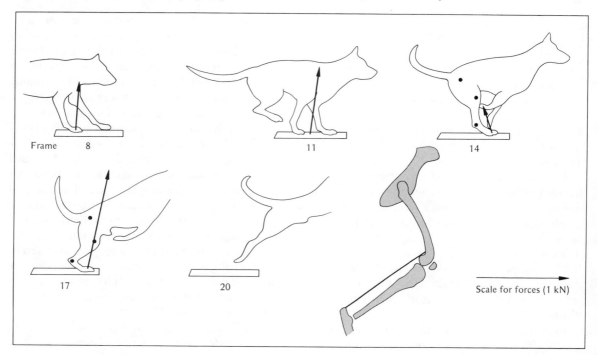

taris is plotted against the change in length of the muscles, calculated from films taken of the jump. The numbers on the graph correspond to the frames on the film, beginning at an instant after the hindlegs hit the platform. The next frames show the increase in force as the muscles are being stretched (15 and 16); then the muscles shorten again (17 through 20), and the force reaches zero as the hindlegs leave the platform.

The record is extremely interesting. During the extension of the muscles, work is being done on the muscles, the work being force times distance. During the takeoff phase, this work is being re- turned, and because it follows exactly the same path, the work (force × distance) during the recoil must be exactly the same as the work done on the muscle during the stretching. Thus, the gastroc- nemius and plantaris muscles with their tendons have behaved as passive elastic bodies. From the amount of stretching of the tendons and their elas- tic properties (the force being known from the records), it can be calculated that most of the elas- tic energy probably is stored in the tendons.

If no work is done by these muscles, what brings the dog over the hurdles? Only a part is attributable to kinetic energy accumulated during the run and recovered elastically. The extensors of the hip shorten and perform work throughout the period of contact of the hindfeet with the ground. The mass of these muscles is substantial, and they contribute a major part of the work that enters into the long jump (Alexander 1974).

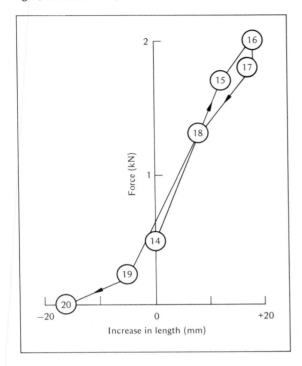

FIGURE 11.22 Force exerted by the combined gastroc-nemius and plantaris muscles during the jump of a 35-kg dog, plotted against the changes in muscle length. Numbers correspond to film frames, as in Figure 11.21. A force of 1 kN equals 102 kgf ; thus the maximum force of about 2 kN (frame 16) equals about 200 kgf. [Alexander 1974]

Kangaroos jump more than most animals; what can we learn from their ways? When kangaroos move slowly with the use of all four feet and the tail, their oxygen consumption increases steeply with increasing speed (Figure 11.23). However, at higher speeds, when they jump on the hindlegs only, there is no further increase in the rate of oxygen consumption as the speed increases from 8 km per hour to 22 km per hour, almost a threefold increase (Dawson and Taylor 1973).

In the context of our discussion of elasticity, it is interesting to examine the kangaroos' frequency of jumping. The number of jumps per minute does not change much in spite of a threefold increase in speed; the increase in speed is almost entirely the result of an increase in the length of each hop.

What does this signify? If the duration of a single hop does not increase, the vertical height of the hop must be unchanged, and only the length is increased. That is, no increase in vertical force is necessary, and only the forward vector of the force is increased to lengthen the hop. The longer hop signifies an increase in the kinetic energy of the kangaroo as it hits the ground, and if a major fraction of this kinetic energy can be taken up in elastic elements, it will constitute a great savings in energy required for the increased speed.

Measurements from force platforms and calculations of the forces in the long Achilles tendon of a kangaroo suggest that the savings attributable to elastic storage in the tendon amounts to 40% of the positive work the leg muscles must do during contact of the hindfeet with the ground (Alexander and Vernon 1975).

More about jumping

Fleas and grasshoppers can make jumps that are perhaps 50 times the length of their bodies, and this has given rise to the popular pastime of calculating how far a man should be able to jump, could he jump in the same proportion. In North America he would jump to the top of the Empire State Building, and in France he would clear the top of the Eiffel Tower. Let us instead examine this question realistically.

Assume that a grasshopper weighs 1 g and can make a standing jump of 50 cm, both figures being quite realistic. Assume a larger animal of isometric build with all linear dimensions increased exactly 10-fold. What is the effect of scaling on the expected height of a jump? The large animal will have a mass 1000 times that of the small one, but the cross section of its jumping muscles (which determines the force the muscle can exert) will be increased only 100-fold. Relative to its larger mass,

FIGURE 11.23 Records from two kangaroos (18 and 28 kg) moving on a treadmill: *a*, oxygen consumption; *b*, stride frequency; *c*, stride length. The left part of each graph refers to movement on all four legs and supported by the tail ("pentapedal" movement); the right part refers to hopping on the powerful hindlegs with the heavy tail used as a balancing device. The dashed line in (*a*) represents oxygen consumption for an 18-kg mammal running quadrupedally. [Dawson and Taylor 1973]

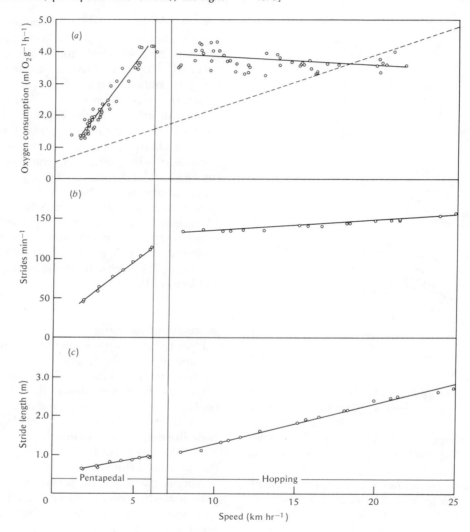

the large animal therefore has only one-tenth the force available to accelerate the body.

Since acceleration equals force/mass, the acceleration during takeoff is also one-tenth that in the small animal. However, as all linear dimensions are increased 10 times, acceleration continues throughout a takeoff distance that is 10 times the takeoff distance of the small animal. The net result is therefore that the takeoff speed is the same for the two animals. After takeoff both ani-

TABLE 11.5 Mechanical performance of four animals in standing jumps. The height of the jumps are all within the same order of magnitude, although the body masses differ by a factor of more than 100 million. (The flea is so small that air resistance is significant, and to reach the indicated height it must in fact have a higher than calculated initial acceleration, estimated at 320 g.)

	Flea (Pulex)[a]	Click beetle (Athous)[b]	Locust (Schistocerca)[c]	Man (Homo)[d]
Body mass, m	0.49 mg	40 mg	3 g	70 kg
Height of jump, h	20 cm	30 cm	59 cm	60 cm
Acceleration distance, s	0.075 cm	0.077 cm	4 cm	40 cm
Takeoff speed, $v = \sqrt{h \cdot 2g}$	190 cm s^{-1}	240 cm s^{-1}	340 cm s^{-1}	343 cm s^{-1}
Time of acceleration, $t = 2s/v$ (takeoff time)	0.00079 s	0.00064 s	0.00235 s	0.233 s
Acceleration, $a = v/t$	241 000 cm s^{-2}	374 000 cm s^{-2}	14 500 cm s^{-2}	1471 cm s^{-2}
Acceleration relative to gravitational acceleration (g)	245 g	382 g	15 g	1.5 g

[a] Bennet-Clark and Lucey (1967). [b] Evans (1972, 1973). [c] Hoyle (1955), Alexander (1968).
[d] Estimate based on center of mass lowered to 60 cm above ground at beginning of jump, accelerated over 40 cm to 100 cm at takeoff, and lifted to 160 cm. World record for standing jump is about 165 cm above ground.

mals are slowed by the effect of the force of gravity. Because both the kinetic energy at takeoff and the retarding force of gravity are proportional to the body mass, the deceleration attributable to gravity permits the two animals to rise to equal heights. (Here we have disregarded the additional loss of speed caused by air resistance, which for a very small animal such as a flea becomes important.)

We can now conclude that if the muscles of the small and large animals exert the same force per cross-sectional area, a small and a large animal of isometric build should be able to jump to exactly the same height.

Consider now the actual jumping performance of a variety of animals (Table 11.5). (We should consider only standing jumps, for a running jump utilizes the kinetic energy of the animal to increase the height of the jump. We will therefore disregard the world record high jump for man or a running horse, which both are about 2 m.) In a standing jump a man can clear about 1.6 m. His center of mass, however, is not lifted over this distance; it is lifted over less distance, for it is not at ground level when he jumps but rather at about 1 m (see note to Table 11.5). Thus, man and grasshopper raise their center of mass by roughly the same amount, although the difference in their body mass is between 10 000- and 100 000-fold.

Of course, in reality animals are not isometric. Nevertheless, it is amazing how similar are the jumping records for a variety of animals. It is a matter of simple physics that man and other mammals do not jump in some proportion to their body lengths and that this indeed would be impossible.

The conclusion that similar animals irrespective of body size should be able to jump to equal height can also be stated as follows. Assume that the jumping muscles make up the same fraction of body mass. The muscle force is proportional to the cross-sectional area, and the shortening is proportional to the initial length of the muscle. The cross section times the length is the volume of the muscle, and the energy output of a single contraction is the product of force and distance. Energy output is therefore proportional to the muscle mass, and, in turn, to body mass. Because a jump uses only a single contraction of the jumping muscles, the work performed during takeoff and used for acceleration is the same relative to body mass. The conclusion is that isometrically built animals of different mass should all jump to the same height, provided their muscles contract with the same force.

The galago. How is it possible, then, for some animals to jump much higher? The record for a standing jump probably belongs to the lesser galago, a small tropical primate that weighs about 250 g. Under well-controlled conditions a galago has jumped 2.25 m (Hall-Craggs 1965). This is more than three times the height of a standing jump for man. Unless the muscles can produce more force per square centimeter cross section (which is unlikely), the galago's superior performance can be explained only by the animal's having a larger muscle mass (more energy from the takeoff contraction) and possibly a more favorable mechanical structure of the limbs.

The galago indeed has large jumping muscles, nearly 10% of the body mass (Alexander 1968) or about twice as much as man. If we assume that all combined mechanical advantages of a highly specialized jumping animal could account for a 50% increase in performance, this in combination with a doubling in the mass of the jumping muscles, could account for a three-fold increase in jumping performance – by no means an unreasonable approximation. As a standing jump this is more impressive than the jump of the locust, but we must remember that the galago is a warm-blooded animal highly adapted to making long jumps in its natural jungle habitat.

The flea. Aside from the fact that animals are not completely isometric, we meet another difficulty. The smaller the animal, the shorter the takeoff distance, and since the takeoff speed should be the same for all, the smaller animal must accelerate its body mass much faster. Because much less time is available for the takeoff, the power output of the muscle must be increased accordingly (i.e., the muscle must contract very fast).

For an animal the size of a flea, the takeoff time is less than 1 ms, and the distance over which acceleration takes place is only 0.75 mm. The average acceleration during takeoff exceeds 2000 m s^{-2}, or roughly the equivalent of 200 g (Rothschild et al., 1972).

Muscles just cannot contract this fast, so how can the flea jump at all? The flea uses the principle of the catapult and stores energy in a piece of elastic material (*resilin*), at the base of the hindlegs. Resilin is a protein with properties very similar to rubber (Weis-Fogh 1960; Andersen and Weis-Fogh 1964). The relatively slow muscles are used to compress the resilin, which returns the total energy again with close to 100% efficiency when a release mechanism is tripped. In this way the elastic recoil works much the same way as a slingshot and imparts the necessary high acceleration to the flea.

The click beetle. A click beetle placed upside down on a smooth, hard surface struggles for a moment to right itself, but when its legs find nothing to hold on to, the beetle stops moving. After a pause, the beetle springs into the air with an audible click. If it falls again upside down, it continues clicking until it does land right side up. This legless mechanism for jumping takes the beetle to heights as much as 30 cm, and the animal may reach a takeoff acceleration of nearly 400 *g*.

The mechanism is simple. The first segment of the thorax has a peg that points backward and fits part way into a pit in the hind section of the body (Figure 11.24). When the peg rests on the lip of the pit, it prevents the large jump muscle in the prothorax from swinging the front half of the body upward. The contracting muscle therefore builds up tension and acts like a spring; when the peg finally slips, the beetle jackknifes and the energy stored in the muscle provides the instant force needed for takeoff. Careful measurements of the

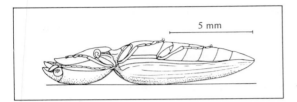

5 mm

jumping action show that during the click the center of mass of the beetle is raised by about 0.6 to 0.7 mm in just over 0.5 ms.

The click beetle, like the flea, is too small for the muscles to provide the necessary acceleration during the short acceleration time. Half a millisecond is not sufficient for muscle contraction, and the best way of obtaining the energy at the necessary speed is to store it in an elastic structure (Evans 1973).

Flying and swimming

Air and water are fluid media, and animals moving through them have similar problems. They do not have a solid substratum to support their weight, and the forces needed to make them move are exerted against a medium that virtually slips away. Air is much less dense and viscous than water, a difference that has substantial effects on animal locomotion in the two media, but there are many fluid dynamics principles in common.

Judging by numbers, flying is extremely successful. Of the roughly 1 million living animal species, more than three-quarters are flying insects. The ability to fly has evolved independently at least four times: in insects, in the extinct large flying reptiles (Pterosaurs), in birds, and in mammals (bats). The size of actively flying animals ranges from about 1 μg for some tiny insects to more than 10 kg for the largest flying birds.

The power needed to propel an animal through the air in forward flight has two components: It must provide lift equal to its weight, and it must provide forward thrust equal to the frictional drag on the animal as it moves through the air. In flying animals lift and thrust are provided by the same surfaces. As we shall see later, this is not always true for animals swimming in water.

The main power for bird flight is provided by the large pectoral muscles, which in all birds make up about 15% of the body mass, irrespective of the size of the bird. The muscles responsible for the upstroke of the wing constitute only about one-tenth of the total mass of the flight muscles and thus cannot provide any major fraction of the power required for forward flight. There are, of course, minor deviations, but as a general rule, in birds the engine makes up the same fraction of the aircraft, irrespective of its size (Greenewalt 1975).

As a group, hummingbirds are different; their flight muscles make up 25 to 30% of the body mass. This is in accord with the greater power needed for hovering than for flapping forward flight (Weis-Fogh 1973). Furthermore, the muscles responsible for the upstroke constitute about one-third of the total mass of the flight muscles. This indicates that for hovering flight, both upstroke and downstroke are important in providing power, and because the bird remains in the same spot, all the power goes to provide lift. This conclusion is confirmed by aerodynamic analysis of the wing movements of hummingbirds (Stolpe and Zimmer 1939).

In spite of the intricate fluid dynamics theory of flight, both bird flight and insect flight are amazingly well understood (Pennycuick 1969; Tucker 1973; Weis-Fogh 1973, 1976; Lighthill 1974). The power required for bird flight is an extremely complex function of the size and shape of the bird, the speed, the size of the wings, and the shape of the wings. The exact analysis is further complicated by the constantly changing speed and shape of the wing during the wing stroke. Aerodynamic

FIGURE 11.25. Power required for bird flight compared with maximum available power. The familiar relationship between resting metabolic rate and body mass of birds is drawn as a regression line with the slope 0.75. A regression line for an assumed maximum power output of 15 times the resting rate is drawn as a parallel line

at a level 15 times higher. The estimated power required for flight is directly proportional to body mass, and a regression line with the slope 1.0 is drawn to intersect maximum power output at 10 kg body mass, the approximate size of the largest flying birds.

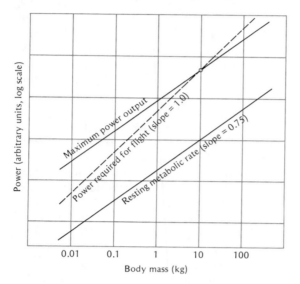

FIGURE 11.25. Power required for bird flight compared with maximum available power. The familiar relationship between resting metabolic rate and body mass of birds is drawn as a regression line with the slope 0.75. A regression line for an assumed maximum power output of 15 times the resting rate is drawn as a parallel line

largest flying birds. It means that a very large bird, such as the ostrich (ca. 100 kg), even if it were anatomically designed to fly, could not be expected to muster enough power for steady-state forward flapping flight.

It is interesting to compare this information with the details of man-powered flight in which the famous Kremer Prize was won by a well-trained athlete weighing 62 kg who flew a plane weighing 32 kg. By cranking the propeller with a bicycle-type drive chain, he flew the plane successfully over a designated course of nearly 2000 m in about 7 minutes (Hirst 1977). This was a peak performance by a well-trained man under ideal conditions; much longer, sustained man-powered flights are unlikely. There is, however, good agreement between our knowledge of power requirements for flapping and propeller-driven flight and actual performance. In fact, a propeller is simply an air foil in which the surface providing thrust has been separated from the surface providing lift.

Swimming is in principle not too different from flying through the air, except that the higher density of water reduces the power needed for providing lift. In fact, some swimming animals have the same density as the water, and no power is used for obtaining lift; all power needed for swimming can be put into thrust. The higher density and viscosity of water are important; fish do not swim as fast as birds fly.

For large animals, such as fish, the drag increases roughly in proportion to the square of the speed, and this puts severe limits on the attainable speed (Lighthill 1974). We have already seen how some of the largest and fastest predatory fish have developed means of keeping their power plant, the muscle mass, at a high temperature so that maximum power output is possible.

The problem of attaining the same density as water, and thus achieving neutral buoyancy and

theory (Lighthill 1974) suggests that the power required for bird flight should increase in proportion to the body weight to the power 1.17. Tucker's analysis, which includes consideration of empirical measurements, gave results which, in a simplified form, suggest that required power input should be proportional to the bird's weight (Tucker 1973). For the following argument it makes little difference whether we use the exponent 1.17 or 1.0.

We know that the resting metabolic rate of a bird increases in proportion to the body mass to the power 0.75 (page 189). Although we do not know for sure, we can assume that the maximum sustainable power output is similarly related to body size. This simply means that available power does not increase as rapidly with body size as the power required for flight. This is shown in Figure 11.25, where the curve for power required for flight is drawn to cross the maximum available power at a body mass of 10 kg, the approximate size of the

avoiding the expenditure of energy to provide lift, has a wide variety of solutions, which we shall examine next.

BUOYANCY

If an animal that swims is heavier than the water, part of its energy expenditure goes to keep from sinking, and only part is available for forward locomotion. If the animal could have the same density as the water (i.e., have neutral buoyancy), it would have more energy available for moving.

The problem of buoyancy is more important for large than for very small animals. Small organisms have a large relative surface and often have extensions of various sorts that reduce the sinking rate in water; the smallest can remain suspended much like dust particles in air and live through their entire life cycle in this way. Surface extensions large enough to act as brakes on sinking become mechanically impossible for large organisms, which must therefore use other means.

A reduction in the density or specific gravity of an organism is the only solution to staying afloat that does not involve some amount of swimming action (i.e., the expenditure of energy). We can list five possible avenues for reducing the tendency to sink and shifting the organism toward neutral buoyancy:

1. Reduction in the amount of heavy substances (e.g., calcium carbonate or calcium phosphate)
2. Replacement of heavy ions (e.g., Mg^{2+} and SO_4^{2-} by Na^+, Cl^-, or even lighter ions, H^+ and NH_4^+)
3. Removal of ions without replacement (i.e., making the organism more dilute)
4. Increase in amount of substances lighter than water, primarily fats and oils.
5. Use of gas floats, such as the swimbladder of fish.

TABLE 11.6 Specific gravity of sea water, fresh water, and substances commonly found as skeletal material in aquatic organisms. [Denton 1961]

Substance	Specific gravity
Sea water	1.026
Fresh water	1.000
Calcium carbonate	
Aragonite	2.9
Calcite	2.7
Calcium phosphate	
Apatite	3.2
Chitin	1.2

We shall discuss these five possibilities: their advantages and disadvantages and the limitations of each. Because sea water is heavier (specific gravity = 1.026) than fresh water (specific gravity = 1.000), it is easier for an organism to achieve neutral buoyancy in the sea; in fact, some ways of producing neutral buoyancy in the sea can never attain this goal in fresh water.

Reduction in heavy substances

Many skeletons consist of very heavy substances, often calcium salts, either calcium carbonate or calcium phosphate. Calcium carbonate is common among invertebrates (the shells of clams and snails, corals, etc.); calcium phosphate is the major component of vertebrate bones, where it is found in the form of the mineral apatite. Some organisms (e.g., some sponges) have skeletons made of silica; others (many arthropods) use chitinous substances, either alone or together with calcium carbonate. The relative contribution of these substances to the weight of an animal in water is evident from their specific gravities (Table 11.6).

It is easy to see that reducing the amount of calcium salts can go far toward reducing the weight of an organism. Accordingly, pelagic nudibranch snails lack shells, whereas opistobranch snails, which are bottom organisms, have well-developed shells. Swimming coelenterates (e.g., jellyfish) completely lack the heavy calcium carbonate skeletons characteristic of many sessile coelenterates such as corals.

It is interesting to note that one group of pelagic organisms, the swimming crabs, has not gone very far toward reducing its heavy exoskeleton (mostly calcium carbonate). This holds for the common blue crab (*Callinectes*), which swims continuously to keep from sinking.

There is, of course, a considerable structural disadvantage to the removal of mechanical support. Therefore, this solution is more common in relatively small organisms. An alternative is replacement of a heavy skeleton by a skeleton built of lighter substances. For example, in the ordinary squid the calcium carbonate skeleton found in its relative, the cuttlefish, has been replaced by a lighter structure made of chitin, the so-called pen of the squid.

Replacement of heavy ions by lighter ions

The best examples of this solution are perhaps found among plants. Two multicellular algae, *Valonia* and *Halicystis*, show what can be achieved by manipulating the ionic composition of the organism. Both are almost neutrally buoyant: The former barely sinks in sea water, and the latter has a slight positive buoyancy. Both have abundant cell sap, which is isotonic with sea water but differs in composition (Table 11.7). Both exclude the heavy sulfate ion almost completely. *Valonia* likewise excludes magnesium, which is present in *Halicystis* in a concentration less than one-third that in sea water. Calcium, another important divalent ion, is also excluded by both, relative to its concentration in sea water.

Of the monovalent ions, K^+ is heavier than Na^+. *Valonia*, which is slightly heavier than sea water, accumulates primarily K^+ in the cell sap, as

TABLE 11.7 Ionic composition of sea water and of the sap of two multinuclear algae that have nearly neutral buoyancy. Ionic molecular concentrations are expressed relative to chloride ($Cl^- = 100$). [Gross and Zeuthen 1948]

	Sea water	Sap of *Valonia*	Sap of *Halicystis*
Cl^-	100	100	100
Na^+	85.87	15.08	92.4
K^+	2.15	86.24	1.01
Ca^{2+}	2.05	0.285	1.33
Mg^{2+}	9.74	Trace	2.77
SO_4^{2-}	6.26	Trace	Trace
Specific gravity	1.0277	1.0290	1.0250
Artificial solution[a]	1.0285	1.0285	1.0252

[a] Containing the ion concentrations listed above.

plants in general do. *Halicystis*, however, which is slightly lighter than sea water, has excluded most of the potassium and uses the lighter sodium as its most important cation.

A very common organism that causes luminescence in the sea is a dinoflagellate, *Noctiluca miliaris*. It is a little less than 1 mm in diameter and tends to accumulate at the surface because it is definitely lighter than sea water. When it is mechanically stimulated, it emits light flashes that are visible as phosphorescence at night. Its light emission has been the subject of extensive studies, but we shall discuss only its buoyancy problems.

The composition of the cell sap of *Noctiluca* is given in Table 11.8. It is isotonic with sea water, and it achieves its positive buoyancy in part by reducing Ca^{2+}, Mg^{2+}, and especially the heavy SO_4^{2-}. Furthermore, it accumulates ammonium ion, which is lighter than the two commonest monovalent cations, Na^+ and K^+.

The hydrogen ion (H^+) is much lighter than any of the other cations. It has been suggested that positive buoyancy could be achieved by increasing the hydrogen ion concentration at the expense of other cations, but this is not feasible. Although the cell sap of *Noctiluca* is rather acid, (i.e., has a high hydrogen ion concentration), it is easy to see that

TABLE 11.8 Ionic composition of sea water and of the cell sap of the dinoflagellate *Noctiluca,* a common source of phosphorescence in the sea. It achieves positive buoyancy by excluding heavy ions and replacing them primarily with NH_4^+. [Kesseler 1966]

	Sap (mEq liter^{-1})	Sea water (mEq liter^{-1})
Na^+	414	418
K^+	34	8.8
NH_4^+	58.5	0
Ca^{2+}	9.5	18.5
Mg^{2+}	15	95
Cl^-	496.5	498.9
SO_4^{2-}	Trace	46.6
$H_2PO_4^-$	13	Trace
pH	4.35	8.2

the effect on the specific gravity must be negligible.

If we take the pH of the *Noctiluca* sap to be 4.0, the hydrogen ion concentration is 10^{-4} mol per liter. In other words, the hydrogen ion concentration is 0.1 mmol per liter, and thus can replace no more than 1/5000 of the total cation content in *Noctiluca*. To replace a substantial amount of cation the pH value would have to be substantially less than 2, which corresponds to a hydrogen ion concentration in excess of 10 mmol per liter. Such high acidity is not known to exist in any cells; it occurs only in a few acid secretions such as the gastric juices of vertebrates and the saliva of some marine snails, which use acid to drill into the shells of their prey.

The principle of replacing heavier ions by light ones can be used just as effectively by large organisms. A striking example is the group of deep-sea squids known as cranchid squids. In these animals the fluid-filled coelomic cavity is very large and makes up about two-thirds of the entire animal. If it is opened and the fluid drained out, the animal loses its neutral buoyancy and sinks. The fluid has a specific gravity of about 1.010 (sea water is 1.026), it is osmotically isotonic with sea water, and the pH is 5.2. The concentration of NH_4^+ is

about 480 mEq per liter and of sodium 80 mEq per liter. The anion is almost exclusively chloride; the heavy SO_4^{2-} ion is excluded (Denton 1960; Denton et al. 1969).

The explanation of the high ammonium concentration is simple. The coelomic fluid is slightly acid, and ammonia, which is the normal metabolic end product of protein metabolism in most aquatic animals, diffuses into the fluid and becomes trapped there. A mechanism that could produce the low pH would be a sodium pump that removes Na^+ from the coelomic fluid, for this would leave an excess of anions and thereby increase the hydrogen ion concentration, thus increasing the acidity. The acid fluid would then serve as a trap to sequester ammonia. It has been calculated that if ammonia is the only end product of protein metabolism in a cranchid squid, about 40% of all the ammonia produced throughout the lifetime of the animal is retained in order to maintain neutral buoyancy (Denton 1971).

The method used by the cranchid squids is quite effective in achieving neutral buoyancy. It has some structural disadvantages, however, for two-thirds of the bulk of the animal is fluid (i.e., the volume of fluid carried around amounts to twice the volume of the living animal itself).

Hypotonicity

Removal of some ions without replacing them by other ions would seem a possible avenue for reducing weight. There would, of course, be osmotic problems, for a dilute solution would be hypotonic to the sea water. In general, invertebrate organisms are in osmotic equilibrium with sea water, and none seems to have used hypotonicity as a buoyancy mechanism.

Teleost fish, however, have ion concentrations much lower than those of sea water, and this contributes toward reducing their weight in water.

This is only of minor importance, however, for teleost fish have skeletons and muscles that contribute sizable amounts of relatively heavy substances. In many deep-sea forms, which have greatly reduced skeleton and muscle masses, however, hypotonicity is of considerable help in balancing the relatively small amounts of heavy substances.

Fats and oils

Many planktonic organisms contain substantial amounts of fats. Fat, of course, is a common form of energy storage, both in animals and in many plants, but the contribution of fats and oils toward buoyancy is also important. Planktonic plants frequently deposit fat, rather than the heavier starch that is a common storage compound in plants in general. All diatoms, for example, deposit only fat.

Elasmobranch fish, as opposed to teleosts, never have swimbladders. But they do have large livers, especially those sharks that are good swimmers. The liver may constitute about one-fifth of the body weight; in a teleost fish it usually is about 1 to 2% of the body weight.

A survey of the black spiny shark (*Etmopterus spinax*) showed that on the average 17% of its body weight was liver, and 75% of the liver was oil (S. Schmidt-Nielsen et al. 1934). (In contrast, mammalian liver contains some 5% fat.) Half of the oil in the shark liver was *squalene*, an interesting unsaturated hydrocarbon ($C_{30}H_{70}$) that derives its name from *Squalus,* or shark.

Most fats and oils have specific gravities of about 0.90 to 0.92, but squalene has a lower specific gravity, 0.86. The difference is not great, but in sea water the buoyancy effect of a given volume of squalene is about 50% higher than the effect of a similar volume of fat or oil. Whether this is a reason for the tremendous accumulation of squalene

in elasmobranch livers is difficult to say; nevertheless, its presence definitely aids in achieving near-neutral buoyancy for the fish. This is particularly effective because elasmobranchs, as opposed to teleosts, have relatively light cartilaginous skeletons that are not weighted down by heavy calcium phosphates.

Interestingly, rays and skates, which are bottom-living animals, have smaller livers with a lower fat content. Five species of rays (*Raja*) averaged a liver size of 7.53% with less than 50% fat in the liver. This makes sense, for rays and skates live at the bottom and are not very good swimmers.

It is worth noting that the relatively low ion concentrations in the body fluids of elasmobranchs, relative to sea water, do not contribute substantially to their buoyancy. The elasmobranchs are isosmotic with sea water, the difference in salt concentration is made up primarily by urea, and urea solutions give only very slight lift over sea water. Trimethylamine oxide (TMAO), which also is found in their body fluids, is appreciably less dense than urea and contributes a minor amount toward giving lift.

Several deep-sea teleosts have swimbladders filled with fat, rather than with air. This is true of some species of *Gonostoma* and *Cyclothone* (Marshall 1960), but since the fatty swimbladders of other fish have a substantial fraction of the fat as cholesterol (up to 49%), which has a higher specific gravity (1.067) than sea water (1.026), the role of the fat-invested swimbladder in buoyancy control is dubious (Phleger 1971).

Gas floats

Compared with an equal volume of water, gas has a very low density, and even a relatively small volume is an excellent solution to the problem of buoyancy. This solution, however, has some limitations and disadvantages. The situation differs de-

PORTUGUESE MAN-OF-WAR. The gas float of this colonial coelenterate (*Physalia pelagica*) is filled with a gas that contains a large proportion of carbon monoxide. The length of the float may exceed 20 cm, and the tentacles, which give extremely painful stings, may extend for more than 10 m. [Courtesy of Charles E. Lane, University of Miami]

pending on whether the gas space is enclosed by soft tissue or by rigid walls.

Soft-walled gas floats

Assume a gas-filled rubber balloon submerged below the surface. The further down it is brought, the greater is the pressure of water on its outside, and as the pressure of gas inside matches the water pressure outside, the balloon becomes increasingly compressed.

The water pressure increases by about 1 atm for each 10 m depth, and for an organism living at considerable depth this causes problems. Consider an organism that lives at 1000 m depth, where the water pressure is about 100 atm. To produce and maintain a volume of gas at that pressure entails three difficulties: (1) the gas must be secreted into the space against a pressure of 100 atm, (2) the *amount* of gas needed to fill the same volume at that pressure is 100 times as great as at the surface, and (3) it is a problem to keep any gas at such a high pressure from diffusing out. The gas tensions in the water at any depth in the ocean are usually not very different from those at the surface, where there is equilibrium with 0.2 atm oxygen and 0.8 atm nitrogen. The gases for the float are obtained from the water, secreted into the high pressure space of the float, and must then be kept from diffusing out.

A gas float with soft walls entails another difficulty: Its volume changes if the organism moves up or down in the water. Say that an organism has enough gas in its float to be in neutral buoyancy at a given depth. If it moves down, the gas is compressed; the animal gets heavier and loses buoyancy. If it moves up, the gas expands and positive buoyancy increases. The latter can be dangerous, for unless the animal actively swims down to the neutral level, the gas will expand faster and faster as the animal rises at increasing speed. Unless it has a mechanism for releasing gas, an organism that has lost control of its rise may even rupture as the gas expands.

If the gas space has rigid walls, the limitations on vertical movements are less strict, but the walls must have sufficient mechanical strength to support changes in pressure. If the walls are made stronger, they also become heavier, and this defeats the purpose of the gas chamber: to make the organism lighter. As we shall see later, the mechanical strength of the float is a limitation on the vertical distribution of the cuttlefish (*Sepia*), which uses rigid gas spaces for flotation.

The swimbladder of fish is undoubtedly the best known type of gas float, but the principle is also used by quite a few other organisms, in particular the colonial siphonophores (Coelenterata). One of these is the conspicuous Portuguese man-of-war (*Physalia*), commonly found along the Atlantic coast of North America. The colony is carried by a "sail," a balloon-like gas chamber, often more than 10 cm long. It floats on the surface of the ocean and carries the colony along as it drifts in the wind.

Because the gas chamber of *Physalia* floats on the surface rather than being submerged, the inside pressure is virtually equal to the atmospheric pressure. This simplifies the problem of filling it with gas. However, the composition of the gas is quite surprising, for it contains up to some 15% carbon monoxide. The oxygen concentration is less than in the atmosphere, the carbon dioxide concentration is negligible, and nitrogen makes up the balance. The gas is produced by a gas gland, which uses the amino acid serine as a substrate ($CH_2OHCHNH_2COOH$) for the formation of carbon monoxide (Wittenberg 1960).

Other colonial siphonophores also carry floats that contain carbon monoxide. One, *Nanomia bi-juga*, has very small floats with a volume of 0.14 to 0.46 μl; in these the carbon monoxide concentration is from 77 to 93% (Pickwell et al. 1964). Carbon monoxide is, of course, highly toxic to man and many other higher animals, and usually occurs in nature only in trace amounts. It is found in the hollow stems of some marine algae, where it may constitute several percent of the total gas. In minute amounts carbon monoxide is found in the exhaled air of man, even nonsmokers. It is not formed by bacteria in the intestine, but from metabolic breakdown of hemoglobin in the liver (Ludwig et al. 1957). The exhaled carbon monoxide does not originate externally from air pollution, for the Antarctic Weddell seal (which presumably breathes entirely unpolluted air) also exhales a measurable amount of carbon monoxide (Pugh 1959).

The presence of gas floats in plankton organisms may explain a rather mysterious phenomenon known as the *deep scattering layer*. Ships that use echo sounding for depth determination often find that sound waves are reflected not only by the bottom, but also by a layer in midwater, which thus gives the impression of a false bottom. This extra layer, which often changes in depth from day to night, is the deep scattering layer. For many years the cause of this sound reflection was unknown; but it is now almost certain that the cause is schools of fish with swimbladders or pelagic siphonophores with gas floats.

Siphonophores have gas floats with an opening so that gas can escape when they rise in the water. A simple experiment has shown that gas can be both given off and secreted by these organisms. Pickwell et al. (1964) put them in a 10-ml syringe and exposed them to vacuum by pulling at the plunger. After gas had been pulled out, the floats were heavier than water, but within 1 hour they were again afloat. They were now filled primarily with carbon monoxide, presumably formed in a manner analogous to that of *Physalia*.

At night the deep scattering layer frequently migrates up toward the surface of the ocean. As gas can escape from the floats of siphonophores, they presumably release bubbles as they ascend, and this can explain the observation that the intensity of the sound reflection of the layer frequently increases as it rises.

If organisms release gas on the way up, we would expect that a similar amount of gas is secreted as they descend again. However, they could avoid repeatedly secreting and releasing gas by keeping a constant amount of gas and putting up with the volume changes during vertical migration. This might be feasible if the animal is in neutral buoyancy at the higher level, but it would hardly be possible if the neutral level is at any depth.

Let us consider an example. Say that a fish has a swimbladder with a volume (at neutral buoyancy) of 5% of the body volume. Descending from a night level of 50 m depth to 300 m would reduce the gas volume to one-fifth, or 1% of the body volume. The fish would now be negatively buoyant, but could keep at the new level by a moderate amount of swimming. However, if the fish were in neutral buoyancy at 300 m depth with the swimbladder constituting 5% of the body volume, rising to 50 m would increase the swimbladder volume fivefold (unless gas were released). A fivefold expansion of the swimbladder volume would involve serious danger of losing control and rising all the way to the surface where, at 1 atm pressure, the swimbladder would be a catastrophic 30 times its original volume. If organisms of the deep scattering layer move vertically without releasing gas, we must therefore expect that they are in neutral

buoyancy at the higher, rather than the lower, level.

Information about volume changes can to some extent be obtained without direct observation. The deep scattering layer reflects sound particularly well at frequencies that correspond to the resonant frequency of the gas bubbles. This in turn depends on the volume of the gas bubbles as well as on their pressure. If the pressure is known (the depth can be read directly from the echo-sounding record) and the resonant frequency is assumed to be the one that is best reflected, the size of the bubbles can be calculated. Now, when the layer moves up, one of two situations may prevail: Either gas is given off and the bubbles maintain constant volume, or the gas is retained and the bubbles expand with the pressure change. If a bubble is kept on constant volume, its resonant frequency should vary with $P^{1/2}$ (P is hydrostatic pressure); if the volume changes with pressure while the mass of gas is kept constant, the resonant frequency should vary with $P^{5/6}$. The sound records are complex and often difficult to interpret, but it appears that organisms of the deep scattering layer may conform to either of these patterns (Hersey et al. 1962).

Rigid-walled gas floats

The problems connected with changes in volume, and thus in buoyancy with changing depth, are eliminated if the gas is enclosed in a rigid space. This solution is used by the cuttlefish (*Sepia officinalis*), a relative of squid and octopus that is common in the upper reaches of coastal water masses. Nearly one-tenth of its body volume is made up of a cuttlebone, a calcified internal structure well known by bird fanciers, who use it to supply calcium to their birds (Figure 11.26). The role of the cuttlebone has been studied by the Brit-

FIGURE 11.26 Diagram of a cuttlefish, a marine cephalopod that uses a gas-filled rigid structure called the cuttlebone to achieve neutral buoyancy in water. The posterior part of the cuttlebone (shown in black above) is filled with liquid. In sea water the cuttlebone gives a net lift of 4% of the animal's weight in air and thus balances the excess weight of the rest of the animal. [Denton and Gilpin-Brown 1961]

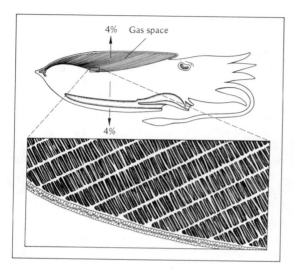

CUTTLEBONE The buoyancy mechanism of the cuttlefish (a relative of the squid and octopus) depends on a gas-filled structure, the cuttlebone, made of calcium carbonate. The cuttlebone consists of thin layers of calcium carbonate, spaced about ⅔ mm apart, and supported by pillars of the same material. [Courtesy of M. L. Blankenship, Duke University]

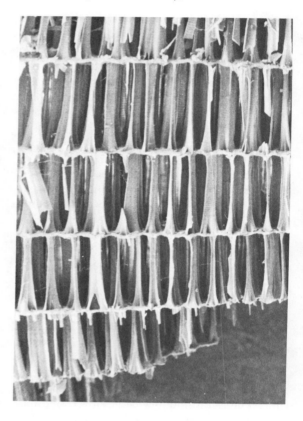

ish investigator Denton (Denton and Gilpin-Brown 1959, 1961).

The volume of the cuttlebone is 9.3% of the body volume, and its specific gravity when freshly removed is in the range of 0.57 to 0.64. This provides sufficient lift to give the animal neutral buoyancy in sea water.

The cuttlebone has a peculiar laminar structure. It is made up of thin layers of calcium carbonate, reinforced by chitin and held about ⅔ mm apart by a large number of "pillars" of the same material. In a fully grown animal the number of parallel layers may be about 100, and most of the space between the layers is filled with gas.

If a cuttlebone is removed from an animal and punctured under water, no gas escapes. This shows that the gas it contains is not under pressure. If the cuttlebone is crushed, gas escapes and can be collected for analysis. The gas is nitrogen under 0.8 atm pressure, plus a small amount of oxygen.

Along the caudal surface of the cuttlebone, some liquid is present between the calcified layers.

This liquid can be collected under paraffin oil by exposing the cuttlebone to a vacuum. It contains mostly sodium chloride and is hyposmotic to sea water and the body fluids (which are isosmotic with sea water). The higher osmotic pressure outside the cuttlebone, therefore, tends to withdraw water from the cuttlebone. The osmotic withdrawal of water is opposed by the hydrostatic pressure, which tends to press water back into the cuttlebone. The maximum force of water withdrawal is limited by the concentration difference. If all ions were removed from the fluid within the cuttlebone, the concentration difference between the cuttlebone fluid and blood would be about 1.1 osmolar. This corresponds to 24 atm pressure; that is, if the outside hydrostatic pressure exceeds 24 atm, fluid will be pressed into the cuttlebone in spite of the osmotic forces acting in the opposite direction. At a depth of about 240 m, the pressure is 24 atm, and below this depth it is impossible to withdraw fluid osmotically from the cuttlebone. However, the cuttlefish live in the upper reaches of the water masses and probably do not descend below about 200 m. Therefore, the osmotic forces suffice to keep water from being forced into the cuttlebone.

The gas pressure inside the rigid walls of the cuttlebone is about 0.8 atm, and the cuttlebone must be sufficiently strong to support the hydrostatic pressure of the surrounding water. This is indeed the case, for if a cuttlebone is covered with a thin plastic bag (so that sea water cannot penetrate into it) and is exposed to high pressures, it withstands more than 25 atm before it is crushed by the pressure.

The advantage of a rigid system is, of course, that the buoyancy is virtually unaffected by changes in depth. This gives the cuttlefish a great deal of freedom in moving up and down without making adjustments in the buoyancy mechanism. In addition to making the cuttlefish relatively independent of depth, the cuttlebone has the further advantage that it serves as a skeleton.

The mechanism responsible for filling the shell of the pearly nautilus, or chambered nautilus, is in principle the same as in the cuttlefish (Figure 11.27). As the nautilus grows, it adds new air chambers to its shell, one by one. A newly formed chamber (next to the animal) is initially filled with liquid in which sodium chloride is the major solute. Sodium is removed from this fluid by active transport; water is therefore withdrawn osmotically in the same way as in the cuttlefish, and gas dif-

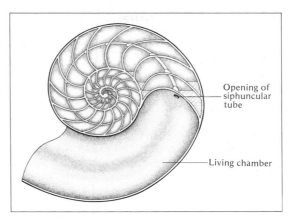

FIGURE 11.27 The shell of the chambered nautilus. Section of the shell of *Nautilus macromphalus*, oriented in its natural position. The chambers are gas filled, and new chambers are added as the animal grows. A new chamber is initially filled with fluid, which is withdrawn osmotically and replaced by gases. A small amount of liquid remains in the most recently formed chambers. [Denton and Gilpin-Brown 1966]

Opening of siphuncular tube

Living chamber

fuses into the chamber to replace the water. Nitrogen by diffusion reaches 0.8 atm, the same tension as in the water and the animal's body fluids. Oxygen, however, is lower and does not reach the concentration in the atmosphere, for the animal uses oxygen and its body fluids therefore have lower than atmospheric oxygen tension. The gas pressure in the shell is always about 0.9 atm, irrespective of the depth at which the animal is caught. Like the cuttlefish, nautilus has a freedom of vertical movement that is limited only by the pressure its rigid shell can support.

We have now discussed examples of the different mechanisms animals can use to improve their buoyancy. Each has advantages and disadvantages, outlined in Table 11.9. It is evident that gas is one of the most favorable solutions to the problem of providing lift and has other advantages as well. But if contained in a soft structure, it has one major disadvantage: It is extremely sensitive to pressure changes. Teleost fish use this mechanism in the form of an air-filled, soft-walled swimbladder, which is the subject we shall discuss next.

TABLE 11.9 Advantages and disadvantages of various mechanisms used by aquatic organisms to improve buoyancy.

Mechanism	Approx. volume needed (% of remaining body)	Buoyancy effectiveness		Pressure independence	Structural qualities	Energy requirement
		Sea water	Fresh water			
Reduction of heavy structures	0	Fair	Poor	Excellent	Poor	No maintenance
Replacement of heavy ions	200	Good	Poor	Excellent	Fair	Needs continuous maintenance
Fat	50	Good	Good	Excellent	Good	High initial cost, no maintenance
Squalene	35	Good	Good	Excellent	Good	High initial cost, no maintenance
Soft-walled gas float	5	Excellent	Excellent	Poor	Excellent	Needs continuous maintenance
Rigid-walled gas float	10	Excellent	Excellent	Good	Excellent	Probably needs maintenance

The swimbladder of fish

Many teleost fish have a *swimbladder* or *air bladder* that provides the necessary lift to give them neutral buoyancy. Whether a particular fish has a swimbladder is related more to its life habits than to its systematic position. Some bottom-living fish lack swimbladders, and it seems reasonable that neutral buoyancy has no particular advantage for a fish that is bound to the bottom. Many pelagic and surface forms do have swimbladders, but others lack this structure. The Atlantic mackerel, for example, has no swimbladder. For a fast and powerful predator this is a definite advantage, for as we have seen, a soft-walled swimbladder places severe constraints on the freedom of moving up and down in the water masses. No elasmobranch (sharks and rays) has a swimbladder.

The swimbladder is a more or less oval, soft-walled sac, located in the abdominal cavity, just below the spinal column (Figure 11.28). The shape varies a great deal, but the volume is rather constant from species to species, mostly about 5% of the body volume in marine fish and 7% in fresh-water species. The reason is that the specific gravity of a fish without the swimbladder is about 1.07. The volume of gas needed to give it neutral buoyancy in fresh water is therefore about 7% of the body volume, and in sea water (sp. gr. = 1.026) about 5%.

A swimbladder can provide a fish with perfect neutral buoyancy, and the fish can thus avoid expending energy to keep from sinking. There is, however, a serious disadvantage: The fish is in equilibrium or neutral buoyancy at only one particular depth. If it swims below this depth, the swimbladder is compressed by the increased water pressure, the buoyancy is decreased, and the fish must swim actively to keep from sinking further. This is not a very serious problem, for even if the swimbladder is almost completely compressed, the fish will not be much heavier than the water.

If the fish rises above the level of neutral buoyancy, however, the situation can become catastrophic. The pressure is reduced, the swimbladder expands, and the fish gains increased lift. It must now swim actively down, or it will continue to rise, and as the swimbladder expands more and more it will eventually lose control.

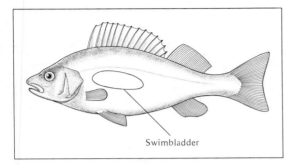

Swimbladder

TABLE 11.10 Composition of gas secreted by the gas gland of codfish. The gland was covered by a thin plastic film and stimulated to secrete gas. Bubbles ca. 0.5 μl in size were collected and analyzed. [Scholander 1956]

	O_2 (%)	CO_2 (%)	N_2 (%)
	81.1	8.6	10.3
	70.3	5.0	24.7
	76.6	15.8	7.6
	49.5	6.9	43.6
	61.1	6.8	32.1
	66.5	5.9	27.6
	30.1	13.9	56.0
	69.2	7.2	23.6
Mean	63.1	8.8	28.1

Some fish have a connection from the swimbladder to the esophagus that permits gas to escape, but others have a completely closed swimbladder. If such a fish is brought up from deep water; it arrives at the surface with the expanded swimbladder everted through the mouth or even ruptured.

The swimbladder is embryologically formed as an evagination from the digestive tract. Those fish that maintain a connection between the swimbladder and the esophagus are known as *physostome* fish (Greek *physa* = bladder; *stoma* = mouth). These fish can fill their swimbladders by gulping air at the surface. For fish living at depth this would be impractical or impossible, for very large volumes of air would have to be taken into the bladder at the surface to achieve neutral buoyancy at depth where the pressure is many times greater.

In many fish the duct degenerates, and there is no connection from the swimbladder to the outside. In these fish, called *physoclist* fish (Greek *kleistos* = closed), the gases in the swimbladder must originate in the blood and be secreted into the swimbladder at a pressure equal to the depth at which the fish live. Fish with swimbladders have been caught below 4000 m depth. Because the swimbladder wall is a soft structure, the swimbladder gas at that depth must be under a pressure in

excess of 400 atm. This poses the problems mentioned earlier in this chapter.

The gases found in the swimbladder are the same as those in the atmosphere (oxygen, nitrogen, and carbon dioxide), but in different proportions. If we withdraw some gas from the swimbladder by puncturing it with a hypodermic needle, the fish is induced to secrete new gas to replace what was lost. In most fish the newly secreted gas is rich in oxygen (Table 11.10).

This is not true for all fish; the whitefish (*Coregonus albus*), for example, fills its swimbladder with virtually pure nitrogen. Because it lives in lakes at depths of 100 m or so, nitrogen must be secreted at a pressure of some 10 atm. Nitrogen gas is highly inert, and it has been difficult to understand how an inert gas can be secreted. Also, the truly inert gases (such as argon) are likewise enriched in the swimbladder gas in about the same proportion to nitrogen as they occur in the atmosphere (Wittenberg 1958). This indicates that the secretion mechanism cannot involve chemical reactions; it must be of a physical nature.

Even when the swimbladder gas is rich in oxygen, the nitrogen may still be more concentrated than in the surrounding water. Say, for example, that we obtain a fish from a depth of 500 m and that a sample of the swimbladder gas shows 80%

BUTTERFLY FISH An x-ray photograph of a long-nosed butterfly fish (*Chelmon rostratus*). The location of the swimbladder is clearly visible as the light area just below the vertebral column. The length of this specimen is about 10 cm. [Courtesy of John Lundberg, Duke University]

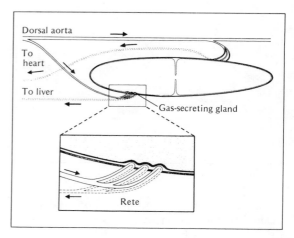

FIGURE 11.29 Diagram of the circulation to the swimbladder of a fish. Blood may reach the swimbladder either through a set of parallel capillaries, the rete, which supplies the gas gland, or through a vessel to the posterior portion where fine blood vessels that spread over the wall can serve to absorb gases rapidly. When gas is being secreted, the gas-absorption vessels are closed and carry no blood.

O_2 and 20% N_2 (and a trace of carbon dioxide). At the depth where the fish was caught, the gases were under 50 atm total pressure, so that their partial pressures were 40 atm for oxygen and 10 atm for nitrogen. Relative to the gas tensions in the surrounding water, 0.2 atm O_2 and 0.8 atm N_2, oxygen in the swimbladder gas was thus enriched 200-fold and nitrogen 12.5-fold. Before we discuss how gases can be secreted at these high pressures, we will examine the simpler question of how the gases are retained and kept from being lost from the swimbladder.

How is gas kept in the swimbladder?

As always, to understand function we must know a bit about structure. In the wall of the swimbladder of most fish there is a gas gland, which stands out because of its bright red color. If the gas gland were supplied with ordinary arterial blood, which comes from the gills and therefore has gas tensions in equilibrium with the water, the blood would tend to dissolve the swimbladder gases and carry them away. How can this be prevented?

The gas gland is supplied with blood through a peculiar structure known as the *rete mirabile* (plural, *retia mirabilia*). The form of the rete and the gas gland varies from species to species, but in principle it is as follows. Before the artery reaches the gas gland, it splits into an enormous number of straight, parallel capillaries, which may unite again into a single vessel before reaching the gas gland (Figure 11.29). The vein from the gland likewise splits into a similar number of parallel capillaries, which run interspersed between the arterial capillaries, and then unite to form a single vein.

Let us see how this structure can help to keep gases inside the swimbladder, once they have been deposited there. The system is analogous to the countercurrent heat exchanger in the flipper of the whale, only it deals with dissolved gases. Assume that the swimbladder contains gas under a pressure of 100 atm. As venous blood leaves the swimbladder, it contains dissolved gases at this high pressure. When this blood enters the venous capillary, only the thin capillary wall separates it from blood in the arterial capillaries. Gases therefore diffuse across the wall to the arterial blood. As the blood runs along the venous capillaries, it loses more and more gas by diffusion to the arterial capillaries. At the end of the venous capillaries, as the blood is about to leave the rete, the surrounding capillaries contain arterial blood, which comes directly from the gills and is not yet enriched with additional gas. The venous blood can therefore lose gases until it is in diffusion equilibrium with the incoming arterial blood (i.e., the venous blood that leaves the rete contains no more gas than the incoming arterial blood).

In this way the rete serves as a trap to retain the gases in the swimbladder and avoid loss of gas to the circulating blood. The rete is a typical countercurrent exchange system that depends on passive diffusion between two liquid streams that run in opposite directions. The exchange of gas is aided by a large surface (i.e., the large number of capillaries), by a short diffusion distance (i.e., the thin capillary wall which is only a fraction of 1 μm thick), and by the length of the capillaries (which are exceptionally long).

Some measurements are informative in this regard. The rete from a common eel weighed about 65 mg, and had some 100 000 arterial capillaries and about the same number of venous capillaries. The capillaries were about 4 mm long, and this gives a total length of 400 m of each kind of capillary. Their diameter was, as most capillaries, some 7 to 10 μm, and simple arithmetic then gives their total wall surface area as being in excess of 100 cm² for each kind, all within the volume of a water drop (Krogh 1929). This large area is thus available for gas exchange between the arterial and venous blood.

The capillaries in the swimbladder rete are exceptionally long, often several millimeters. Muscle capillaries, which are otherwise among the longest anywhere, are only about 0.5 mm long. There is a clear correlation between the length of the rete capillaries and the gas pressure in the swimbladder. Those fish that live at the greatest depth and secrete gases under the highest pressures have the longest capillaries, the record being held by a deep-sea fish, *Bassozetus taenia*, which has rete capillaries with the enormous length of 25 mm (Marshall 1960).

The effectiveness of the rete has been estimated by Scholander, who found that its dimensions would suffice for maintaining pressures in excess of 4000 atm in the swimbladder (Scholander 1954). Fish that have swimbladders almost certainly do not occur below 5000 m (500 atm), and we must therefore conclude that the rete is more than adequate as a device to keep the swimbladder gas from being carried away, dissolved in the blood. Because the rete depends entirely on diffusion and requires no energy, it provides a rather advantageous solution to the problem.

The other way gas could escape from the swimbladder is by diffusion through the swimbladder wall. Because the total pressure of gas in the swimbladder is nearly identical to the hydrostatic pressure in the water where the fish finds itself, the partial pressures are inevitably higher than in the surrounding water. Typically, the partial pressure of nitrogen in the water (and in the blood) is 0.8 atm. For oxygen the partial pressure in the water, and therefore in the blood, does not normally exceed 0.2 atm. In contrast, the partial pressure in the swimbladder may be several hundred atmospheres.

The water content of the swimbladder wall is as high as in other tissues, and we might therefore expect that oxygen would readily diffuse out. However, the diffusibility for oxygen is unexpectedly low. This is apparently attributable to multiple layers of very thin platelets of crystalline guanine, each sheet having a thickness of only about 0.02 μm. These thin sheets form a barrier to diffusion of gases out of the swimbladder and thereby reduce the metabolic energy needed for secretion of gases into the swimbladder to maintain its volume (Lapennas and Schmidt-Nielsen 1977).

How is gas secreted into the swimbladder?

The problem of depositing gas into the swimbladder against a high pressure is more formidable than the problem of keeping the gas in, once it is there. We will first discuss oxygen, for this is the gas most commonly secreted into the swimbladder. Let us base this discussion on a swimbladder that already contains oxygen at a high pressure, say 100 atm (corresponding to 1000 m depth) and assume that that additional gas is to be secreted against this enormous pressure. Because the oxygen tension in the arterial blood that comes from the gills at best equals that of the surrounding water, the arterial oxygen tension is no more than 0.2 atm.

Let us consider the oxygen *content* of the blood, rather than the gas tension. If oxygen is being removed from the blood and secreted into the swimbladder, it follows that the venous blood leaving the rete must contain less oxygen than the entering arterial blood. As an arbitrary example, say that the arterial blood contains 10 ml O_2 per 100 ml blood and that the venous blood contains 9 ml O_2 per 100 ml when leaving the rete. In Figure 11.30, in which for simplicity the entire rete structure is represented by a single loop, these two numbers are placed at the far left of the diagram. The difference between the arterial and the venous blood, 1 ml O_2 per 100 ml blood, is the oxygen that is being deposited into the swimbladder.

Assume now that the gas gland begins to produce lactic acid (which in reality it does when gas secretion takes place). The lactic acid enters the blood and reduces its affinity for oxygen. This ef-

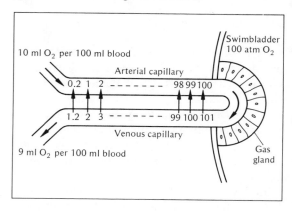

FIGURE 11.30 Diagram of the countercurrent multiplier system of the fish swimbladder. The numerous capillaries of the rete are represented by a single loop. As the gas gland produces lactic acid, the oxygen tension increases in the venous capillary, and gas diffuses across to the arterial capillary, thus remaining within the loop. The venous blood, as it exits, contains less oxygen than the incoming arterial blood.

fect of acid is much more pronounced in fish blood than in mammalian blood and is known as the *Root effect*. The lactic acid drives oxygen off the hemoglobin, and the oxygen tension in the blood leaving the swimbladder is therefore increased (Figure 11.31). The oxygen tension now is higher in the venous than in the arterial capillary, and oxygen diffuses across to the arterial capillary. This continues as long as lactic acid is added to the venous blood, which thus leaves the rete with less oxygen than is contained in the incoming arterial blood. Owing to the countercurrent flow in the rete, lactic acid also tends to remain in the loop, increasing the effect.

In this system it is important to distinguish clearly between *amount* and *tension*. The amount of oxygen in the venous blood that leaves the rete must be less than in the incoming arterial blood, but for oxygen to diffuse across from the venous to the arterial capillary, the oxygen tension must be higher in the venous than in the arterial capillary. This higher tension is caused by the acid produced by the gas gland. In this way oxygen is constantly returned to the arterial capillary and gradually accumulates in the loop, where it can reach very high concentrations. How the gas gland itself func-

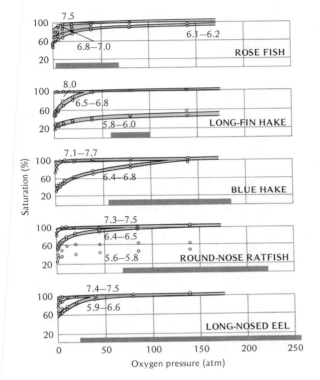

FIGURE 11.31 Oxygen dissociation curves of fish blood demonstrate the pronounced effect of acid, known as the Root effect. The upper curve in each diagram represents untreated blood at 4 °C; in the lower curve of each diagram lactic acid has been added to give the pH stated for each curve. The solid bar on the abscissa gives the depth range at which the fish lives. Note that the abscissa gives the oxygen pressure in atmospheres. [Scholander and Van Dam 1954]

tions is not known in detail, but the system described explains how the blood that enters the gland can have very high oxygen concentrations.

The mechanism described has received experimental verification. The Norwegian investigator, John Steen, has succeeded in taking minute blood samples from the artery and the vein of the rete in eels that were in the process of secreting oxygen. He did indeed find that the venous blood had a lowered oxygen *content*, but an increased oxygen *tension*, attributable to the production of lactic acid and carbon dioxide in the gas gland (Steen 1963).

We can now see that the rete, when it secretes, is a *countercurrent multiplier* system. A small dif-

ference in oxygen tension between venous and arterial blood is multiplied along the length of its capillaries. Thus, the longer the capillaries, the higher the multiplication effect. Again, this is in accord with the much longer rete capillaries found in fish from greater depths. The multiplication requires energy, as does any process that transports a substance against a gradient; in this case the lactic acid released by the gas gland provides the necessary driving force.

The system described depends on the use of acid to decrease the affinity of hemoglobin for oxygen. However, acid is not the only substance that has an effect on the blood gases; any solute added to the blood decreases the solubility for gases. This we could call a *salting-out effect*. Thus, lactic acid, in addition to its Root effect, also has a salting-out effect on the blood gases – not only on oxygen, but also on the other gases, including nitrogen.

Because the addition of solute decreases the solubility for nitrogen, the rete acts as a multiplier for this gas as well as for oxygen, and we now can explain how inert gases can accumulate in the swimbladder. Until the salting-out effect was understood, it was very difficult to explain how any inert gas could be secreted. The possibility was pointed out as early as 1934 by Henri Koch, but received little attention until the late 1960s when sufficient experimental work had accumulated to yield a clear understanding of how the rete acts as a countercurrent multiplier system.

The salting-out effect, however, does not seem to be as effective as the Root effect. Careful estimates indicate that the salting-out effect of an increase in the blood salts of 0.02 mol per liter could result in a nitrogen concentration of 25 atm, but due to the Root effect a lactic acid concentration of only 0.005 mol per liter (as observed in actively secreting eels) could potentially produce some 3000 atm of oxygen (Kuhn et al. 1963).

One further finding seems to put the final piece of the puzzle in place. As acid is added to fish blood and oxygen is forced off the hemoglobin, the effect is spoken of as the *Root-off reaction*. The reverse, that oxygen is again bound to the hemoglobin when the pH is increased, is known as the *Root-on effect*. The interesting finding is that the reaction rates for the Root-off and Root-on effects are quite different. The half time for the off reaction is 50 ms (at 23 °C); for the on reaction it is in the order of 10 to 20 seconds. Blood in the arterial capillary of the rete, as acid diffuses into it, rapidly gives off its oxygen. But as the same blood runs in the venous capillary and the lactic acid diffuses out and into the arterial capillary, the oxygen does not immediately enter into combination with hemoglobin; it remains as uncombined oxygen. In other words, a high oxygen tension persists. Thus, the slow Root-on effect permits a large fraction of the hemoglobin to leave the venous capillary in the deoxygenated state, although the tension of oxygen may still be quite high (Berg and Steen 1968).

Although the function of the rete as a counter-current multiplier system is clear, the function of the secretory epithelium of the gas gland itself has not been well studied. One of the unsolved problems is how gas is moved from the capillary into the gas phase of the bladder itself. The role of the secretory epithelium in this process is not understood.

We have now seen how animal movement is achieved with the aid of various force-producing mechanisms. The underlying principle, the sliding between adjacent microfilaments, appears to be a universal mechanism for the production of force. In any but the smallest animals, location depends on forces that act on skeletal elements, either a hydraulic skeleton or a rigid skeleton that can be internal (e.g., vertebrates) or external (e.g., arthropods). Animal adaptations to locomotion

show many interesting principles of biomechanics, including how aquatic animals achieve neutral buoyancy and thus avoid the expenditure of energy to keep from sinking.

REFERENCES

Abbott, B. C., and Lowy, J. (1953) Mechanical properties of *Mytilus* muscle. *J. Physiol.* 120:50P.

Alexander, R. McN. (1968) *Animal Mechanics.* London: Sidgwick & Jackson. 346 pp.

Alexander, R. McN. (1974) The mechanics of jumping by a dog (*Canis familiaris*). *J. Zool., Lond.* 173:549–573.

Alexander, R. McN., and Vernon, A. (1975) The mechanics of hopping by kangaroos (Macropodidae). *J. Zool., Lond.* 177:265–303.

Allen, R. D., Francis, D., and Zeh, R. (1971) Direct test of the positive pressure gradient theory of pseudopod extension and retraction in amoebae. *Science* 174:1237–1240.

Andersen, S. O., and Weis-Fogh, T. (1964) Resilin: A rubber-like protein in arthropod cuticle. *Adv. Insect Physiol.* 2:1–65.

Baguet, F., and Gillis, J. M. (1968) Energy cost of tonic contraction of a lamellibranch catch muscle. *J. Physiol.* 198:127–143.

Barber, V. C. (1974) Cilia in sense organs. In *Cilia and Flagella* (M. A. Sleigh, ed.), pp. 403–433. London: Academic Press.

Bennet-Clark, H. C., and Lucey, E. C. A. (1967) The jump of the flea: A study of the energetics and a model of the mechanism. *J. Exp. Biol.* 47:59–76.

Berg, H. C., and Anderson, R. A. (1973) Bacteria swim by rotating their flagellar filaments. *Nature, Lond.* 245:380–382.

Berg, T., and Steen, J. B. (1968) The mechanism of oxygen concentration in the swim-bladder of the eel. *J. Physiol.* 195:631–638.

Blum, J. J., and Lubliner, J. (1973) Biophysics of flagellar motility. *Annu. Rev. Biophys. Biomed. Engin.* 2:181–219.

Bone, Q. (1966) On the function of the two types of myotomal muscle fibre in elasmobranch fish. *J. Mar. Biol. Assoc. UK* 46:321–349.

Brokaw, C. J., and Gibbons, I. R. (1975) Mechanisms of movement in flagella and cilia. In *Swimming and Flying in Nature*, vol. 1 (T. Y.-T. Wu, C. J. Brokaw, and C. Brennen, eds.), pp. 89–126. New York: Plenum Press.

Buller, A. J., Eccles, J. C., and Eccles, R. M. (1960) Interactions between motoneurons and muscles in respect to the characteristic speeds of their responses. *J. Physiol.* 150:417–439.

Buller, A. J., and Lewis, D. M. (1964) The rate of rise of tension in isometric tetani of cross-innervated mammalian skeletal muscles. *J. Physiol.* 170:67P–68P.

Burke, R. E., Levine, D. N., Zajack, F. E., III, Tasiris, P., and Engel, W. K. (1971) Mammalian motor units: Physiological-histochemical correlation in three types in cat gastrocnemius. *Science* 174:709–712.

Casella, C. (1950) Tensile force in total striated muscle, isolated fibre and sarcolemma. *Acta Physiol. Scand.* 21:380–401.

Cavagna, G. A., and Citterio, G. (1974) Effect of stretching on the elastic characteristics and the contractile component of frog striated muscle. *J. Physiol.* 239:1–14.

Cavagna, G. A., Dusman, B., and Margaria, R. (1968) Positive work done by a previously stretched muscle. *J. Appl. Physiol.* 24:21–32.

Cavagna, G. A., Saibene, F. P., and Margaria, R. (1964) Mechanical work in running. *J. Appl. Physiol.* 19:249–256.

Cooper, S., and Eccles, J. C. (1930) The isometric responses of mammalian muscles. *J. Physiol.* 69:377–385.

Dawson, T. J., and Taylor, C. R. (1973) Energetic cost of locomotion in kangaroos. *Nature, Lond.* 246:313–314.

Denton, E. (1960) The buoyancy of marine animals. *Sci. Am.* 203:119–128.

Denton, E. J. (1961) The buoyancy of fish and cephalopods. *Prog. Biophys. Biophys. Chem.* 11:178–234.

Denton, E. J. (1971) Examples of the use of active transport of salts and water to give buoyancy in the sea. *Philos. Trans. R. Soc. Lond. B.* 262:277–287.

Denton, E. J., and Gilpin-Brown, J. B. (1959) On the buoyancy of the cuttlefish. *Nature, Lond.* 184:1330–1332.

Denton, E. J., and Gilpin-Brown, J. B. (1961) The distribution of gas and liquid within the cuttlebone. *J. Mar. Biol. Assoc. UK* 41:365–381.

Denton, E. J., and Gilpin-Brown, J. B. (1966) On the buoyancy of the pearly Nautilus. *J. Mar. Biol. Assoc. UK* 46:723–759.

Denton, E. J., Gilpin-Brown, J. B., and Shaw, T. L. (1969) A buoyancy mechanism found in cranchid squid. *Proc. R. Soc. Lond. B.* 174:271–279.

Eckert, R. (1972) Bioelectric control of ciliary activity. *Science* 176:473–481.

Evans, M. E. G. (1972) The jump of the click beetle (Coleoptera: Elateridae): A preliminary study. *J. Zool., Lond.* 167:319–336.

Evans, M. E. G. (1973) The jump of the click beetle (Coleoptera: Elateridae): Energetics and mechanics. *J. Zool., Lond.* 169:181–194.

Gibbons, I. R. (1977) Structure and function of flagellar microtubules. In *International Cell Biology* (B. R. Brinkley and K. R. Porter, eds.), pp. 348–357. New York: Rockefeller University Press.

Gibbons, I. R., and Rowe, A. J. (1965) Dynein: A protein with adenosine triphosphatase activity from cilia. *Science* 149:424–426.

Gordon, A. M., Huxley, A. F., and Julian, F. J. (1966) The variation in isometric tension with sarcomere length in vertebrate muscle fibres. *J. Physiol.* 184:170–192.

Gray, J., and Lissmann, H. W. (1938) Studies in animal locomotion. 7. Locomotory reflexes in the earthworm. *J. Exp. Biol.* 15:506–517.

Greenewalt, C. H. (1975) The flight of birds: The significant dimensions, their departure from the requirements for dimensional similarity, and the effect on flight aerodynamics of that departure. *Trans. Am. Philos. Soc.* (new series) 65(4):1–67.

Gross, F., and Zeuthen, E. (1948) The buoyancy of plankton diatoms: A problem of cell physiology. *Proc. R. Soc. Lond. B.* 135:382–389.

Hall-Craggs, E. C. B. (1965) An analysis of the jump of the lesser galago (*Galago senegalensis*). *J. Zool.* 147:20–29.

Haxton, H. A. (1944) Absolute muscle force in the ankle flexors of man. *J. Physiol.* 103:267–273.

Hersey, J. B., Backus, R. H., and Hellwig, J. (1962)

Sound-scattering spectra of deep scattering layers in the western North Atlantic Ocean. *Deep Sea Res.* 8:196–210.

Hirst, M. (1977) America's man-powered prizewinner. *Flight Int.* 112:1253–1256.

Hoyle, G. (1955) Neuromuscular mechanisms of a locust skeletal muscle. *Proc. R. Soc. Lond. B.* 143:343–367.

Hoyle, G. (1957) *Comparative Physiology of the Nervous Control of Muscular Contraction.* Cambridge: Cambridge University Press. 147 pp.

Hoyle, G. (1964) Muscle and neuromuscular physiology. In *Physiology of Mollusca*, vol. 1 (K. M. Wilbur and C. M. Yonge, eds.), pp. 313–351. New York: Academic Press.

Huxley, H. E. (1969) The mechanism of muscular contraction. *Science* 164:1356–1366.

Huxley, H. E. (1973) Muscular contraction and cell motility. *Nature Lond.* 243:445–449.

Jahn, T. L., Landman, M. D., and Fonseca, J. R. (1964) The mechanism of locomotion of flagellates. 2. Function of the mastigonemes of *Ochromonas. J. Protozool.* 11:291–296.

Johnson, W. H., and Twarog, B. M. (1960) The basis for prolonged contractions in molluscan muscles. *J. Gen. Physiol.* 43:941–960.

Kesseler, H. (1966) Beitrag zur Kenntnis der chemischen und physikalischen Eigenschaften des Zellsaftes von *Noctiluca miliaris. Veröff. Inst. Meeresforsch. Bremerhaven* 2:357–368.

Koch, H. (1934) L'émission de gaz dans la vésicule gazeuse des poissons. *Rev. Quest. Sci.* 26:385–409.

Krogh, A. (1929) *The Anatomy and Physiology of Capillaries*, 2nd ed. New Haven, Conn.: Yale University Press. 422 pp.

Kuffler, S. W., and Williams, E. M. V. (1953a) Small-nerve junctional potentials: The distribution of small motor nerves to frog skeletal muscle, and the membrane characteristics of the fibres they innervate. *J. Physiol.* 121:289–317.

Kuffler, S. W., and Williams, E. M. V. (1953b) Properties of the "slow" skeletal muscle fibres of the frog. *J. Physiol.* 121:318–340.

Kuhn, W., Ramel, A., Kuhn, H. J., and Marti, E.

(1963) The filling mechanism of the swimbladder: Generation of high gas pressures through hairpin countercurrent multiplication. *Experientia* 19:497–511.

Lapennas, G. N., and Schmidt-Nielsen, K. (1977) Swimbladder permeability to oxygen. *J. Exp. Biol.* 67:175–196.

Lighthill, J. (1974) *Aerodynamic Aspects of Animal Flight.* Fluid Science Lecture, British Hydromechanics Research Association. 30 pp.

Lowy, J., and Millman, B. M. (1962) Mechanical properties of smooth muscles of cephalopod molluscs. *J. Physiol.* 160:353–363.

Ludwig, G. D., Blakemore, W. S., and Drabkin, D. L. (1957) Production of carbon monoxide and bile pigment by haemin oxidation. *Biochem. J.* 66:38P.

Machin, K. E. (1958) Wave propagation along flagella. *J. Exp. Biol.* 35:796–806.

Marshall, N. B. (1960) Swimbladder structure of deep-sea fishes in relation to their systematics and biology. *Discovery Rep.* 31:1–122.

Naitoh, Y., and Kaneko, H. (1972) Reactivated triton-extracted models of paramecium: Modification of ciliary movement by calcium ion. *Science* 176:523–524.

Parry, D. A., and Brown, R. H. J. (1959a) The hydraulic mechanism of the spider leg. *J. Exp. Biol.* 36:423–433.

Parry, D. A., and Brown, R. H. J. (1959b) The jumping mechanism of salticid spiders. *J. Exp. Biol.* 36:654–664.

Peachey, L. D. (1965) The sarcoplasmic reticulum and transverse tubules of the frog's sartorius. *J. Cell Biol.* 25:209–231.

Pennycuick, C. J. (1969) The mechanics of bird migration. *Ibis* 111:525–556.

Phleger, C. F. (1971) Pressure effects on cholesterol and lipid synthesis by the swimbladder of an abyssal *Coryphaenoides* species. *Am. Zool.* 11:559–570.

Pickwell, G. V., Barham, E. G., and Wilson, J. W. (1964) Carbon monoxide production by a bathypelagic siphonophore. *Science* 144:860–862.

Pollard, T. D. (1977) Cytoplasmic contractile proteins. In *International Cell Biology* (B. R. Brinkley and K. R. Porter, eds.), pp. 378–387. New York: Rockefeller University Press.

Pringle, J. W. S. (1949) The excitation and contraction of the flight muscles of insects. *J. Physiol.* 108:226–232.

Pringle, J. W. S. (1957) *Insect Flight*. Cambridge: Cambridge University Press. 132 pp.

Pugh, L. G. C. E. (1959) Carbon monoxide content of the blood and other observations on Weddell seals. *Nature, Lond.* 183:74–76.

Pugh, L. G. C. E. (1971) The influence of wind resistance in running and walking and the mechanical efficiency of work against horizontal or vertical forces. *J. Physiol.* 213:255–276.

Rayner, M. D., and Kennan, M. J. (1967) Role of red and white muscles in the swimming of the skipjack tuna. *Nature, Lond.* 214:392–393.

Ridgway, E. B., and Ashley, C. C. (1967) Calcium transients in single muscle fibers. *Biochem. Biophys. Res. Commun.* 29:229–234.

Rikmenspoel, R. (1965) The tail movement of bull spermatozoa: Observations and model calculations. *Biophys. J.* 5:365–392.

Rikmenspoel, R., Sinton, S., and Janick, J. J. (1969) Energy conversion in bull sperm flagella. *J. Gen. Physiol.* 54:782–805.

Roeder, K. D. (1953) *Insect Physiology*. New York: Wiley. 1100 pp.

Rüegg, J. C. (1971) Smooth muscle tone. *Physiol. Rev.* 51:201–248.

Satir, P. (1968) Studies on cilia. 3. Further studies on the cilium tip and a "sliding filament" model of ciliary motility. *J. Cell. Biol.* 39:77–94.

Satir, P. (1974) The present status of the sliding microtubule model of ciliar motion. In *Cilia and Flagella* (M. A. Sleigh, ed.), pp. 131–142. London: Academic Press.

Schmidt-Nielsen, S., Flood, A., and Stene, J. (1934) On the size of the liver of some gristly fishes, their content of fat and vitamin A. *Kongelige Norske Videnskabers Selskab Forhandlinger* 7:47–50.

Scholander, P. F. (1954) Secretion of gases against high pressures in the swimbladder of deep sea fishes. 2. The rete mirabile. *Biol. Bull.* 107:260–277.

Scholander, P. F. (1956) Observations on the gas gland in living fish. *J. Cell. Comp. Physiol.* 48:523–528.

Scholander, P. F., and Van Dam, L. (1954) Secretion of gases against high pressures in the swimbladder of deep sea fishes. 1. Oxygen dissociation in blood. *Biol. Bull.* 107:247–259.

Seymour, M. K. (1971) Burrowing behaviour in the European lugworm *Arenicola marina* (Polychaeta: Arenicolidae). *J. Zool.* 164:93–132.

Sleigh, M. A. (1968) Patterns of ciliary beating. *Symp. Soc. Exp. Biol.* 22:131–150.

Sleigh, M. A. (1974) *Cilia and Flagella*. London: Academic Press. 500 pp.

Sleigh, M. A. (1977) Fluid propulsion by cilia and flagella. In *Comparative Physiology: Water, Ions and Fluid Mechanics* (K. Schmidt-Nielsen, L. Bolis, and S. H. P. Maddrell, eds.), pp. 255–265. Cambridge: Cambridge University Press.

Sotavalta, O. (1953) Recordings of high wing-stroke and thoracic vibration frequency in some midges. *Biol. Bull.* 104:439–444.

Squire, J. M. (1975) Muscle filament structure and muscle contraction. *Annu. Rev. Biophys. Bioeng.* 4:137–163.

Steen, J. B. (1963) The physiology of the swimbladder in the eel *Anguilla vulgaris*. 3. The mechanism of gas secretion. *Acta Physiol. Scand.* 59:221–241.

Stolpe, M., and Zimmer, K. (1939) Der Schwirrflug des Kolibri im Zeitlupenfilm. *J. Ornithol.* 87:136–155.

Taylor, C. R., Caldwell, S. L., and Rowntree, V. J. (1972) Running up and down hills: Some consequences of size. *Science* 178:1096–1097.

Thys, H., Cavagna, G. A., and Margaria, R. (1975) The role played by elasticity in an exercise involving movements of small amplitude. *Pflügers Arch.* 354:281–286.

Thys, H., Faraggiana, T., and Margaria, R. (1972) Utilization of muscle elasticity in exercise. *J. Appl. Physiol.* 32:491–494.

Trueman, E. R. (1966) Observations on the burrowing of *Arenicola marina* (L.). *J. Exp. Biol.* 44:93–118.

Trueman, E. R., and Packard, A. (1968) Motor performances of some cephalopods. *J. Exp. Biol.* 49:495–507.

Tucker, V. A. (1973) Bird metabolism during flight: Evaluation of a theory. *J. Exp. Biol.* 58:689–709.

Weber, A., and Murray, J. M. (1973) Molecular control mechanisms in muscle contraction. *Physiol. Rev.* 53:612–673.

Weber, H. H., and Portzehl, H. (1954) The transference of the muscle energy in the contraction cycle. *Prog. Biophys. Biophys. Chem.* 4:60–111.

Weis-Fogh, T. (1960) A rubber-like protein in insect cuticle. *J. Exp. Biol.* 37:889–907.

Weis-Fogh, T. (1973) Quick estimates of flight fitness in hovering animals, including novel mechanisms for lift production. *J. Exp. Biol.* 59:169–230.

Weis-Fogh, T. (1976) Energetics and aerodynamics of flapping flight: A synthesis. In *Insect Flight* (R. C. Rainey, ed.), pp. 48–72. New York: Wiley.

Wiersma, C. A. G. (1961) The neuromuscular system. In *The Physiology of Crustacea*, vol. II (T. H. Waterman, ed.), pp. 191–240. New York: Academic Press.

Wigglesworth, V. B. (1972) *The Principles of Insect Physiology*, 7th ed. London: Chapman & Hall. 827 pp.

Wilson, J. A. (1972) *Principles of Animal Physiology*. New York: Macmillan.

Wittenberg, J. B. (1958) The secretion of inert gas into the swimbladder of fish. *J. Gen. Physiol.* 41:783–804.

Wittenberg, J. B. (1960) The source of carbon monoxide in the float of the Portuguese man-of-war, *Physalia physalis* L. *J. Exp. Biol.* 37:698–705.

ADDITIONAL READING

Alexander, R. McN. (1968) *Animal Mechanics*. Seattle: University of Washington Press. 346 pp.

Blake, J. R., and Sleigh, M. A. (1974) Mechanics of ciliary locomotion. *Biol. Rev.* 49:85–125.

Bülbring, E., Brading, A. F., Jones, A. W., and Tomita, T. (1970) *Smooth Muscle*. London: Arnold. 676 pp.

Carlson, F. D., and Wilkie, D. R. (1974) *Muscle Physiology*. Englewood Cliffs, N.J.: Prentice-Hall. 170 pp.

Close, R. I. (1972) Dynamic properties of mammalian skeletal muscles. *Physiol. Rev.* 52:129–197.

Cold Spring Harbor Laboratory (1973) The mechanism of muscle contraction. *Cold Spring Harbor Symp. Quant. Biol.* 37, 706 pp.

Gray, J. (1968) *Animal Locomotion*. London: Weidenfeld & Nicolson. 479 pp.

Hill, A. V. (1965) *Trails and Trials in Physiology: A Bibliography 1909–1964; with Reviews of Certain Topics and Methods and a Reconnaissance for Further Research*. London: Edward Arnold. 374 pp.

Hubbard, J. I. (1973) Microphysiology of vertebrate neuromuscular transmission. *Physiol. Rev.* 53:674–723.

Karpovich, P. V., and Sinning, W. E. (1971) *Physiology of Muscular Activity*, 7th ed. Philadelphia: Saunders. 374 pp.

Lighthill, M. J. (1969) Hydromechanics of acquatic animal propulsion. *Annu. Rev. Fluid Mech.* 1:413–446.

Margaria, R. (1976) *Biomechanics and Energetics of Muscular Exercise*. London: Oxford University Press. 146 pp.

Nachtigall, W. (1974) Locomotion: Mechanics and hydrodynamics of swimming in aquatic insects. In *The Physiology of Insecta*, 2nd ed., vol. 3 (M. Rockstein, ed.), pp. 381–432. New York: Academic Press.

Pedley, T. J. (ed.) (1977) *Scale Effects in Animal Locomotion*. London: Academic Press. 545 pp.

Rainey, R. C. (ed.) (1976) *Insect Flight*. New York: Wiley. 287 pp.

Sleigh, M. A. (ed.) (1974) *Cilia and Flagella*. London: Academic Press. 500 pp.

Squire, J. M. (1975) Muscle filament structure and muscle contraction. *Annu. Rev. Biophys. Bioeng.* 4:137–163.

Wainwright, S. A., Biggs, W. D., Currey, J. D., and Gosline, J. M. (1976) *Mechanical Design in Organisms*. New York: Wiley. 423 pp.

Westfall, J. A., et al. (1973) Symposium on invertebrate neuromuscular systems. *Am. Zool.* 13:233–445.

Wu, T. Y.-T., Brokaw, C. J., and Brennen, C. (eds.) (1975) *Swimming and Flying in Nature*, vols. 1 and 2. New York: Plenum Press. 1005 pp.

12

CHAPTER TWELVE

Information and senses

In the preceding chapters we have discussed some aspects of the environment that are important to animals – oxygen, food, temperature, water. This chapter deals with the questions of how animals obtain information about their environment and how this information is used. We shall be concerned with (1) what kind of information is available and (2) how information is processed and passed on to the central nervous system. The next and final chapter will discuss how information from outside as well as from within the body is used and how various physiological functions are controlled and integrated.

Virtually all animals depend on information about their surroundings, whether for finding food and mates or escaping from predators. They must find their way about and also assess important qualities of the environment (e.g., temperature).

Most information about the environment is obtained through specialized sensory organs. Traditionally, sense organs are separated into *exteroceptors*, which respond to stimuli coming from outside, such as light and sound, and *proprioceptors*, which refer to internal information, such as the position of the limbs. This separation does not have much inherent meaning and, at best, is a matter of convenience. Another traditional classification of senses is based on the five most obvious senses of humans – vision, hearing, taste, smell, and touch. In reality our sensory equipment is not nearly so limited.

A list of external stimuli to which at least some animals respond is quite extensive (Table 12.1). The listed categories are not discrete, and the separation is somewhat arbitrary. However, for convenience we shall follow this sequence in our discussion of the possibilities and limitations that apply to the use of the various kinds of available information. The information naturally falls into three major categories: electromagnetic and thermal en-

TABLE 12.1 Environmental stimuli significant in sensory perception.

Electromagnetic and thermal energy
Light
Infrared radiation
Thermal: "heat," "cold"
Electric
Magnetic
Mechanical energy and force
Sound and sonar
Touch and vibration
Pressure
Gravity
Inertia
Chemical agents
Taste
Smell
Humidity

ergy, mechanical energy and mechanical force, and chemical agents.

In this context the question of whether information can reach the central nervous system via other avenues outside the sensory organs (extrasensory perception, or ESP) is irrelevant. In this chapter we are dealing with measurable physical quantities that can be recognized, described, and manipulated in controlled ways. Although some sensory mechanisms may still be unknown (this is true of the response to magnetic fields), the word extrasensory by definition means that no sensory structure is involved.

The wide variety of stimuli and the striking structural differences in the sense organs seem confusing until we realize that the general sequence of events is much the same for all of them (Figure 12.1). The external stimulus impinges on an accessory structure, which may be highly complex such as the eye or the ear, or much simpler such as the touch receptors of the skin. Through these accessory structures the external stimulus reaches one or more sensory neurons.

These have many properties in common, but they differ from organ to organ in that each type is particularly sensitive to one specific kind of stimulus. The sensory neurons of the retina are highly sensitive to light, those of the ear to vibrations at frequencies in the range we identify as "sound," and so on. However, all sensory neurons respond in the same way; they translate the stimulus into nerve impulses that, via the appropriate sensory nerves, are transmitted to the central nervous system.

We now come to the intriguing fact that the nerve impulses carried in the different sensory nerves are all of the same fundamental nature. For example, the optic nerve carries the same kinds of nerve impulses as the auditory nerve, and the sorting out of what the original stimulus was is carried out in the central nervous system. If the auditory nerve is stimulated artificially, the induced nerve impulses are perceived by the central nervous system as "sound," and artificial stimulation of the optic nerve is perceived as "light." Most of us have experienced this kind of interpretation; we know that mechanical pressure on the eyeball is perceived as light and that a sharp blow on the eye makes us "see stars," although no light is involved.

Returning to the sequence in Figure 12.1, we say that the sensory neurons are the *transducers* that receive the external information and *encode* it as impulses in the sensory nerves.* In the central nervous system the nerve signals are again *decoded* and the pertinent information is integrated and utilized. The most interesting events are the trans-

---

* A *transducer*, in a general sense, is a device that transforms or "translates" energy supplied in one form into another form. For example, a microphone is a transducer that translates sound energy into electric energy. A transducer often requires an auxiliary source of energy and thus, strictly, is an energy controller. Furthermore, the output is often at a higher energy level than the input (i.e., the transducer is also an amplifier).

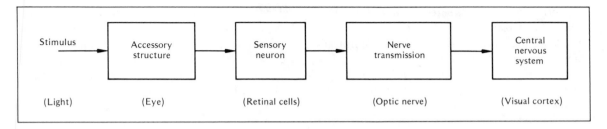

ducing and encoding of information on the one hand, and the decoding and processing on the other. The sensory nerves are transmission lines and in this context the least interesting part of the system.

In this chapter we shall first consider what kinds of information can be obtained and the possibilities and limitations inherent in the nature of the information. The sensory information is encoded and transmitted as nerve signals and again decoded in the central nervous system. The astounding fact is that these various operations are all carried out by nerve cells that function in the same fundamental way, based on the electric potential across the cell membrane. We shall therefore discuss the membrane potential of nerve cells and how impulses are generated and transmitted. We can then return to the consideration of how the wealth of available information is screened, sorted, and processed.

SENSORY INFORMATION: POSSIBILITIES AND LIMITATIONS

Determination of direction and distance

The direction from which a stimulus comes is of utmost importance, for this permits an animal to move toward the source of the stimulus or to evade it, as need may be. Without knowledge of direction, this is impossible. Distance is also important, although usually not by itself but in combination with information about direction. Based on the nature of the stimulus, there are strict physical limitations to what is possible and what the accessory structure can help to achieve in this regard.

The most directional sense is vision, to a majority of animals probably the most important single sense. The reason for the directionality of vision is, of course, that light rays travel in straight lines. In addition to information about the location, the sense of vision gives much information about the nature of various objects. Although vision is the most highly directional sense, information about direction can also be obtained with other senses, but there are certain inevitable limitations.

In principle, there are only three ways in which directional information can be obtained:

1. The sense organ itself can be directional, which means that it must give different signals if turned toward or away from the source of the stimulus, as the eye does.
2. Signals obtained from a pair of similar sense organs can be compared. Directional information about sound is obtained in this way by the use of two ears, although the hearing organ itself is not very directional.
3. Signals can be compared in space and time; this is achieved by successively assessing the condition at a series of points in space.

FIGURE 12.2 The eyes of vertebrates and some invertebrates have a lens and function on the same principle
as a camera. Insects have compound eyes in which the
single elements combine to form the image. Four types
of eyes are shown: a, mammal; b, octopus, c, insect
eye with a single ommatidium shown in detail, d, annelid.

Information about distance is more difficult to
obtain. Here again vision is superior, and with the
aid of two eyes and the differences in the images
they receive, distance can be evaluated. Most
other sense organs can achieve only a rough estimate of distance, based on the attenuation of the
signals. Certain auditory mechanisms can, however, be used for extremely precise information
about distance as well as direction, as we shall see
later.

Light and vision

Undoubtedly vision can give more detailed information than any other sense. With the aid of
suitable accessory structures (eyes), light can form
a detailed image of the environment, both near
and distant.

Information obtained with the aid of light depends on differences in *intensity*, for a uniformly
luminous environment conveys no information of
interest. Much additional information is contained
in differences in *wavelength* or *color*. Finally, for
animals that have appropriate accessory organs the
plane of polarization of the light can also be highly
informative. As far as animals are concerned, the
transmission of light is so fast (300 000 km per second) that it involves no delay. This means that
changes in the environment can be perceived instantaneously, except for the time used in trans-
ducing the signal and transmitting it to the central
nervous system. Compare this, for example, with
the slowness with which a change in an olfactory
signal is perceived.

Image-forming eyes

Sensitivity to light is extremely widespread; even
many unicellular organisms are sensitive to light.
Multicellular animals usually have this sensitivity
concentrated in certain spots (eyespots). If these are
shielded from light on one side, they become
direction-sensitive as well.

More highly organized animals have increasingly complex light-sensitive structures that, in
their ultimate form, are excellent image-forming
eyes. Well-developed image-forming eyes are
found in four different animal phyla: worms, molluscs, arthropods, and vertebrates. The various
eyes differ in design and development, and it must
be assumed that they have evolved independently
in these animals.

An image-forming eye can be based on one of
two different principles: either a multifaceted eye
as in insects, or a single-lens, camera-like structure
as in vertebrates (Figure 12.2). The multifaceted
eye has the same angle of resolution whether an
object is distant or close by, but the lens-type eye
must have some device for focusing if it is to perform equally well at different distances. It is interesting that among the few worms that have well-
developed, image-forming eyes, both single-lens
and multifaceted eyes occur.

Wavelength sensitivity

What man perceives as light lies within a very
narrow wavelength band, 380 to 760 nm, out of
the wide spectrum of electromagnetic radiation
that ranges from the extremely short gamma rays to
long-wave radio waves (Figure 12.3).

All other animals' visual sensitivity lies within or
very close to the same range of wavelengths as
man's. The remarkable fact is that not only animals, but also plants, respond to light within this
same range. This includes both photosynthesis and
phototropic growth of plants. The reason for this
universal importance of a very narrow band of
electromagnetic radiation is simple. The energy
carried by each quantum of radiation is inversely
related to the wavelength. Therefore, the longer
wavelengths do not carry sufficient energy in each
quantum to have any appreciable photochemical
effect, and the shorter wavelengths (ultraviolet and
shorter) carry so much energy that they are de-

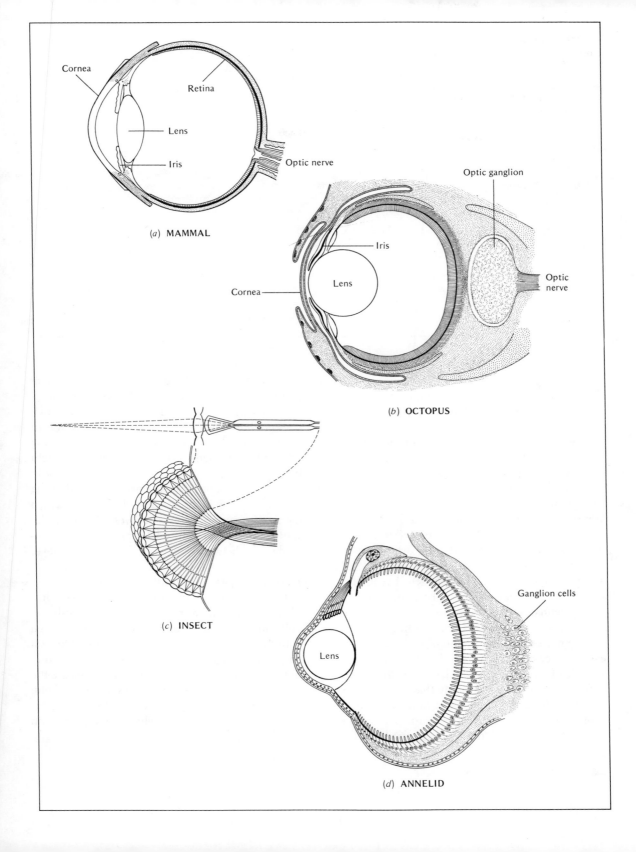

(a) MAMMAL

Cornea

Retina

Lens

Iris

Optic nerve

(b) OCTOPUS

Optic ganglion

Iris

Cornea

Lens

Optic nerve

(c) INSECT

(d) ANNELID

Lens

Ganglion cells

FIGURE 12.3 The spectrum of electromagnetic radiation ranges from the shortest cosmic rays to long-wave radio waves. Only a narrow band between about 380 and 760 nm is perceived as light. The energy carried by each quantum of electromagnetic radiation increases 10-fold for each 10-fold decrease in wavelength. V, violet; B, blue; G, green; Y, yellow; O, orange; R, red.

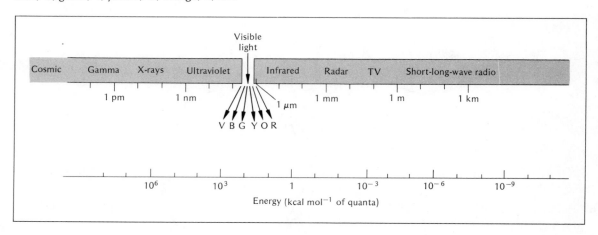

structive to organic materials. The universal biological use of what we know as "light" is a result of the unique suitability of these particular wavelengths.

The range is not exactly the same for all animals. The vision of insects, for example, extends into the near ultraviolet range, to slightly shorter wavelengths than the vertebrate eye. This is evident from the ability of honeybees to distinguish any spectral color between 313 and 650 nm from white light, an ability that is unrelated to light intensity and therefore must depend on wavelength discrimination (i.e., "color" vision) (Kühn 1927). The mammalian retina would be sensitive to ultraviolet light, but these wavelengths do not penetrate to the retina, primarily because of a slight yellowness of the lens, which acts as a filter.

This narrowing of the band of wavelengths that can be perceived is probably advantageous. A lens, if made of a uniform material, refracts short wavelength radiation more strongly than longer wavelengths. This means that the various wavelengths cannot be brought into focus simultaneously. In man-made lenses this difficulty is known as chromatic aberration, and is corrected for by the use of composite lenses consisting of several elements with different refractive indexes. For an eye that does not have color correction, the simplest way of reducing the difficulty of simultaneously focusing different wavelengths is to narrow the wavelength band that is permitted to enter. It is then best to eliminate the shortest wavelengths, for which the refraction error is the greatest. In insects, which have nonfocusing composite eyes, this is irrelevant, for the resolution is determined by the angular distance between the single elements of the eye.

Light absorption: retinal pigments

For the sensory neurons of the retina to be stimulated by light and respond, a sufficient number of light quanta must be absorbed. This is achieved with the aid of a light-absorbing pigment. The best known such pigment is *rhodopsin*, which is found in the rods of the vertebrate eye. (The rods are the retinal elements responsible for vision in dim light; the cones, which will be discussed below, are involved in vision in bright light and the perception of color.)

INSECT VISION What a bee can perceive of the color pattern of a flower is remarkably different from what the human eye sees. To the human eye the marsh marigold (*Caltha palustris*) appears as an even, bright yellow flower (top photograph); an ultraviolet photograph (bottom) reveals a strong pattern that serves as a nectar guide leading bees to the source of food. [Courtesy of Thomas Eisner, Cornell University]

Rhodopsin can be isolated from retinas of animals whose eyes are fully dark-adapted. It is light-sensitive, and when exposed to light breaks down to *retinene*, a molecule closely related to vitamin A, and a protein called *opsin*. This process is the initial step in the perception of light by the rods of the vertebrate eye.

FIGURE 12.4 The relative sensitivity of the dark-adapted human eye at different wavelengths compared with the absorption spectrum of the visual pigment. [Crescitelli and Dartnall 1953]

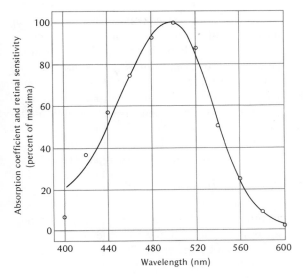

The strongest argument for the role of rhodopsin in the visual process is that, if the rhodopsin in the retina is depleted by exposing the eye to strong light, the eye is insensitive to dim light until the rhodopsin has regenerated. Another strong argument for the role of rhodopsin comes from its absorption spectrum, which has its maximum at 497 nm. This maximum and the detailed absorption spectrum coincide with the maximum sensitivity and the sensitivity spectrum of the human eye in dim light (Figure 12.4).

It has been more difficult to establish the role of pigments in the perception of light by the cones, primarily because of difficulties in extracting appropriate pigments. It has long been understood that the perception of color requires more than one pigment, and that three pigments with absorption peaks in blue, green, and red are sufficient to explain the perception of all colors we see (the trichromatic vision theory).

The difficulties have been overcome by making measurements of the absorption spectra directly in the human eye. By using an ingenious instrumentation, measurements on single cones have revealed that there are three kinds of cones: blue-sensitive cones with an absorption maximum about 450 nm, green-sensitive cones with an absorption maximum about 525 nm, and red-sensitive cones with an absorption maximum at 555 nm. Presumably, these three types of cones are responsible for human color vision in bright light (Brown and Wald 1964).

The spectral sensitivity of the human eye is in accord with the three-pigment theory and the measured absorption spectra. Furthermore, what is known about deficiencies in the color vision of man (often called color blindness) is satisfactorily explained on the basis of the existence of these same three pigments (Wald 1964; Rushton 1972).

Polarized light

One quality of light that man is unaware of is its plane of polarization.* For this reason, the information carried by the polarization of light and its importance for animals remained unnoticed until the German zoologist Karl von Frisch discovered

* In polarized light the vibrations of the propagated wave (ray) are all in one plane, perpendicular to the direction of the ray. In ordinary unpolarized light the vibrations are in all directions (yet perpendicular to the ray). Unpolarized light may become polarized by reflection from water and other nonmetallic surfaces and by transmission through certain materials. The plane of polarization can carry information not contained in unpolarized light. It appears that, in nature, the most important information animals can obtain in this way is attributable to the fact that the light from the blue sky is polarized, the polarization at any point in the sky depending on its position relative to the sun. Indirectly, this tells the location of the sun, information that is most important in animal orientation and navigation.

that honeybees use the plane of polarization as a directional cue.

Von Frisch studied communication among bees and how information is relayed from a bee that has found a good food source to other bees. Usually the bees indicate the direction of the food source relative to the direction of the sun (i.e., they use the sun as a compass), and they have an internal clock that compensates for the movements of the sun across the sky.

It was found, however, that the bees could orient correctly, even in the absence of a visible sun, provided a small piece of the blue sky could be seen. The light from the blue sky is polarized to a degree that gives information about the actual position of the sun, and this is used by the bees. One way to show the dependence on the plane of polarization of the light is to reflect the light by a mirror. This makes the bees orient in the opposite of the correct direction, as expected from the reversal of the plane of polarization caused by the mirror. Also, by imposing filters that change the direction of polarization, any degree of deviation from the correct orientation can be imposed on the bees (von Frisch 1948).

After the importance of polarized light had been demonstrated, it was found that insects in general are sensitive to polarized light. Surprisingly, several aquatic animals are also sensitive to the polarization of light. This was first observed in the horseshoe crab, *Limulus* (Waterman 1950). Many other aquatic animals, including octopus and fish, are also sensitive to polarized light (Waterman and Forward 1970).

A substantial amount of evidence indicates that the ability to perceive the plane of polarization of light is important in orientation and navigation, not only for terrestrial insects, but for a large number of aquatic animals as well (Waterman and Horch 1966).

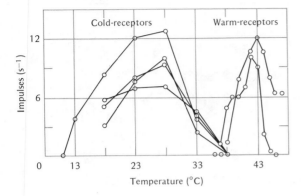

FIGURE 12.5 Response curves obtained from single warm-receptors and cold-receptors in the scrotal skin of rats. [Iggo 1969]

Cold-receptors   Warm-receptors

Impulses (s⁻¹)

Temperature (°C)

Temperature

We think we are able to sense the thermal condition of the environment, and our behavior often seems to indicate that this is so. What is sensed, however, is not the environmental temperature, but the temperature of the skin at the depth of the appropriate receptors. A good example of our inability to judge air temperature is provided by the kind of electric hand drier that is installed in some public washrooms. The drier blows a stream of hot air (as high as 100 °C) over the hands, and as long as the skin is wet, the air feels moderate, but as soon as the skin is dry, the airstream becomes painfully hot. .

Heat and cold are perceived by different sense organs in the skin. This can be shown by an examination of the activity in the individual nerve fibers from the temperature-sensitive skin receptors. The fibers from a heat-sensitive receptor show no activity until the temperature increases above a certain point; from then on the activity increases, roughly in proportion to the temperature rise, until a certain upper limit is reached, above which the nerve activity again decreases.

Records obtained from the scrotal skin of the rat show exactly this (Figure 12.5). The warm-recep-

tors are "silent" below approximately 37 °C. Above this temperature their output increases rapidly with increasing temperature but above 43 °C their output falls off. Below 37 °C the warm-receptors remain silent, and cold-receptors show an increasing response, again roughly proportional to the decrease in temperature. Below a certain level, the cold-receptors show a decreased activity that eventually falls to zero.

Within the narrow neutral range, which corresponds to a normal scrotal skin temperature for a rat in heat balance, there is a low level of activity from both warm- and cold-receptors. A minor change in either direction can therefore be perceived with rapidity and precision. This provides the makings of a sensitive control system based on obtaining accurate information about slight changes in skin temperature and hence on changes in heat transfer between the skin and the environment.

Infrared radiation

Cutaneous temperature receptors of the kind we have just discussed are sensitive to the local skin temperature. They give information about the environment only indirectly, for skin temperature is a result of heat flow from the core of the animal as well as the conduction and radiation to the environment. They discover radiation only through its effect on skin temperature, as we know when we can feel the radiation from a hot stove or other hot object. To some extent we can also sense a loss of heat by radiation. For example, a cold brick wall in a house may give a definite feeling of "cold"; we can feel that the chill comes from the wall and we often interpret it as a "cold draft," although obviously no air flows through the wall.

Infrared radiation can be perceived directly by a few animals that have specialized sense organs which respond to this type of radiation. The so-called *facial pits* or *pit organs* on the head of some snakes are such specialized infrared receptors.

When a rattlesnake strikes, the direction of the strike seems to be guided by the infrared radiation from its prey. A rattlesnake strikes only at warm-blooded prey, and when the prey is dead and at room temperature, the snake will not strike. However, a blindfold snake strikes correctly at a moving dead rat, provided the rat is warmer than the surroundings. Blindfold, the snake cannot be guided by vision. Nor is it guided by the sense of smell, for it will strike correctly even at a moving, cloth-wrapped electric bulb. The pit organ is evidently involved in sensing the location of warm objects. All snakes that have pit organs feed preferentially on warm-blooded prey, and this further supports the view that these organs are infrared sensors. In the rattlesnake the pit organ is located, one on each side, between the nostril and the eye; it is richly innervated, and this in itself suggests a sensory role for the organ.

The sensitivity of the facial pit has been examined by recording the activity in the nerve leading from the organ. A variety of stimuli, such as sound, vibration, or light of moderate intensity (with the infrared part of the spectrum filtered out), has no detectable effect on the activity in the nerve. However, if objects of a temperature different from the surroundings are brought into the receptive field around the head, there is a striking change in nerve activity, regardless of the temperature of the intervening air (Bullock and Cowles 1952; Barrett et al. 1970).

How is the infrared radiation sensed? The pit is covered by a thin transparent membrane (Figure 12.6), and it has been suggested that a rise in temperature in the chamber behind the membrane could cause an expansion of the gas with a consequent deformation of the membrane, which in

FIGURE 12.6 The pit organs on the head of a rattlesnake are specialized receptors for infrared radiation used to locate warm-blooded prey. The two pits are located, facing forward, one on each side of the head, between the eye and the nostril. The pit has an outer and an inner chamber, separated by a thin membrane. [Gamow and Harris 1973. Copyright © 1973 by Scientific American, Inc. All rights reserved.]

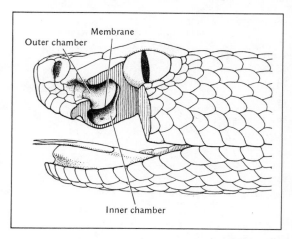

turn could be sensed by a suitable receptor. This hypothesis is highly improbable, for a cut in the membrane that opens the chamber to the outside air causes no loss in responsiveness. This result is incompatible with the hypothesis that a pressure change is sensed.

We are left with two other possibilities to consider: Either the infrared radiation is absorbed by a specific compound, analogous to the light-sensitive pigments in the eye, or the pit organ is sensitive to the slight temperature rise caused when infrared radiation reaches it.

The infrared radiation emitted from a mammalian body has its peak around 10 000 nm, and the low quantum energy in this wavelength makes any photochemical effect on a pigment extremely unlikely. Pure infrared radiation can be produced by a laser, and experiments with such radiation of known wavelength provide strong evidence that the mode of reception in the facial pit organ is entirely thermal (Harris and Gamow 1971).

Can the pit organs be used for stereoscopic perception of an infrared source, the way our two eyes are used for stereoscopic vision? This seems highly likely, not only from observations of the precision with which a snake can strike, but also from studies of its brain activity. When infrared radiation falls on the facial pit organ, electric potentials can be recorded from the optic tectum, the part of the brain with which the optic nerve is connected. This in itself is interesting; although the nerve from the pit organ (a branch of the trigeminal nerve) is completely separate from the optic nerve, the same part of the brain seems to handle visual and infrared information.

Many of the neurons in the tectum respond to stimulation of the pit organ on the opposite side of the head. This is reminiscent of the way information from the eyes is handled; the crossover of the optic nerve in the optic chiasma is essential for stereoscopic vision for interpretation of distance. Information from the two pit organs is apparently coordinated and interpreted in a similar way, a conclusion in agreement with recorded changes in the neural activity in the tectum when the position of the infrared source is changed. The response is particularly noticeable when the source is in a position to irradiate both pits at once. It therefore appears that the facial pits indeed provide stereoscopic perception and thus substantially aid in the precision of estimating the location of prey (Goris and Terashima 1973).

It is unlikely that specialized infrared-sensitive organs are of widespread importance among animals. In particular, it is almost unthinkable that any similar organ could be important for aquatic animals, partly because of the very low penetration of infrared radiation in water, and also because the direct contact of the body surface with a medium of high thermal conductivity and thermal capacity would make it impossible to perceive the small amounts of heat involved.

Animal electricity

It is well known that some fish can produce strong electric shocks. Ancient peoples such as the Greeks and Egyptians knew and wrote about the powerful shocks delivered by the electric ray (*Torpedo*) and the electric catfish (*Malapterurus*), but the phenomenon must have appeared utterly mysterious until the nature of electricity became known.

Powerful electric discharges serve obvious offensive and defensive purposes. The strongest shocks are produced by the South American electric eel (*Electrophorus*). It can deliver discharges of between 500 and 600 V, which are powerful enough to kill other fish and possibly animals the size of a man. Such strong electric discharges can be delivered only by a few species of fish. These are not particularly closely related: Some are teleosts; others are elasmobranchs; and they occur both in fresh water and in the sea.

As electric fish have been studied more closely, it has become apparent that even weak discharges – too weak to have any direct effect on other fish – serve a variety of useful purposes. Obviously, very weak discharges cannot stun prey and are useless as weapons of offense or defense. However, it is now clear that they can be used to obtain information about the environment in general. Electric discharges are particularly suitable for distinguishing between nonconducting objects and good conductors, such as another animal. A freshwater animal, say a fish, has a much higher electric conductivity (because of the salt content of its body fluids) than the poorly conducting water. Weak electric discharges can be used for navigation in murky water, as well as for the location of predators and prey. Furthermore, electric discharges are used in communication between individuals, and in some cases a fish will respond with

FIGURE 12.7 Feeding responses of a small shark (*Scyliorhinus canicula*). (*a*) When a hungry shark passes in the vicinity of a flounder completely buried in the sand, it detects the flounder and immediately attacks it. (*b*) To exclude olfactory cues, the flounder is covered with an agar chamber perfused with water that exits at some distance (broken arrow). The shark nevertheless attacks the correct location of the prey. (*c*) When pieces of fish are placed in the agar chamber, the shark searches for food where the perfusing water exists. (*d*) If the agar chamber is covered with a plastic film, the signal is attentuated and the shark passes by without noticing. (*e*) An artificial electric field of the same magnitude as generated by the breathing movements of the flounder excites the shark, which immediately attacks this "prey." [Kalmijn 1971]

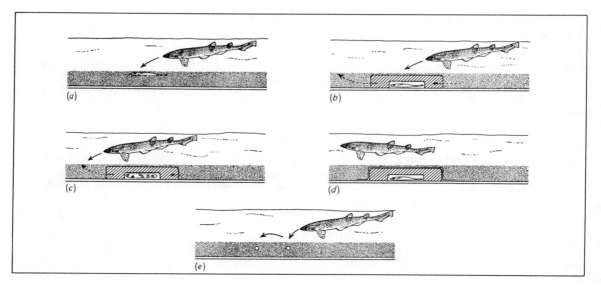

sexual behavior only to the electric signals from the opposite sex (Black-Cleworth 1970; Hopkins 1972).

Finally, it has recently become known that many fish which cannot produce electric signals themselves are quite sensitive to the weak electric activity that results from the ordinary muscle function of other organisms. In this way sharks and rays are able to locate another fish, even if it is at rest. A hungry dogfish (a shark, *Scyliorhinus*) will even respond to a flounder that is resting at the bottom of an aquarium, completely covered by sand. If the shark passes within less than 15 cm of the flounder, it makes a sudden turn toward the hidden prey, removes the sand by sucking it up and expelling it, and seizes the flounder (Figure 12.7).

Careful experiments have shown that this ability to locate the flounder is not associated with the sense of smell. Apparently the dogfish responds to the minute electric potentials produced by the breathing movements of the hidden flounder, for it will react to an artificial electric discharge of 4μA (the same order of magnitude as that produced by a living flounder) and try to dig out this imaginary prey (Kalmijn 1971).

To summarize, electric fish can use their discharges to stun prey, to obtain information about their environment (much more common), and for communication. Furthermore, nonelectric fish can use their extremely sensitive electroreceptors to locate prey by means of the action potentials caused by normal muscle activity.

Production of the electric discharges

In nearly all electric fish the discharges are produced by discrete electric organs, which consist of modified muscle. These organs have been most carefully studied in the electric eel, which can produce discharges of over 500 V. The large electric organs of the electric eel run along most of the body, one large mass on each side, and make up about 40% of the animal's volume.

FIGURE 12.8 The electric organs of fish consist of thin, waferlike cells (electroplaques) stacked in columns of several thousand. When a plaque is inactive and at rest (top), both faces are positively charged, the outsides being at+84 mV. Therefore, no potential difference exists between the two outside faces. During discharge (bottom) the potential on the posterior face of the plaque is reversed and becomes −67 mV on the outside. The po-

tential difference between the two outside faces therefore is 84 + 67 mV = 151 mV. [Keynes and Martins-Ferreira 1953]

FIGURE 12.9 Location of sense organs on the head of the dogfish *Scyliorhinus*. Openings of the ampullae of Lorenzini are shown by black dots. The open circles indicate the pores of the lateral line system and the black lines its location. [Dijkgraff and Kalmijn 1963]

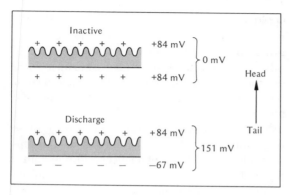

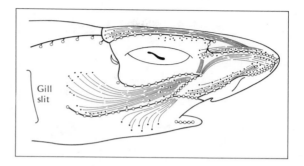

These organs consist of thin, waferlike cells, known as *electroplaques* or *electroplates*. The electroplaques are stacked in columns that may contain between 5000 and 10 000 plaques, and there are some 70 such columns on each side of the body. The two faces of each electroplaque are markedly different; one face is innervated by a dense network of nerve terminals, and the other is deeply folded and convoluted.

When the organ is at rest, the two surfaces of the electroplaque are positively charged on the outside, each being +84 mV relative to the inside (Figure 12.8). The overall potential from the outside of one membrane to the outside of the other therefore is zero. During a discharge, however, the potential on the innervated surface is reversed, and the total voltage across the single electroplaque becomes about 150 mV. With the serial arrangement of the electroplaques, the voltage adds up as when we connect a number of batteries in series. With several thousand electroplaques arranged in this way, the electric eel can reach several hundred volts. (Because of internal losses the voltage does not reach the full magnitude of an ideal serial connection of the electroplaques.)

The arrangement of electroplaques in parallel columns increases the current flow. In the electric

eel the current may be about 1 A. The giant electric ray (*Torpedo*) lives in sea water, which has a much lower resistance than fresh water. Its voltage is lower (some 50 V), but because it has some 2000 columns in parallel in each electric organ, the total current is extraordinarily high, up to 50 A. A current of 50 A at 50 V corresponds to a power of 2500 W during the discharge (Keynes and Martins-Ferreira 1953).

Electroreceptors

Most fish have a large number of sensory organs in the skin, mostly around the head and along the lateral line. One kind, the ampullae of Lorenzini, has a characteristic small flask-shaped appearance. These ampullae are found on the heads of all marine elasmobranchs (Figure 12.9), and similar organs are found in a variety of teleosts. Appropriate nerve sections and behavior experiments have shown that the ampullae of Lorenzini of the dogfish (*Scyliorhinus*) are electroreceptors (Dijkgraaf and Kalmijn 1963). The electroreceptors of electric fish are of several different types; the commonest has a duct leading from the surface of the skin to an enlarged ampulla situated deeper in the skin (Figure 12.10).

The electric reception has been studied in detail in the African electric fish, *Gymnarchus niloticus* (Machin and Lissman 1960). This fish, which

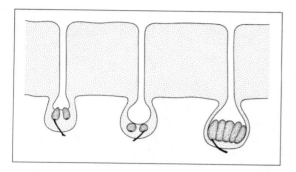

lacks a common name, discharges a continuous stream of pulses at a frequency of about 300 to 400 per second. During each discharge the tip of the tail is momentarily negative with reference to the head, so that an electric current flows into the surrounding water. The configuration of the electric field depends on the conductivity of the surroundings and is distorted if an object with higher or lower conductivity than the water is introduced into the field. The lines of current flow will converge on an object of higher conductivity than water and diverge around a poorer conductor.

The sense organs are sensitive to such changes in the field. However, they do not respond to single pulses, but to the average current over a period of about 25 ms, a time sufficient for the emission of some 7 to 10 discharges. The total current needed to excite an individual electric receptor is extremely small, about 3×10^{-15} A (Machin and Lissmann 1960).

Many electric fish live in murky water where visibility is poor, and they often have poorly developed eyes. The value of an electric sense that permits scanning of the environment and location of prey is therefore obvious. And such a sense has the advantage above vision that the system is independent of the day-and-night light cycle. A disadvantage, compared with light, is the limited range.

What about interference? If many individual electric fish live in the same area, how can they avoid confusing their own signals with those of others? If an electric fish is subjected to an artifical electric pulsing of the same frequency as its own, it responds by shifting its frequency away from the artificial signal. This presumably can prevent the fish from confusing its own signals with those originating from other fish. This conclusion is supported by the tendency of an electric fish in nature to shift and find its own individual private frequency when approached by another fish with the same frequency (Bullock et al. 1972).

Magnetic sense

It can no longer be doubted that many animals respond to magnetic fields, although innumerable experiments with artificial magnets have failed to yield clear results. However, the fact that a magnetic field has some effect on an animal does not mean that, in nature, the animal senses a magnetic field or makes any use of it.

Apparently one reason many experiments have failed to show an effect is that experimenters have used very strong magnetic fields. Fields of the same magnitude as the natural magnetic field of the earth systematically influence the direction in which mud snails (*Nassarius*) and planarians (*Dugesia*) move. When the strength of the artificial magnetic field is decreased substantially, response also becomes minimal. It seems that the magnetic response (compass orientation) is exquisitely adjusted to the strength of the natural geomagnetism, and this may explain the lack of success in so many earlier experiments (Brown et al. 1964).

Many other invertebrates also react to weak magnetic fields, but the meaning of their responses is often difficult to establish. We can assume that, if the response is more than incidental, an influence on direction and orientation is important, as has been established for honeybees. When bees communicate to other worker bees the direction of a food source, the waggle dance they perform on a vertical honeycomb is affected by the earth's magnetic field (Lindauer and Martin 1968).

It seems that in the orientation of birds, the role of a magnetic sense is beyond doubt. When pigeons are homing from unknown sites, they use a number of different cues to obtain navigational information. One cue is the sun, which is used to establish compass direction. Even on overcast days, however, pigeons can carry out successful homing, but if they carry small magnets, they become disoriented, whereas control birds that carry similar but nonmagnetic pieces of metal show no disorienting effects. It seems that magnetic cues are only one of several kinds of information used by homing pigeons and therefore do not fully explain the birds' excellent orientation abilities (Keeton 1971).

In laboratory experiments, where single environmental cues can be better isolated and controlled, the effect of magnetic fields can be studied more accurately. The European robin (*Erithacus rubecula*) displays a period of migratory restlessness in the spring and again in the fall. If it is placed in a circular cage, it tends to move toward the direction that would be its natural migratory direction for that time of year.

The preferred flight direction is maintained in a completely closed room without any optical reference points or other known cues that would make the bird orient correctly. However, if the test cage is moved to an all-steel chamber, which provides a complete shield to magnetic fields, the birds can no longer find the natural migratory direction (Fromme 1961). Furthermore, artificially generated magnetic fields influence the directional choice made by the bird. Caged indigo buntings orient during their migration period according to the natural geomagnetic field, orienting toward the north. If the horizontal component of the magnetic field is artificially deflected 120° clockwise, the birds orient accordingly toward east-southeast (Emlen et al. 1976). There are also reports that free-flying nocturnal migrants are affected by natural disturbances in the geomagnetic field (Moore 1977).

It appears that the magnetic compass of these birds is not sensitive to the polarity of the field, but that the inclination of the axial direction of the field lines is used as a cue for deriving information about north–south polarity (Wiltschko and Wiltschko 1972).

What organ or structure responds to magnetic fields is completely obscure. No information is available about any sensory organs that could be responsible for a magnetic sense, and there are no convincing hypotheses about mechanisms that could be involved.

Sound and hearing

In trying to define sound, we easily end up with a circular definition. We know that sound consists of regular compression waves that we perceive with the ears; in turn, an ear can be defined as an organ sensitive to sound. Because compression waves can be transmitted in air, in water, and in solids, we cannot restrict the word hearing to perception in air only.

In describing human hearing, we meet no great conceptual difficulties. Our ears are sensitive to regular compression waves in the air in a range from approximately 40 to 20 000 Hz (cycles per second). However, a dog can perceive higher frequencies that are completely inaudible to humans, up to 30 000 or 40 000 Hz. Bats can perceive frequencies as high as 100 000 Hz, and we still consider it hearing. There are two good reasons. One is that dogs and bats have ears very similar to ours. The other is that the frequency range perceived is continuous; our own ear is just designed so that it is insensitive to the very high frequencies many other animals can hear.

On the other hand, we know that compression waves at frequencies below those we perceive as sound can often be perceived as a vibration (i.e., we still sense frequencies lower than those that we consider hearing).

When a fish senses footsteps at the edge of the water, is this hearing? The vibration from the steps is transmitted through ground and water, and these low-frequency vibrations are probably perceived by sensory organs in the lateral line, rather than in the ear. Is it then correct to say that the fish "hears" or "feels" the approaching steps? In general, we tend to speak about hearing when animals have specially developed organs that are sensitive to what we consider to be sound, but this is obviously not a good way to define hearing.

Defining hearing becomes much more difficult in relation to invertebrate animals, in which the organs responsible for perception of vibration are very different from the vertebrate ear and in many cases have not been identified at all.

What kinds of information can be obtained from sound waves? Man can perceive sound waves and their temporal pattern, determine intensity, distinguish different frequencies (pitch), and carry out very complex frequency analysis. He is also able to determine the direction from which sound comes. In addition, some animals can use self-generated sound for obtaining detailed information about the physical structure of the environment.

The complexity of frequency analysis of which man is capable is astonishing if we consider that what he hears basically is nothing more than a minute in-and-out motion of the eardrum. In spite of this fundamental physical simplicity, the average musical ear can distinguish individual instruments in an orchestra with a large number of simultaneously performing instruments.

Insects can produce and perceive a wide variety of sounds, which are used extensively for communication, often with the opposite sex. The hearing organs of insects vary greatly in structure and may be located in various parts of the body. Apparently, insect hearing organs are not sensitive to pitch the way our ear is; instead, information is transmitted mainly as changes in intensity, duration, and pattern of the sound.

Records of sound produced by male insects to attract females can be played back with so much frequency distortion that they are unrecognizable to the human ear, and still attract the female insects. The meaningful information seems to be carried in the pattern of pulses, not in pitch and tone quality. The pattern of insect sound is relatively fixed and is highly species-specific. The human ear is inherently unsuited for perceiving the important features of insect sound, but as electronic equipment for recording and analysis has become available, the pattern of insect song can be transcribed to visual patterns that we understand better and can evaluate more intelligently.

Directional information

Determining the direction of sound depends mainly on having two ears, separated in space. Some information is derived from the fact that the signals arriving at the two ears often differ in intensity, but more important is the difference in time of arrival of the sound from a given source.

For a complex signal the interpretation of direction may be brought about by a simultaneous comparison of intensity, frequency spectrum, phase, and time of arrival of the signals at the two ears. The comparisons must be made by the central nervous system, which must also be capable of separating the important signal from other noise in the environment (Erulkar 1972).

For man the ability to locate sound is much less precise than for predators that normally work in the dark, such as cats or owls. The barn owl, for example, can locate prey in total darkness, using only the sense of hearing, with an error of less than 1° in both the vertical and the horizontal planes. For this accurate location of the source of a sound, the owl depends on frequencies above 5000 Hz, which are more directional than lower frequencies (Payne 1971).

Sonar

Owls use sounds coming from the prey to determine its location, but some vertebrates obtain information from the faint reflections or echoes of sound they themselves produce. This *echo location* or *animal sonar* is particularly well developed in bats, but exists also in whales, dolphins, shrews, and a few birds.

It has long been known that bats can fly unhindered around a completely darkened room, and it makes no difference whether or not their poorly developed eyes are covered.

One of the best studied bats is the big brown bat (*Eptesicus fuscus*). As this bat cruises in the dark, it emits pulses of sound that are inaudible to man, for the frequency is between some 25 000 and 50 000 Hz. Each pulse or click lasts about 10 to 15 ms and may be given off at a rate of some five single clicks per second. The bat obtains information about the environment from the echoes and is able to avoid all sorts of obstacles and can even fly between wires strung across the room.

If the bat approaches a particularly "interesting" object, such as a flying insect, the number of clicks increases and the duration of the single click is shortened, so that the rate may be as high as 200 clicks per second, each click lasting less than 1 ms.

One advantage of using a high-frequency sound (short wavelength) is that the directional precision is much better than for low-frequency sound, which spreads too widely and gives diffuse reflections unsuited for pinpointing accurately the location of objects. Furthermore, in order to reflect sound waves an object must be above a certain size, relative to the wavelength. The shorter the wavelength, the smaller the size of object that gives a distinct reflection, and the highest frequencies therefore permit the detection of the smallest objects.

The bat's impressive skill at avoiding fine wires can be tested in a long room where the animal can fly back and forth. If the center of the room is divided by vertical wires, spaced 30 cm apart, a bat flying without information about the wires, has a 35% random chance of missing a wire. Wires several millimeters in diameter are successfully avoided, but if the diameter of the wires is reduced to less than 0.3 mm, the flying bat hits or touches the wires more often. If the wires are very thin, 0.07 mm, the number of hits and touches increases and the number of misses decreases to the chance value of 38%, the same as in experiments with bats made temporarily deaf by plugging their ears (Figure 12.11).

The shortest wavelength used by the little brown bat, which was used in these experiments, is about 3 mm (100 000 Hz). This means that it detects objects no more than about 0.1 wavelength in diameter.

Acoustic orientation is, of course, particularly useful for animals that are active in the dark, and it is also of importance in deep or murky waters. Both dolphins and whales use acoustic echoes to avoid colliding with objects and with the ocean

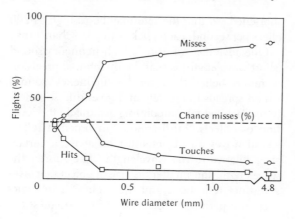

FIGURE 12.11 Obstacle avoidance of the little brown bat (*Myotis lucifugus*) when flying in darkness through a row of vertical wires spaced 30 cm apart. The statistical chance of missing the wires was 35%. The ability to avoid the wires deteriorated when the diameter of the wires was decreased to less than 0.3 mm. [Griffin 1958].

bottom, as well as for finding food. When they swim in open water, there are few obstacles, but at night and at depths where little or no light penetrates, their sonar helps them avoid collision with the bottom. It is, of course, very valuable to be able to locate food both in the dark and when the visibility is low.

How dolphins use acoustic information has been tested in captive animals. A trained dolphin can find a dead fish thrown into the water and reach it within seconds from the opposite end of the tank, even if the water is so murky that a man cannot see the fish at all. If a panel of clear plastic is used to block its path, the dolphin can immediately find an opening, which it swims through without hesitation (Kellogg 1958; Norris et al. 1961).

Other mammals that use acoustic orientation or echo location include the shrews, which employ sound pulses of short duration and high frequency (to over 50 000 Hz) (Gould et al. 1964).

At least two species of birds use echo location: the oilbirds (*Steatornis*) of South America and the cave swiftlets (*Colocalia*) of Southeast Asia. They are not closely related, but both live and nest in

deep caves. The best known are the oilbirds or guacharos, which can fly around freely in dark caves without hitting the walls or other obstacles. They use a sonar system, much like bats, except that their sounds have a frequency of only about 7000 Hz. Therefore, they are fully audible to man, and their sound has been described as something like the ticking of a typewriter. If the birds' ears are plugged, they lose their sense of orientation in the dark, but they can still fly about in a lighted room and evidently use their eyes for orientation (Griffin 1953, 1954).

Other mechanical senses

We have seen that vision and hearing can provide a great deal of detailed information about the environment – hearing, in fact, far more than we initially would think. There is no doubt that those animals that use echo location obtain a three-dimensional impression of their immediate environment, somewhat analogous to what humans obtain through vision, except that the range is more limited.

Other senses provide additional information, but mostly of a less comprehensive nature. These include a variety of mechanical senses about which we have only rudimentary information.

Sensitivity to *touch*, which actually discovers a deformation of the skin surface, is widespread and gives information about objects in immediate contact with the body. The sensation of vibrations is similar in nature. The sensations picked up by the lateral line of fish belong in the same category. The lateral line is, in fact, a highly complex organ that is used for a multiplicity of purposes, among others to sense the rate of flow of water over the fish, thus giving information about speed (Cahn 1967).

Sensitivity to *hydrostatic pressure* has been demonstrated in a variety of aquatic invertebrates.

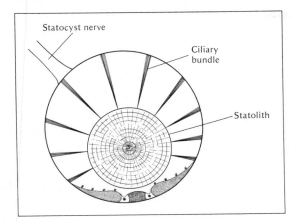

FIGURE 12.12 The statocyst of a pelagic snail. Within the statocyst, a heavy body of calcium carbonate, the statolith, rests on sensitive sensory cells. These respond to changes in the position of the statolith as the animal changes its position in the field of gravity. [Tschachotin 1908]

Statocyst nerve

Ciliary bundle

Statolith

Some are quite sensitive to small pressure changes, but the mechanism involved remains obscure. It is easy to understand how an animal that carries a volume of gas can be sensitive to pressure changes, for the resulting distortion can easily be sensed with the aid of mechanoreceptors. However, the majority of plankton organisms that are pressure-sensitive carry no bodies of air that can explain their high sensitivity to small pressure changes (Enright 1963).

Sensitivity to the force of *gravity* is extremely widespread and gives information about what is up and down. More complex animals have gravity receptors, known as *statocysts*, which are uniformly built on the same principle (Figure 12.12). A heavy body, the statolith, often of calcium carbonate, rests on sensory hairs. Any displacement relative to the force of gravity causes the statolith to stimulate the ciliary bundles of other sensory cells, which then send information to the central nervous system.

Angular acceleration is sensed in a similar way; fluid within a suitable structure is set in motion and deflects the ciliary cells in a direction that depends on the direction of acceleration. In ver-

tebrates the organ responsible for sensation of angular acceleration is located in the inner ear and consists of the three semicircular canals, oriented in three planes at right angles to each other. They are filled with fluid, and any angular movement causes the inertia of the fluid to stimulate the sensory cells in the plane of the movement. With the canals oriented in three different planes, any angular acceleration can be perceived.

The receptor cells involved in hearing, in perception of position and equilibrium, and in the detection of water movements by the lateral line of a fish are all of the same basic type. These receptor cells possess ciliumlike processes, and movements of these processes lead to appropriate sensory outputs. Let us, as an example, examine the typical receptor cell in the lateral line of fish, a system used in the perception of movements relative to the surrounding water.

Each sensory receptor cell has two types of processes: a group of relatively short *stereocilia* and one longer *kinocilium*. The kinocilium contains two central and nine peripheral filaments (i.e., the characteristic structure of a cilium).

These sensory cells are usually placed together in a group, and the protruding cilia are covered by a gelatinous projection, the *cupula* (Figure 12.13). When the cupula is deformed by movements in the adjacent water, the cilia are bent, and the movement excites the sensory cell.

The kinocilium is always placed to one side of the group of stereocilia (Figure 12.14), and because of this orientation, the response of the receptor cell is polarized. The sensory cell is excited when the cilia are bent toward the side where the kinocilium is located. In the lateral line adjacent hair cells are oriented in opposite directions: Some are therefore stimulated by the movement of water in one direction; other hair cells are stimulated by movement in the opposite direction.

FIGURE 12.13 Structure of the lateral line organs of fish. Top: several sensory cells, each carrying a sensory hair, are grouped together with the protruding sensory hairs covered by a jellylike cupula. Bottom: longitudinal section of lateral line. Movement of water causes deformation of the cupulae (dotted). [Dijkgraaf 1963]

FIGURE 12.14 The common sensory cells of mechanoreceptors, known as hair cells, have a longer kinocilium placed at one side of a group of smaller stereocilia. Deformation of the protruding cilia gives rise to action potentials in the sensory cell. [Flock 1971]

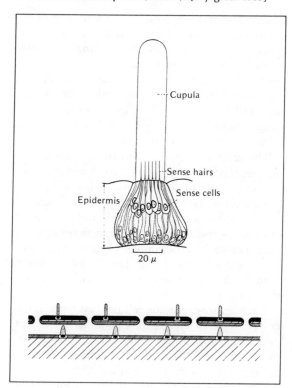

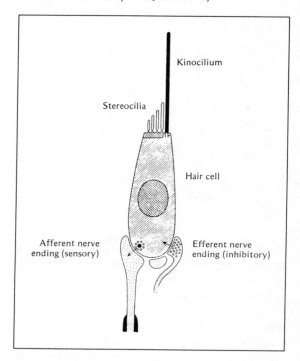

Chemical senses: taste and smell

To a human the sense of taste refers to materials in contact with the mouth; smell refers to gaseous substances that reach the nose via the air. The separation is highly subjective, for the sense of taste, located in the mouth, separates only the four basic qualities of salt, sweet, sour, and bitter. When we speak about the "taste" of food, most of the information is obtained through the nose.* When we

---

* The texture of the food is also an important quality in "taste"; if familiar foods have been homogenized, they often are very difficult to recognize.

extend the use of the terms taste and smell to other air-breathing vertebrates, the analogy is justified, for their sense organs have a great deal of structural similarity to ours.

For aquatic animals a distinction between taste and smell rapidly becomes meaningless. For example, some fish have chemical senses of unbelievable acuity, but should we say that they "smell" or "taste" the water? In common language it is often convenient to use the term smell for objects at a distance and taste for materials in direct contact with the animal, in particular with the mouth. However, a catfish has sensitive chemoreceptors located all over the body, and we find it natural to call these receptors tastebuds (Figure 12.15).

The chemoreceptors of fish are highly sensitive

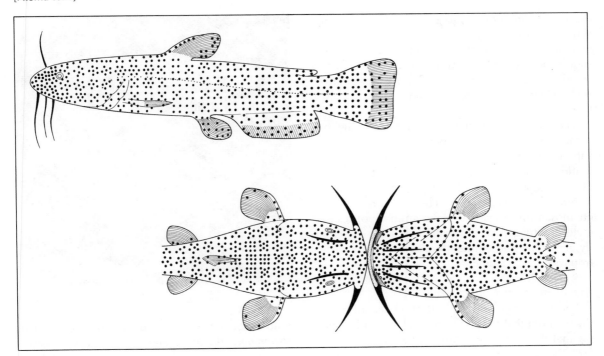

to the amino acids, and this is undoubtedly of importance in finding food. Electrophysiological tests of the responses of the chemoreceptors on the barbels of the channel catfish (*Ictalurus punctatus*) show different sensitivities for different amino acids with estimated threshold concentrations between 10^{-9} and 10^{-11} mol per liter (1 to 100 μg per liter). These unbelievably low concentrations are the lowest electrophysiological thresholds reported for taste in any vertebrate (Caprio 1975).

The chemical senses cannot be used for scanning the environment, the way light and sound are used, for the chemical compounds to be sensed must first arrive at the animal's sensory organ, either by diffusion or by mass movement of the medium. Light has, for physiological purposes, the

quality of instantaneous arrival, and sound arrives very rapidly, but with such a delay that the delay itself carries valuable information (as in directional hearing and echo location). The limitations on the use of the chemical senses are much more severe, and detailed information is usually obtained only over extended periods of time, often in combination with the animal's moving about.

The most useful information obtained with the chemical senses concerns finding food or mates and locating enemies. The chemical senses must be considered primarily as useful alerting or warning devices, allowing the efforts of other senses to be concentrated.

Some kind of chemical sense seems to be universally present among animals; even very simple organisms and protozoans respond to chemical stimuli. Obviously, it is of interest to an ameba whether or not a particle is digestible and thus should be ingested. However, discrete chemical sensory organs do not occur in the simplest animals. Their highest development is reached in vertebrates and in arthropods.

The presence of a chemical sense in less highly organized animals is easily demonstrated by observation of their behavior, but virtually all our information about the physiological events in the chemoreceptor organs has been obtained from arthropods and vertebrates. The reason is that the chemosensitive organs in these animals form discrete structures that, especially in insects, are easily accessible and eminently well suited for electrophysiological studies.

Insect taste and olfaction

Let us separate taste and olfaction according to the medium that carries the chemical stimulus and say that the transfer medium for taste is water and for olfaction is air. We can then say that taste and olfactory receptors in the insect are distinctly separate structures. They can be located almost anywhere – on the mouth parts, the antennae, the feet. The taste organs are usually bristles or hairs, a fraction of a millimeter long, open at the tip, and with one or more sensory neurons at the base. The olfactory organs may be hair- or bristlelike, or may consist of thin-walled pegs or pits in a wide variety of forms and shapes.

The taste organs usually require the stimulating molecules to be in much higher concentrations than the olfactory receptors. The latter in some cases are unbelievably sensitive to specific odoriferous molecules that are important for locating food, prey, and mates.

Substances used to find other individuals of the same species, as well as for a variety of social com-

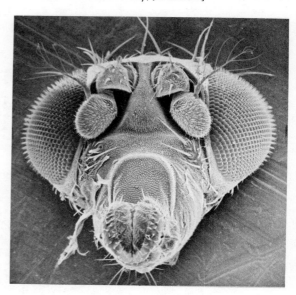

FRUIT FLY The head of a fruit fly (*Drosophila*) carries numerous sensory hairs. The large bulging areas on each side of the head are the two compound eyes. The mouth parts, with a number of sensory hairs attached, are seen at the bottom of the photograph. [Courtesy of R. Falk, Hebrew University, Jerusalem]

munication activities in colonial insects, are called *pheromones* (Greek *pherein* = to carry). Pheromones constitute a chemical language that is of great importance in laying trails, in recognizing individuals from the same nest, in marking the location of a food source, as alarm substances, and so on.

The sensory hairs on the proboscis of the ordinary blowfly (*Phormia regina*) make an excellent experimental model. Severing the head from the fly and attaching it to a fine micropipette filled with saline provides a way of mounting it, as well as an electrical connection to the inside. By penetrating the wall of a sensory hair, at some distance from its tip, with a microelectrode, the electrical activity in the hair can be recorded. A minute glass tube, filled with a solution to be tested, can then with the aid of a micromanipulator be brought into contact with the open tip of the hair, and if the solution contains a suitable stimulant, it causes

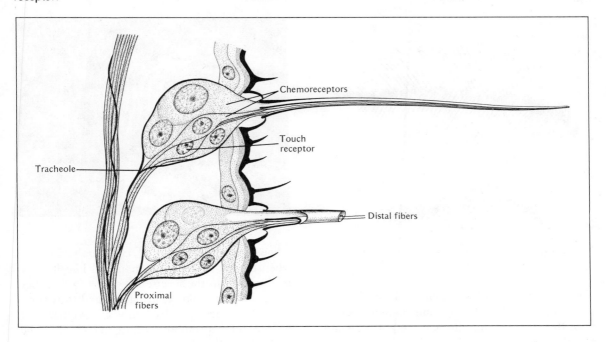

FIGURE 12.16 The proboscis of the blowfly carries a large number of sensory hairs; these hairs can sense various chemicals and thus are taste receptors. At the base of the hollow hair are several sensory neurons, each more or less specialized in the range of compounds it responds to, and one of them always a touch receptor.

Chemoreceptors

Touch receptor

Tracheole

Distal fibers

Proximal fibers

electric activity, which the microelectrode records (Morita and Yamashita 1959).

A typical *taste hair* (Figure 12.16) may have five sensory neurons at its base. One of these is always a mechanoreceptor, sensitive to deflection of the hair; of the others, one is usually sensitive to sugar, one to water, and one or two to salts and various other compounds (Hodgson 1961). By recording directly from the taste hair, it is possible to determine which substances stimulate the receptor and what concentration is necessary.

In contrast to the taste receptors, which are sensitive to a very small range of compounds, the olfactory receptors of insects are highly sensitive and respond to a wide variety of stimuli. They are more complex than the taste receptors and often contain a large number of sensory neurons.

The olfactory receptors are often, but not al-

ways, located on the antennae. The ability of certain male moths to locate females of the same species, even at distances of several kilometers, is well known. The female gives off a sex attractant, and the male is equipped with giant antennae that have a primary role in sensing the sex attractant in immensely dilute concentrations (Figure 12.17). In the male polyphemus moth (*Telea polyphemus*) each antenna carries about 70 000 sensory organs with about 150 000 sensory cells. The female, in contrast, has much smaller antennae, each with only about 14 000 sensory organs and 35 000 sensory cells (Boeckh et al. 1960). Approximately two-thirds of the receptor cells of the male are specialized for sensing the female sex attractant. The shape of the antennae makes them serve as sieves, almost combing through the air for the appropriate molecules. With the aid of radioactively la-

FIGURE 12.17 The large antennae of the male polyphemus moth (*Telea polyphemus*) are sensory organs that are highly sensitive to the sex attractant of the female moth. Their shape serves to attain contact with the largest possible volume of air.

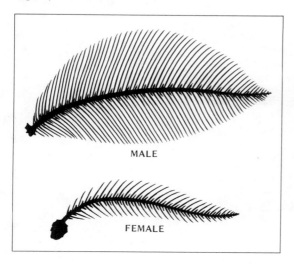

MALE

FEMALE

ANTHERAEA MOTH The large antennae of the male moth (*Antheraea pernyi*) are used in locating the female of the same species by means of the sex attractant she gives off. [Courtesy of Muriel V. Williams, Harvard University]

beled sex attractant, it has been demonstrated that more than 25% of the sex attractant molecules from the air streaming through the antennae are absorbed (Schneider et al. 1968).

Experiments with male silk moths (*Bombyx*) have provided an estimate of how many molecules are necessary to cause a reaction in the moth. If the air contains 1 molecule of the sex attractant per 10^{15} air molecules, and the air flows at a speed of 60 cm per second for a total time of 2 seconds, the male responds. The number of molecules that reaches the antennae corresponds approximately to a level where each of the 10 000 receptors has received a maximum of one molecule of the attractant during the 2 seconds (Schneider 1969).

A male moth that senses a female from a distance of several kilometers cannot possibly use concentration gradients in the air to discover the direction in which he can find the female. The typical behavior of the male is to take off upwind, and this gets him started in the right direction. He then follows the aerial trail of the attractant in a characteristic zigzag flight across it. Animals that

follow an odoriferous trail commonly do this; when they pass out of the trail and lose it, they tend to turn in the reverse direction, so that they enter the trail again. Leaving the trail thus automatically brings them back in again, so that their path becomes a sinuous crossing in and out of the trail. When they have reached the immediate vicinity of their goal, the final localization can take place with the aid of concentration gradients.

The localization of the source of a chemical signal at a long distance (such as the sex attractant) requires a long-lasting, relatively stable molecule. For alarm substances, which should act briefly and in a localized area, the molecule should be relatively unstable to avoid long-lasting residual effects that would be confusing. When the odoriferous substance is unstable and decays rapidly, it results in steep gradients in the vicinity of the source, which therefore is more easily pinpointed.

TRANSDUCTION AND TRANSMISSION OF INFORMATION

Sensory information, such as light, sound, and so on, must be transformed into nerve impulses, and these must be conveyed to the central nervous

FIGURE 12.18 Neurons can have various forms, but every neuron has a cell body and a number of long extensions. These schematic drawings represent (a) an arthropod motor neuron, (b) a mammalian spinal sensory neuron, (c) a bipolar neuron from the vertebrate retina, (d) a neuron from the nerve net of a coelenterate, and (e) a basket cell from the mammalian cerebellum. [Aidley 1971]

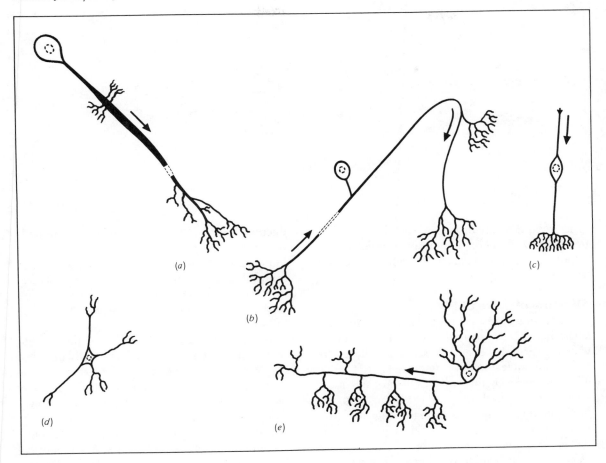

(a)

(b)

(c)

(d)

(e)

system, which coordinates and uses the information in directing appropriate responses. In the preceding section we were concerned with the nature of the information and how it is received by the sensory organs. We shall now examine the effect of the stimulus on the receptor organs, how the stimulus is transduced and encoded as nerve signals, and in what form the information is transmitted to the central nervous system.

To understand this, we must know how nerve cells function. Because of the astounding uniformity inherent in the function of nerve cells, it is relatively unimportant what animal we deal with. A snail or a squid or a frog or a fish is equally relevant to the understanding of the principles on which nerves and sense organs operate.

We shall now discuss the nerve cell, the nature of nerve impulses, and the encoding and decoding

FIGURE 12.19 The mantle of the common squid is supplied with nerves of exceptionally large diameter. These so-called giant axons are fast-conducting and serve to obtain a nearly simultaneous contraction of the entire mantle musculature. [Keynes 1958. Copyright © by Scientific American, Inc. All rights reserved.]

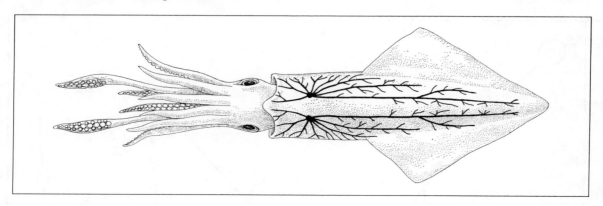

of sensory information. In Chapter 13 we shall return to nerve function and examine in detail how impulses are communicated from one nerve cell to another.

Structure of nerve cells

All nervous systems consist of a large number of single nerve cells (*neurons*). Neurons can have a wide variety of shapes and sizes, but they have certain important features in common (Figure 12.18). There is a *cell body* that contains the nucleus and a large number of thin fibers extending from it. Each neuron usually has a single long fiber (the *axon*), which in a large animal may be several meters long, and a large number of shorter fibers (*dendrites*), which are heavily branched and for the most part are less than 1 mm long. In vertebrates the cell body of the neuron is usually quite small, often less than 0.1 mm in diameter and the fibers less than 0.01 mm thick. The entire cell, including all the fibers, is surrounded by a thin membrane (the *nerve membrane*). A complex nervous system contains an immense number of neurons. In man, for example, the brain alone contains around 10 000 000 000 cells.

The long fibers, the axons, are the main conduction lines in the body. What is commonly known as a nerve, or nerve trunk, consists of hundreds or thousands of axons, each originating from a different neuron. A nerve has no cell bodies in it; these are found in the central nervous system, in special aggregations known as *ganglia*, and in the sensory organs.

The points where nerve cells and their extensions make contact with other nerve cells are called *synapses*. A single nerve cell may, through the synapses, be connected to hundreds of other neurons. The most important feature of the synapse is that it functions as a one-way valve. Transmission of an impulse can take place in one direction only, from the axon to the next cell, and not in the opposite direction (for details, see Chapter 13). As a result, conduction in an axon anywhere in the nervous system takes place in one direction only.

An axon in itself is perfectly capable of conducting impulses in either direction, but in the integrated nervous system all conductance in any given axon is always in the same direction. In a nerve trunk, however, some axons may conduct in one direction, and others in the opposite direction. For example, in a nerve to a given muscle, the impulses stimulating the muscle to contract are transmitted from the central nervous system to the muscle, and at the same time sensory information from the muscle is carried in other axons back to the central nervous system.

How the nerve cell functions

It has been known for about two centuries that electric phenomena are associated with the transmission of nerve impulses and muscular contraction. The detailed understanding of the nature of the electric phenomena has come about only during the last few decades. The single most important event was the discovery that certain nerve fibers in squid consist of a single axon of giant dimensions. Most axons measure from less than 1 μm to about 10 μm in diameter; the *giant axons* of squid may be nearly 1000 μm in diameter (Figure 12.19). Because of their exceptional size, giant axons can be used for many experiments that would be impossible on smaller axons. However, the general principles revealed in studies of giant axons apply to the function of other neurons and also to certain other tissues, such as muscle.

The normal neuron, including its axon, shows a potential difference between the inside and the outside of the cell membrane. In the inactive or resting neuron this membrane potential is known as the *resting potential*. In the giant axon of squid the resting membrane potential is about − 70 mV (inside negative). During activity this can change to + 55 mV (inside positive), giving a total change, an *action potential* of 125 mV.

Membrane resting potential

The most important information to consider in relation to the membrane potential is the concentration of ions inside and outside the axon and the permeability of the axon membrane to these ions. The inside axoplasm has a relatively high concen-

TABLE 12.2 Concentrations (in millimoles per liter) of major ions in axoplasm of giant axons from squid, in squid blood, and in sea water. [Data mainly from Hodgkin and Huxley 1952]

Ion	Axoplasm	Blood	Sea water
Potassium	400	20	10
Sodium	50	450	470
Chloride	100	570	550
Calcium[a]	0.3×10^{-3}	10	10
Magnesium	10	55	54

[a] The concentration of ionized calcium in axoplasm is uncertain, but extremely low (Baker et al. 1971).

tration of potassium and a low concentration of sodium; the outside fluid, in contrast, is low in potassium and high in sodium (Table 12.2). Chloride is distributed according to a Donnan equilibrium and plays no major role in our considerations.

The nerve membrane, when at rest, is permeable to potassium ions, and its permeability to sodium is so low that for the moment it can be disregarded. Potassium ions therefore tend to diffuse out of the axon to the lower concentration outside, giving rise to an electric potential that is negative inside the axon and positive outside. The negative potential inside the membrane counteracts the outward flow of positive ions, and at equilibrium this potential exactly balances the concentration gradient that drives potassium out. The magnitude of the membrane potential depends on the potassium concentrations inside and outside and can be calculated from the laws of physical chemistry.

The potential difference (E) caused by the unequal distribution of ions inside and outside a permeable membrane is described by the Nernst equation (see Appendix E). By introducing into the equation the proper numerical values for the constants, it is reduced (at 20 °C) to:

$$E = 58 \log \frac{C_o}{C_i}$$

FIGURE 12.20 The axoplasm can be squeezed out of a giant axon, and the axon can then be filled with an artificial solution of desired composition. An action potential recorded from the extruded axon filled with isotonic potassium sulfate (a) is virtually indistinguishable from an action potential recorded from an intact axon filled with its natural axoplasm (b). [Baker et al. 1962]

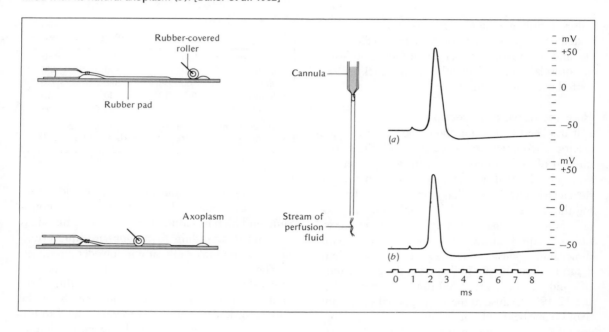

where C_o and C_i are the outside and inside concentrations of the ion under consideration.

The ratio of potassium concentrations in blood and axoplasm of the giant axon is 1:20, and the calculated resting potential should be -75 mV, with the inside negative relative to the outside. The membrane potential can be measured by inserting a tiny microelectrode into the axon. The observed resting potential is very close to the predicted potential, but because of a slight sodium leakage, it is usually a few millivolts less, about -70 mV.

The fundamental correctness of the conclusion that the membrane potential is caused by the potassium gradient can be verified by changing the potassium gradient across the membrane. By increasing or decreasing the potassium concentration in the outside medium, the transmembrane potential changes as predicted from Nernst's equation. If the outside potassium concentration is increased to equal the inside concentration, the transmembrane potential declines to zero, as predicted (Hodgkin and Keynes 1955).

The most dramatic experiment to show that the nerve membrane is the essential element, and that the axoplasm plays no direct role, consists in removing the entire content of the axon and replacing it by an artificial salt solution (Figure 12.20) (Baker et al. 1961, 1962). Such a perfused axon behaves amazingly like a normal axon; it can remain excitable for many hours and able to give several hundred thousand action potentials.

These elegant experiments, which are possible because of the large size of the giant axons, show two important things. First, the axoplasm is not needed for the generation of membrane potentials;

FIGURE 12.21 Entry of radioactive sodium into a squid axon placed in sea water containing radioactive sodium as a tracer. When the axon has been stimulated for 5 minutes to produce action potentials at a rate of 50 per second, a considerable amount of sodium has entered the axon. After stimulation ceases, the sodium concentration inside the axon slowly declines as active trans- port removes sodium to the outside. When the sodium pump is poisoned by the inhibitor DNP, the sodium content remains constant. Renewed stimulation of the axon causes more sodium to enter, and when the in- hibitor DNP is removed, sodium is again extruded by active transport. [Hodgkin and Keynes 1955]

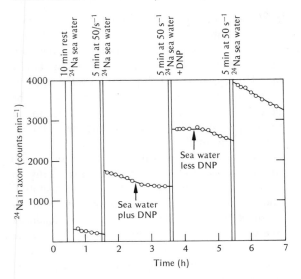

it can be replaced by an artificial salt solution. Sec- ond, if the internal potassium concentration is changed from the normal, the membrane poten- tial changes as predicted from Nernst's equation, and if the internal potassium concentration equals the outside concentration, the resting potential is abolished. The membrane potential can even be reversed by making the inside potassium concen- tration lower than the outside (Baker et al. 1961, 1962). However, changing the concentration of sodium or chloride does not produce similar alter- ations in the resting potential.

Action potential

The next subject to consider is the action poten- tial. When we discussed the resting potential, we assumed that the permeability to sodium was zero and showed that, because of the potassium perme- ability, the inside potential must be negative. The situation is, therefore, that the high outside so- dium concentration as well as the inside negative potential tend to drive sodium into the axon. Let us now disregard potassium and suppose that there is a momentary change in the membrane that for an instant makes it highly permeable to sodium ions. Both concentration gradient and membrane potential drive sodium in, making the inside posi- tive until the potential (now positive) prevents fur- ther entry of sodium ion. The potential will, ac- cording the Nernst's equation, be $+55$ mV (inside positive). Thus, the membrane potential has changed from -70 mV (potassium potential) to $+55$ mV (sodium potential). In other words, the membrane potential of the axon can be changed by 125 mV, merely by altering the relative perme- abilities to sodium and potassium ions.

If, an instant later, the membrane again be- comes impermeable to sodium, the permeability to potassium reestablishes the resting potential. As we shall see below, the permeability to sodium ions does indeed change as outlined here.

During the brief change in permeability, some sodium has entered the cell, and this sodium must be removed to keep the system from running down. This is achieved through active transport of sodium out of the axon. The amount of sodium that entered during the action potential can be calculated from theoretical considerations and it can be measured experimentally. The experiment described in Figure 12.21 shows what happens when a giant axon is immersed in sea water that contains radioactive sodium ions. When the axon is stimulated repeatedly to produce action poten- tials, sodium ions enter the axon and the inside becomes radioactive. When stimulation ceases and the axon is at rest, the sodium is slowly re- moved from the inside. To show that this removal is an active transport, the sodium pump was poi- soned by the inhibitor dinitrophenol (DNP); after the pump was poisoned, sodium extrusion ceased and the inside concentration remained constant.

After renewed stimulation of the axon, more sodium entered the axon, again showing that the action potentials and sodium entry are coupled.

The amount of sodium that enters the axon during a single action potential is extremely small. Giant axons in which the sodium pump has been inactivated with DNP or cyanide can continue to show normal excitability and action potentials for several hours. On one occasion such a giant axon continued to conduct impulses at a rate of 50 per second for a total of 70 minutes (i.e., a total of 210 000 impulses) before the system ran down due to inside accumulation of sodium (Hodgkin and Keynes 1955).

So far we have disregarded the chloride ion. Chloride ions are distributed according to the electrical potential to satisfy Nernst's equation, and for our purposes we need not consider their movements. Their effect is small, and, most importantly, chloride permeability does not change during the action potential.

We have now established that the membrane potential, whether the axon is at rest or active, is determined by the membrane permeability and that its magnitude can be calculated from basic principles. By selectively changing the permeabilities to sodium and potassium, the membrane potential can shift anywhere between -75 mV and $+55$ mV.

The permeability changes are extremely fast and take place in microseconds. The experimental difficulties involved in measuring the magnitude of the currents carried by the ions flowing through the membrane in such a short time have been solved by a technique known as the voltage clamp method. By maintaining the membrane potential constant ("clamped") at any particular desired value, one can determine the current flow carried by each ion across the membrane. From this one can estimate the magnitude of the permeability of the membrane to ions as a function of the membrane potential.

The most important result of such studies, primarily executed by the British investigators Hodgkin, Huxley, and Katz, is to establish the ion permeability of the membrane as a function of the potential across the membrane. A decrease in the resting potential (*depolarization*) increases sodium permeability of the membrane. This change increases the rate of inward flow of sodium, which further reduces the membrane potential. This, in turn, further increases the sodium permeability, and so on. The rapidly increasing change in sodium permeability is a form of positive feedback (see Chapter 13) that in an extremely short time causes sodium permeability to reach its maximum. A complete depolarization in turn deactivates the system that permits the sodium ion to cross the membrane, thus restoring its initial low permeability (Hodgkin and Huxley 1952).

Depolarization of the membrane also causes the potassium permeability to change, but more slowly. If the permeability to both sodium and potassium were increased simultaneously, there would be no action potential, and the membrane potential would arrive at some intermediate value. However, because the potassium permeability changes after sodium permeability has declined, the effect is to restore the original state and drive the voltage back to its initial value (known as *repolarization* of the membrane (Figure 12.22).

Nerve impulses

Until now we have described only the local events at a given point of the nerve membrane. However, we are in possession of the information needed to understand how an action potential is propagated along the axon (i.e., how a nerve impulse is conducted along the axon).

We have seen that a local depolarization of the

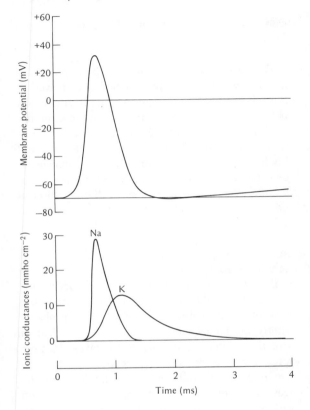

FIGURE 12.22 Calculated changes in sodium and potassium conductances during a propagated action potential in a squid giant axon. The main increase in sodium conductance occurs during the rising phase of the action potential; the increase in potassium conductance is slower and serves to reestablish the resting potential after the sodium conductance has declined. [Hodgkin and Huxley 1952]

nerve membrane increases the inflow of sodium and through positive feedback leads to an action potential. The immediately adjacent area of the nerve membrane will be unable to retain its full resting potential and will be partly depolarized; its sodium conductance increases, causing further inflow of sodium, and so on, until it reaches a full action potential. The depolarization therefore spreads from area to area and travels as a rapidly propagated action potential along the axon.

A very weak stimulus does not result in an action potential. For this to happen, the stimulus must have a certain strength, referred to as the *threshold* strength. Below this threshold there is no action potential, but if the stimulus exceeds the threshold value, an action potential results. The size of the action potential, however, is not influenced by the magnitude of the stimulus. Let us assume that we double, or triple, the strength of the stimulus above the threshold; the resulting action potential remains the same. The reason, of course, is that the action potential is caused by the concentration of ions on the two sides of the membrane. Therefore, the stimulus either causes a full-strength action potential, or none at all. In physiology, this type of response is called an *all-or-none response*.

Once the axon membrane has been depolarized and an action potential generated, it travels as a rapidly propagated action potential along the axon. Because the action potential results from the local concentrations of ions, it spreads along the axon without change in magnitude. This is called *conduction without decrement* and is a fundamental characteristic of the axon, different from conduction in, for example, an electric conductor.

If an action potential is always of the same magnitude and uninfluenced by the strength of the stimulus, how can an axon then convey information about the strength of a stimulus – information that is obviously of the greatest interest?

The answer is that a change in frequency of the action potentials in the axon can be used as an indicator of stimulus strength. This is illustrated in Figure 12.23, which shows that a pressure-sensitive skin receptor cell in the finger of a man responds to an increased stimulus strength with increasing frequency of the action potentials in its axon. This particular receptor had a threshold of about 0.5 g, and therefore showed no response to a force 0.2 g. A force of 0.6 g, however, gave a clear response, which increased in frequency with the application of increasing forces. Thus, we can

FIGURE 12.23 Impulses recorded from a single pressure-sensitive receptor on the human hand. A plastic rod 1 mm in diameter was pressed against the skin with the force indicated on each record. This receptor did not respond to temperature changes (i.e., it was a specific mechanoreceptor). [Hensel and Boman 1960]

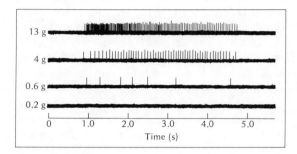

say that the magnitude of a sensory stimulus is coded and transmitted as a frequency-modulated signal.

Some receptors show *spontaneous activity*; they produce action potentials and nerve impulses in the absence of stimulation. At first glance this may seem meaningless and counterproductive; in reality, it has certain important advantages.

The most important consequence is that it increases the *sensitivity* of the receptor. A very small stimulus – too small to cause a depolarization by itself – increases the rate of impulse discharge. This means that there is no such thing as a subthreshold stimulus; any slight increase in stimulus changes the discharge frequency.

The other advantage of spontaneous receptor activity is that a change in either direction can be encoded. Consider, for example, the infrared receptor of the rattlesnake. In a completely uniform environment the receptor shows spontaneous activity, and a small change in the environmental radiation in either direction can be encoded as a frequency change. The sensitivity of the system is high enough so that it will respond every time the snake suddenly faces a suitable surface with a temperature 0.1 °C above or below the previous (Bullock and Diecke 1956). Thus spontaneous activity not only brings about a highly sensitive system, but

also permits determination of the polarity of a change.

We have now established some basic principles in sensory physiology. Information is conveyed in the sensory nerves as action potentials; the action potentials in all sensory nerves are of the same nature; the magnitude of the action potentials is constant and uninfluenced by stimulus intensity; and finally, information about stimulus intensity is coded as a frequency modulation of the action potentials. We can now proceed to the question of how sensory information is further processed.

Sorting and processing sensory information

If all information available to the sensory organs were transmitted to the central nervous system, the mass of signals would be formidable and probably utterly unmanageable. However, a great deal of screening, filtering, and processing takes place before the signals are passed on, beginning at the sensory neuron and continuing at several levels on the way to the brain. The filtering networks pass on only selected portions of the information they receive; furthermore, they carry out certain steps of processing that improve on the information that is transmitted to higher levels. We shall discuss these principles using the processing of visual information as an example.

In spite of the complexity of the visual system, the processing of visual signals is better understood than that of other complex senses. There are several reasons. One is that artificial stimuli (light) can be directed, timed, and quantified with great accuracy. Another reason is the use of the arthropod eye as experimental material, because the structure of the compound eye permits ready

FIGURE 12.24 Action potentials recorded simultaneously from two adjacent ommatidia, A and B, in the compound eye of the horseshoe crab *Limulus*. The discharge rate is indicated at the right of each record. Top: illumination of ommatidium A alone. Center: illumination of A and B, resulting in lateral inhibition (see text). Bottom: illumination of B alone. [Hartline and Ratliff 1957]

FIGURE 12.25 Diagram showing that lateral inhibition causes an enhanced edge effect. For details, see text.

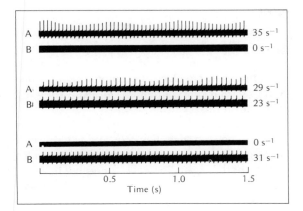

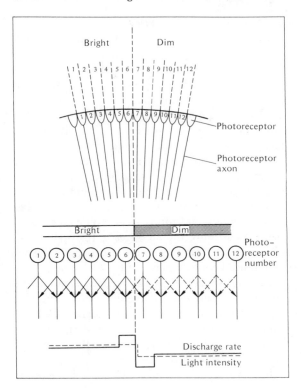

access to single sensory receptor units. However, even the complex signal processing in the vertebrate eye is amazingly well understood.

Lateral inhibition

The horseshoe crab, *Limulus*, has a compound eye in which individual receptor units can readily be stimulated by a fine lightbeam. It is also possible to record the impulses from that fiber of the optic nerve which connects to the particular receptor in question. It turns out that the signals in the axon do not completely represent the stimulus; the pattern also depends on the amount of light falling on other receptor units. This is because each visual receptor is connected to its neighbors and inhibits their activity. This characteristic, known as *lateral inhibition*, has the effect of enhancing the contrast between the amounts of light falling on two adjacent receptors.

This may need some further explanation. Figure 12.24 shows the records from two adjacent receptor units of the *Limulus* eye. The top record shows the regularly spaced discharges when a single receptor (A) was stimulated. The rate of discharge was 35 impulses per second. When the adjacent receptor (b) was also stimulated, the discharge rate in unit A decreased to 29. Unit B was likewise influenced by its neighbor A, for when the stimulus was removed from A, the discharge rate from receptor B increased from 23 to 31 impulses per second. Thus, there is a mutual lateral inhibition between these two units (Hartline and Ratliff 1957).

The effect of lateral inhibition can be understood from the diagram in Figure 12.25. Consider 12 units that are stimulated by uniform light of two intensities, there being a sharp transition between units 6 and 7. In this system, units 2, 3, 4, and 5 are all bordered by units receiving bright illumination and are therefore subject to lateral inhibition. Unit 6, however, is not subjected to the same degree of lateral inhibition from unit 7, which is within the dimly stimulated zone. As a result, its discharge frequency is higher. Next, consider the dimly lit units, of which receptors 8, 9, 10, and 11 receive the same small amount of lateral inhibition. Unit 7, however, is inhibited by its brightly lit neighbor, receptor 6, and therefore discharges at a lower frequency than the other dimly lit receptors. The overall effect is that the transition between brightly and dimly lit receptors is emphasized. As a result, the messages in the optic nerve give a correct picture of the edge, but with an emphasized contrast between the two zones.

Another aspect of visual reception should be emphasized before we move on to the vertebrate eye. It has been observed in the scallop (*Pecten*), which has well developed image-forming eyes along the edge of the mantle, that the retina has two layers. One layer, as expected, responds to light as a normal stimulus. The other layer of the retina, however, behaves in a very different way; it does not respond to increased light, but is sensitive only to a decrease in the intensity of illumination. This phenomenon is common in the eyes of vertebrates, and we shall return to it shortly. The importance of this phenomenon can easily be imagined. When a shadow suddenly falls on an animal, it often means the approach of a predator, and in this case, a decrease in stimulus intensity is far more important to the animal than any other information.

We are now familiar with two relatively simple but important aspects of visual systems: the amplification of edge effects and the importance of dark areas in the visual field.

Information processing

The frog's eye. The frog's retina has no central fovea,* and rods and cones are uniformly distributed so that the structure of the retina is much the same from place to place. The photoreceptors connect to several kinds of neurons located within the retina at various levels.

We shall be concerned only with those neurons that are known as *ganglion cells* and whose axons make up the optic nerve. There are about half a million of these ganglion cells, a number that corresponds approximately to the number of fibers in the optic nerve (Maturana 1959). However, there are more than 1 million receptor cells (rods and cones) in the retina. Clearly, the optic nerve cannot carry a complete point-by-point picture of the image that falls on the retina; the analogy between the retina and the photographic film in a camera is not fully valid, for there must be some discrimination or processing before the signals are sent on to the brain.

In the optic nerve there are five different kinds of fibers that originate from the ganglion cells. They have been clearly identified because they respond differently to specific kinds of stimulation of the retina. Some fibers respond only to the onset of illumination; these are called *on fibers*. Other fibers respond only to the termination of a light stimulus and are called *off fibers*. A third kind of fiber responds to either onset of a light stimulus or its termination; these are known as *on–off fibers*. Their response to the movement over the retina of a linear shape is marked, and they could therefore also be called moving-edge detectors.

---

* In the mammalian eye a small, central area of the retina, the *fovea centralis*, contains only cones. It is the area that provides the most acute vision in strong light, but because of the absence of rods the fovea is insensitive to dim light.

FIGURE 12.26 In the eye of a frog a high degree of processing and filtering of information takes place in the retina. The various fibers in the optic nerve carry different kinds of information. Certain fibers do not respond to movement of large objects (a), to movement of the whole visual field (b), or to on and off light. They do respond to small moving objects if these are darker than the background (c), but not if they are lighter (d). The response is independent of the general level of illumination (e), and there is no response to movement of objects with fuzzy edges (f). [Bullock and Horridge 1965. Courtesy of T. H. Bullock and G. A. Horridge and W. H. Freeman and Company. Copyright © 1965.]

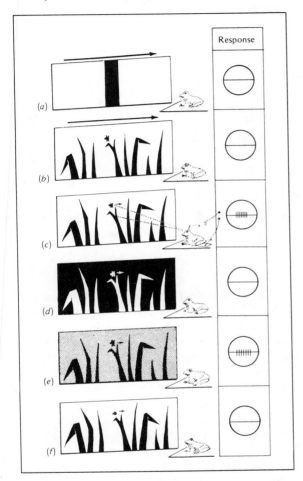

objects. The two last categories do not react to changes in the general intensity of the light; even switching a light on or off does not affect them (Figure 12.26).

We can now see that the frog's retina can carry out a great deal of analysis of the visual signals before information is transmitted to the optic nerve, and if we examine this information in view of what may be important to the frog, the system makes sense. From the frog's viewpoint the insects on which it feeds are some of the most relevant objects in its life. Therefore, a small dark object is important, particularly if it moves, and the frog is equipped to respond specifically to this stimulus through the bug-detector fibers. Small stationary objects, such as spots, shadows, or pebbles, are of no special interest, and information about these does not reach that part of the system designed for a rapid response to live prey. The most meaningful processing of signals related to food has therefore already taken place in the retina and is transmitted as specialized information to the central nervous system (Lettvin et al. 1961). We shall not discuss the details of the other types of retinal detection systems but instead move on to the mammalian visual system.

The mammalian eye. The eye of a cat, and of many other mammals, has about 100 million receptor cells in the retina. The optic nerve carries about 1 million axons. This immediately tells us that the brain cannot receive separate information from each individual receptor cell; a great deal of sorting and processing takes place before the information is sent to the central nervous system.

In spite of the complexity, what occurs in the mammalian optical system is reasonably well known. In a simplified way we can regard the transmission as taking place over six levels: Three

There are also fibers that respond to the presence of a sharp edge in the visual field, whether it is stationary or moving, and these *edge receptors* differ distinctly from the on–off fibers. Finally, there are fibers that could be called *bug detectors* because they respond to small, dark, moving objects, but not to large, dark objects or to stationary

FIGURE 12.28 Simplified diagram of the connections between the receptor cells (R), the bipolar cells (B), and the ganglion cells (G) in the retina of the eye. Also shown are horizontal cells (H), amacrine cells (A), and a glial or supporting cell (GL). [Courtesy of John E. Dowling, Harvard University]

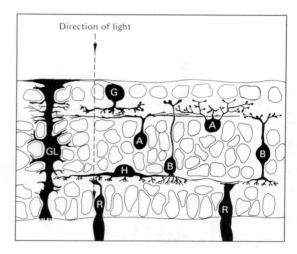

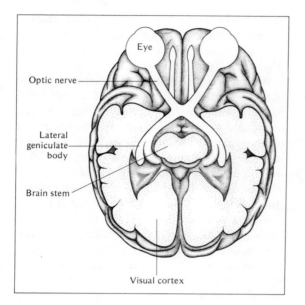

in the retina of the eye, one in the lateral geniculate body of the brain, and two in the visual cortex of the brain (Figure 12.27).

Let us follow these levels and examine the processing that takes place before information reaches the visual cortex, which in turn is connected with other parts of the central nervous system.

The light *receptor neurons* in the retina (Figure 12.28) connect to a layer of nerve cells known as the *bipolar cells.** These in turn are connected to a layer of retinal *ganglion cells* whose axons make up the optic nerve. The connections between these three types of cells are very complex. A receptor cell may connect to more than one bipolar cell, and several receptors may be connected to one and the same bipolar cell. The same holds for the con-

* For simplicity the discussion of amacrine and horizontal cells is omitted.

nections between the bipolar cells and the ganglion cells. Because there are more than 100 receptor cells for each ganglion cell, the ganglion cell must evidently receive information from a large number of receptor cells. Together, such a group of receptor cells makes up a *receptive field*, which designates that area of the retina with which the particular ganglion cell is connected.

Without going into detail, let us remember that retina cells may have both "on" and "off" characteristics and that the synaptic connections may be excitatory as well as inhibitory (for a further discussion of excitation and inhibition, see Chapter 13).

One of the most striking characteristics of the glion cells. Since there are more than 100 relation, they pour impulses into the optic nerve at a steady rate of about 20 to 30 per second. Even more surprising is the fact that subjecting the entire retina to illumination does not have any prominent effect on the number of impulses from the ganglion cells. However, if a small spot of light falls on one receptive field, it has a very marked, although highly complex effect on the corresponding ganglion cell.

It was discovered by Kuffler (1953) that the receptive fields consist of two concentric areas, one with "on" receptors and the other with "off" receptors. The receptive fields are of two kinds: the *on-center*" fields, in which the central area consists of "on" receptors surrounded by a ring of "off" receptors, and the "*off-center*" fields, have a central "off" region, surrounded by a circle of "on" receptors. If a small spot of light shines on the center of an "on-center" field, the corresponding ganglion cell shows increased activity. Shining two spots at adjacent points within the "on" region increases the response, but if one of the spots falls on the "on" region, and the other on the surrounding "off" region, the response is greatly reduced. This explains why uniform illumination of the entire retina gives rise to very little response, for the various "on" and "off" regions to a great extent cancel each other.

The meaning of this signal evaluation or processing is obvious. It is of no particular interest to record and inform the brain about the general intensity of uniform light; the details of light and dark contrast are of much greater importance. What we have is a case of *selective destruction of information,* carried out in the eye. The eye selects what kind of information is to be transmitted in the optic nerve and thus reduces the information load on the brain.

The fourth level in the processing of visual stimuli is in the brain, in the *geniculate body.* The cells of the geniculate body correspond to the receptive fields of the retina; they have some characteristics in common with the retinal ganglion cells, and they function to increase the contrast between small spots of light and changes in diffuse lighting. For our purposes, however, we shall regard the geniculate body merely as a way-station, consisting of synaptic connections between the optic nerve and the visual cortex of the brain.

Of the many different kinds of cells in the *visual cortex,* we shall discuss only two: simple cells and complex cells. *Simple cells* respond primarily to contrasting lines, such as light bars on a dark background or dark bars on a light background, and to sharp light – dark edges. Whether or not a given simple cell responds to such a bar depends on the orientation and position of the bar within the receptive field of the retina. For example, if a vertical bar gives a response in a given simple cell of the visual cortex, the cell will not respond if the bar is rotated some 10 or 15° or if the bar is displaced slightly to one side or the other. A careful testing of the retina with minute light spots has shown that the simple cells respond to arrangements such as are shown in Figure 12.29.

Because the response of a given simple cell depends on the orientation as well as the position on the retina of a straight line, it constitutes a built-in filter that removes or destroys a great deal of information and transmits only specialized information about certain kinds of contrasting lines. Each simple cell thus has a specific function: It corresponds to one restricted part of the retina and responds only to one particular kind of information falling on this spot. There is no evidence that in the cat's retina there is any preferred orientation; the receptive fields may occur in all possible orientations with no preference for vertical, horizontal, or oblique orientations.

The *complex cells* of the visual cortex respond, like simple cells, to the orientation of a given stimulus. If a complex cell responds to a certain orientation of a light bar on the retina, rotation of the bar causes the cell to cease its activity. If the bar is moved over the retina with unchanged orientation, however, the complex cell continues to be active. This can most readily be explained by as-

FIGURE 12.29 Various arrangements of receptive fields in the visual cortex. Areas giving an "on" response are indicated by crosses; those giving an "off" response by triangles. The orientation of the receptive fields, whether of the "on" or "off" type, is indicated by straight lines through the center of the field. [Hubel and Wiesel 1962]

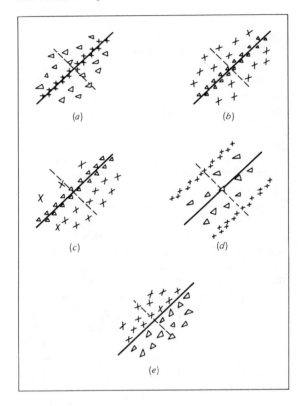

(a) (b) (c) (d) (e)

ranged in fields that readily process information about contrast between light and dark, rather than about the general level of illumination. This takes place with the aid of two levels of transmitting neurons within the retina, which select the most pertinent information and reject less meaningful information.

The selected information is transmitted to the geniculate body, which accentuates the information obtained from the "on-center" and "off-center" receptive fields. The visual cortex, finally, rearranges the information from the geniculate body so that lines and contours are selected for further processing. Here the first step (in the simple cells) is concerned with the orientation of lines; then the information converges on the complex cells, from which the information is transmitted to other parts of the brain for integration with the general functioning of the body.

The processing and sorting of sensory signals carried out before the signals are passed on to the central nervous system remove much irrelevant information and greatly reduce the load on the control systems responsible for adequate and proper responses. We shall now pass on to the question of how these control systems work and coordinate the many functions and activities of the animal body.

suming that the complex cell receives information from a large number of simple cells, all of which are of the same general type and have the same orientation of their fields. This requires an enormous complexity of the connections in the visual cortex, but all the evidence that has been obtained indicates that the cells are indeed located in such anatomically well-defined groups.

We can now summarize the main steps in the processing of visual information in the mammalian eye. The receptors can be "on"- or "off" type receptors. They do not individually send information to the central nervous system, but are ar-

REFERENCES

Aidley, D. J. (1971) *The Physiology of Excitable Cells.* Cambridge: Cambridge University Press. 468 pp.
Atema, J. (1971) Structures and functions of the sense of taste in the catfish (*Ictalurus natalis*). *Brain Behav. Evol.* 4: 273–294.
Baker, P. F., Hodgkin, A. L., and Ridgway, E. B. (1971) Depolarization and calcium entry in squid giant axons. *J. Physiol.* 218: 709–755.
Baker, P. F., Hodgkin, A. L., and Shaw, T. I. (1961) Replacement of the protoplasm of a giant nerve fibre with artificial solutions. *Nature, Lond.* 190: 885–887.

Baker, P. F., Hodgkin, A. L., and Shaw, T. I. (1962) Replacement of the axoplasm of giant nerve fibres with artificial solutions. *J. Physiol.* 164:330–354.

Barrett, R., Maderson. P. F. A., and Meszler, R. M. (1970) The pit organs of snakes. In *Biology of the Reptilia*, vol. II (C. Gans and T. S. Parsons, ed.), pp. 277–300. London: Academic Press.

Black-Cleworth, P. (1970) The role of electrical discharges in the nonreproductive social behaviour of *Gymnotus carapo* (Gymnotidae, Pisces). *Anim. Behav. Monogr.* 3:1–77.

Boeckh, J., Kaissling, K.-E., and Schneider, D. (1960) Sensillen und Bau der Antennengeissel von *Telea polyphemus* (Vergleiche mit weiteren Saturniden: *Antheraea, Platysamia* und *Philosamia*). *Zool. Jahrb., Abt. Anat.* 78:559–584.

Brown, F. A., Jr., Barnwell, F. H., and Webb, H. M. (1964) Adaptation of the magnetoreceptive mechanism of mud-snails to geomagnetic strength. *Biol. Bull.* 127:221–231.

Brown, P. K., and Wald, G. (1964) Visual pigments in single rods and cones of the human retina. *Science* 144:45–52.

Bullock, T. H., and Cowles, R. B. (1952) Physiology of an infrared receptor: The facial pit of pit vipers. *Science* 115:541–543.

Bullock, T. H., and Diecke, F. P. J. (1956) Properties of an infra-red receptor. *J. Physiol.* 134:47–87.

Bullock, T. H., Hamstra, R. H., Jr., and Scheich, H. (1972) The jamming avoidance response of high frequency electric fish. 1. General features. *J. Comp. Physiol.* 77:1–22.

Bullock, T. H., and Horridge, G. A. (1965) *Structure and Function in the Nervous System of Invertebrates*, vols 1 and 2. San Francisco: Freeman. 1719 pp.

Cahn, P. H. (ed.) (1967) *Lateral Line Detectors*. Bloomington: Indiana University Press. 496 pp.

Caprio, J. (1975) High sensitivity of catfish taste receptors to amino acids. *Comp. Biochem. Physiol.* 52A:247–251.

Crescitelli, P. and Dartnall, H. J. A. (1953) Human visual purple. *Nature, Lond.* 172:195–197.

Dijkgraaf, S. (1963) The functioning and significance of the lateral-line organs. *Biol. Rev.* 38:51–105.

Dijkgraaf, S., and Kalmijn, A. J. (1963) Untersuchungen über die Funktion der Lorenzinischen Ampullen an Haifischen. *Z. Physiol.* 47:438–456.

Emlen, S. T., Wiltschko, W., Demong, N. J., Wiltschko, R., and Bergman, S. (1976) Magnetic direction finding: Evidence for its use in migratory indigo buntings. *Science* 193:505–508.

Enright, J. T. (1963) Estimates of the compressibility of some marine crustaceans. *Limnol. Oceangr.* 8:382–387.

Erulkar, S. D. (1972) Comparative aspects of spatial localization of sound. *Physiol. Rev.* 52:237–360.

Flock, A. (1971) Sensory transduction in hair cells. In *Handbook of Sensory Physiology*, vol. 1, *Principles of Receptor Physiology* (W. R. Lowenstein, ed.), pp. 396–441. Berlin: Springer-Verlag.

Von Frisch, K. (1948) Gelöste und ungelöste Rätsel der Bienensprache. *Naturwissenschaften* 35:38–43.

Fromme, H. G. (1961) Untersuchungen über das Orientierungsvermögen nächtlich ziehender Kleinvögel (*Erithacus rubecula, Sylvia communis*). *Z. Tierpsychol.* 18:205–220.

Gamow, R. I., and Harris, J. F. (1973) The infrared receptors of snakes. *Sci. Am.* 228:94–101.

Goris, R. C., and Terashima, S. (1973) Central response to infra-red stimulation of the pit receptors in a crotaline snake, *Trimeresurus flavoviridis*. *J. Exp. Biol.* 58:59–76.

Gould, E., Negus, N. C., and Novick, A. (1964) Evidence for echo location in shrews. *J. Exp. Zool.* 156:19–39.

Griffin, D. R. (1953) Acoustic orientation in the oil bird, *Steatornis. Proc. Natl. Acad. Sci. U.S.A.* 39:884–893.

Griffin, D. R. (1954) Bird sonar. *Sci. Am.* 190:78–83.

Griffin D. R. (1958) *Listening in the Dark: The Acoustic Orientation of Bats and Men.* New Haven, Conn.: Yale University Press. 413 pp.

Harris, J. F., and Gamow, R. I. (1971) Snake infrared receptors: Thermal or photochemical mechanism? *Science* 172:1252–1253.

Hartline, H. K., and Ratliff, F. (1957) Inhibitory interaction of receptor units in the eye of *Limulus*. *J. Gen. Physiol.* 40:357–376.

Hensel, H., and Boman, K. K. A. (1960) Afferent impulses in cutaneous sensory nerves in human subjects. *J. Neurophysiol.* 23:564–578.

Hodgkin, A. L., and Huxley, A. F. (1952) The dual effect of membrane potential on sodium conductance in the giant axon of *Loligo. J. Physiol.* 116:497–506.

Hodgkin, A. L., and Keynes, R. D. (1955) Active transport of cations in giant axons from *Sepia* and *Loligo. J. Physiol.* 128:28–60.

Hodgson, E. S. (1961) Taste receptors. *Sci. Am.* 204:125–144.

Hopkins, C. D. (1972) Sex differences in electric signaling in an electric fish. *Science* 176: 1035–1037.

Hubel, D. H., and Wiesel, T. N. (1962) Receptive fields, binocular interaction and functional architecture in the cat's visual cortex. *J. Physiol.* 160:106–154.

Iggo, A. (1969) Cutaneous themoreceptors in primates and sub-primates. *J. Physiol.* 200:403–430.

Kalmijn, A. J. (1971) The electric sense of sharks and rays. *J. Exp. Biol.* 55:371–383.

Keeton, W. T. (1971) Magnets interfere with pigeon homing. *Proc. Natl. Acad. Sci. U.S.A.* 68:102–106.

Kellogg, W. N. (1958) Echo ranging in the porpoise. *Science* 128:982–988.

Keynes, R. D. (1958) The nerve impulse and the squid. *Sci. Am.* 199:83–90.

Keynes, R. D., and Martins-Ferreira, H. (1953) Membrane potentials in the electroplates of the electric eel. *J. Physiol.* 119:315–351.

Kuffler, S. W. (1953) Discharge patterns and functional organization of mammalian retina. *J. Neurophysiol.* 16:37–68.

Kühn, A. (1927) Ueber den Farbensinn der Bienen. *Z. vergl. Physiol.* 5:762–800.

Lettvin, J. Y., Maturana, H. R., Pitts, W. H., and McCulloch, W. S. (1961) Two remarks on the visual system of the frog. In *Sensory Communication* (W. A. Rosenblith, ed.), pp. 757–776. New York: M.I.T. Press and Wiley.

Lindauer, M., and Martin, H. (1968) Die Schwereorientierung der Bienen unter dem Einfluss des Erdmagnetfeldes. *Z. Physiol.* 60:219–243.

Lissmann, H. W., and Mullinger, A. M. (1968) Organization of ampullary electric receptors in Gymnotidae (Pisces). *Proc. R. Soc. Lond. B.* 169:345–378.

Machin, K. E., and Lissmann, H. W. (1960) The mode of operation of the electric receptors in *Gymnarchus niloticus. J. Exp. Biol.* 37:801–811.

Maturana, H. R. (1959) Number of fibers in the optic nerve and the number of ganglion cells in the retina of anurans. *Nature, Lond.* 183:1406–1407.

Moore, F. R. (1977) Geomagnetic disturbance of the orientation of nocturnally migrating birds. *Science* 196:682–684.

Morita, H., and Yamashita, S. (1959) Generator potential of insect chemoreceptors. *Science* 130:922.

Norris, K. S., Prescott, J. H., Asa-Dorian, P. V., and Perkins, P. (1961) An experimental demonstration of echo-location behavior in the porpoise, *Tursiops truncatus* (Montagu). *Biol. Bull.* 120:163–176.

Payne, R. S. (1971) Acoustic location of prey by barn owls. *J. Exp. Biol.* 54:535–573.

Rushton, W. A. H. (1972) Pigments and signals in colour vision. *J. Physiol.* 220:1–31P.

Schneider, D. (1969) Insect olfaction: Deciphering system for chemical messages. *Science* 163:1031–1037.

Schneider, D., Kasang, G., and Kaissling, K.-E. (1968) Bestimmung der Riechschwelle von *Bombyx mori* mit Tritium-markiertem Bombykol. *Naturwissenschaften* 55:395.

Tschachotin, S. (1908) Die Statocyste der Heteropoden. *Z. Wissensch. Zool.* 90:343–422.

Wald, G. (1964) The receptors of human color vision. *Science* 145:1007–1016.

Waterman, T. H. (1950) A light polarization analyzer in the compound eye of *Limulus. Science* 111:252–254.

Waterman, T. H., and Forward, R. B., Jr. (1970) Field evidence for polarized light sensitivity in the fish *Zenarchopterus. Nature, Lond.*, 228:85–87.

Waterman, T. H., and Horch, K. W. (1966) Mechanism of polarized light perception. *Science* 154:467–475.

Wiltschko, W., and Wiltschko, R. (1972) Magnetic compass of European robins. *Science* 176:62–64.

ADDITIONAL READING

Aidley, D. J. (1971) *The Physiology of Excitable Cells.* Cambridge: Cambridge University Press. 468 pp.

Autrum, H., et al. (editorial board) (1971–1978) *Hand-*

book of Sensory Physiology; vol. 1, *Principles of Receptor Physiology*; vol. 2, *Somatosensory System*; vol. 3, *Enteroceptors*; vol. 4, *Chemical Senses*; vol. 5, *Auditory System*; vol. 6, *Vestibular System*; vol. 7, *Photochemistry of Vision*; vol. 8, *Perception*; vol. 9, *Development of Sensory Systems*. New York: Springer-Verlag.

Bench, R. J., Pye, A., and Pye, J. D. (eds.) (1975) *Sound Reception in Mammals*. London: Academic Press. 340 pp.

Bennett, M. V. L. (1970) Comparative physiology: Electric organs. *Annu. Rev. Physiol.* 32:471–528.

Catton, W. T. (1970) Mechanoreceptor function. *Physiol. Rev.* 50:297–318.

Cold Spring Harbor Laboratory. (1965) Sensory receptors. *Cold Spring Harbor Symp. Quant. Biol.* 30:1–649.

Davson, H. (1972) *The Physiology of the Eye*, 3rd ed. New York: Academic Press. 643 pp.

Daw, N. W. (1973) Neurophysiology of color vision. *Physiol. Rev.* 53:571–611.

de Vries, H. (1956) Physical aspects of the sense organs. *Prog. Biophys.* 6:207–264.

Erulkar, S. D. (1972) Comparative aspects of spatial localization of sound. *Physiol. Rev.* 52:237–360.

Griffin, D. R. (1958) *Listening in the Dark: The Acoustic Orientation of Bats and Men*. New Haven, Conn.: Yale University Press. 413 pp.

Hodgkin, A. L. (1964) The ionic basis of nervous conduction. *Science* 145:1148–1154.

Horridge, G. A. (ed.) (1975) *The Compound Eye and Vision of Insects*. Oxford: Clarendon Press. 595 pp.

Huxley, A. F. (1964) Excitation and conduction in nerve: Quantitative analysis. *Science* 145:1154–1159.

Roeder, K. D. (1967) *Nerve Cells and Insects Behavior*, rev. ed. Cambridge, Mass.: Harvard University Press. 238 pp.

Rushton, W. A. H. (1965) Chemical basis of colour vision and colour blindness. *Nature, Lond.* 206:1087–1091.

Sales, G., and Pye, D. (1974) *Ultrasonic Communication by Animals*. London: Chapman & Hall. 281 pp.

13

CHAPTER THIRTEEN

Control
and integration

In this book we have often mentioned regulation and control, but without discussing details of the process or mechanism of regulation. To regulate means to adjust an amount, a concentration, a rate, or some other variable, usually in order to attain or keep it at some desired level. An example is the regulation of respiration. We take for granted that respiration should be adjusted to provide oxygen at the rate it is used by the organism (i.e., the rate of oxygen uptake in the lung should be coordinated with its use in the tissues). Similarly, all the various physiological processes should be controlled, regulated, and integrated.

Integration means putting parts together. In physiology we use the word to describe the process of controlling all the functional components so that they merge into a smoothly operating organism in which no single process or function is permitted to run wild or proceed at its own independent rate.

There are several ways to control physiological functions. We have seen that carbon dioxide is an important element in the control of respiration in air-breathing animals, but carbon dioxide does not directly influence the breathing muscles. In a mammal the contraction of the diaphragm during inspiration is controlled by a nerve from the respiratory center in the brain; this center, in turn, is sensitive to the carbon dioxide level in the blood. The respiratory movements are thus under nervous control.

We have learned that the secretion of digestive juice from the pancreas is stimulated by a hormone (secretin), but in addition the pancreas is supplied with nerves that can also stimulate secretion. Some processes, such as the release of stored sugar from the liver, seem to be primarily under endocrine control. Evidently both nerves and hormones can control physiological processes; the two can work together in a well-coordinated fashion, and it may be difficult to delineate where one influence terminates and the other takes over.

Although there is much overlap between nervous and hormonal control, two important differences stand out. One relates to the *speed of action*; the other, to the *size of the target*.

When a quick response is required, such as the contraction of skeletal muscles, the rapid conduction in a nerve is necessary for fast action. Nerve impulses move at speeds up to about 100 m per second, and the delay in transmission of a message therefore need not be more than milliseconds. A process regulated by a hormone requires that the hormone reach the target organ before there is any effect. The speed of transmission is limited by the transport of the hormone via the blood. The minimum response time will therefore be in the magnitude of seconds.

The second important difference between nervous and hormonal control is that high precision can be obtained in the spatial distribution of nervous control, whereas hormonal control is often more diffuse. The single axon in a motor nerve connects only to a limited number of muscle fibers, and this permits the separate stimulation (or inhibition) of a single muscle without affecting other muscles; it even permits the stimulation of a small fraction of a single muscle. Hormones, in contrast, affect all sensitive cells they reach via the circulation; whole organs or organ systems are affected. For example, noradrenaline affects the sugar release from the entire liver. Control of the degree of effect is still possible, however, through the amount of the hormone released into the blood.

Hormones usually control processes in which the response is slow (e.g., the secretion of digestive juices, the control of urine concentration and volume, the excretion of sodium) as well as some very slow processes such as the development of the gonads and the growth of the body. However, as already mentioned, we shall find no sharp separation between nervous and hormonal control; the nervous system not only contributes to the regulation of endocrine function but is itself important in the production of hormones.

CONTROL AND CONTROL THEORY

The control mechanisms responsible for the maintenance of steady states in living organisms constitute a major chapter in physiology, and it is helpful to examine their performance in the light of simple control theory. For example, birds and mammals maintain nearly constant body core temperatures in spite of wide variations in external temperature and in internal heat production. This regulation of body temperature is carried out with the aid of a complex control system.

Theoretical considerations

In engineering, control systems are so important that control theory and design constitute an independent branch of engineering with its own terminology and theoretical approach. For physiology this is important in two respects: (1) many ill-defined and hazy old terms have been replaced by exact and well-defined concepts, and (2) the theoretical approach to control theory has led to a more precise definition of the components of physiological mechanisms and a better understanding of the relations between the component parts.

Feedback

Let us examine a familiar control system: the thermostat that regulates the temperature in a house or a water bath. This and all other control systems operate with a *controlled variable* (in this case, temperature), which is kept within a more or less narrow range around a desired value. A mea-

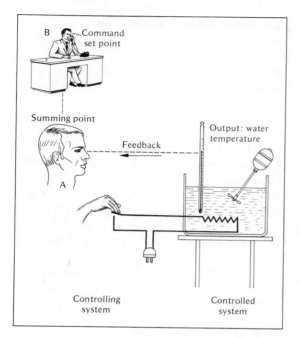

FIGURE 13.1 Diagram of a temperature-controlled water bath that includes a controlling system and a command. See text for details.

B — Command set point

Summing point

Feedback

Output: water temperature

A

Controlling system

Controlled system

perature has reached the desired level, and then break the contact again.

We can, of course, replace A by an automatic mechanism, a thermostat, similar to that which controls the central heating system in a house. We can then represent the entire system by the block diagram shown in Figure 13.2a. Information about the output of the system, in this case water temperature, is fed back into the thermostat so that appropriate action is taken to correct any deviation in water temperature and keep it at the desired level. This is known as *feedback*, a term used when we compare the condition of the output of the control system with the set point. In this case, a rise in water temperature is corrected by a decrease in the heat input. This is *negative feedback*, a term used when a deviation is offset by a corrective action in the opposite direction.

For the further discussion it is useful to expand the diagram slightly and introduce standardized terms used by control system engineers (Figure 13.2b).

The typical negative feedback system is known as a *closed-loop control system*, a term whose meaning is obvious from the diagram. The signal from the controlled variable is fed back into the system, forming a closed loop. For completeness we should also briefly discuss *open-loop systems*, although they are less important in physiological regulation. Assume that a house furnace has a variable fuel supply arranged so that a drop in outside temperature increases the flow of fuel to the furnace. We can carefully adjust the system so that a decrease in the outside temperature gives precisely the amount of fuel needed to keep the room temperature constant. In this example the input is the outside temperature, and the output is the heat supplied by the furnace. However, if a disturbance enters into the carefully calibrated system (e.g., a strong wind), more heat is carried away from the

surement of the value of the controlled variable is compared with the desired value, the *set point*. This is done by an *error detector*, which delivers a signal that in turn activates a control mechanism that results in the necessary correction.

If we want to keep a water bath at a "constant" temperature, we can use the arrangement shown in Figure 13.1, which is nearly self-explanatory. Person A can decide to add heat to the water bath by throwing an electric switch, thus causing the temperature to rise. A thermometer tells him the temperature of the water. Person B, who is in charge, has already told A what temperature he wants maintained in the water bath. If the water temperature falls below the set point, A can throw the switch to raise the water temperature. By looking at the thermometer, he can see when the tem-

FIGURE 13.2 Diagrams of (a) the control system represented in Figure 13.1 and (b) a control system that includes the terms most commonly used in control theory.

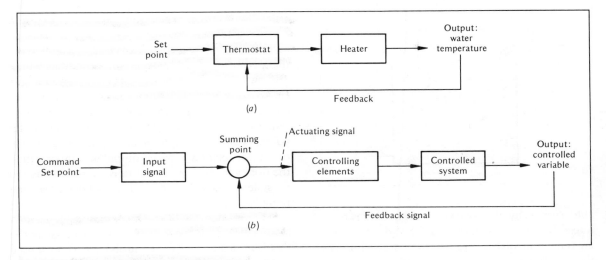

house, but the furnace supplies no more heat and the room temperature falls. In this system there is no feedback; it is an *open-loop control system.*

Negative feedback serves to reduce the difference between the output and the desired value; a thermostat is a good example. What about *positive feedback?* Does it exist? Does it have any importance in biology?

The following situation has been suggested by my colleague, Steven Vogel, as an illustration of positive feedback. Assume that a husband and wife have separate electric blankets, each thermostatically controlled through negative feedback. Say that the husband prefers a rather cool blanket, and the wife a higher temperature. Let us now assume that the thermostats inadvertently get interchanged. The husband sets "his" thermostat at his preferred low temperature, and his wife, who now finds her blanket cooler than she likes, turns up "her" thermostat. The husband soon finds his blanket much too warm and turns "his" thermostat further down, whereupon the wife turns "hers" up

even more. This is *positive feedback,* in which a deviation leads to an ever-increasing augmentation of the deviation.

Obviously, positive feedback is no good at all for control purposes, for the system will proceed to some extreme condition. However, positive feedback systems can be useful in some biological situations.

Negative feedback is used to maintain a steady state; positive feedback makes a system change more and more rapidly toward some extreme state. We discussed an example of positive feedback in connection with the generation of a nerve impulse: A decrease in the nerve membrane potential increases the permeability to sodium ions, sodium entry in turn further decreases the membrane potential, and so on, until a full action potential is generated. Positive feedback thus can serve to amplify a small signal and bring about a full response.

Positive feedback often is useful also in synchronizing events. A familiar example is the positive feedback during the process of mating. When suit-

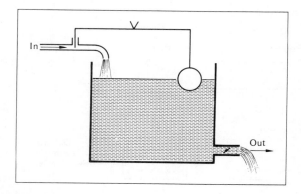

able partners meet, progress toward mating is reinforced by positive feedback; as the partners get increasingly involved, mutual response reinforces the appetite for sex, and continued positive feedback leads to copulation and completion of mating.

On–off, proportional, integral, and derivative control

The thermostatic control of a house furnace is an *on–off system*. The furnace is either on or off and therefore gives a discontinuous supply of heat. This inevitably leads to oscillations in temperature. In order to start the on cycle, a certain deviation from the set point is needed. Similarly, when the heater is on, there is usually an overshoot in temperature before the heater is cut off. The overshoot can be reduced by making the system more sensitive, but there is no way to eliminate oscillations entirely from an on–off system.

A more constant output of the controlled variable can be achieved with other types of control systems. One important system, known as *proportional control*, perhaps is best illustrated by a mechanical analogy (Figure 13.3). The water level in the tank is controlled with the aid of a float, which gives continuous instead of on–off control. If the

outflow from the tank for some reason increases, the water level sinks, thus opening the valve to allow increased inflow. The more the level sinks, the greater the increase in the inflow. Should the outflow be impeded, the water level rises and causes the inflow valve to close. This system has continuous control, and the degree of control action is directly related to the deviation from the set point. (The term *proportional* implies a continuous linear relation between output and input with the origin at zero, but the term is commonly used for systems that do not meet this strict definition of proportionality.)

A proportional controller has an interesting characteristic that is important. Assume a system that is in steady state and controls exactly at the set point. If a disturbance is introduced (e.g., a change in outflow), the system readily reaches a new steady state, but it is *not possible to attain the original set point*. This may need explanation.

Assume a given steady state, which is the original set point of the water level. Let us now decrease the outflow and examine the corrective action. The increased water level lifts the float and decreases the inflow. The new steady state involves a higher water level and the float must remain at this higher level to keep the inflow equal to the reduced outflow (i.e., the new level represents an error from the original set point). The corrective action cannot return the fluid level exactly to the original set point (zero error), for the float would then be in its original position and inflow would not match the decreased outflow. Now assume that a different kind of disturbance is introduced by adding a second inflow. The fluid level again rises and reduces the regulated inflow, but the new steady state is again above the set point.

A physiological system that represents proportional control is the control of mammalian respiration (ventilation of the lungs) by the carbon dioxide level in the blood. When a mammal at rest has an arterial P_{CO_2} of 40 mm Hg, we can consider this as the set point for the arterial P_{CO_2}. If we increase the carbon dioxide content of the inspired air, there is an increase in arterial P_{CO_2} that results in an increase in ventilation, and the amount of increase is directly related to the increase in P_{CO_2} (Chapter 2).*

As we have seen, a steady state with zero error cannot be achieved through simple proportional control. This is possible, however, by using *integral control*. In this system the total error over time is used as an input to the controller. Thus the output is proportional to the time integral of the input (i.e., the rate of change of output is proportional to the input). If a disturbance is introduced into such a system, and the disturbance remains constant, the error in the output will tend toward zero with time. Whatever the value of the disturbance, if it remains constant the integral action can achieve a steady state with zero error.

An additional type of control action is known as *derivative control* (or *rate control*) because the time derivative of a signal, or the rate of change in the signal, is used to anticipate the amount of corrective action required. Because this control action responds only to a change, and not to the signal itself, it is usually useful only in combination with other control action. In physiology such combination is particularly valuable, for it can be used to

---

* Respiratory control is in reality more complex. We know that in exercise carbon dioxide production is increased; if there were a simple proportional control system, the arterial P_{CO_2} should be slightly increased in exercise to provide the actuating signal for the increased ventilation. Actually, the situation is the opposite: During exercise arterial P_{CO_2} is slightly decreased. This is often "explained" as a result of a "resetting" of the set point in exercise. The real reason is that the respiratory center receives and integrates several different control inputs, including nerve impulses from the working muscles.

take appropriate action when there are temporary transients in a controlled variable.

Physiological mechanisms

Let us now return to the regulation of body temperature. We used the analogy with the thermostatic control of a water bath through a negative feedback loop. Central heating in a house can be regulated in the same way, and if in summer the house tends to get too warm, we can add an airconditioning (cooling) system. If the house temperature rises unduly, the thermostat starts the cooling cycle, an action that decreases the deviation from the desired set point (i.e., there is negative feedback). Thus regulation of both heating and cooling depends on negative feedback control.

The analogy to body temperature control is obvious. If body temperature tends to fall, heat production is increased, mainly through involuntary muscle contractions (shivering). If body temperature increases because of a heat load, whether external or internal, cooling is achieved through sweating or panting. Thus both heating and cooling of the body depend on negative feedback control systems.

The temperature regulation center is located in the hypothalamus of the brain. This can be demonstrated in various ways. For example, if the blood in the carotid artery of a dog is heated, it causes the dog to pant. This shows that the regulation takes place in the head, because as the experimental animal is caused to pant, it loses too much heat and its core temperature drops. Conversely, if the carotid blood is cooled, the dog begins to shiver, and the body core temperature rises. The exact location of the heat regulation center can be pinpointed by heating or cooling small areas in the hypothalamus.

The heat regulation center can be regarded as the thermostat and the "normal" body temperature as the set point. The system is not so simple, however, for there is no constant set point and there are multiple inputs. To begin with, the body temperature fluctuates with a daily cycle, even if the external temperature and internal heat production remain constant. This means that the set point undergoes a diurnal cycle. Also, we learned that during exercise the core temperature is reset and regulated at a higher level than at rest. Of the many inputs to the heat regulation center, the warm and cold receptors in the skin provide important information, but many other inputs provide additional information, one of them being the temperature of the arterial blood that reaches the center.

We mentioned that regulation of body temperature depends on both heating and cooling (i.e., control action takes place in either direction). This is an important principle in physiological regulation and much more common than a superficial examination might indicate.

The heart is another example. Its rhythmic contraction is an inherent characteristic of the heart muscle. Contraction starts at the sinus node, spreads throughout the heart muscle, and is followed by relaxation; whereupon, a new contraction begins. The rate of contraction is under the control of two nerves: the *accelerator nerve* speeds the heartbeat; the *decelerator nerve*, a branch of the vagus nerve, slows the rate. The heart rate is thus determined by the balance between two antagonistic nerves, one stimulating and the other inhibiting.

We shall see that a balance between stimulation and inhibition is very common in physiological control systems. An understanding of nerve cells and how they transmit excitatory and inhibitory impulses will help to clarify this common principle, and this is what we will turn to next.

NERVOUS CONTROL SYSTEMS

Nerve cells or neurons are the basic components of all nervous systems. We shall now focus our attention on two of their most important functional parts: the long fibrous extensions, the *axons*, and the connections between cells, the *synapses*. The single neuron has only one axon but may have hundreds or even thousands of synaptic connections, and their role in integration is extremely important. The axons function as cables and the synapses as highly complex contact or switching devices.

We shall first describe some important characteristics of axon physiology and then discuss how synapses handle the information that impinges on the neuron. As an example of integration we shall then describe how simultaneous stimulatory and inhibitory control by the central nervous system is used to control the muscles of a limb used in locomotion.

Axons

Action potentials

The nerve impulse, or action potential, was discussed in Chapter 12. It consists of a rapid transient change in membrane permeability accompanied by ion movement and a change in membrane potential. Although there is some minor variation among different animal species, the magnitudes of both resting and action potentials are similar throughout the animal kingdom. The resting membrane potential of both axon and cell body is mostly about 60 to 90 mV (outside positive).

The action potential, which consists of a localized drop in membrane potential, is mostly about -80 to -120 mV relative to the resting potential. No regular variations are associated with the size of the animal, the size of the axon, or whether the

TABLE 13.1 Resting and action potentials in neurons from a variety of animals. [Bullock and Horridge 1965]

Animal	Fiber or cell	Resting potential (mV)	Action potential peak (mV)	Spike duration (ms)
Squid (*Loligo*)	Giant axon	60	120	0.75
Earthworm (*Lumbricus*)	Median giant fiber	70	100	1.0
Crayfish (*Cambarus*)	Median giant fiber	90	145	2
Cockroach (*Periplaneta*)	Giant fiber	70	80–104	0.4
Shore crab (*Carcinus*)	30-μm leg axon	71–94	116–153	1.0
Frog (*Rana*)	Sciatic nerve axon	60–80	100–130	1.0
Sea slug (*Aplysia*)	Visceral ganglion	40–60	80–120	10
Land snail (*Onchidium*)	Visceral ganglion	60–70	80–100	9
Crayfish (*Cambarus*)	Stretch receptor cell	70–80	80–90	2.5
Puffer fish (*Sphaeroides*)	Supramedullary cell	50–80	80–110	3
Toad (*Bufo*)	Dorsal root ganglion	50–80	80–125	2.8
Toad (*Bufo*)	Spinal motor neuron	40–60	40–84	2
Rabbit (*Oryctolagus*)	Sympathetic cell	65–82	75–103	4–7
Cat (*Felis*)	Spinal motor neuron	55–80	80–110	1–1.5

animal is an invertebrate or a vertebrate. The time course of the action potential, on the other hand, is much more variable (Table 13.1).

Conduction speed

Speed of conduction is an entirely different matter; it varies tremendously from nerve to nerve and from animal to animal. A few typical values are listed in Table 13.2. Vertebrate motor nerves have a much higher speed of conduction than the regular motor nerves of invertebrates. This is not because the invertebrates live at lower temperatures, for the cold-blooded vertebrates also have relatively high conduction speeds. If the values for the cold-blooded vertebrates were recalculated to the body temperature of mammals, using a Q_{10} of 1.8 (which is characteristic of nerve fibers), they would all be within the mammalian range. For the invertebrates this would not be true. Some invertebrates, however, have a few fast-conducting axons, *giant axons*, which are much larger than ordinary axons, up to nearly 1 mm in diameter. In these giant axons the speed of conduction is around 10 times as fast as in ordinary axons from the same animal.

The biological role of high-speed conduction is obvious. It is always related to a quick response mechanism the animal uses in locomotion, mostly to avoid predators. One of the fastest such responses is that of the cockroach, which within 25 ms reacts to an air puff on the tip of the abdomen (which is equipped with receptor hairs). Another example is the squid, which has giant fibers that run the length of the mantle. Because of the rapid conduction in these giant axons, the entire mantle musculature can contract almost simultaneously, which is needed when the squid swims by jet propulsion. If conduction to the more distant parts of the mantle took appreciably longer, contraction would spread slowly and not give the necessary sudden forceful jet of water. In an earthworm, which on a moist morning reaches halfway out of its burrow, giant fibers permit almost instantaneous withdrawal in response to a mechanical disturbance, in sharp contrast to the relatively slow locomotion otherwise characteristic of this animal.

TABLE 13.2 Conduction velocities (in meters per second) in nerves from various animals. [Data from Bullock and Horridge 1965]

Animal	Regular motor nerves	Giant axons
Vertebrates		
Cat	30–120	
Snake	10–35	
Frog	7–30	
Fish	3–36	
Invertebrates		
Cockroach	2	10
Squid	4	35
Earthworm	0.6	30
Crab	4	
Snail	0.8	
Sea anemone	0.1	

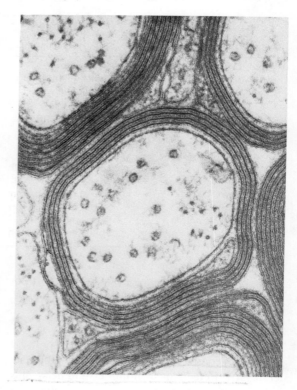

In general, there is a direct relationship between conduction velocity and axon diameter (Hodgkin 1954). It can be derived on theoretical grounds, based on cable theory, that the speed of conduction (u) should be approximately proportional to the square root of the fiber diameter (d), or:

$$u = k \sqrt{d}$$

A great deal of observational material indicates that this relationship is approximately correct, but the value of the constant k varies from animal to animal. The equation therefore tells us that, for a given type of fiber, we can expect a 10-fold increase in conduction speed for a 100-fold increase in fiber diameter. The giant axons mentioned above may be 50 or even 100 times thicker than the ordinary axons in the same animals, and their high conduction velocities therefore fall within the expected range.

Vertebrate myelinated fibers

Vertebrates do not have giant axons; yet, speed of conduction in their motor nerves is very high – the highest found anywhere in the animal kingdom. The reason mammalian axons conduct rapidly, although they are very thin, is their peculiar structure. The entire axon is covered with a thin sheath of a fatlike substance, *myelin*, which is interrupted at short intervals to expose the nerve membrane. The exposed sites are known as *nodes*, and the distance between the nodes is from a fraction of a millimeter up to a few millimeters (Figure 13.4). The myelin sheath is formed from glial (or supporting) cells, which grow into a many-layered wrapping from which the protoplasm disappears so that multiple layers of the glial cell membrane remain.

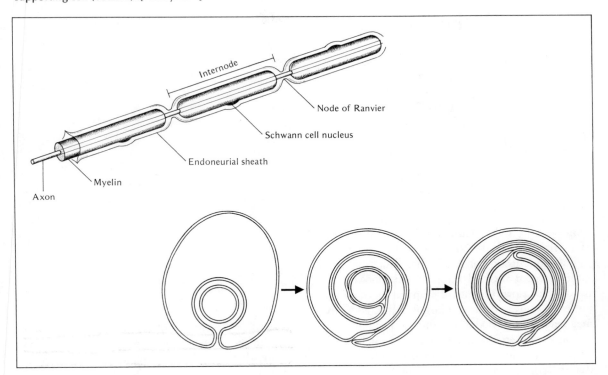

Let us examine the conduction of an impulse in a myelinated axon. At a node the action potential is exactly like any other action potential (i.e., there is a local depolarization of the membrane). This means that this node, relative to the neighboring node, appears negative (the usual positive membrane charge at our node has disappeared). This instantaneously sets up an electric current between this and the neighboring node, sufficient to initiate depolarization and trigger an action potential at the neighbor. This is a virtually instantaneous process, so that the action potential appears to jump from one node to the next. The depolarization at the node itself is less rapid, but as it develops, it in turn causes depolarization at the next node, and so on. The rapid transmission from node to node is known as *saltatory conduction* (from Latin *saltare* = to dance or jump), and because it takes place almost without delay, the result is a rapid transmission of action potentials in fibers whose small diameter would otherwise give a very low conduction velocity.

The conduction velocity in a myelinated frog axon has been accounted for as follows. A myelinated axon of 10 μm diameter has a conduction velocity of 20 ms^{-1}, and the delay at each node is approximately 0.06 ms. Because the nodes in a frog axon of this size are spaced about 1.6 mm apart, it can be seen that most of the conduction time is accounted for by the delay at the nodes; transmission

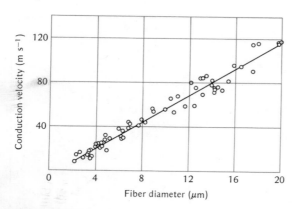

FIGURE 13.5 The speed of conduction in myelinated fibers of the cat, plotted against fiber diameter. [Hursh 1939]

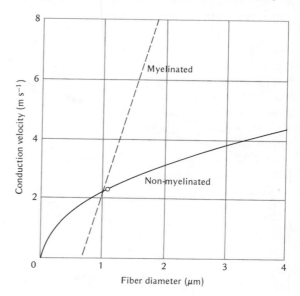

FIGURE 13.6 The conduction velocities in myelinated and nonmyelinated fibers are different functions of fiber diameter. Below a certain diameter the myelinated fiber therefore conducts more slowly than a nonmyelinated fiber of the same diameter. The broken line is an extrapolation from data in Figure 13.5; the solid line is a theoretical curve based on the fastest observed nonmyelinated C-fibers from the cat. [Rushton 1951]

from node to node occupies only a small fraction of the time (Tasaki 1959).

This conclusion can be further supported by a rather elegant method that involves cooling the nerve. This slows the rate of propagation of an impulse. However, if the cooling is restricted to the area between two nodes, the duration of the transmission between the nodes is nearly unchanged. This result is again consistent with the hypothesis that the conduction between the nodes is an electric phenomenon and that the potential change at the node itself is of the same nature as the action potential in a nonmyelinated axon (Hodler et al. 1951).

The conduction velocity in myelinated axons still depends on the fiber diameter, but in these fibers the conduction velocity is directly proportional to fiber diameter (Figure 13.5). The reason the conduction velocity is higher in the larger-diameter myelinated axon seems to be that the distance between the nodes increases with the diameter of the axon. Because virtually all delay is at the nodes, a smaller number of nodes permits a faster overall propagation of the action potential (Rushton 1951).

The fact that, in myelinated fibers, velocity is

linearly related to the fiber diameter, but in nonmyelinated fibers is proportional to the square root of the diameter, has the interesting consequence that for very small diameters, less than about 1 μm, myelinated axons conduct more slowly than nonmyelinated axons (Figure 13.6). This coincides with the lower limit for the actual size of myelinated fibers in the organism, which seem never to be smaller than 1 μm in diameter. The so-called C-fibers of the sympathetic nervous system are nonmyelinated, and the fastest and largest C-fibers have a diameter of 1.1 μm and conduct at 2.3 meters per second. This point is marked with a circle on Figure 13.6.

The greatest advantage of myelinated axons comes from their small size, which allows a highly complex nervous system with high conduction velocities without undue space occupied by the conduits. Let us say that we wish to increase the conduction velocity 10-fold in a given nonmyelinated fiber. This would require a 100-fold increase in its diameter, and the volume of nerve per unit length would in turn be increased 10 000-fold. Obviously, if this avenue were used for increasing conduction velocity, a nerve trunk that should contain hundreds or thousands of nonmyelinated axons would be unreasonably voluminous. Imagine the size of the optic nerve in humans if a high conduction velocity were to be achieved without myelination. This nerve has a diameter of 3 mm; if it were to contain the same number of fibers without myelination and conduct at the same speed, it would require a diameter of 300 mm.

Invertebrate myelinated nerves

We have seen that vertebrates achieve rapid nerve conduction with the aid of a myelin sheath and saltatory conduction. Many invertebrates achieve rapid conduction by increasing the fiber diameter (i.e., they employ giant axons when rapid conduction is essential, such as in escape reactions). This, however, is not the full story.

Invertebrate axons show two kinds of structural modification related to fast conduction. First, as we have already discussed, there are giant axons whose rapid conduction is directly related to their cable properties (Hodgkin 1954). Second, some invertebrate nerve fibers are covered with multiple layers of sheaths that are strikingly similar to the myelination of vertebrate nerves. Such nerves covered by "myelin," have been found in insects, earthworms, crabs, and prawns.

The speed of conduction in these fibers is much greater than in other invertebrate nerves of similar diameter. In fibers from prawns (*Palaemonetes vulgaris*) with a diameter of 26 μm the speed is as high as 18 to 23 meters per second (Holmes et al. 1941), far faster than the speed recorded in nerve fibers of such small diameter in other invertebra-

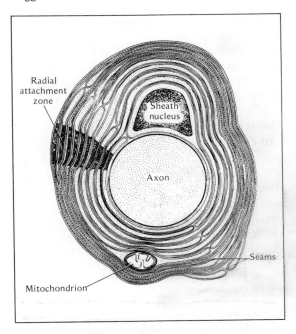

tes, although it is not as great as the conduction velocity in myelinated axons of most vertebrates.

The sheaths of the prawn nerve are somewhat different from vertebrate myelination (Figure 13.7). The layers are arranged more loosely, and there is an extracellular space between the inner layer of the sheath and the axon. Because of the loose arrangement, fewer membranes are present in a prawn sheath of a given thickness than in a vertebrate myelin sheath of the same thickness (Heuser and Doggenweiler 1966).

Another difference is that the spaces form a reservoir for ions that increases the capacitance of the nerve sheath. This influences the cable properties and offers a reasonable explanation for the slower rate of conduction in prawn fibers than in vertebrate fibers. Also, the spacing of nodes in prawn fibers seems to be shorter than is usually found in vertebrates, and if the greatest delay in saltatory conduction is at the node, the greater number of nodes explains the lower conduction velocity. Nevertheless, the structure is strikingly similar to vertebrate myelinated axons, and for a given fiber diameter the conduction velocity is strikingly high.

The synapse: excitation, inhibition, and computation

Information that travels in axons is transmitted to other neurons at the synapses. The transmission at the synapse is of two distinct kinds: *electrical* or *chemical*. Electrical transmission is not as widespread as chemical transmission, but it has characteristics that in certain situations confer considerable biological advantage.

Electrical synapses

At electrical synapses the terminal end of an axon comes so close to the next neuron that the space between the two membranes is no more than about $0.002\ \mu$m. This space forms a low-resistance pathway that provides a shunt for current to flow from the axon terminal into the next cell. In addition to a low-resistance pathway between the interior of one neuron and the next, electric transmission also requires a high resistance that prevents sideways spread of the current at the site of contact.

Electrical transmission takes place without measurable delay and therefore is not subject to the characteristic time lag that occurs at chemical synapses. This is important in certain synapses in the nervous systems of both vertebrates and invertebrates.

Electrical synapses were first demonstrated in the abdominal nerve cord of crayfish (Furshpan and Potter 1959) and has since been found in several arthropods, annelids, and molluscs. In fish, electrical transmission occurs at certain large nerve cells, known as Mauthner cells, where rapid trans-

mission is important in escape reactions (see later in this chapter); electrical transmission is probably common in the vertebrate central nervous system.

Many electrical synapses conduct equally well in both directions, but in others the contact area permits current to flow only from the presynaptic to the postsynaptic site, and not in the reverse direction. Thus, some but not all electrical synapses show rectifying properties; they allow impulses to pass in only one direction, like chemical synapses, which also permit conduction in one direction only.

Chemical synapses

The structure of chemical synapses is important for the understanding of their function (Figure 13.8). The terminal end of the axon spreads out and forms the *axon knob*, which in turn makes contact with a dendrite or a cell body of another neuron. The axon and the other cell do not fuse; there remains a narrow space or gap, the *synaptic cleft*, which has a width of about 20 nm. The knob is referred to as *presynaptic*; the dendrite or neuron is the *postsynaptic* structure. Both the appearance of the cleft and its width are surprisingly similar throughout the animal kingdom, but the cleft is so small that information about its contents has remained virtually unknown. It has been suggested that the relatively constant width of the cleft can best be explained by assuming that it is not merely a fluid-filled space, but contains a structure of oriented molecules arranged between the two nerve membranes.

The synaptic knob has an important structural characteristic: It contains a large number of small vesicles, which are usually about 20 to 100 nm in diameter. These vesicles are closely packed near the presynaptic membrane. It is well established that the transmission of an impulse from the presynaptic knob to the postsynaptic neuron takes

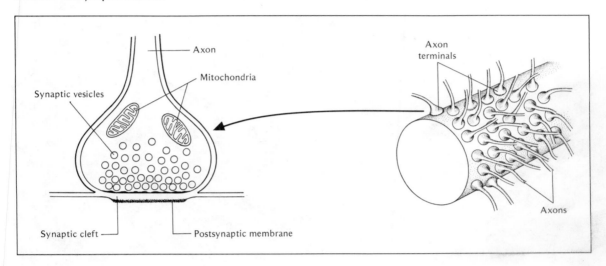

place through the release of a *chemical transmitter substance* from the synaptic vesicles that diffuses across the synaptic cleft and affects the postsynaptic membrane. The occurrence of vesicles is so typical of all synapses that their presence is used by electron microscopists as a criterion for the presence of a synapse.

There are several known synaptic transmitter substances, and one of them, *acetylcholine* (ACh), has been identified with greater certainty and studied in more detail than any other. Acetylcholine is also the transmitter substance released at the motor nerve end plates of striated muscle. As we shall see later, other transmitter substances are known, but there are great difficulties in identifying such substances with certainty.

Acetylcholine is released from the presynaptic knob in small discrete quantities, or "packages," corresponding to a certain number of synaptic vesicles. The release is dependent on the presence of calcium ions, and in a low-calcium medium little or no transmitter is released. Experiments with

variations in the calcium concentration have permitted the detailed study of the release mechanism and an estimate of how many single packages (vesicles) of acetylcholine are released in the normal transmission process (del Castillo and Katz 1954). At the neuromuscular junction, about 300 such packages are released for each nerve impulse. The precise relation between the presynaptic membrane potential and the amount of transmitter released has also been worked out in the stellate ganglion of squid (Katz and Miledi 1967).

We can now see why chemical synapses can transmit impulses only in one direction. Because transmission depends on the release of a transmitter substance that is present only on the presynaptic side, there is no means for transferring impulses in the opposite direction.

The diagram in Figure 13.8 makes the transfer of information to a neuron appear much simpler than it actually is. Most axons are highly branched and connect to a large number of other neurons, as we saw in the discussion of the structure of the ret-

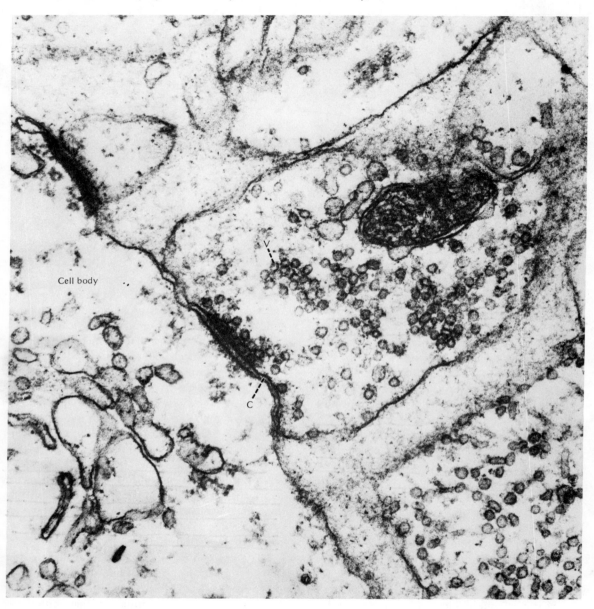

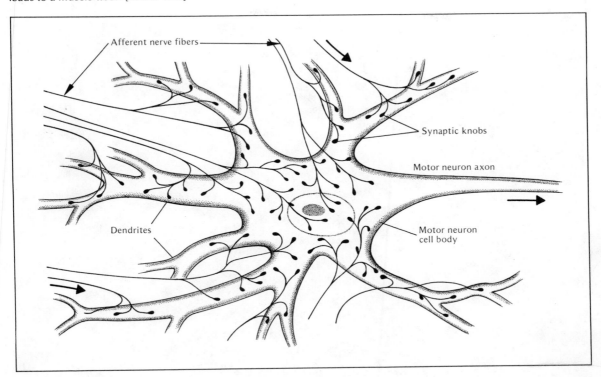

Afferent nerve fibers

Synaptic knobs

Motor neuron axon

Dendrites

Motor neuron cell body

ina (Chapter 12). Although the single neuron gives off only one axon, this axon may branch widely and connect to a large number of other neurons. As a consequence, each single neuron must also receive a large number of axon branches that terminate on the neuron or its dendrites (Figure 13.9).

There are often hundreds of synapses on a single neuron in the central nervous system, and motor neurons in the vertebrate spinal cord may have over 1000 synapses. Some specialized neurons are so densely covered with synaptic knobs that it has been estimated that a single cell may have about 10 000 synaptic connections. This, viewed in combination with the enormous number of neurons (estimated at 10 000 000 000 in the central nervous system of man), should give an idea of the complexity of the central nervous system. It must be emphasized, however, that the connections are by no means random; they are highly specific and form precisely functioning tracts within the central nervous system. These have been studied and described in great detail by neuroanatomists as well as neurophysiologists.

Postsynaptic potentials

The understanding of synaptic transmission was revolutionized in the 1950s because methods were developed to record electrically from single neurons. Extremely fine glass pipettes with tip di-

SYNAPTIC KNOBS This scanning electron micrograph shows the multitude of synaptic knobs that impinge on the surface of a single neuron in the nervous system of *Aplysia,* a marine snail. The diameter of each disclike terminal knob is about 1 μm. [Courtesy of Edwin R. Lewis, University of California, Berkeley]

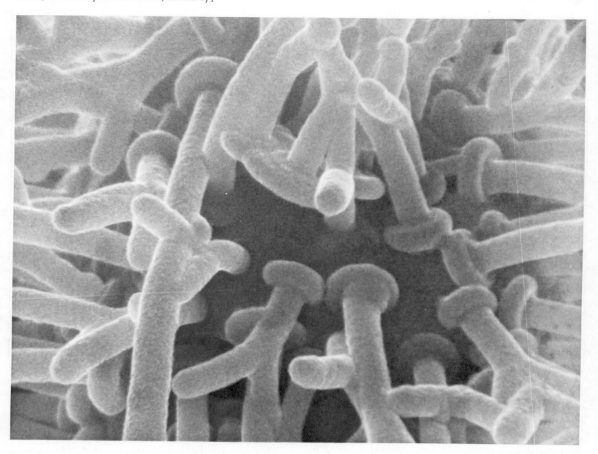

ameters of less than 1 μm are prepared. These are filled with a salt solution (usually concentrated potassium chloride), making microelectrodes that can be used to record any potential change between the tip and a common "ground." As such a microelectrode is slowly moved toward and inserted into a neuron, it first shows the same potential as the ground, but the moment it penetrates the nerve membrane, it shows a negative potential relative to the ground. The membrane seems to

seal around the glass, and it is possible to record from a neuron for several hours while it continues to function apparently normally. By using pipettes with two or more channels, instead of a single one, it is even possible to apply minute amounts of various chemicals at the site from which the recording is being made.

With this technique we can study the events at the synapse. As an impulse arrives at the presynaptic membrane, there is a delay of a fraction of a

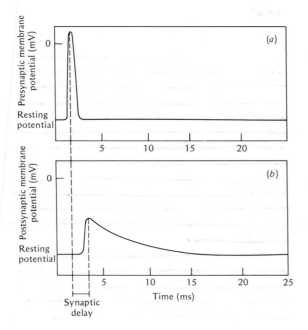

FIGURE 13.10 The arrival of an action potential at the terminal knob (*a*) gives rise to a potential at the postsynaptic membrane, the postsynaptic potential (*b*), which is smaller in magnitude but has a much longer duration. [Stevens 1966]

milliseconds before a change occurs in the potential at the postsynaptic membrane. This latter potential is the *post-synaptic potential* or *PSP* (Figure 13.10). The PSP initially rises rapidly, and then decays, but at a much slower rate.

PSPs differ from action potentials in two important respects: They are usually much smaller in amplitude, and they have a much longer duration, sometimes as much as 10 or 100 times as long. These characteristics have two important consequences: A single PSP is rarely (if ever) sufficient to cause an action potential in the postsynaptic neuron, and the long duration permits a great deal of interaction with other PSPs in the same neuron, both at the same synapse and at neighboring synapses on the same neuron. This interaction is very important and requires a more detailed discussion of the postsynaptic events.

What causes the PSP? We will assume that the PSP is caused by a transmitter substance released from the presynaptic membrane and that the delay corresponds to the diffusion time across the synaptic cleft.

The best known and understood transmitter substance is *acetylcholine*, whose function in synapses and at neuromuscular junctions is beyond doubt. In the synapse, the magnitude of the PSP seems to be directly related to the amount of acetylcholine released. The system could not function, however, unless the acetycholine were rapidly removed again, for otherwise it would gradually accumulate and maintain a continuous PSP. The enzyme *acetylcholinesterase* is always present at the synapse and serves to hydrolyze and thus remove the transmitter. As the transmitter substance disappears, the PSP gradually decays.

Summation. If a second impulse arrives at the same axon terminal before the preceding PSP has fully decayed, the amount of transmitter substance is increased, and the PSP is greater. We therefore observe a summation of the two impulses, and because this is a summation in time, it is referred to as *temporal summation*.

Assume that a long train of impulses arrives at an axon terminal at a constant rate. The PSPs are summed, giving an overall PSP that is directly related to the frequency of impulse arrival. If the frequency is increased, the PSP increases, and a new steady-state PSP is established at a level where the decay rate of the transmitter equals the release rate from the axon terminal. The magnitude of the PSP, therefore, is a direct expression of the impulse frequency in the axon; in other words, the PSP is a *frequency modulated* potential. This clarifies how a series of action potentials, each an all-or-none event of constant magnitude, can be used to transmit information about a changing signal. For example, an increase in the magnitude of a signal from a sensory neuron is encoded and transmitted in its axon as an increase in the frequency of action potentials; on arrival at the synapse it is then decoded as a PSP of a magnitude directly related to the magnitude of the original signal.

Let us return to the postsynaptic neuron and remember that there are numerous synapses on its surface. The PSP at any one synapse is not completely restricted to the area under the axon knob; it spreads to the immediate neighborhood with a decreasing magnitude as the distance increases. This spatial spread means that there is a slight change in membrane potential in the neighborhood of the synapse, and should an impulse arrive at another synapse within this area, the new PSP is added to the existing potential. Such summation in space is known as *spatial summation*.

We have now seen that various inputs to a neuron, arriving at different synapses, influence each other and can be summated through both temporal and spatial summation. These two types of summation form the basis for computations in each neuron and thus in the entire nervous system.

Before we discuss these integration processes further, we must know what effect postsynaptic potentials have on the neuron. We said that the PSPs are usually insufficient to cause an action potential in the postsynaptic neuron. There is, however, an area of the neuron, the *"hillock,"* near the origin of its axon, which fires action potentials if it is sufficiently stimulated by depolarization of the membrane. If many impulses arrive at various synapses on a neuron, a combination of temporal and spatial summation may lead to a PSP sufficiently great to cause the passive spread to reach and depolarize the *axon hillock*, thus setting up an action potential in the axon.

Excitation and inhibition

Until now we have not been concerned with the polarity of the PSP. A PSP can, in fact, either decrease or increase the normal membrane potential. If the post synaptic membrane potential is slightly decreased, or depolarized, by the PSP, the change is in the direction that normally leads to an action potential. This is referred to as an *excitatory postsynaptic potential* or (EPSP). If the PSP causes an increase in the normal membrane potential, or a hyperpolarization, its effect is the opposite of that which would lead to an action potential. This is known as an *inhibitory postsynaptic potential* (IPSP) because it tends to inhibit the generation of an action potential.

Current evidence indicates that any one synapse is always of one kind or the other and that a given axon causes the same type of PSP at all its synaptic terminals. One particular synapse is always either inhibitory or excitatory and does not change.

Except for the polarity, the two kinds of postsynaptic potentials are of the same nature. The EPSPs transform an arriving train of impulses into a sustained depolarization; the IPSPs convert a similar train of impulses into a sustained hyperpolarization. A given neuron responds differently to arriving information, depending on what information it has recently received or is receiving from other sources. Under circumstances when a neuron is already partly depolarized, an additional excitatory potential may be sufficient to fire an action potential from the hillock, although at other times the same EPSP would be far from sufficient.

These characteristics explain how a single neuron can carry out extensive integration of information received from various sources. Although the synapse is a one-way valve, the amount of integration and computation that results from excitatory and inhibitory synapses, in combination with spatial and temporal summation, makes the single neuron a formidable device in the computation processes in the nervous system.

Presynaptic inhibition

We must consider one further characteristic of neurons. Without going into details, we should mention that there is an important process known as *presynaptic inhibition*. An axon terminal may, for a time, remain slightly hyperpolarized, or slightly depolarized (without any action potential). An impulse that arrives during a period of slight depolarization causes the release of a smaller than usual amount of transmitter and therefore causes a reduced PSP. Since the reason for the smaller PSP is located in the axon terminal, this type of inhibition is presynaptic.

It may seem trivial whether inhibition is pre- or postsynaptic, but in fact it is important. Postsynaptic inhibition works by subtracting from excitatory PSPs that arrive at the neuron; it therefore is a nonselective inhibition. Presynaptic inhibition is highly selective, for it affects only signals arriving at that particular synapse. It also has another characteristic: It is not a simple additive inhibition (as is postsynaptic summation), for it influences the transmission of each impulse in a larger train of an arriving signal. Thus, presynaptic inhibition increases both the specificity and the complexity of the integration that can take place at the neuronal level.

Escape: a flip of the tail

The normal escape reaction of a fish is a sudden flip of the tail, followed by ordinary undulatory swimming. The reaction is directed from two remarkable large nerve cells, the *Mauthner cells*, located in the brain of teleost fish. Their role is to integrate and relay information from sense organs and the brain to the motor nerves that drive the body muscles.

The Mauthner cells receive a rich supply of nervous inputs, many forming electric rather than chemical synapses. The main output from the Mauthner cell is through a large axon that from each Mauthner cell crosses to the opposite side and runs down along the spinal cord, where it makes synapses with motor neurons that innervate the main swimming muscles (Diamond 1971).

The importance of the Mauthner cell is in its reaction to sensory information, particularly information produced by mechanical disturbances in the water. In response, the Mauthner cell discharges a single impulse to the large axon, which crosses to the opposite side and causes a rapid, vigorous muscle contraction (i.e., a fast flip of the tail, followed by normal swimming movements). This escape reaction can readily be observed in aquarium fish by tapping on the glass side.

What would happen if both Mauthner cells fired impulses simultaneously? If the muscles on both sides contract at the same time, there will be no movement. This is where the importance of an inhibitory input comes in. It is virtually impossible for a sensory input to be completely bilaterally symmetrical, both in amplitude and timing, and one Mauthner cell is therefore usually activated an instant ahead of the other. Each Mauthner cell sends inhibitory connections to the Mauthner cell on the opposite side, and as one fires an impulse, the other cell is inhibited. This is where the speed of electrical synaptic transmission is important, for it makes the transmission nearly instantaneous. A chemical synapse would involve a delay, and the inhibition of the second Mauthner cell might not occur soon enough to prevent it from firing an impulse.

The superiority of electric transmission at the synapse is its speed; the delay characteristic of chemical synapses is eliminated. This saving of time explains the importance of the mechanism for fast processing of signals in the nervous system. When a Mauthner cell fires, it almost instantaneously inhibits the other Mauthner cell, thus preventing a paralyzing contraction on both sides simultaneously.

Except where there is a great need for speed, chemical transmission appears to be superior. The greatest advantage is probably in regard to the duration of a change, whether it is excitatory or inhibitory, for the presence of a chemical transmitter permits integration over much longer periods. The preponderance of chemical transmission implies some functional superiority, and we can assume that the greater refinement in temporal interactions is a rational explanation.

Control of muscle function

In Chapter 11 we saw that crustacean muscles are controlled by a small number of axons (often three to five). Each axon branches to most or all of the muscle fibers within a muscle. Vertebrate muscles, in contrast, are innervated by a large number of axons, each branching to a small number of muscle fibers (*a motor unit*). This difference in innervation causes much of the control of crustacean muscle to take place peripherally, whereas vertebrate muscle is essentially under central control, directed by a highly complex central nervous system.

The difference is readily explained. Myelinated axons permit the fast transmission of impulses without undue increase in axon diameter; the necessary large number of connections can be carried in a nerve of moderate size. For the crustacean muscle, in contrast, rapid conduction requires a substantial increase in fiber size, and a finely

graded central control that depends on a large number of fibers in the motor nerve is therefore not feasible. We have already described how crustacean muscles nevertheless are under precise control, although they may receive no more than two or three stimulatory and one or two inhibitory axons. We shall now examine the control of vertebrate muscle.

Vertebrate muscle

The contraction of vertebrate muscle is under precise central control, and there is no need for inhibitory axons to reach the muscle and modify the degree of contraction.* The vertebrate muscle system acts under the direction of a central computer that receives a great deal of information from various sources, including a very important part that comes as feedback directly from the muscles themselves. We shall use the admirable performance of this system as an example of what can be achieved through central control and integration.

The nerve to a vertebrate muscle has fibers that carry the impulses for stimulation of contraction; in addition, it contains other fibers that carry sensory information from the muscle to the central nervous system. There are two major kinds of such sensory fibers (Figure 13.11). Some come from small sensory units located in the tendons, the *tendon organs*; others come from a specialized type of muscle fiber, known as *muscle spindles*. The ten-

* Although vertebrate skeletal muscle apparently has no inhibitory innervation, this is not true for other kinds of vertebrate muscle. The heart, which is a striated muscle, has double innervation, both stimulatory and inhibitory. All smooth muscle receives both adrenergic and cholinergic fibers from the autonomic nervous system. For some smooth muscle (e.g., bronchial muscle) the adrenergic fibers are inhibitory; for others (vascular smooth muscle) they are stimulatory. Because cholinergic fibers are always antagonistic to the adrenergic fibers, they may likewise be either inhibitory or stimulatory.

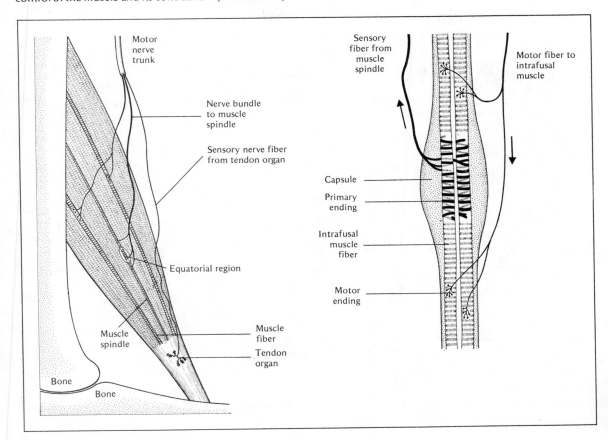

FIGURE 13.11 A vertebrate muscle is equipped with sensory organs that transmit information to the central nervous system. The tendon organs provide information about the force of contraction; the muscle spindles, primarily about the length of the muscle. This information is used as important feedback signals in the precise control of the muscle and its contraction. [Merton 1972]

Labels (left diagram): Motor nerve trunk; Nerve bundle to muscle spindle; Sensory nerve fiber from tendon organ; Equatorial region; Muscle spindle; Muscle fiber; Tendon organ; Bone; Bone

Labels (right diagram): Sensory fiber from muscle spindle; Motor fiber to intrafusal muscle; Capsule; Primary ending; Intrafusal muscle fiber; Motor ending

don organs seem to be used for sensing the deformation produced by tension in the tendon and give information about the force of muscle contraction. The muscle spindles, which we shall discuss in greater detail, are used for obtaining information about length, but as we shall see, they are even more useful because of the way their information is used in the feedback control of muscle contraction. In fact, it has been stated that the muscle spindles, next to the eye and the ear, are the most complex sensory organs in the body.

Let us use a highly simplified diagram (Figure 13.12). A person can easily, in response to a constant load, maintain a constant force in a muscle. If an additional load is unexpectedly added, it causes a stretching of the muscle that almost instantly is compensated for by an increased force of contraction in the muscle. This is caused by a reflex that originates in the muscle spindles. When stretched, the spindles send impulses to the spinal cord where the sensory fibers make synaptic contact with motor neurons, which in turn cause suf-

FIGURE 13.12 Feedback control of muscle contraction. A constant load on a muscle is balanced by a constant force (a). A sudden increase in the load (b) stretches the muscle; the muscle spindles send signals back to the spinal cord, where additional motor neurons are excited and send impulses back in the motor nerve, causing increased contraction to balance the load (c). [Merton 1972]

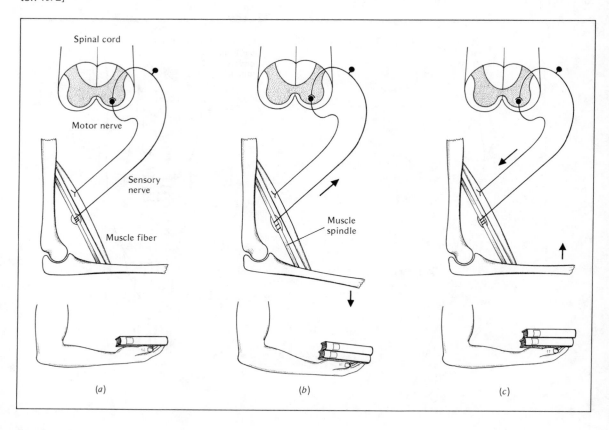

ficient contraction in the muscle to compensate for the initial stretching.

In reality, more than one synapse is usually involved, and there are connections from the spinal cord that carry information to different levels of the central nervous system, including the brain; but we need not be concerned with this complexity. The important matter is that the muscle spindles can serve to maintain a constant length of the muscle in the face of variations in the load – an obvious advantage in coordinating movements of all sorts.

Another important characteristic of the muscle spindles is that these are themselves contractile and innervated by motor axons that are separate from the axons that cause contraction of the regular muscle fibers. The detailed structure of the muscle spindles is important (Figure 13.11). They consist of a bundle of a few modified muscle fibers, called the *intrafusal muscle fibers* (Latin *fusus* = spindle). These fibers are contractile, but in the middle of the fiber the contractile apparatus is absent and the fiber has sensory nerve endings wrapped around it.

Let us say that an intrafusal fiber is stimulated to

contract; this stretches its central region, which sends to the spinal cord a volley of impulses. These in turn stimulate the motor neurons, which respond by sending action potentials in the motor nerve, causing contraction in the muscle. When the contraction in the muscle has reached the same level as the contraction in the intrafusal fiber, the equatorial region is no more stretched and is therefore silenced.

This elegant system in a way is analogous to a mechanical servosystem. For example, in the power steering of a heavy automobile a small movement of the steering wheel is used to direct a motor that performs the heavy work of turning the wheels on the road. A servomechanism is no more than an automatic feedback control mechanism in which the controlled variable is the position of a mechanical device. Such systems are used widely in the design of all sorts of mechanical and industrial control processes.

The characteristic of intrafusal fibers just described provides a clever mechanism for maintaining a graded speed of contraction against a variable load. If the contraction of a given muscle is directed by a certain slow contraction of the intrafusal fibers, an increase in the load on the muscle only momentarily slows its rate of contraction. The intrafusal fibers continue to contract at a set rate, but if the muscle follows too slowly because of the increased load, the equatorial region is stretched and immediately sends additional impulses to the spinal cord, which in turn increases the contraction of the muscle. The result is an automatic compensation for increases or decreases in the load that occur during muscle contraction.

When we turn to the question of how the muscles are used in the movements of the limbs, we find a need for additional control. The muscles that serve in maintaining posture normally maintain some degree of contraction; if they did not, the

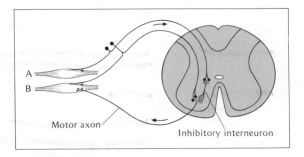

FIGURE 13.13 When two antagonistic muscles (A and B) participate in the movement of a limb, the contraction of muscle A causes impulses to be sent to the spinal cord. Here an inhibitory interneuron, by inhibiting the appropriate motor neuron, causes relaxation of the antagonistic muscle (B).

body would collapse, as happens when a person faints. Therefore, to move a limb, it is not sufficient that a given set of muscles contract; it is also necessary that the antagonistic muscles relax. This is achieved through a relatively simple reflex mechanism shown in schematic form in Figure 13.13. Suppose that stimulation of the muscle spindles in muscle A evokes contraction of this muscle. Information about this contraction is received by certain neurons in the spinal cord, which are called *interneurons* * and act by inhibiting other neurons. The pathways are arranged in such a way that these *inhibitory interneurons* cause the inhibition of the motor neurons of the antagonistic muscle (B). The reciprocal action of antagonistic muscles, with one relaxing as the other contracts, is functionally very useful and could not take place without the aid of inhibitory neurons or other inhibition.

In addition to the control of antagonistic muscles, inhibition serves other important purposes. For example, if a contracted muscle is stretched very forcefully, the tendon organs are activated and initiate impulses that reach inhibitory interneurons in the spinal cord. This causes inhibition of contraction in the muscle that is being

* Neurons whose axons do not extend outside the central nervous system are known as *interneurons*. They may be excitatory or inhibitory.

TABLE 13.3 Important vertebrate hormones.

Source	Hormone[a]	Major functions
Hypothalamus	Releasing and release-inhibiting hormones acting on adenohypophysis	Hormones delivered via portal circulation to adenohypophysis; for functions, see Table 13.6
Hypothalamus (via neurohypophysis)	Oxytocin	Stimulates contraction of uterine muscle; releases milk
	Antidiuretic hormone (ADH)[b]	Stimulates water reabsorption in kidney
Adenohypophysis	Adrenocorticotropic hormone (ACTH)	Stimulates adrenal cortex
	Thyrotropic hormone (TSH)	Stimulates thyroid
	Follicle-stimulating hormone (FSH)	Stimulates ovarian follicle development; seminiferous tubule development in testes
	Luteinizing hormone (LH)	Stimulates conversion of ovarian follicle to corpus luteum; stimulates progesterone and testosterone production
	Prolactin	Stimulates milk production
	Melanocyte-stimulating hormone (MSH)	Stimulates dispersion of melanin in amphibian skin pigment cells
	Growth-stimulating hormone (GSH)	Stimulates growth (acts via liver, see below)
Liver	Somatomedin	Stimulates growth (is effector for hypophyseal GSH)
Adrenal cortex	Glucocorticoids (corticosterone, cortisone, hydrocortisone, etc.)	Regulate carbohydrate metabolism
	Mineralocorticoids (aldosterone, deoxycorticosterone, etc.)	Regulate sodium metabolism and excretion
	Cortical androgens, progesterone	Stimulate secondary sexual characteristics, predominantly male
Ovary	Estrogens	Initiate and maintain female secondary sexual characteristics; initiate periodic thickening of uterine mucosa; inhibit release of FSH
	Progesterone	Cooperates with estrogens in stimulating female secondary characteristics; supports and glandularizes uterine mucosa; inhibits release of LH and FSH
Testis	Testosterone	Initiates and maintains male secondary sexual characteristics

[a] The terms *tropic hormone* and *tropin* are equivalent (e.g., thyrotropic hormone is also known as thyrotropin). In analogy with other uses of the suffix *-tropin* (from *tropein* = to turn or change), the preferred spelling is *tropin*, rather than *trophin* (from *trophos* = feeder) (cf. *phototropic* and *hypertrophy*) (Stewart and Li 1962).
[b] ADH is also known as vasopressin, a name it received because in high concentration it stimulates contraction of vascular smooth muscle and causes increased blood pressure.

Thyroid	Thyroxine, triiodothyronine	Stimulate oxidative metabolism; stimulate amphibian metamorphosis; inhibit release of TSH
	Calcitonin[c]	Inhibits excessive rise in blood calcium
Parathyroid	Parathormone	Increases blood calcium
Stomach	Gastrin	Stimulates secretion of gastric juice
Duodenum	Secretin	Stimulates secretion of pancreatic juice
	Pancreozymin[d]	Stimulates secretion of pancreatic enzymes
	Cholecystokinin[d]	Stimulates release of bile by gallbladder
	Enterogastrone	Inhibits gastric secretion
Pancreas	Insulin	Reduces blood glucose; stimulates formation and storage of carbohydrates
	Glucagon	Increases blood glucose by mobilization of glycogen from liver
Adrenal medulla	Adrenaline, noradrenaline	Augment sympathetic function (vasodilation in muscle, liver, lungs; vasoconstriction in many visceral organs); increase blood sugar

[c] In mammals, from the C-cells of the thyroid; in lower vertebrates, from the ultimobranchial bodies.
[d] Pancreozymin and cholecystokinin are believed to be identical.

stretched, and the active resistance to stretching is decreased. This reflex contributes to the protection of the muscles from rupture when they are suddenly loaded. Without this protection there would be far more "pulled muscles" than we now see.

The integration of muscular movement in locomotion is more complex than described in the examples mentioned here. It depends on various commands from the brain and on information received from sensory organs such as eyes, ears, skin, and so on. The central nervous system is a highly complex control system that acts on information and feedback signals from a variety of sources.

We shall now turn to the endocrine system and to the close interaction between the endocrine and the nervous systems. In fact, we shall see that the central nervous system not only is a control system, based on the transmission and integration of nerve impulses, but is a major producer of hormones as well.

HORMONAL CONTROL SYSTEMS

Numerous physiological functions are under hormonal control. Often this is referred to as chemical control, in contrast to nervous control, but the term is both unfortunate and misleading. We have already seen that synaptic transmission of nerve impulses usually is of a chemical nature, and we shall soon see that the nervous system is a major factor in hormone production and control. Furthermore, some chemical substances that are normally formed in the body (e.g., carbon dioxide) have pronounced physiological effects and are important in regulation, but they are not hormones and have nothing to do with endocrine function.

The term *hormone* can best be defined as a substance that is released from a well-defined organ or structure and has a specific effect on some other discrete structure or function. The most important vertebrate hormones and their major functions are listed in Table 13.3.

Until recently, endocrinology and hormone research were empirical sciences centered around a fairly uniform approach. An organ suspected of producing a hormone was removed, depriving the organism of the normal source of this hormone. The results were observed and described, and if the symptoms could be relieved by injections of extracts of the organ, further work consisted of purification of the extract, chemical isolation of the active component, and eventually its chemical synthesis. Another important method of replacement therapy consisted of transplantation of hormone-producing tissue.

These general methods are quite successful in research on organs that can be removed without hazard to life, such as the gonads. However, many hormone-producing organs have functions that are essential in other respects, and their removal leads to serious difficulties. The liver and kidneys, for example, produce hormones, but their removal rapidly leads to fatal effects unconnected with their role as sources of hormones.

The physiological functions that are under mainly endocrine control can conveniently be divided into a five broad areas (Table 13.4). The first three categories have to do with metabolic functions and maintenance of steady states; the remaining two with reproduction, growth, and development. Many of these hormones are similar or identical throughout all vertebrate classes; others have specific functions that differ characteristically

TABLE 13.4 Major functional areas under endocrine control with some well-known hormones listed as examples.

> *Digestion and related metabolic functions*
>   Secretin
>   Gastrin
>   Insulin
>   Glucagon
>   Noradrenaline
>   Thyroxine
>   Adrenal corticoids
> *Osmoregulation, excretion, water and salt metabolism*
>   Vasopressin
>   Prolactin
>   Aldosterone
> *Calcium metabolism*
>   Parathormone
>   Calcitonin
> *Growth and morphological changes*
>   Growth hormone
>   Adrenocortical androgens
>   Thyroxine (amphibian metamorphosis)
>   Melanocyte-stimulating hormone (amphibian color change)
> *Reproductive organs and reproduction*
>   Follicle-stimulating hormone
>   Luteinizing hormone
>   Estrogen
>   Progesterone
>   Prolactin
>   Testosterone

from group to group. One example is prolactin, which in mammals stimulates milk secretion, in pigeons stimulates the formation of the crop "milk," and in fish affects renal function and osmotic permeability of the gills.

Chemical nature of vertebrate hormones

In spite of the large number of vertebrate hormones and the great variety in their actions, in regard to structure and chemical nature they fall in three distinct groups: (1) tyrosine-derived hormones, (2) steroids, and (3) peptides and proteins (Table 13.5). These groupings are important, for

TABLE 13.5 Classification of vertebrate hormones on the basis of structure and chemical nature. These groups are important for the cellular mechanism of hormone action.

Tyrosine-derived hormones	
Catecholamines	
Noradrenaline	
Adrenaline	
Thyroid hormones	
Thyroxine	
Triiodothyronine	
Steroid hormones	
Testosterone	
Estrogen	
Progesterone	
Corticosteroids	
Vitamin D_3	
Peptide and protein hormones	
Peptides	
Hypothalamic hormones	(3 to 14 amino acids)
Angiotensin	(8 amino acids)
Gastrin	(17 amino acids)
Secretin	(27 amino acids)
Glucagon	(29 amino acids)
Calcitonin	(32 amino acids)
Insulin	(51 amino acids)
Parathormone	(84 amino acids)
Larger proteins	
Growth hormone	
Prolactin	
Luteinizing hormone	
Follicle-stimulating hormone	
Thyrotropic hormone	

the structure of a hormone is closely related to the process through which its action is communicated to the cells of its target organ (see later in this chapter).

Tyrosine-derived hormones. The structures of the two common catecholamines, noradrenaline and adrenaline, are in Figure 13.19. These two hormones differ only in a methyl group (—CH_3) that is absent in noradrenaline and present in adrenaline.

The formation of the catecholamines begins with the amino acid tyrosine and in a few steps leads to the formation of noradrenaline and adrenaline. Tyrosine is also the raw material for the syntheses of the thyroid hormones triiodothyronine and thyroxine. These hormones are not catecholamines, but form a group by themselves; they are formed from tyrosine by the condensation of two six-membered carbon rings, which after iodination become the active hormones.

Steroid hormones. All the steroid hormones are derived from cholesterol, which has a characteristic basic structure consisting of three six-membered and one five-membered carbon rings. Modifications in this basic structure give rise to a whole series of important hormones in which a slight change can lead to radical changes in physiological effects. Compare, for example, the small structural differences between estradiol and testosterone, which are sex hormones with diametrically differing effects. Hormones of this group include adrenocortical androgens, estrogens, progesterones, and corticosteroids such as cortisone and aldosterone.

Peptide and protein hormones. The peptide hormones include the hypothalamic control hormones, which contain from 3 to 14 amino acids. The thyrotropin releasing hormone (T-RH) contains only 3 amino acids, growth hormone releasing hormone (GH-RH) contains 10, and growth hormone release-inhibiting hormone (GH-RIH) contains 14. Several other peptide and protein hormones are also listed in Table 13.5.

The major hormones originating from the adenohypophysis are protein in nature and may contain several hundred amino acids. Some are glycoproteins, which in addition to characteristic peptide chains also contain carbohydrate components. As proteins go, they are not very large; some have molecular weights around 30 000, but it is often difficult to say whether active components that have been isolated from the gland are in fact identical

C_8H_17

HO

Cholesterol

OH

HO

Estradiol, a major estrogen

OH

O

Testosterone

with the functional hormone in the living organism.

Integration of endocrine and nervous control

There is a constant mutual interaction between the endocrine organs and the central nervous system. In the following discussion we shall see how the brain, directly or indirectly, influences and controls endocrine function. But we should not forget that the reverse is also true: Hormones have profound effects on the function of the central nervous system. Just consider the simple fact that a female dog in heat accepts the mating behavior of a male dog, although at other times the same signals from the male leave her indifferent or antagonistic. The fact that the same signal can lead to completely different behaviors depends on hormonal effects, and these effects can be mimicked by the injection of appropriate hormones.

The central nervous system in turn has an essential role in the control of endocrine function. Present evidence shows that the region of the brain known as the *hypothalamus* plays a completely dominant role in this control. Adjacent parts of the brain, the supraoptic and paraventricular nuclei, are part of this system.

The hypothalamic control system

The hypothalamus is located at the base of the brain, immediately above the hypophysis (pituitary gland),* just posterior to the optic chiasm, where it forms the floor of the third ventricle (Figure 13.14).

The hypothalamus is the seat of several nervous control functions, notably temperature regulation and the regulation of intake of water and of food. Control of body temperature has already been discussed as a feedback system. The role in regulation of food intake can be demonstrated by destruction of certain parts of the hypothalamus and by electric stimulation, which, if properly located, make animals ingest huge quantities of food and grow abnormally obese. Regulation of water intake is shown similarly; electric stimulation or the injection of small amounts of hypertonic salt solution into certain areas of the hypothalamus causes animals to drink excessively. Goats, for example, have

* The terms *hypophysis* and *pituitary* are equivalent. The organ was known to the ancient Greeks as the hypophysis, which refers to its location under the brain. The name pituitary came into use later because it was mistakenly believed that this gland secretes the mucus of the nose (Latin *pituita* = mucus or slime). The use of the terms *anterior* and *posterior* in connection with the hypophysis is unfortunate, for the relative position of these two major parts of the gland differs in different vertebrates. The *adenohypophysis*, also called the anterior lobe, is that part which is of glandular origin, formed by an invagination from the digestive tract (Greek *aden* = gland). The *neurohypophysis*, also called the posterior lobe, is that part of the hypophysis which is of neural origin. Further confusion is caused by the term *intermediate lobe*, a part that arises with the adenohypophysis but is included with the neurohypophysis in the posterior lobe.

FIGURE 13.14 The location of the hypophysis and the hypothalamus in the human brain. The adenohypophysis is controlled by neurohormones originating in the hypothalamus; the neurohypophysis is a neurohemal organ that releases hormones produced by neurons in the adjoining parts of the brain.

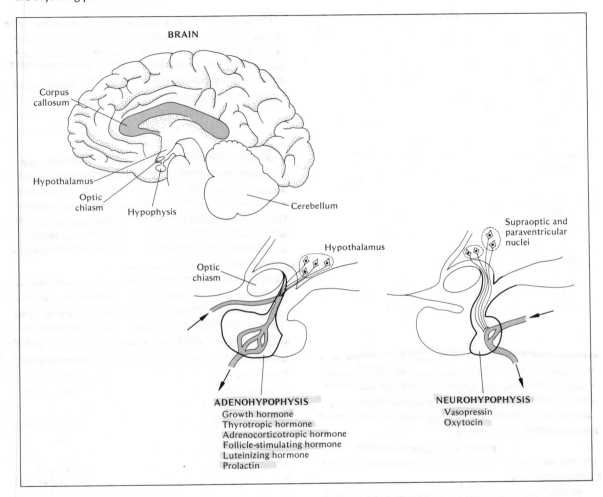

in this way been induced to drink, within minutes, 40% of their body weight in water.

The hypothalamus is of major importance in the endocrine system, for it controls the function of the hypophysis, which has been called the master gland of the endocrine system. This control is mediated to the *neurohypophysis via neural connections* and to the *adenohypophysis via special blood vessels*, known as the *portal circulation*.

The *neurohypophysis* contains two well-known hormones, vasopressin and oxytocin. It has been known for more than half a century that the neurohypophysis contains substances that affect the reabsorption of water in the kidney and are neces-

TABLE 13.6 Hypothalamic hormones that control the release of hormones from the adenohypophysis. [Schally et al. 1973]

Hormone	Abbreviation
Growth hormone releasing hormone	GH-RH
Growth hormone release-inhibiting hormone	GH-RIH
Prolactin releasing hormone	P-RH
Prolactin release-inhibiting hormone	P-RIH
Melanocyte-stimulating hormone releasing hormone	MSH-RH
Melanocyte-stimulating hormone release-inhibiting hormone	MSH-RIH
Corticotropin (ACTH) releasing hormone	C-RH
Thyrotropin releasing hormone	T-RH
Luteinizing hormone releasing hormone	LH-RH
Follicle-stimulating hormone releasing hormone	FSH-RH

sary for the formation of a concentrated urine. It is now known that the mammalian *antidiuretic hormone*, ADH, is identical with *vasopressin*, which, if injected in large and unphysiological amounts, causes a marked rise in blood pressure owing to the constriction of arterioles. Another substance in the neurohypophysis, *oxytocin*, causes contraction of the smooth muscle of the uterus in the pregnant female at term.

Vasopressin and oxytocin are octapeptides (i.e., they consist of eight amino acids). Both are formed in nerve cells near the hypothalamus and are transported along axons to the nerve endings in the neurohypophysis, from where they are released into the blood. The neurohypophysis therefore serves as a storage and release organ (a *neurohemal organ*) for the hormones that are produced in the adjacent parts of the brain.

The *adenohypophysis*, in contrast, produces hormones, and their release into the bloodstream is regulated by the hypothalamus. This control is achieved by hormones that are produced in the hypothalamus and reach the adenohypophysis via the portal circulation. It is now well established that there are at least 9 and probably 10 hypothalamic regulating hormones involved in the system that controls the hypophysis (Table 13.6).

Three of the hormones from the adenohypophysis – *growth hormone* (GH), *prolactin* (P), and *melanocyte-stimulating hormone* (MSH) – are under dual hypothalamic control, one inhibitory and one stimulatory. The release of these three hormones thus is not regulated by simple feedback systems, although feedback signals undoubtedly are involved in their control.

The regulation of release of the four other hormones seems to depend on a regular negative feedback system. *Corticotropin* (ACTH), *thyrotropin* (TSH), *luteinizing hormone* (LH), and *follicle-stimulating hormone* (FSH) have as their targets the adrenal cortex, the thyroid, and the gonads, respectively. These glands when stimulated release the appropriate hormones into the blood. These hormones (corticosteroids, thyroxine, and sex steroids) in turn inhibit, by negative feedback, the secretion from the adenohypophysis of the tropic hormones. There is now strong evidence that this inhibition acts via the hypothalamus (except that thyroxine may have a shorter feedback loop and act via the adenohypophysis).

The hypothalamic control system is summarized in Figure 13.15, which shows the central role of the hypothalamus in endocrine function, controlling the adenohypophysis through a series of regulating hormones. The adenohypophysis in turn controls the secretion from the thyroid, gonads, and adrenal cortex through a negative feedback system; three other functions are regulated by

FIGURE 13.15 Diagram of the central role of the hypothalamus in endocrine control. The neurohypophysis is controlled via neural pathways. The release of some hormones from the adenohypophysis is under dual (stimulatory and inhibitory) control (indicated by ±); the release of other hormones is controlled by negative feedback from the target organ (indicated by dashed lines and minus signs).

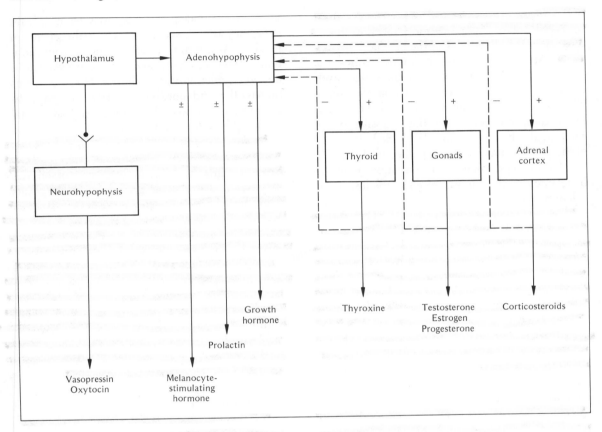

dual sets of hormones, one stimulating and the other inhibiting.

This central role of the hypothalamus prompts the question of how this important organ in turn is controlled. We know that there are extensive nerve connections to the hypothalamus from other parts of the brain and that these are influenced by many environmental as well as emotional factors, by light cycles, seasons, and so on. We can therefore conclude that the major portion of the entire endocrine system is under nervous control, acting through the central role of the hypothalamus in the control system.

Endocrine glands outside direct hypothalamic control

Parathyroids and thyroid

The parathyroid glands, which are located near or embedded in the thyroid gland, produce a hormone, *parathormone*, which causes the calcium ion concentration in the blood plasma to increase.

An antagonistic hormone, *calcitonin*, which lowers the plasma calcium ion concentration, is released from the C-cells of the thyroid in mammals. An excess of parathormone causes the plasma calcium to increase to an abnormal level, the calcium being taken primarily from the bones. This in turn results in an increased renal excretion of calcium, and because of the excessive loss of calcium the bones become decalcified. Removal of the parathyroids has the opposite effect, leading to a fall in blood calcium concentration, followed by tetanic muscle cramps caused by the low blood calcium level.

The normal secretion from the parathyroids is apparently controlled by a simple feedback mechanism, because a high plasma level of calcium inhibits release of the hormone from the parathyroids, and a low level stimulates release. The effect of the calcium level on the release of calcitonin is the opposite; a high calcium level stimulates hormone release and a low level inhibits release. The calcium ion concentration in the blood plasma is thus controlled by the balance between two negative feedback systems.

Digestive organs

Various organs of the digestive tract secrete hormones that control the secretory activity of the digestive glands. Their release is to a great extent under nervous control. In addition, the pancreas (specifically the small groups of endocrine cells known as the *islets of Langerhans*) produces two important hormones that have no direct role in digestion but are extremely important in carbohydrate metabolism. These are *insulin* and *glucagon*.

The most obvious effect of insulin is that it decreases the glucose concentration in the blood by stimulating the formation and deposition of glycogen in the cells from the blood glucose. In the disease diabetes mellitus, the production or release

of insulin is inadequate (there may be an insulin deficiency or a tissue resistance to the effects of insulin). As a result the blood glucose concentration remains too high, and glucose is excreted in the urine, often in vast quantities. However, insulin has other and more complex roles in intermediary metabolism, and there are several additional serious effects of insulin deficiency.

Insulin is produced by the *beta cells* of the islets of Langerhans; glucagon is produced by the *alpha cells*. If the glucose concentration in the blood rises above the normal level (e.g., after ingestion of carbohydrates), insulin is released, glucose uptake in the muscles and the formation of glycogen is stimulated, and blood sugar falls. The release of insulin is therefore a negative feedback control system.

If the blood glucose level falls too low, release of glucagon causes mobilization of glucose from the liver and an increase in blood glucose concentration. The glucagon effect is opposite to the effect of insulin; the normal blood glucose level thus is regulated by two antagonistic hormones, one inhibitory and one stimulating, each acting through its own negative feedback loop.

Adrenal medulla

The cells of the adrenal medulla are of neural origin and belong to the sympathetic nervous system. The adrenal medulla secretes two similar hormones, *noradrenaline* and *adrenaline* (Figure 13.19). Adrenaline causes acceleration of the heartbeat, increased blood pressure, increase in blood sugar through conversion of glycogen to glucose, vasodilation and increased blood flow in heart muscle, lungs, and skeletal muscle; but causes vasoconstriction and decreased blood flow in smooth muscle, digestive tract, and skin. These effects of adrenaline are well known as the fight or flight syndrome, which occurs in response to fear, pain, and anger. These reactions help mobilize the physical resources of the body in response to an emergency situation.

Noradrenaline has very similar effects, and the differences between the two hormones are mostly quantitative. Adrenaline, for example, has a more potent effect on the heart rate, and in some organs noradrenaline causes more vasoconstriction. The two hormones act through receptor sites known as alpha receptors (most affected by noradrenaline) and beta receptors (most affected by adrenaline); many organs have both alpha and beta receptors, and in addition each receptor type is to some degree sensitive to both hormones. This explains why it is difficult to give a definite listing of similarities and differences between the two hormones — there is too much overlap in their effects.*

The cascade effect

The control of metabolic functions by the endocrine system can lead to a cascade of step-by-step amplification that makes it possible to control a final process through minute amounts of the initial hormone. An example is shown in Figure 13.16, where the approximate amounts are given for a series of hormones that lead to deposition of glycogen. A minute amount, 0.1 μg, of the hypothalamic corticotropin-releasing hormone (C-RH) causes a series of subsequent events in which the final step is the formation of 5600 μg of glycogen in the liver. The initial amount of C-RH required is so small that the final response represents an amplification of 56 000 times.

Hormonal interactions with target cells

We have discussed the origin, release, and transport of hormones, but we have not discussed details of how they affect a target organ. For a hor-

---

* Noradrenaline has been found in the nervous systems of insects and annelids, but it seems to be absent in crustaceans, molluscs, echinoderms, and many other phyla.

FIGURE 13.16 An example of the biological amplification possible in the endocrine system. The release of a minute amount of hypothalamic hormone leads to the deposition of glycogen in the liver, with a total amplification of 56 000 times. [Data from Bentley 1976.]

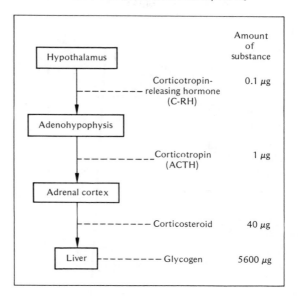

mone to have an effect on a specific target, and not on other organs exposed to the same concentration of the hormone in the blood, the responding cells must possess on their surfaces specific receptor sites that interact with the hormone, and other cells must lack these receptors.

In regard to the effect on target cells, hormones can be classified in two categories: (1) those, the catecholamines and peptide hormones, that act by controlling the formation of the cyclic nucleotide *3′,5′-adenosine monophosphate (cyclic AMP or cAMP)*; and (2) those, including the steroid and thyroid hormones, that penetrate into the cell and exert their effects directly on the cell nucleus and the mechanism for cellular protein synthesis.

The discovery of the role of cyclic AMP was made during studies of the action of adrenaline on the liver, where it causes the conversion of glycogen to glucose (Sutherland 1972). This process depends on a series of enzymes, of which one is

FIGURE 13.17 The adrenaline-induced release of glucose from the cells of the liver. Adrenaline interacts
with a receptor site in the cell membrane, which sets off
the release of cyclic AMP as a "second messenger." The
last step shown, the rate-limiting phosphorylase a, includes two additional enzymes, phosphoglucomutase
and glucose-6-phosphatase. [Sutherland 1972]

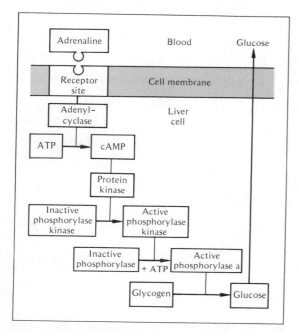

phosphorylase, which is the rate-limiting step in
the process (Figure 13.17). The active enzyme,
phosphorylase a, is formed from a precursor, phosphorylase b, which through the action of a phosphorylase kinase reacts with adenosine triphosphate (ATP) to form the active enzyme. The whole
process is initiated by the binding of adrenaline to
the receptor site in the cell membrane. This causes
the release of the enzyme *adenylcyclase,* which in
turn catalyzes the formation of cyclic AMP from
ATP.

Cyclic AMP has been called a "second messenger" in the process of hormone action, and cAMP
and adenylcyclase have been found in several different tissues from vertebrates as well as invertebrates from many different phyla and in bacteria
(Sutherland et al. 1962). The initiating event in
the action of those hormones that act via cAMP is
always the release of adenylcyclase from the hormone receptor site in the cell membrane.

The *steroid hormones,* which include the male
and female sex hormones and hormones secreted
by the adrenal cortex, act through a different
mechanism. As we saw before, the hormones affect those cells that contain the appropriate receptor sites and leave other cells unaffected. Estradiol
binds to receptors in the uterus, testosterone to
receptors in the prostate, progesterone to the ovi-
ducts of birds, and so on. However, at the cell surface these hormones form a complex with a receptor protein, and this complex is rapidly moved
from the cell membrane to the cell nucleus, where
its presence stimulates or induces the expression of
genetic information (Jensen et al. 1971).

Recent studies of the receptor sites for steroid
hormones have revealed some fascinating information. Steroid hormones secreted by the gonads of a
newborn rat can be traced to target cells in specific
areas of the brain. The hormones interact with
these receptor areas and induce the development
of nervous circuits that later control whether the
adult animal shows male or female behavior. This
sexual differentiation of nerve circuits in the brain
lays down a morphological basis for the activation
of certain types of behavior and the suppression of
others (McEwen 1976). During development the
functioning of the central nervous system is already modulated by hormones, and as we have
seen, the central nervous system is in turn a major
regulator of endocrine function throughout the
body.

Chemical transmission and transmitter substances

The effects of adrenaline and noradrenaline released into the blood from the adrenals resemble
the effects of a general stimulation of the sympa-

thetic nervous system.* This similarity is easy to understand when we realize that the cells of the adrenal medulla are of neural origin and analogous to the large sympathetic ganglia located along the spinal column. The similarity is even more clearly understood when we learn that noradrenaline is the terminal transmitter substance of the sympathetic nervous system as a whole.

We have already seen that most synaptic transmission is accomplished through the release of a transmitter substance, acetylcholine, and that acetylcholine is released from the terminals of the motor nerves at the neuromuscular junction. The autonomic nervous system also exerts its effects through transmitter substances.

The sympathetic nerves release noradrenaline from their nerve terminals, and the parasympathetic nerves release acetylcholine. These two substances have antagonistic effects; when one stimulates, the other inhibits. One of the best known examples of such antagonistic action is the control of the heartbeat. The heart is innervated by two nerves. One is a branch of the splanchnic nerve (sympathetic), which when stimulated releases noradrenaline and speeds the heartbeat; the other is a branch of the vagus nerve, which is parasympathetic and when stimulated releases acetylcholine from its nerve endings and causes the heart to slow.

The effect of acetylcholine in slowing the heart is a striking example of the fact that the response to

* The major components of the peripheral vertebrate nervous system are sensory nerves, motor nerves, and the autonomic nervous system. The autonomic nervous system in turn is divided into the *sympathetic* and the *parasympathetic* nervous systems, which jointly control the function of internal organs such as heart, stomach, glands, intestine, kidneys, and so on (i.e., functions outside voluntary control). These organs are innervated by both sympathetic and parasympathetic nerves, which act antagonistically, one being stimulatory and the other inhibitory.

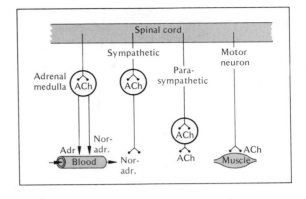

FIGURE 13.18 Diagram of the actions of peripheral transmitter substances in vertebrates. For details, see text. ACh, acetylcholine; Adr, adrenaline; Noradr, noradrenaline.

a given substance depends on the target organ rather than on the nature of the transmitter substance. Ordinary striated muscle is stimulated to contract through the release of acetylcholine; the effect of acetylcholine on the heart is to reduce the frequency of contraction.

The major aspects of transmitter action in the peripheral nervous system of a vertebrate are summarized in Figure 13.18. We can now see that acetylcholine is the common transmitter substance, not only for the terminals of the motor nerves and synapses within the parasympathetic nervous system, but also in sympathetic ganglia and for the nerves that innervate the adrenal medulla. Thus, in the nerve-to-nerve transmission outside the brain and spinal cord, acetylcholine seems to be the universal transmitter substance in vertebrates. Within the central nervous system the situation is less clear.

Although acetylcholine is so widespread in peripheral synaptic transmission, there is strong evidence that other substances function in transmission, particularly within the vertebrate central nervous system. For invertebrates several other substances, most of them amines, seem to be important transmitters. The structures of a few of these substances are shown in Figure 13.19. The

FIGURE 13.19 Structural formulas of some compounds known or believed to be transmitter substances.

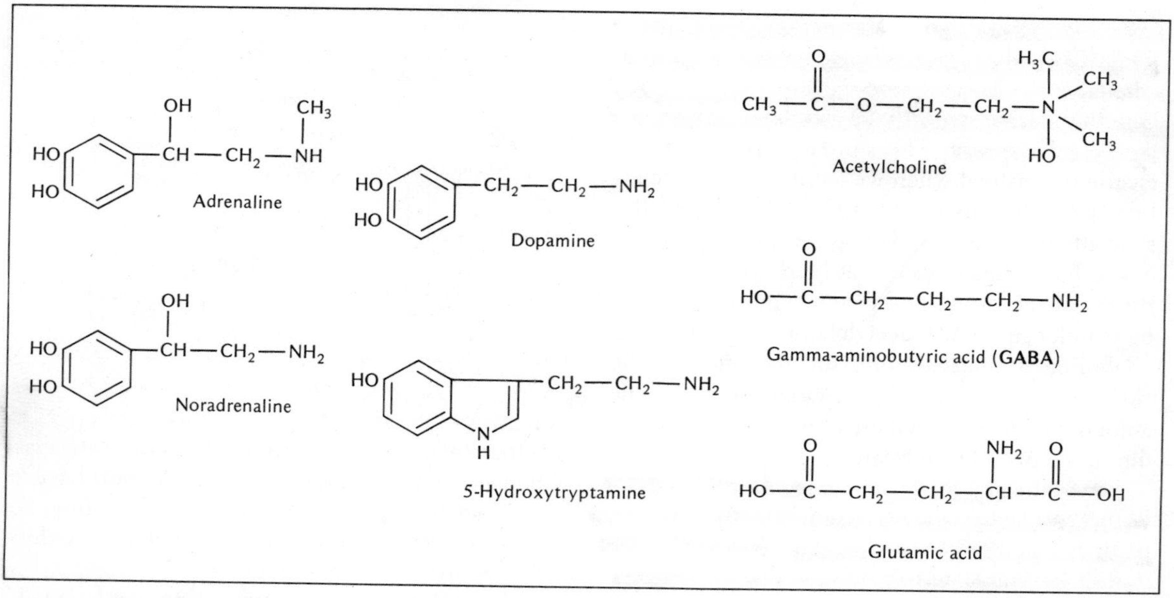

reason the information about transmitter substances is so uncertain is the extreme difficulty of establishing with certainty that a given substance has a role as a transmitter.

The greatest problem is that many substances that occur in the organism in small amounts have rather profound physiological effects, particularly on the nervous system. The fact that a substance causes a reaction that is similar to some normally occurring nervous phenomenon does not establish this substance as a transmitter. A number of criteria should be fulfilled before this can be established; among these criteria are:

1. The substance or a precursor should be present in the neuron from which the suspected transmitter is released.
2. The substance should be present in the extracellular fluid in the region of the activated synapse.
3. When applied to the postsynaptic structure, the substance should mimic the action of the transmitter.
4. A mechanism for removal or inactivation of the substance should be present, either through enzymatic inactivation or through specific uptake or reabsorption.

Although pharmacological agents have helped clarify mechanisms by interfering with synthesis, release, removal, inactivation, or target, they do not provide direct evidence for transmitter function.

The crucial demonstration of a transmitter substance consists in proving that, on the arrival of an action potential at the presynaptic ending, the substance is released in sufficient quantity to produce the observed physiological effect on the postsynaptic structure.

It is indeed difficult to meet all these criteria, and very often the technical obstacles have been insurmountable. As a result much work on suspected transmitter substances, especially in invertebrate animals, has been unsatisfactory because it has not progressed much beyond the demonstration of an effect of some suspected agent. Therefore, present information about transmitter substances is insufficient for generalizations. Nevertheless, two substances, 5-hydroxytryptamine and gamma-aminobutyric acid, should be mentioned here because they are invariably recognized as being highly active in many invertebrates.

5-Hydroxytryptamine, often written 5HT, is found in many invertebrate neurons, notably of molluscs and arthropods. 5HT is not restricted to nervous structures, but the concentration in other tissues is usually lower. Large amounts are found in certain animal toxins, such as those from octopus, the snail *Murex*, and the venom of wasps.

The heart of bivalve molluscs is quite sensitive to 5HT, and there is good evidence that this substance is a normal transmitter for acceleration of the mollusc heart. The effect of 5HT on vertebrate blood pressure (which it increases) was known before the substance was chemically identified. It was then called *serotonin*, a name that is still in use. 5HT is also believed to be a transmitter within the vertebrate central nervous system, but whether it has a normal role in vertebrates remains uncertain.

Gamma-aminobutyric acid, often written GABA, may be a transmitter substance in the central nervous system of vertebrates, and it almost certainly is a transmitter at inhibitory synapses of crustaceans. The terminals of the inhibitory nerve to crustacean muscle release GABA, and the effect can be mimicked by the direct application of GABA (Otsuka et al. 1966; Takeuchi and Takeuchi 1967).

GABA is found in rather high concentration in certain regions of the mammalian brain, but this does not prove it has a normal function as a trans-mitter substance. A great deal of research is currently underway to clarify the role of GABA and other putative neurotransmitters in the central nervous system and especially the brain.

Some amino acids, notably *glutamic acid* and *aspartic acid*, have also been suggested as transmitter substances. They occur in both vertebrate and invertebrate nervous tissues, but substantial evidence for their normal role is still lacking. One major difficulty in establishing amino acids as transmitter substances is that they also occur normally in the animals, and their mere presence therefore carries no particular significance. Another problem is the extreme difficulty of establishing an inactivation mechanism specifically for amino acids as related to their suspected role as transmitter substances, for all the amino acids are regular participants in normal metabolic pathways.

Other neuroactive substances

It has become increasingly evident that the central nervous system produces more chemical substances with specific physiological actions than anybody imagined a few years ago. A series of remarkable discoveries has come from studies of the action of such well-known opiate drugs as *morphine* and *heroin*. These are used to suppress pain, but they also cause euphoria and, as everybody knows, are addictive.

The most important basic discovery grew out of a search for specific receptors in the brain that would react with opiate drugs and thus provide a basis for further study of their characteristic effects. If such specific receptor sites could be found in the central nervous system, one would know where to begin the further search for the mechanism of action of the opiates (Pert and Snyder 1973).

The search was successful and set the stage for the next step. Because opiates do not occur naturally in animals, it was reasonable to assume that the receptors must have some other normal physiological role and that there might be endogenous materials that normally act at these sites. This hypothesis has led to the tracking down of a whole series of endogenous substances that participate in the normal functioning of the nervous system – substances that may be involved not only in the suppression of pain, but also in changes in mood and emotions, and possibly also in certain types of mental disease.

The naturally occurring substances that interact with the opiate receptors are called *endorphins*. Among them are two pentapeptides, known as *enkephalins*. Several active polypeptides with somewhat longer chains are also known, and new active substances are constantly being found. Some of them have the exact same amino acid sequence as certain fragments from longer peptide hormones, such as β-lipotropin, which serves to mobilize fat from adipose tissue and is a chain of 91 amino acids (Kosterlitz et al. 1977). Because of the demonstrated action of smaller and larger chains, it is difficult at the present time to say which specific compounds are normally active in the nervous system and which are precursors or breakdown products.

There is no reason to believe that these developments are in any way the end of neurochemical discoveries; on the contrary, we can expect that exciting new principles of interaction will be exposed in the field of neurochemistry.

We will now turn to a few examples of endocrine function among invertebrates. Among these animals the role of the nervous system in the production of hormones and control of endocrine function is even more prominent than among vertebrates.

CONTROL AND INTEGRATION IN INVERTEBRATES

In the more highly organized invertebrates physiological function is under endocrine as well as nervous control. As in vertebrates, the nervous system serves for rapid communication, which is essential for actions related to escape, feeding, mating, and so on. As in vertebrates, an endocrine system produces hormones that control slower processes, such as growth, maturation, and many other metabolic functions. The nervous system has a direct and primary role in hormone production, and the connection between the nervous and endocrine systems is even closer than in vertebrates.

Function and role of nervous systems

We have seen that in vertebrates the central nervous system coordinates a large number of separately controlled functions, both nervous and endocrine. The apparent contrast between nervous control (usually rapid) and hormonal control (usually slow to very slow) breaks down under closer scrutiny. Not only is the transmission from neuron to neuron, and from neuron to effector, usually of a chemical nature, but the central nervous system itself is a major producer of hormones and exerts much of its overall regulatory function in this way.

When we examine the role of nervous systems in the less highly organized invertebrates, we will need a more precise definition of nervous system than we used for vertebrates, where the meaning of the term is self-evident.

A nervous system can be defined as an assembly of neurons specialized for repeated and organized transmission of information from sensory receptor sites to neurons, or between neurons, or from neurons to effectors (e.g., muscles, glands, etc.). This definition precludes the existence of a nervous system in unicellular animals (protozoans), although functions within such cells may be coordinated with the aid of conducting organelles.

It is, on the whole, reasonably easy to recognize a neuron or nerve cell, based on its membrane potential and the ability to produce repeated action potentials. However, it may be difficult to recognize whether a "neurosecretory" cell is truly a modified nerve cell. Its location in association with nerve cells is helpful, and further information can be gained from recognition of action potentials, structural characteristics, and the function of similar structures in related species.

Invertebrate neurons have the same characteristics as vertebrate neurons. They are able to produce *action potentials* that consist of changes in membrane permeability and potential. Such action potentials are of an *all-or-none* nature and can be propagated in a nerve fiber, or axon. Typical neurons are able to produce repeated brief action potentials that last from a fraction of a millisecond to several milliseconds, and such neurons occur in all major phyla except protozoans and sponges.

The transfer of information from one neuron to another takes place at *synapses*, which usually show evidence of *chemical transmission*. As in vertebrates, the nerve membranes of the two neurons most often remain spatially separated, and transmission is achieved through the release of an intermediary *chemical transmitter substance*. It appears, however, that there may be direct electric transmission at certain synapses where the nerve membranes of the two cells become contiguous, a situation that also prevails at some vertebrate synapses.

Typical *postsynaptic potentials* seem to be universally present in at least four major phyla: molluscs, annelids, arthropods, and chordates. These may be *excitatory* or *inhibitory postsynaptic poten-tials*. In general, the excitatory postsynaptic potentials consist in a decrease in membrane polarization (in the direction of complete depolarization), and inhibitory potentials consist in an increase in membrane potential. These general characteristics are in principle similar in invertebrate and vertebrate neurons.

Neurosecretion

The importance of specialized neurosecretory cells and organs was recognized in invertebrates long before their importance was generally established in vertebrates. Neurosecretory cells are difficult to define and characterize so that they are easily and unequivocally recognized; the major approach for locating and pinpointing such cells still depends on the use of certain stains and other substances that have a special affinity for neurosecretory cells.

Neurohemal organs

Neurosecretory organs are groups of neurons that are the source of secretion. These cells have nerve fibers in which the secreted agent is transported and which usually terminate in close association with a vascular structure. Here they form a *neurohemal organ*, where the secreted product is stored and released. Neurohemal organs are very widespread, but it is not certain that they are present in connection with all neurosecretory organs.

In all the more highly organized animal phyla neurosecretory systems are central to the control and operation of endocrine mechanisms. We have already seen the importance of neurosecretion in the control of vertebrate endocrine organs; among invertebrates neurosecretion seems to be even more important.

We shall discuss later in this chapter examples

of endocrine function in some insects, where neurosecretion and neurohemal organs have a central role in the control systems. Neurosecretory systems are also important in annelids and crustaceans, in which they are involved in control of reproduction, metabolism, molting, pigmentation, and so on.

Although some of the endocrine glands of vertebrates as well as invertebrates are outside direct neuroendocrine control, the normal function of most or all glands ultimately depends on regulation from the nervous system, either directly through nerve pathways or through hormonal mechanisms. It seems that control by neurosecretory mechanisms is the general rule in all animals in which endocrine mechanisms play a major role.

Hormones and endocrine function

Endocrine function is best understood and has been most extensively studied in those invertebrates that are morphologically highly organized. There are two reasons for this. One is that a highly organized animal such as an insect needs more detailed control and integration than a less organized animal such as a sea anemone. The other reason is that in the more highly organized animals many functions are delegated to specialized organs. This permits experiments such as removal of organs, reimplantation, extraction, and other procedures that are difficult or impossible in less highly organized animals.

Most invertebrate hormones are different from vertebrate hormones, both in chemical constitution and in effects. Many vertebrate hormones that profoundly affect vertebrate growth, development of gonads, metabolic processes, and so on, have no effect whatever if introduced into invertebrate animals.

Endocrine function has been demonstrated in many invertebrates, but the mere fact that some

extract has a physiological effect does not prove that the extract contains a substance that has a normal role as a hormone in this animal. More careful study is required, and the necessary criteria include demonstration of a normal occurrence of the substance in question, demonstration of where it originates, examination of the effects of removing its source or blocking its secretion, isolation and purification of the substance, and eventually its synthesis, including demonstration that the synthetic product is effective in similar concentrations to the natural product. As a result, detailed knowledge of invertebrate endocrinology is restricted to a relatively few forms.

The groups in which organs of internal secretion have been clearly demonstrated and are reasonably well understood are molluscs (notably cephalopods), annelids, crustaceans, insects, and tunicates. All these are highly organized animals where the complexity of function and the specialization of organs facilitate endocrine research. In particular, insects have been studied in great detail for reasons that include their easy accessibility, the ease of collecting and breeding them, their tolerance to drastic surgical procedures, and their economic importance.

Insect endocrinology

Hormones play a major role in the physiology of insects, especially in growth, molting, pupation, and metamorphosis into the mature adult form. These phenomena have been studied in many different species, but considering that there may be around 1 million different insect species, the fraction is very small indeed. However, there is a striking similarity in the endocrine function of different insects. For example, the *hemimetabolic* insects*,

---

* The preferred form is hemimetabolic, rather than hemimetabolous. The latter is less desirable because it mixes a Greek prefix with a Latin ending.

which go through a number of molts as they change from the newly hatched form and gradually develop into adults, employ the same hormones as the *holometabolic insects*, which remain in the completely larval form through a number of molts, and then transform into pupae from which the fully formed adults emerge. We shall discuss both types.

As an example of a hemimetabolic insect we shall use the South American bloodsucking bug, *Rhodnius*, a relative of the ordinary bedbug. *Rhodnius* hatches from the egg as a tiny bloodsucking bug, a nymph, which through five stages (*instars*) gradually develops into the adult form. In each stage the nymph must obtain a meal of blood before it can develop into the next stage by shedding its old cuticle. After it has sucked blood, about 4 weeks pass by; the nymph then sheds the old cuticle and increases in size by filling the tracheal system with air before the new cuticle hardens. It is then ready to suck blood again, but if the opportunity does not occur, it can survive for many months. However, if it obtains blood, it molts again about 4 weeks later.

The molt of *Rhodnius* is stimulated by a hormone, *ecdysone* (also known as molting hormone), which is secreted after the blood meal has been ingested. This hormone is secreted by the two *prothoracic glands*, located in the thorax of the insect. The prothoracic glands, in turn, are stimulated by a hormone secreted by specialized neurosecretory cells in the brain.

The adult *Rhodnius* has wings and mature gonads, and differs from the larval forms in other respects as well. What directs this development? The lack of adult characters in the earlier nymphal stages is attributable to another hormone, the *juvenile hormone*, which is secreted by the *corpus allatum*, a tiny cluster of cells just behind the brain.

The juvenile hormone determines that the new cuticle will have nymphal characteristics, and its presence prevents the formation of adult characters. Thus, in all the early nymphal stages, when juvenile hormone is present, ecdysone causes molting that results in a larger nymph. In the fourth nymphal stage the amount of juvenile hormone declines, and the fifth nymphal stage shows the beginning of the development of wings. In the very last molt the juvenile hormone is absent, and an adult emerges (Figure 13.20).

Let us examine the evidence for endocrine control of these events. If a fifth-stage *Rhodnius* is decapitated shortly after a blood meal, the final molt does not take place, although the headless animal may live for more than a year. This makes it appear that the molting hormone, ecdysone, is produced in the head. However, if the decapitation takes place after a certain "critical period," which occurs around the seventh day after the meal, the animal will molt and develop into a headless adult. Thus, the head is needed only early in the period and not for the molt itself.

The difference can be shown to result from endocrine factors, for if blood from an insect decapitated after the "critical period" is transferred to an insect decapitated before the "critical period", the latter is induced to molt. The explanation is that during the critical period the brain releases a hormone, the *brain hormone*,[*] which stimulates the prothoracic gland. If the decapitation occurs early, the prothoracic gland is never stimulated; if decapitation is later, the brain hormone has taken effect and is not needed, and ecdysone is released and induces molting in due time.

The simplest way of performing a blood trans-

---

[*] The brain is the source of several hormones, but it is conventional among insect endocrinologists to use the term *brain hormone* for the agent that specifically activates the prothoracic gland and leads to release of ecdysone. It is also called the *prothoracicotropic hormone*.

FIGURE 13.20 (top) The bloodsucking bug *Rhodnius* develops into an adult through five molts. Each molt is caused by release of the hormone ecdysone from the prothoracic glands. In the first four molts the juvenile hormone (from the corpora allata) causes the new cuticle to be of the larval type, thus preventing the formation of an adult. In the last molt the absence of juvenile hormone results in a fully developed adult.

FIGURE 13.21 (bottom) The transfer of hormones via the blood can be demonstrated by joining together two animals. In this case a fourth-stage *Rhodnius* nymph (*left*) with only the tip of the head removed (leaving the brain intact) has been joined to a headless fifth-stage nymph (*right*). [Wigglesworth 1959]

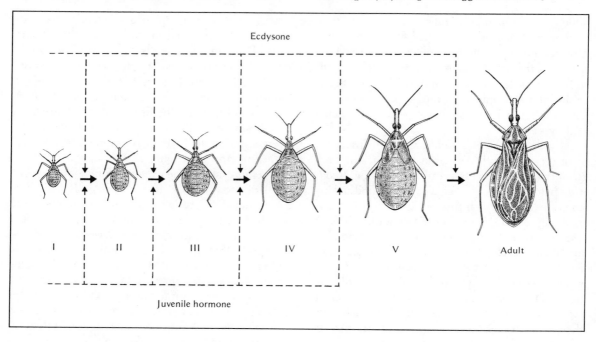

Ecdysone

I II III IV V Adult

Juvenile hormone

fusion in these insects is to connect them with a short glass capillary that can be fixed into the neck of the decapitated insect with a droplet of wax (Figure 13.21). If we join together a fifth-stage nymph that has had a blood meal and has passed through the critical period, and a small first-stage nymph recently emerged from the egg and not yet fed, the small nymph will develop adult characteristics such as wings and genital organs and appear as a midget adult. If the experiment is repeated with an older first-stage nymph, this animal is again induced to molt, but does not acquire adult characteristics because by this time it has produced and released enough juvenile hormone to molt into a juvenile nymph.

Ecdysone is obviously needed for each molt, and its effects are merely modified by the juvenile

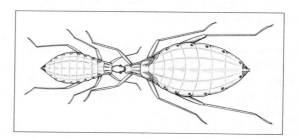

hormone, which is secreted later. Therefore, if decapitation is carried out after the brain hormone has been released and has stimulated the prothoracic gland, molting is induced; the type of molt depends on whether the decapitation occurred before or after juvenile hormone was secreted. By accurately timing the decapitation in relation to the

FIGURE 13.22 Endocrine organs that control development of the cecropia moth. Each molt is caused by the hormone ecdysone, which is released from the prothoracic gland. This gland in turn is controlled by the brain hormone, produced by neurosecretory cells in the brain. The juvenile hormone, secreted from the corpora allata, causes the new cuticle to be of the larval type; in the last larval molt less juvenile hormone is present, and a pupa with a heavier pupal cuticle is formed; in the final molt juvenile hormone is absent and an adult moth emerges. [Schneiderman and Gilbert 1964.]

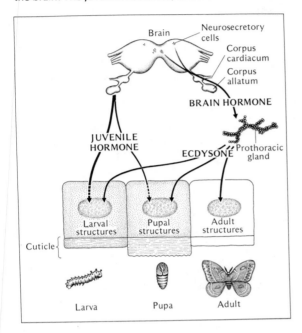

pupa). The larval tissues break down and are transformed into pupal structures, and after a final molt, when no juvenile hormone remains, a fully developed adult moth emerges (Figure 13.22).

The simplest way to confirm the function of the corpus allatum is to remove this organ in the young silkworm during one of the early instars. If this is done, molting takes place as usual (the brain–prothoracic system is untouched), but instead of developing another larva, the molt results in a diminutive pupa, from which, in turn, a tiny adult emerges (Figure 13.23). There is no other effect of the removal of the corpus allatum (and thus the juvenile hormone) so that, except for size, a normal-looking adult develops from the pupa.

The role of the corpus allatum can be confirmed by transplantation. If several corpora allata from early larvae are implanted into the last larval instar, the next molt does not produce a pupa, but instead an oversized larva. This larva may continue to grow, and by introducing additional corpora allata it is possible to produce a giant larva that can molt into a giant adult. There is a limit to this development, and it is unlikely that an animal will survive beyond an artificial seventh instar.

The chemical structure of the hormones we have discussed should be mentioned for comparison with vertebrate hormones. Ecdysone is a steroid (Karlson and Sekeris 1966) that was originally isolated and crystallized by extraction of 500 kg silkworm pupae, which yielded 25 mg of the crystalline hormone. The full chemical structure of the hormone was finally established on material extracted from a batch of 1000 kg dried silkworm pupae (corresponding to about 4 ton fresh weight). It is interesting that ecdysone is a steroid because of the widespread occurrence of hormones of steroid nature in vertebrates.

The juvenile hormone, on the other hand, is entirely different and is similar in structure to ter-

blood meal and the critical period, it has been possible to obtain small, headless *Rhodnius* with adult morphological characteristics.

In holometabolic insects the events are similar. Endocrine function has been extensively studied in several moths, including the silk moth (*Bombyx mori*) and the cecropia moth (*Hyalophora cecropia*). The larva of the cecropia moth goes through four molts (five stages), all initiated by the brain hormone, which causes the prothoracic gland to release ecdysone.

In the younger larval stages the *corpora allata* release juvenile hormone, so that the larva molts into another larva, which because of its loose and wrinkled skin can grow and increase in size as it feeds. At the end of the fifth larval stage, the corpora allata cease to secrete juvenile hormone, and at the next molt, a harder cuticle is formed (i.e., a

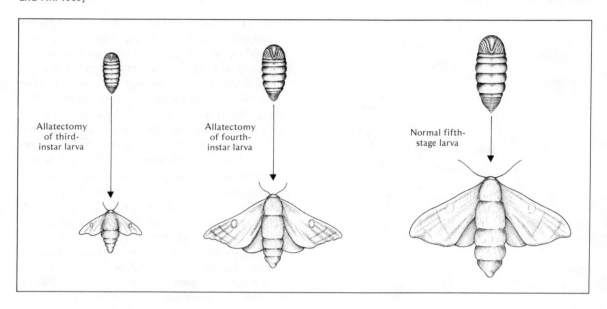

Allatectomy of third-instar larva

Allatectomy of fourth-instar larva

Normal fifth-stage larva

penes. A large number of compounds with juvenile hormone activity have been isolated, and many have been synthesized, some apparently far more potent than the naturally occurring ones. The major natural hormone from cecropia has the following structure (Röller et al. 1967):

The synthesis of compounds with juvenile hormone activity is very challenging because they may be potential insecticides. These compounds are effective in extremely small amounts, and if they are applied at suitable periods during the life history of insects, they can prevent normal adult development and thus reproduction. The advantage of juvenile hormones as insecticides is that insects are less likely to develop immunity to substances on

which they normally depend than to the various poisonous substances now in use. Because some of the artificial juvenile hormones act more specifically on some insects than on others, it may be feasible to develop insecticides that are more specific for one insect species than for another – a far more attractive prospect than the use of highly toxic substances such as DDT, which kill desirable and undesirable insects alike. Also, as juvenile hormone has little or no effect on many other animals, at least not on vertebrate predators, such substances probably will have less disastrous ecological effects.

An accidental discovery of a naturally occurring juvenile hormone substance suggests that plants in fact already use such substance for their own protection. The discovery was made when a Czech investigator, Dr. Karel Sláma, spent a year at Harvard University and brought with him his stock of the bug *Pyrrhocoris*. In the laboratory these bugs

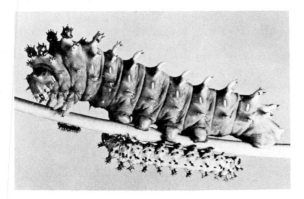

SILK WORM The larva of the silk moth, *Hyalophora cecropia,* in the first, third, and fifth instars. After the fifth instar the larva pupates and goes through a complete metamorphosis, whereupon the adult moth emerges. [Courtesy of Charles Walcott, State University of New York, Stony Brook]

are reared on linden seeds with paper towels as a climbing surface. At Harvard they would not develop into adults after the fifth instar, but developed into oversized sixth or even seventh instar immature forms.

It appeared as if the bugs had unintentionally been exposed to contamination with juvenile hormone, and a systematic comparison of all possible differences between conditions in the United States and in Europe revealed that the paper towels used at Harvard contained a substance with juvenile hormone activity (Sláma and Williams 1965, 1966). It turned out that all American-made paper, including newsprint, contained the substance. It was not added during manufacture, but originated in the wood from fir trees used for paper manufacture in the United States. Paper of Japanese or European origin had no such effect. The substance, called the *paper factor*, has been isolated and has some chemical resemblance to the juvenile hormone. This suggests that many terpenes, which occur commonly in plant material, especially in evergreen trees, may be natural protective substances that have evolved as a countermeasure to insect pests.

The insect hormones discussed so far are all concerned with the control of growth and develop-

ment, but many other physiological processes are also under endocrine control. As an example, we shall briefly describe a hormone concerned with the regulation of water excretion in a bloodsucking insect, which, because of its feeding habits, is periodically subjected to a heavy water load.

The bloodsucking bug *Rhodnius* is faced with this problem each time it takes a blood meal. Immediately after feeding it excretes a large volume of dilute urine, eliminating most of the excess water. This sudden surge in the excretion of water is regulated by a hormone that stimulates water excretion.

The action of this hormone has been studied on isolated Malpighian tubules of *Rhodnius*. These organs can continue to secrete urine if they are placed in blood from *Rhodnius* or in a suitable saline solution. If they are placed in the blood from a recently fed *Rhodnius,* the tubules secrete a copious urine at a much higher rate than if they are placed in blood from an unfed animal. This suggests that the blood contains a hormone which is directly responsible for stimulating the secretion of urine (i.e., a *diuretic hormone*).

Extracts of various known hormone-producing organs of *Rhodnius* have no effect on urine secretion, but extracts of the ganglia from the first abdominal segment are effective. These ganglia contain several groups of large neurosecretory cells, and 97% of the diuretic activity is contained in the most posterior group of these.

After a *Rhodnius* has sucked blood, urine production increases within less than a minute, showing that the hormone must be released very rapidly, undoubtedly stimulated by nervous reflexes from the feeding process or from the abdominal expansion caused by the feeding (Maddrell 1963).

These few examples of insect hormones give merely an indication of the complexity of endocrine regulation. The importance of the nervous system in hormone production is remarkable, not only in insects, but in many other invertebrates. An increasing number of hormones have been isolated, chemically identified, and finally synthesized, but the field of invertebrate endocrinology is so vast that this area of physiological research will continue to expand and remain an area of fruitful research for many years to come.

REFERENCES

Aidley, D. J. (1971) *The Physiology of Excitable Cells.* Cambridge: Cambridge University Press. 468 pp.

Bentley, P. J. (1976) *Comparative Vertebrate Endocrinology.* Cambridge: Cambridge University Press. 415 pp.

Bullock, T. H. and Horridge, G. A. (1965) *Structure and Function in the Nervous Systems of Invertebrates,* vols. I and II. San Francisco: Freeman. 1719 pp.

del Castillo, J., and Katz, B. (1954) Quantal components of the end-plate potential. *J. Physiol.* 124:560–573.

Diamond, J. (1971) The Mauthner cell. In *Fish Physiology,* vol. 5, *Sensory Systems and Electric Organs* (W. S. Hoar and D. J. Randall, eds.), pp. 265–346. New York: Academic Press.

Eccles, J. (1965) The synapse. *Sci. Am. 212:*56–66.

Furshpan, E. J., and Potter, D. D. (1959) Transmission at the giant motor synapses of the crayfish. *J. Physiol.* 145:289–325.

Heuser, J. E., and Doggenweiler, C. F. (1966) The fine structural organization of nerve fibers, sheaths, and glial cells in the prawn, *Palaemonetes vulgaris. J. Cell. Biol.* 30:381–403.

Highnam, K. C., and Hill, L. (1969) *The Comparative Endocrinology of the Invertebrates.* London: Edward Arnold. 270 pp. 2 ed. (1977), 368 pp.

Hodgkin, A. L. (1954) A note on conduction velocity. *J. Physiol.* 125:221–224.

Hodler, J., Stämpfli, R., and Tasaki, I. (1951) Ueber die Wirkung internodaler Abkühlung auf die Erregungsleitung in der isolierten markhaltigen Nervenfaser des Frosches. *Pflügers Arch.* 253:380–385.

Holmes, W., Pumphrey, R. J., and Young, J. Z. (1941) The structure and conductance velocity of the medullated nerve fibres of prawns. *J. Exp. Biol. 18:*50–54.

Hursh, J. B. (1939) Conduction velocity and diameter of nerve fibers. *Am. J. Physiol.* 127:131–139.

Jensen, E. V., Numata, M., Brecher, P. I., and Desombre, E. R. (1971) Hormone-receptor interaction as a guide to biochemical mechanism. In *The Biochemistry of Steroid Hormone Action* (Biochemical Society Symposium No. 32) (R. M. S. Smellie, ed.), pp. 133–159. London: Academic Press.

Karlson, P., and Sekeris, C. E. (1966) Ecdysone, an insect steroid hormone, and its mode of action. *Recent Prog. Horm. Res.* 22:473–502.

Katz, B., and Miledi, R. (1967) A study of synaptic transmission in the absence of nerve impulses. *J. Physiol.* 192:407–436.

Kosterlitz, H. W., Hughes, J., Lord, J. A. H., and Waterfield, A. A. (1977) Enkephalins, endorphins, and opiate receptors. *Soc. Neurosci. Symp.* 2:291–307.

Maddrell, S. H. P. (1963) Excretion in the blood-sucking bug, *Rhodnius prolixus* Stål. *J. Exp. Biol.* 40:247–256.

McEwen, B. S. (1976) Interactions between hormones and nerve tissue. *Sci. Am. 235:*48–58.

Merton, P. A. (1972) How we control the contraction of our muscles. *Sci. Am. 226:*30–37.

Otsuka, M., Iversen, L. L., Hall, Z. W., and Kravitz, E. A. (1966) Release of gamma-aminobutyric acid from inhibitory nerves of lobster. *Proc. Natl. Acad. Sci. U.S.A.* 56:1110–1115.

Pert, C. B., and Snyder, S. H. (1973) Opiate receptor: Demonstration in nervous tissue. *Science* 179:1011–1014.

Röller, H., Dahm, K. H., Sweely, C. C., and Trost, B. M. (1967) The structure of the juvenile hormone. *Angew. Chem., (Engl.)* 6:179–180.

Rushton, W. A. H. (1951) A theory of the effects of fibre size in the medullated nerve. *J. Physiol.* 115:101–122.

Schally, A., Arimura, A., and Kastin, A. J. (1973) Hypothalamic regulatory hormones. *Science* 179:341–349.

Schneiderman, H. A., and Gilbert, L. I. (1964) Control of growth and development in insects. *Science* 143:325–333.

Sláma, K., and Williams, C. M. (1965) Juvenile hormone activity for the bug *Pyrrhocoris apterus*. *Proc. Natl. Acad. Sci. U.S.A.* 54:411–414.

Sláma, K., and Williams, C. M. (1966) The juvenile hormone. 5. The sensitivity of the bug, *Pyrrhocoris apterus*, to a hormonally active factor in American paperpulp. *Biol. Bull.* 130:235–246.

Stevens, C. F. (1966) *Neurophysiology: A Primer*. New York: Wiley.

Stewart, J., and Li, C. H. (1962) On the use of -tropin or -trophin in connection with anterior pituitary hormones. *Science* 137:336–337.

Sutherland, E. W. (1972) Studies on the mechanism of hormone action. *Science* 177:401–408.

Sutherland, E. W., Rall, T. W., and Menon, T. (1962) Adenyl cyclase. 1. Distribution, preparation, and properties. *J. Biol. Chem.* 237:1220–1227.

Takeuchi, A., and Takeuchi, N. (1967) Anion permeability of the inhibitory post-synaptic membrane of the crayfish neuromuscular junction. *J. Physiol.* 191:575–590.

Tasaki, I. (1959) Conduction of the nerve impulse. In *Handbook of Physiology*, sect. 1; *Neurophysiology*, vol. 1 (J. Field, H. W. Magoun, and V. C. Hall, eds.), pp.75–121, Washington, D.C.: American Physiological Society.

Wigglesworth, V. B. (1959) *The Control of Growth and Form: A Study of the Epidermal Cell in an Insect*, Ithaca, N.Y.: Cornell University Press. 140 pp.

ADDITIONAL READING

Aidley, D. J. (1971) *The Physiology of Excitable Cells*. Cambridge: Cambridge University Press. 468 pp.

Barrington, E. J. W. (1975) *An Introduction to General and Comparative Endocrinology*, 2nd ed. Oxford: Clarendon Press.

Barth, R. H., and Lester, L. J. (1973) Neuro-hormonal control of sexual behavior in insects. *Annu. Rev. Entomol.* 18:445–472.

Benson, G. K., and Phillips, J. G. (eds.) (1970) *Hormones and the Environment*. London: Cambridge University Press. 629 pp.

Bentley, P. J. (1976) *Comparative Vertebrate Endocrinology*. Cambridge: Cambridge University Press. 415 pp.

Bullock, T. H., and Horridge, G. A. (1965) *Structure and Function in the Nervous Systems of Invertebrates*, vols. 1 and 2. San Francisco: Freeman. 1719 pp.

Cooke, I., and Lipkin, M., Jr. (1972) *Cellular Neurophysiology: A Source Book*. New York: Holt. 1039 pp.

Eccles, J. C. (1973) *The Understanding of the Brain*. New York: McGraw-Hill. 238 pp.

Field, J., Magoun, H. W., and Hall, V. C. (eds.) (1959–1960) *Handbook of Physiology*, sect. 1, *Neurophysiology*: vol. 1, pp. 1–780 (1959); vol. 2, pp. 781–1440 (1960); vol. 3, pp. 1441–1966 (1960). Washington, D.C.: American Physiological Society.

Ganong, W. F., and Martini, L. (eds.) (1978) *Frontiers in Neuroendocrinology*, vol. 5. New York: Raven Press. 416 pp.

Gershenfeld, H. M. (1973) Chemical transmission in invertebrate central nervous systems and neuromuscular junctions. *Physiol. Rev.* 53:1–119.

Gilbert, L. I. (ed.) (1975) *The Juvenile Hormones*. New York: Plenum Press.

Greep, R. O., and Astwood, E. B. (eds.) (1972–1976). *Handbook of Physiology*, sect. 7, *Endocrinology*, vols. 1–7. Washington, D.C.: American Physiological Society.

Highnam, K. C., and Hill, L. (1977) *The Comparative Endocrinology of the Invertebrates*, 2nd ed. Baltimore: University Park Press. 368 pp.

Holmes, R. L., and Ball, J. N. (1974) *The Pituitary Gland: A Comparative Account*. Cambridge: Cambridge University Press. 397 pp.

Kuffler, S. W., and Nicholls, J. G. (1976) *From Neuron to Brain*. Sunderland, Mass.: Sinauer. 486 pp.

Roeder, K. D. (1967) *Nerve Cells and Insect Behavior*, rev. ed. Cambridge, Mass: Harvard University Press. 238 pp.

Sutherland, E. W. (1972) Studies on the mechanism of hormone action. *Science* 177:401–408.

Turner, C. D. and Bagnara, J. T. (1976) *General Endocrinology*, 6th ed. Philadelphia: Saunders. 596 pp.

Wigglesworth, V. B. (1970) *Insect Hormones*. Edinburgh: Oliver & Boyd. 159 pp.

APPENDIXES

Measurements and units

TABLE A.1 Base units.

Physical quantity	SI base unit	Symbol
length	meter	m
mass	kilogram	kg
time	second	s
thermodynamic temperature	kelvin	K
amount of substance	mole	mol
electric current	ampere	A
luminous intensity	candela	cd

The purpose of physiological measurement is to determine the magnitude of certain physical quantities. To express such measurements, we need generally accepted units. Such units, their symbols, and the mathematical signs connected with their use, are standarized and internationally adopted through a long-standing permanent international body, *The General Conference on Weights and Measures* (Conférence Générale des Poids et Mésures). The system of units is known as *The International System of Units*, or commonly the *SI System* (=Système Internationale).

This appendix provides (1) names of units and their symbols, (2) a brief list of conversion factors, and (3) a list of some physical quantities and the recommended symbols for their designation.

UNITS

The SI System uses seven *base units* for seven dimensionally independent basic physical quantities; each base unit is represented by a *symbol*. Use of these symbols is *mandatory* (Table A.1).

The SI System recognizes several *derived units* (as well as some supplementary units). Although some of these have special symbols, they are derived from and defined in terms of the base units. A few of particular interest in physiology are listed in Table A.2.

529

TABLE A.2 Derived units.

Physical quantity	SI derived unit	Symbol for unit	Definition in terms of base units
force	newton	N	$m \ kg \ s^{-2}$
energy	joule	J	$m^2 \ kg \ s^{-2}$
power	watt	W	$m^2 \ kg \ s^{-3}$
pressure	pascal	Pa	$m^{-1} \ kg \ s^{-2}$
viscosity	poise	P	$m^{-1} \ kg \ s^{-1} \times 10^{-1}$
electric potential difference	volt	V	$m^2 \ kg \ s^{-3} \ A^{-1}$
electric resistance	ohm	Ω	$m^2 \ kg \ s^{-3} \ A^{-2}$
electric charge	coulomb	C	$s \ A$
frequency	hertz	Hz	s^{-1}

The following should be noted about the use of these units and symbols:

1. The symbols are printed in roman (upright) letters. The symbols are not abbreviations and are not followed by a period, except at the end of a sentence.
2. The plural form of a symbol is unchanged, and the ending -*s* should not be used.
3. Certain units are derived from proper names (e.g., Volta); the symbols for these units are capital roman (upright) letters (V), but the unit names are not capitalized (volt).

There is frequently need for smaller and larger decimal multiples of the units. These are obtained by the use of the prefixes listed in Table A.3.

These prefixes may be attached to any SI base or derived unit. (Large multiples of seconds, however, are rarely expressed this way.) The combination of prefix and symbol is regarded as a single symbol, and compound prefixes are not allowed.

The decimal sign is usually printed as a point (.) in English, but as a comma (,) in French and German. In the SI system a point or comma should be used only for the decimal sign. To facilitate the reading of long numbers the digits may be divided into groups of three, but not separated by commas.

Example: 212 489.143 but not 212,489.143

TABLE A.3 Prefixes.

Prefix	Symbol	Equivalent
tera	T	10^{12}
giga	G	10^{9}
mega	M	10^{6}
kilo	k	10^{3}
deci	d	10^{-1}
centi	c	10^{-2}
milli	m	10^{-3}
micro	μ	10^{-6}
nano	n	10^{-9}
pico	p	10^{-12}
femto	f	10^{-15}
atto	a	10^{-18}

When the decimal sign is placed before the first digit of a number, a zero should always be placed before the decimal sign.

Example: 0.143 872 but not .143 872

Products or quotients of two or more units are treated according to the rules of mathematics. Note that the use of more than one solidus(/) is ambiguous. Thus, a specific rate of oxygen consumption can be written:

ml/(kg s), or ml (kg s)$^{-1}$, or ml kg^{-1} s^{-1}
but *not* ml/kg/s

Several circumstances make it impractical, at the present time, to express all physiological measurements in SI base and derived units. One reason is the large body of existing information, including widely known and used tables of physical and chemical constants, which uses other units (e.g., calorie). Another reason is that methods in common use make it simpler to use certain traditional units (e.g., gram-force, rather than the SI unit, the newton). This practice should progressively be abandoned in favor of SI units.

USEFUL CONVERSION FACTORS

Traditional units are commonly used in those fields of physiology that concern energy (e.g., in the area of metabolic rates) and pressure (e.g., in the area of gas exchange).

Force
1 kilogram-force = 9.807 N
1 newton = 0.102 kgf

Energy, work (force × distance)
1 cal = 4.184 J
1 joule = 0.239 cal

1 liter O_2 (at 0 °C, 760 mm Hg), when used in oxidative metabolism, is often equated with 4.8 kcal; thus 1 liter $O_2 \approx 20.083$ kJ

Power (work per unit time)
1 cal s^{-1} = 4.184 J s^{-1} = 4.184 W
1 kcal h^{-1} = 1.163 W
1 watt = 0.239 cal s^{-1} = 14.34 cal min^{-1}
 = 860.4 cal h^{-1}

Pressure (force per unit area)
1 atm = 1.013 × 10^5 N m^{-2} = 101.3 kPa
1 mm Hg at 0 °C (1 torr) = 1.333 × 10^2 N m^{-2} = 133.3 Pa
1 kN m^{-2} = 7.50 mm Hg = 9.869 × 10^{-3} atm

Volume
1 liter is (since 1964) defined as exactly equal to

1 dm^3 (0.001m^3); thus, ml = cm^3 and μl = mm^3
1 U.S. gallon = 3.785 liter
1 U.S. quart = 0.946 liter

Length
1 inch = 25.4 mm
1 foot = 304.8 mm = 0.3048 m
1 yard = 0.914 m
1 mile = 1609.3 m
1 ångström (Å) = 10^{-10} m = 0.1 nm

Mass
1 pound (avoirdupois) = 0.4536 kg
1 ounce = 0.0283 kg

Temperature
The temperature unit of 1 degree Celsius (°C)* equals the temperature unit of 1 kelvin (K).
The Celsius temperature is defined as the excess of the thermodynamic temperature over 273.15 K.
For practical purposes, the melting point of ice at 1 atm pressure is 0 °C; the triple point of ice, water, and water vapor is 0.01 °C; and the boiling point of water at 1 atm pressure is 100 °C.

PHYSICAL QUANTITIES

We often need symbols to represent physical quantities. This is especially true when relationships are expressed in the form of equations. An example is Poiseuille's law, which states that the rate of flow ($\dot{Q}$) of a fluid through a tube is proportional to the pressure (p) and inversely proportional to the resistance (R), or $\dot{Q} = p/R$.

The names of some physical quantities and *recommended* symbols are given in Table A.4.

The symbol for a physical quantity should be a

* Also known as "centigrade."

TABLE A.4 Physical quantities and recommended symbols. Alternative symbols are separated by commas.

Physical quantity	Symbol
Length	l
Height	h
Radius	r
Area	A
Volume	V
Mass	m
Density (m/V)	ρ
Pressure	p
Time	t
Velocity	u, v
Frequency	f
Force	F
Work	W
Energy	E, W
Power	P
Temperature	T

single letter of the Latin or Greek alphabet and should be printed in sloping (italic) type. When necessary, subscripts (and/or superscripts) are used to indicate a specific meaning of a symbol. For example: body temperature, T_b; blood volume, V_{bl}; rate of oxygen consumption, $\dot{V}_{O_2}$.

Equations that represent physical or physiological relationships use these and similar symbols for physical quantities. Symbols for units (m, kg, s, etc.) should *never* be used in such equations.

REFERENCES

Mechtly, E. A. (1973) *The International System of Units. Physical Constants and Conversion Factors* (second revision). NASA SP-9012. Washington, D.C.: National Aeronautics and Space Administration. 21 pp.

Symbols Committee of the Royal Society (1975) *Quantities, Units, and Symbols*, 2 ed. London: The Royal Society. 55 pp.

U.S. Department of Commerce (1977). *The International System of Units (SI)*. NBS Special Publication 330. Washington, D.C.: U.S. Government Printing Office. 41 pp.

B

Diffusion

Diffusion is a fundamental process in the movement of materials. Those diffusion processes that are of physiological importance take place over short distances, a fraction of a millimeter, a few micrometers, or less. Examples are the diffusion of respiratory gases across the respiratory membranes or the diffusion of ions across the nerve membrane during the action potential. Over longer distances, diffusion is physiologically unimportant, and transport takes place by mass movement (convection) (e.g., protoplasmic streaming or circulation of blood). Why is diffusion limited to short-distance transport? The reason is that diffusion time increases with the square of the diffusion distance.

Let us consider the time it takes for particles to travel a given distance by diffusion.

The average distance (R) covered by a diffusing substance depends on the average length (l) of each track in the random thermal motion of the particles (the mean free path) and on the number of individual tracks (n). The relationship is $R = l \sqrt{n}$. Because the number of individual tracks is proportional to time (t), we can write $R \propto l \sqrt{t}$, or $t \propto R^2 l^{-2}$. This means that diffusion time (t) increases with the square of the average distance (R) covered by the diffusing substance.

What does this mean for diffusion in the tissues of an organism? Suppose we impose a stepwise increase in oxygen concentration at a given point, A. Oxygen will diffuse into the surrounding tissues and will attain an average diffusion distance of 1 μm in about 10^{-4} s. An average diffusion distance 10 times as long will require 100 times the time, and so on. Thus, for an average diffusion distance of 10 μm the time required will be 10^{-2} s, for 1 mm it will be 100 s, for 10 mm, 10^4 s (nearly 3 h), and for 1 m, the time will be 10 000 times as long again, or more than 3 years. The distance from the lungs of a man to his extremities is in the order of

533

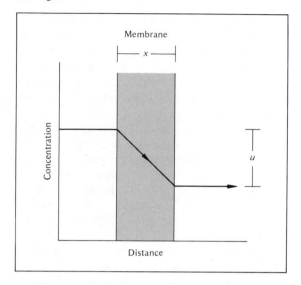

in which Q is the quantity of a substance diffused, A is the area through which diffusion takes place, u is the concentration at point x (du/dx thus being the concentration gradient), t is time, and D is the diffusion coefficient. The minus sign is a convention that makes the flux positive in the direction in which the gradient is negative. Diffusion coefficients for various substances can be found in standard tabular works.*

From this equation we can see that the amount of a substance diffusing is proportional to elapsed time. This also makes sense, intuitively. Consider a physiological example. Alveolar air has a partial pressure of oxygen of 100 mm Hg, and mixed venous blood arriving at the lung may have an oxygen partial pressure (tension) of 40 mm Hg. Normally, these concentrations remain very constant, and oxygen therefore diffuses from the air through the alveolar wall at a constant rate. It is now obvious that in 2 minutes twice as much oxygen diffuses as in 1 minute, in 10 minutes 10 times as much, and so on. Thus, if a constant gradient is maintained, the amount of the substance diffusing is directly proportional to time.

Another consequence of Fick's equation is that if the diffusion distance (x) is increased, the amount of the substance diffusing (Q) decreases. If x is doubled, Q is reduced to half; if x is 10 times as large, Q is one-tenth, and so on. The amount diffusing is thus inversely proportional to the diffusing distance, but again, this applies only when a steady state is maintained. This requires that we must have, on one side of the membrane, a constant supply or an infinite pool of the diffusing substance, and on the other side a constant lower concentration maintained by the continous removal of, or an infinite sink for, the diffusing substance.

* At 20 °C the diffusion coefficient D, for oxygen in air is $D = 0.196$ cm^2 s^{-1}, and for oxygen in water $D = 0.183 \times 10^{-4}$ cm^2 s^{-1}.

1m; clearly, transport by diffusion alone is impossibly slow and mass movement is necessary.

It is important to realize that we have discussed the time needed for diffusion over a certain distance when we impose a change in a concentration of the diffusing substance at one point and observe the effect of this change at another point at a certain distance. Thus, the diffusion of the neurotransmitter acetylcholine across the synaptic cleft (about 0.02 μm) is very fast; in contrast, if the hormone adrenaline were to reach its target somewhere in the body by diffusion from its source, the adrenal gland, the process would be so slow as to be useless.

Let us now consider a steady-state diffusion over an unchanged gradient (as we find in many physiological process).

The diffusion over a constant concentration gradient (Figure B.1) is expressed by Fick's equation:

$$dQ = - D\, A \frac{du}{dx} dt$$

This, for practical purposes, applies to our example of the lung.

The diffusion constant. The diffusion constant (K) which was used for the calculation on page 16, differs from the diffusion coefficient (D) used in Fick's equation. The need for the use of the diffusion constant in physiology arose in connection with the study of diffusion of gases between air and tissues, notably in the mammalian lung. For this purpose it is practical to refer the diffusion rate of a gas, say oxygen, to the partial pressure of the gas in the alveolar air. The diffusion coefficient (D) cannot be used in this case, for it requires knowledge of the concentration gradient (du/dx) in the tissue. The concentration in the tissue is difficult to determine directly and cannot be calculated unless the solubility for the particular gas in the particular tissue is known. The diffusion constant circumvents this problem by relating the rate of diffusion directly to the partial pressure of the gas in the gas phase.

The diffusion constant (K) is defined as the number of cubic centimeters of a gas which diffuses in 1 minute through an area of 1 cm² when the pressure gradient is 1 atm per cm. These units can be written out and reduced to a simpler form as follows:

$$\frac{cm^3}{min \times cm^2 \times atm \times cm^{-1}} = cm^2\ atm^{-1}\ min^{-1}$$

Diffusion constants have been determined for only a small number of animal tissues; some examples are listed in Table B.1. These constants are empirically determined and include the solubility of gas in the tissues in question. The figures show that the diffusion in those tissues that have a relatively high water content (e.g., muscle and con-

TABLE B.1 The diffusion constant (K) for oxygen at 20 °C in various materials. Note that these figures are not diffusion coefficients. [Krogh 1919]

Material	K (cm² atm⁻¹ min⁻¹)
Air	11
Water	34×10^{-6}
Gelatin	28×10^{-6}
Muscle	14×10^{-6}
Connective tissue	11.5×10^{-6}
Chitin[a]	1.3×10^{-6}
Rubber	7.7×10^{-6}

[a] The diffusion constant for chitin often is given incorrectly as 10 times the value listed here. This is the result of a misprint in a book by Krogh (1941) which has been copied by many later authors.

nective tissue) is somewhat lower than for water, but of the same order of magnitude. The diffusion through chitin (the hard cuticle of insects), on the other hand, is lower by an order of magnitude.

It is worth noting that the rate of diffusion of oxygen through rubber is considerable, about one-half the rate for muscle. This is of importance in physiological experimentation, for if rubber tubing is used for connections, oxygen can diffuse through the walls of the tubing, and carbon dioxide can diffuse even faster because of its high solubility. The actual rate of diffusion depends on the kind of rubber or plastic used; some kinds are more permeable and others much less permeable.

REFERENCES

Krogh, A. (1919) The rate of diffusion of gases through animal tissues, with some remarks on the coefficient of invasion. *J. Physiol.* 52:391-608.

Krogh, A. (1968) *The Comparative Physiology of Respiratory Mechanisms* (reprint ed.). New York: Dover. 172 pp.

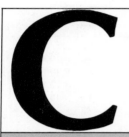

APPENDIX C

Logarithmic and exponential equations

LOGARITHMIC EQUATIONS

Logarithmic equations are of the general form:

$$y = b \cdot x^a \tag{1}$$

The logarithmic form of this equation is:

$$\log y = \log b + a \log x \tag{2}$$

Equation 2 shows that $\log y$ is a linear function of $\log x$; that is, by plotting $\log y$ against $\log x$ we obtain a straight line with the slope a.

Example: Rate of oxygen consumption ($\dot{V}_{O_2}$) plotted against body mass (M_b) (Figure 6.12, page 188).

$$\dot{V}_{O_2} = k \cdot M_b{}^a \tag{3}$$

The independent variable, body mass (M_b), is customarily plotted on the abscissa. The rate of oxygen consumption per unit mass is obtained by dividing both sides of equation (3) by M_b:

$$\frac{\dot{V}_{O_2}}{M_b} = k \frac{M_b{}^a}{M_b} = k \, M_b{}^{(a-1)} \tag{4}$$

Example: Figure 6.10, page 185.

EXPONENTIAL EQUATIONS

Exponential equations are of the general form:

$$y = b \cdot a^x \tag{5}$$

The logarithmic form of this equation is:

$$\log y = \log b + x \log a \tag{6}$$

Equation 6 shows that $\log y$ is a linear function of x, and plotting $\log y$ against x gives a straight line. Use of semilog graph paper (linear abscissa and logarithmic ordinate) for plotting y against x gives the same result.

Example: Rate of oxygen consumption plotted against temperature (Figure 7.4, page 209).

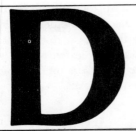

Thermodynamic expression of temperature effects

TABLE D.1 The change in Q_{10} for temperature intervals between 0 °C and 50 °C, calculated for a constant activation energy (μ) of 12 286.

T_1 (°C)	T_2 (°C)	Q_{10}
0	10	2.22
10	20	2.10
20	30	2.00
30	40	1.91
40	50	1.84

The term Q_{10} (defined on page 207) is widely used in biology, for it gives a simple and easily understood expression of the effect of temperature on rates (e.g., rate of oxygen consumption). In physical chemistry, however, temperature effects are expressed differently, based on thermodynamic considerations of the frequency of molecular collision as a function of temperature. The equation, as developed by Arrhenius, is usually given in the following form:

$$K_{T_2} = K_{T_1} \cdot e^{-\mu/R[(T_2-T_1)/T_2 T_1]} \tag{1}$$

in which K stands for rate constant, T for thermodynamic temperature (absolute temperature, in kelvin), μ for activation energy, R for the universal gas constant, and e is the base of the natural logarithm.

We can obtain a thermodynamically correct expression for Q_{10} (the ratio of two rates 10 °C apart) by solving equation (1) for K_{T_2}/K_{T_1} when $T_2 - T_1 = 10$. We then find:

$$Q_{10} = \frac{K_{T+10}}{K_T} = e^{-\mu/R\{10/[T(T+10)]\}} \tag{2}$$

When we examine equation 2, we find that Q_{10} is a function of and changes with temperature (T) and thus cannot be a constant. R, e, and (for a given process) μ are constants, and Q_{10} therefore decreases with increasing temperature. An example of this change in Q_{10} with temperature is given in Table D. 1 for $\mu = 12\ 286$ (a value of μ chosen to correspond to $Q_{10} = 2.00$ for the temperature interval between 20 and 30 °C). If $\mu = 19\ 506$, Q_{10} is 3.0 for the interval between 20 and 30 °C, but will be higher at low temperatures and lower at high temperatures. The use of Q_{10} to describe temperature effects therefore lacks a theoretical foundation in thermodynamics; however, Q_{10} is a highly descriptive term that is convenient and much used in biology.

537

APPENDIX E

Solutions
and osmosis

Water is the universal solvent for virtually all biological reactions and related phenomena. The following brief summary therefore is restricted to aqueous solutions. For a more extensive treatment, the reader is referred to an introductory chemistry or physical chemistry text.

SOLUTIONS AND CONCENTRATIONS

A 1 *molar* solution contains 1 mol of solute per liter of solution. Thus a 1 molar solution of sodium chloride contains 58.45 g NaCl per liter solution. In biological work it is common to refer to a one-thousandth of a 1 molar solution, a millimolar solution. This helps to eliminate small decimal fractions.

A 1 *molal* solution contains 1 mol of solute per kilogram water. For example, a solution of 58.45 g NaCl in 1 kg of water is a 1 molal solution. Because the volume of this solution is more than 1 liter, its concentration is less than 1.0 molar.

The molarity designation is often used in biological work and is convenient because it is easy to determine the volume of a given sample of a biological fluid, whereas it is cumbersome or impossible to determine its exact water content. For example, the volume of a blood sample is easily measured, and its content of sodium can readily be determined. Referring the sodium concentration to the amount of water in the sample is more difficult, for both plasma and red cells contain many other solutes, including substantial amounts of proteins. In physical chemistry, the molality concept is more useful; this includes the consideration of osmotic phenomena.

Concentrations are sometimes given as *percent* (%). Unfortunately, this expression is ambiguous and there is no accepted convention for how it should be used. This often leads to confusion. For example, does a 1% solution of sodium chloride contain 1 g NaCl and 99 g water? Or does it contain 1 g NaCl in 100 ml solution? Or, as the term *percent* means *per hundred*, does it mean 1 g sodium chloride per 100 g water? Even more confusing is the expression *milligram percent* (mg %), which logically should mean "milligrams per 100 milligrams." Usually it means the number of milligrams of a substance per 100 ml solution. Ambiguities can be avoided by using the appropriate units (e.g., 120 mg glucose per 100 ml blood) or often better, the molar concentration.

Addition of modifiers such as "weight percent" helps little. It commonly, but not always, refers to parts of solute by weight per 100 parts of solution by weight. Accordingly, a 1% solution of sucrose contains 1 g sucrose in 99 g water. However, if a salt that contains water of crystallization is weighed out, both the amount of the salt and the total amount of water may be uncertain.

Strictly, the term *percent* should express a nondimensional fraction. This means that the ratio of the two quantities should be expressed in the same units that cancel out in the division. It should be noted that gas concentrations can properly be expressed in percent. For example, dry atmospheric air contains 20.95% oxygen, whether the quantity of oxygen relative to the total is given in number of molecules, in milliliters, or in pressure units.

OSMOSIS

Osmosis is the movement of water between two solutions with different solute concentrations when the solutions are separated by a semipermeable membrane (i.e., a membrane permeable to water but impermeable to the solute).

Consider the system in Figure E.1. Water will tend to move from compartment B into compartment A, causing a pressure, the osmotic pressure, to develop in A. The movement of water can be

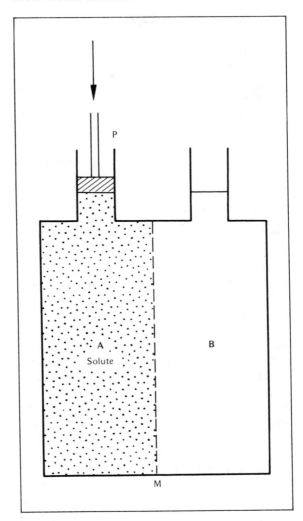

FIGURE E.1 When two solutions are separated by a semipermeable membrane (M), water tends to move from the more dilute solution to the more concentrated solution. If the membrane is rigid, the water movement can be prevented by exerting a pressure on the piston (P), equal to the difference in osmotic pressure between the two solutions.

counteracted by pressure exerted on the piston, and when this pressure exactly balances the osmotic pressure there is no net movement of water.

The force (π) which tends to move water is defined by van't Hoff's equation:

$$\pi V = n R T$$

in which V is volume, n is number of moles of solute particles, R is the universal gas constant, and T is absolute temperature. This equation is similar to the gas equation $PV = nRT$ (in which P is pressure).

A 1 molal solution of an ideal nonelectrolyte has an osmotic pressure of 22.4 atm at 0 °C (273 K). This equals the pressure exerted by 1 mol of an ideal gas when the volume is 1 liter.

Solutes depress the freezing point of water, and a 1 molal solution of an ideal nonelectrolyte has a freezing point of − 1.86 °C. Solutes also reduce the vapor pressure and increase the boiling point of a solution.

The osmotic pressure of a solution ideally depends only on the number of particles present, and not on their size or nature. An electrolyte such as sodium chloride, which in water ionizes into Na^+ and Cl^-, therefore has an osmotic concentration higher than expected from its molality.

Unfortunately, observations of actual osmotic pressures in solutions of electrolytes do not completely conform to the expected results as estimated from the expected number of ions. This is because of interaction between the positive and negative ions and interaction between the ions and the water molecules. The osmotic pressure of a solution of an electrolyte therefore cannot be calculated directly from the molal concentration of the electrolyte, and it is necessary to introduce an empirical *osmotic coefficient* (ϕ).

If sodium chloride were completely ionized in solution, we would expect a 1.00 molal solution to have a freezing point depression of 2×1.86 °C, or 3.72 °C. In fact, a 1 molal solution of NaCl freezes at − 3.38 °C. Its osmotic coefficient therefore is 3.38/3.72 = 0.91. The osmotic coefficient varies with the concentration of the electrolyte and differs from one electrolyte to another.

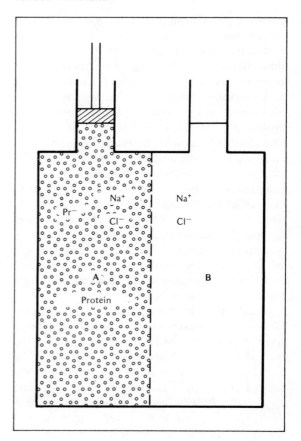

FIGURE E.2 An osmotic system consisting of two chambers, A and B, separated by a rigid membrane which is freely permeable to water and small ions but impermeable to large ions such as a negatively charged protein (Pr^-). The distribution of ions and the osmotic pressure at equilibrium (the Donnan equilibrium) are described in the text.

THE DONNAN EQUILIBRIUM

Assume a system similar to that in Figure E.1, but with the compartments separated by a rigid membrane permeable to water and electrolytes, but not to a large ionized molecule, say a negatively charged protein, Pr^- (Figure E.2). Let the electrolyte be sodium chloride.

If the compartments have rigid walls, pressure on the piston P prevents any net movement of water through the membrane. It can now be shown that at equilibrium:

$$[Na^+]_A \times [Cl^-]_A = [Na^+]_B \times [Cl^-]_B$$

Rearranging the terms, we find that the ratio of the diffusible cations on one side to those on the other is inversely proportional to the ratio of the anions on the two sides:

$$\frac{[Na^+]_A}{[Na^+]_B} = \frac{[Cl^-]_B}{[Cl^-]_A} = r$$

These ratios are known as the *Donnan constant* (r).

Assume that the initial concentration of sodium chloride is the same on the two sides of the membrane, and that compartment A also contains sodium proteinate, NaPr, dissociated as Na^+ and Pr^-. Thus, the total concentration of Na^+ in compartment A is higher than in B, and sodium ions will diffuse into B. This will give an excess of Na^+ ions relative to Cl^- ions in compartment B, and an equal number of chloride ions will therefore cross the membrane into compartment B. However, this tendency is counteracted by the resulting higher chloride concentration in B than in A. At equilibrium there will be unequal concentrations of ions across the membrane and a net negative charge in chamber A relative to chamber B. This equilibrium situation with the unequal distribution of diffusible ions and the resulting electric charge is known as the *Donnan equilibrium*.

We should note that the potential difference is attributable to the presence of the nondiffusible anion in chamber A. Because the system is in equilibrium, the potential is maintained without expenditure of energy and remains indefinitely.

The potential difference (E) can be calculated from Nernst's equation:

$$E = \frac{RT}{zF} \ln \frac{[Na^+]_A}{[Na^+]_B}$$

in which R is the universal gas constant, T is absolute temperature, z is valency of the ion (in this case unity), and F is Faraday's constant (96 500 coulomb per mole).

The pressure applied to the piston in Figure E.2 necessary to prevent osmotic flow of water into compartment A is known as the *colloidal osmotic pressure*. In the event that the pressure on the plunger exceeds the colloidal osmotic pressure, water and electrolytes are forced out through the membrane while the nonpermeable protein is withheld in A. This process is called *ultrafiltration*. The osmotic pressure of the ultrafiltrate will equal the osmotic pressure of the solution in compartment B.

ADDITIONAL READING

Overbeek, J. Th. G. (1956) The Donnan equilibrium. *Progr. Biophys.* 6:57–84.
Hammel, H. T., and Scholander, P. F. (1976) *Osmosis and Tensile Solvent*. New York: Springer-Verlag. 133 pp.

The animal kingdom

An abbreviated classification
of living animals.

PHYLUM PROTOZOA
CLASS FLAGELLATA
Flagellates
CLASS SARCODINA
Protozoans with pseudopods. *Amoeba*
CLASS CILIATA
Ciliates. *Paramecium*

PHYLUM PORIFERA. Sponges

PHYLUM COELENTERATA (or Cnidaria)
CLASS HYDROZOA
Hydrozoans. *Hydra*
CLASS SCYPHOZOA
Jellyfishes
CLASS ANTHOZOA
Sea anemones and corals

PHYLUM CTENOPHORA. Comb jellies

PHYLUM PLATYHELMINTHES. Flatworms
CLASS TURBELLARIA
Free-living flatworms
CLASS TREMATODA
Flukes
CLASS CESTODA
Tapeworms

PHYLUM NEMATODA. Roundworms. *Ascaris*

PHYLUM MOLLUSCA. Molluscs
CLASS AMPHINEURA
Chitons
CLASS GASTROPODA
Snails
CLASS PELECYPODA
Bivalve molluscs, clams, mussels, etc.
CLASS CEPHALOPODA
Squids, octopuses, and *Nautilus*

PHYLUM ANNELIDA. Segmented worms
CLASS POLYCHAETA
  Marine worms, tubeworms, etc.
CLASS OLIGOCHAETA
  Earthworms and many fresh-water annelids
CLASS HIRUDINEA
  Leeches

PHYLUM ARTHROPODA. Arthropods
CLASS XIPHOSURA
  Horseshoe crabs. *Limulus*
CLASS ARACHNIDA
  Spiders, ticks, mites, scorpions, etc.
CLASS CRUSTACEA
  Crustaceans. Lobster, *Daphnia, Artemia*
CLASS CHILOPODA
  Centipedes
CLASS DIPLOPODA
  Millipedes
CLASS INSECTA
  Insects

PHYLUM ECHINODERMATA. Echinoderms
CLASS ASTEROIDEA
  Sea stars
CLASS OPHIUROIDEA
  Brittle stars
CLASS ECHINOIDEA
  Sea urchins and sand dollars
CLASS HOLOTHUROIDEA
  Sea cucumbers

PHYLUM CHORDATA. Chordates
SUBPHYLUM UROCHORDATA (OR TUNICATA). Tunicates

SUBPHYLUM CEPHALOCHORDATA. Amphioxus

SUBPHYLUM VERTEBRATA. Vertebrates
CLASS AGNATHA
  Jawless fishes. Lampreys and hagfish
CLASS CHONDRICHTHYES
  Elasmobranchs. Sharks, skates, rays
CLASS OSTEICHTHYES
  Bony fishes
Subclass Crossopterygii. *Latimeria*
Subclass Dipnoi. Lungfishes
Subclass Actinopterygii. Higher bony fishes
CLASS AMPHIBIA
  Amphibians. Frogs, toads, salamanders
CLASS REPTILIA
  Reptiles
Order Chelonia. Turtles
Order Crocodylia. Crocodiles and alligators
Order Squamata. Lizards, snakes
CLASS AVES
  Birds
CLASS MAMMALIA
  Mammals
Subclass Prototheria
  Order Monotremata. Egg-laying mammals
  (echidna and platypus)
Subclass Metatheria
  Order Marsupialia (Kangaroos, opossum, etc.)
Subclass Eutheria.

  Orders: Insectivores (moles, shrews, etc.),
  bats, primates (lemurs, monkeys, apes, man),
  sloths and armadillos, rabbits and hares, rodents, whales and dolphins, carnivores and
  seals, elephants, manatees and sea cows,
  ungulates.

REFERENCE
Rothschild, L. (1965) A *Classification of Living Animals*, 2nd ed. London: Longmans, Green. 134 pp.

Index

myosin, 380-1, 386, 390-1
Myotis, 454
Mytilus, 400

narcotic effect, inert gases, 173
nasal glands, *see* salt glands
nasal temperature, 325-6
Nassarius, 450
Nautilus, 425
negative feedback, 480ff
nematocyst, 132
Neoceratodus, 38, 41-3, 367
nephron, 350ff
Nephrops, 291
Nernst's equation, 541
nerve
  action potential, 465ff
  myelinated, 486-90
  resting potential, 463
  speed of conduction, 485-90
nerve cells, 461-3
neurohemal organs, 517-18
neurohypophysis, 507
neuron, 461-3
neurotransmitters, 512-16
nickel, nutrition, 150
nitrogen
  diving, 170-3
  narcotic effect, 173
nitrogen excretion, 360-73
nitrogen fixation, 146
Noctiluca, 418-19
noradrenaline, 103, 510-12
  structure, 514
Notonecta, 54
noxious compounds, 152-7
nutrition, 142-52
  mineral, 147-52
  protein, 144-6

Ochromonas, 383
Ocypode, 313
oil bird, echolocation, 454
olfaction, 458-60
Oncorhynchus, 221-2
opiate receptors, 515-16
Opsanus, 355
opsin, 443
organic phosphates, red blood cell, 70ff

ornithine cycle, 364-5
osmoregulation
  aquatic vertebrates, 297-308
  amphibians, 306-8
  cyclostomes, 298-300
  definitions, 288
  elasmobranchs, 300-2
  energy requirement, 294-6
  hagfish, 298-300
  hypersaline environments, 286-97
  invertebrates, 289-97
  lamprey, 298-300
  mechanisms, 293-6
  teleosts, 302-6
osmosis, 538-41
osmotic coefficient, 540
osmotic pressure, colloidal, 110-12
ostrich, respiration, 45-6
oxalic acid, 153-4
oxidation water, 319-20
oxygen
  in atmosphere, 9
  caloric equivalent, 162-3
oxygen consumption, 162ff
  birds, 188-9
  body size, 183-90
  diving, 178-80
  marsupials, 189-90
oxygen diffusion, capillary, 74-5
oxygen dissociation curve, 68-77
  altitude, 72-3
  body size, 73-5
  carbon dioxide, 69-70
  diving, 73
  fetal blood, 71-2
  invertebrates, 76-7
  organic phosphates, 70ff
  pH effect, 69-70
  tadpole, 72
  temperature, 69
oxygen extraction, fish gills, 20
oxygen minimum layer, 169
oxygen toxicity, 173
oxytocin, 507-8

P_{50}, 69ff
pacemaker, heart, 102
Pachygrapsus, 291
PAH, 346